ROUTE MAPS IN GENE TECHNOLOGY

ROUTE MAPS IN GENE TECHNOLOGY

Matthew R. Walker BA, PhD

Department of Clinical Chemistry
Wolfson Research Laboratory
Queen Elizabeth Medical Centre
University of Birmingham

WITH

Ralph Rapley BSc, PhD

School of Natural and Environmental Sciences
Coventry University
Priory Street, Coventry

**Blackwell
Science**

© 1997, by
Blackwell Science Ltd
Editorial Offices:
Osney Mead, Oxford OX2 0EL
25 John Street, London WC1N 2BL
23 Ainslie Place, Edinburgh EH3 6AJ
350 Main Street, Malden
  MA 02148 5018, USA
54 University Street, Carlton
  Victoria 3053, Australia
10, rue Casimir Delavigne
  75006 Paris, France

Other Editorial Offices:

Blackwell Wissenschafts-Verlag GmbH
Kurfürstendamm 57
10707 Berlin, Germany

Blackwell Science KK
MG Kodenmacho Building
7–10 Kodenmacho Nihombashi
Chuo-ku, Tokyo 104, Japan

First published 1997
Reprinted 1999

Set by Excel Typesetters, Hong Kong

ISBN: 978-0-632-03792-6

The Blackwell Science logo is a
trade mark of Blackwell Science Ltd,
registered at the United Kingdom
Trade Marks Registry

DISTRIBUTORS

Marston Book Services Ltd
PO Box 269
Abingdon, Oxon OX14 4YN
(*Orders*: Tel: 01235 465500
            Fax: 01235 465555)

USA
Blackwell Science, Inc.
Commerce Place
350 Main Street
Malden, MA 02148 5018
(*Orders*: Tel: 800 759 6102
                  781 388 8250
            Fax: 781 388 8255)

Canada
Login Brothers Book Company
324 Saulteaux Crescent
Winnipeg, Manitoba R3J 3T2
(*Orders*: Tel: 204 837 2987)

Australia
Blackwell Science Pty Ltd
54 University Street
Carlton, Victoria 3053
(*Orders*: Tel: 3 9347 0300
            Fax: 3 9347 5001)

A catalogue record for this title
is available from the British Library

ISBN 0-632-03792-X

Library of Congress
Cataloging-in-Publication Data

Walker, Matthew R.
  Route maps in gene technology/
  Matthew R. Walker, with Ralph Rapley.
    p. cm.
  Includes bibliographical references and index.
  ISBN 0-632-03792-X
  1. Genetic engineering.   I. Rapley, Ralph.
  II. Title.
  QH442.W345   1997
  660′.65—dc20   96–31611
                                              CIP

For further information on
Blackwell Science, visit our website:
www.blackwell-science.com

CONTENTS

CONTENTS/Contd

PREFACE

There is no doubt that gene-based techniques have revolutionised our understanding of how the inherited messages of life are coded and decoded within cells from organisms as diverse as viruses and humans. Their continued development and application has also had a direct impact on many aspects of our lives, providing new horizons for clinical diagnosis and therapy, and enabling the generation of novel molecules and genetically modified organisms for industrial and agricultural exploitation. Although some of these achievements have been highlighted by the media, the expanding map of the molecular biological world has remained inaccessible to many. This is partly due to the rapidity of technological and intellectual developments, and the often undecipherable nature of modern scientific jargon.

Our aim has been to provide a series of overlapping 'introductions' to the wealth of knowledge contained within the outwardly cryptic and detailed texts furnished by the pioneers and practitioners of recombinant genetic techniques. Furthermore, the book is specifically formatted to allow the reader the flexibility to follow any one of numerous and interlinking paths through molecular biology concepts, principles and key recombinant genetic methods or approaches towards numerous defined destinations or horizons. Indeed, many different routes may be taken to reach the same destination, starting from any point within the book, each taking in more or less detail of the available molecular biology 'scenery'. Accordingly, those needing to rapidly grasp the essential principles may wish to take a direct route, whilst the more curiosity-driven may take an indirect route(s), frequently diverting along new but interconnected intellectual paths. This permits readers of varying background and knowledge to rapidly become conversant with general concepts/principles, and provides glimpses of some of the new frontiers of the expanding molecular biology universe. They may then be prepared, and indeed stimulated, to tackle the existing 'ordnance survey' maps of molecular biology, complete with jargonised signposts and detail (many of which appear in a substantial bibliography section in the latter part of the book).

Route Maps is suitable as an introductory text for undergraduate and graduate students in all areas of life sciences, medical students, qualified clinicans, and anyone interested in finding out more about recombinant DNA technologies, their potential applications, and implications. Although also useful as a revision guide, the principal aim of writing *Route Maps* has been to provide the requisite information in a manner which enables each individual to construct a personalised conceptual map of recombinant DNA technologies within the larger setting of molecular biology as a whole. In other words, by allowing the reader to define the order of topics covered, he/she should be able to develop their own 'neural networks' which serve to bridge the gaps in understanding often evidenced in exam answers. *Route Maps* may therefore also be simultaneously used by lecturers as a framework for course development, and by students to reinforce the interrelatedness of the information presented in individual lectures.

In conclusion I hope that by exploring *Route Map*s, the cumulative knowledge it contains will be optimally assimilated, and will transmit both the excitement and concerns of those actively engaged in the pursuit of recombinant genetic goals.

Matthew R. Walker

ACKNOWLEDGMENTS

This book would not have been possible without the support and assistance of numerous individuals to whom I owe a great debt of thanks.

I am especially grateful to my wife and children for enduring my protracted physical (and mental) absences during the lengthy composition of this book. I also wish to express my thanks and appreciation to my parents and all members of my family (especially Jeremy and Joanne) for their encouragement and practical support during the numerous difficulties encountered along the way. Similarly, my thanks also go to friends and colleagues for their support and useful discussions. An expression of thanks is also due to the many students whose input and forebearance I have greatly appreciated.

Special thanks go to colleagues who have assisted with reading and commenting upon early drafts, and who have provided many useful pieces of advice. Especially noteworthy are John Pound, John Heptinstall and Craig Winstanley.

I am also indebted to Harvey Shoolman and Andrew Robinson at Blackwell Science, and Steve Meyfroidt of Integrated Design, for their initial encouragment, technical guidance, crystallisation and polishing of my original concept. I am also indebted to Simon Rallison and those members of the Blackwell Science Book Production Department whose dedication greatly assisted the difficult process of composing and refining the manuscript.

I would also like to thank Ralph Rapley for his encouragement, our many discussions, his constructive criticisms, meticulous checking for inaccuracies and inconsistencies, and for generally helping to bring the text to life.

ABOUT THE ROUTE MAPS FORMAT

The *Route Maps* format contains several core features (such as 'route prompts', 'biblio-boxes' or 'compass boxes') designed to assist the reader to develop and extend their understanding of gene technology.

In order to facilitate the rapid acquisition of knowledge concerning key aspects and technologies of molecular biology, and stimulate the development of conceptual cross connections (i.e. 'neural networks'), *Route Maps* is divided into numerous short sections (of two or four pages) covering a single topic or subtopics. Each deals with general principles, technologies and their applications in more or less detail in a deliberately overlapping manner. Although there is a certain degree of enforced logic to the arrangement of individual sections within the book, the reader has total freedom in the order in which sections are read, or indeed, which parts of which sections to read at any given time. However, each section follows a fairly standard format in order to maintain a level of continuity and allow selection of individual learning pathways. The *Route Maps* format has therefore been purposefully designed to contain several core features to guide the reader further along a specific intellectual path (or 'route') or to divert the reader onto a new, parallel or interconnected path.

WHY CALL THE BOOK ROUTE MAPS?
The format of *Route Maps* presents the information in a manner analogous to that of a road atlas or A–Z streetmap. In other words, the reader can find their way to specific destinations from any initial page by following the information and prompts to overlapping pages. Indeed, just as the pages of an atlas or A–Z do not necessarily follow in sequence, the pages of *Route Maps* also guides the reader in many directions enabling a personal 'intellectual map' to be developed. Although akin to the formation of 'neural networks' such a title was discounted because of its common use in computer terminology concerning 'artificial intelligence'. The format is however designed to facilitate presentation in electronic versions which may be used in conjunction with new computer-based learning approaches (e.g. as tutorial packages for self teaching, self-assessment, and/ or distance learning.)

1 Introduction Each two or four page section begins with a general introduction aiming to summarise its' coverage within the context of preceeding and following sections. Indeed, it is possible to read only these portions of each section before tackling the more detailed portions of text (i.e. use the introductions to select the order of sections to read to suit your individual needs or preferences).

2 'Summary boxes' The top right hand corner of each left hand page also contains a short summary statement indicating the content and coverage. This is intended to both facilitate the selection of sections to read in detail and to serve as a form of 'memory key'.

3 'Compass boxes' The first page of each section also contains a box of information which is intended to serve as an information 'bite' which links that section to another related section or extends into another related topic for further reading. These boxes may also provide points intended as stimuli for discussion or further investigation and are therefore complemented by the next core feature.

4 'Biblio-boxes' Each section also contains one or two oblong shaded boxes which directs the reader to key references relating to the information within that particular section. These are numbered according to their order within the three bibliographies located at the end of the book containing original articles (**OAs**), review articles (**RAs**), and books for suggested further reading (**SFR**). It is hoped that these will provide easy access to further detailed texts (i.e. as a launch pad for further investigation of the subject) .

OAs: 1. 115–8, 287–95, 336–8, 445, 569–83,1005–19. **RAs:** 35, 46–8, 77–91, 233–4. **SFR:** 32, 35, 56–9, 117, 119–25.

5 'Route prompts' The last feature which appears at the top right of each odd-numbered page is a listing of other related sections which may be used in combination to follow particular intellectual or technical concepts throughout the book. These are neither an exhaustive list, nor intended to be strictly adhered to. They are merely intended to suggest those section(s) containing information on particular related topics/techniques and which may therefore be used to construct a specific route or path. They may consequently be used to move both backwards and forwards

through the book to either fill in gaps in knowledge or to extend an individual's knowledge of a specific topic. For example, the route prompts in Chapter 44 covering restriction fragment length polymorphisms (RFLPs) may direct the reader backwards to sections covering restriction endonucleases or the organisation of genes within chromosomes, etc., and forwards to the sections covering some applications of RFLPs e.g. in genetic fingerprinting, prenatal diagnosis, or genome mapping.

1. THE CONCEPT OF GENES IS DEVELOPED

The origins of molecular biology can be traced to Mendel's experiments which allowed the concept of discrete units of inheritance and the basic principles dictating their transmission to be formulated.

Humankind has long been aware that physical characteristics or traits are inherited from parents to offspring. Selective breeding exploited this to improve the quality of domestic plants and animals. The practice of selective breeding was however unpredictable, with many undesirable traits (e.g. sterility or disease susceptibility) unexpectedly occurring or recurring in offspring. The explanation for these events awaited the careful analysis of the inheritance of traits in offspring from well-controlled matings between organisms expressing well-defined traits. Initial insights were provided by Gregor Mendel's painstaking pea breeding experiments performed in the early 1860s. These laid the cornerstones of modern genetics, and subsequently led to the development of what we now term recombinant genetic technologies. Mendel's observations, mathematical analyses and interpretations first established the discreteness of units of inheritance (now termed genes), and that each simple trait studied was under the control of two such units; one inherited from the male parent and one from the female. This led him to formulate some of the basic rules of genetics such as gene segregation, independent assortment and dominance. Following their rediscovery in the 1900s, these became known as Mendel's Principles. The conceptualisation of transmissable genes can therefore be viewed as the starting point for molecular genetics, which led to the identification of DNA as the genetic material and the determination of the biochemical structure of genes. This allowed the way in which DNA structures store and regulate the flow of genetic information to be determined, rapidly followed by the development of methods by which genetic information/material could be analysed, manipulated or exploited.

THE FATHER OF GENETICS

Now regarded as the father of genetics, Gregor Mendel (1822–1884) was an Augustinian monk and substitute science teacher (having repeatedly failed the required examinations to be a full time teacher). His painstaking pea plant (*Pisum sativum*) breeding experiments were performed in the monastery garden at Brno, Czechoslovakia. Over timescales not possible in science today, Mendel applied mathematical analyses to the results obtained from over 21 000 hybrid plants. Despite presenting his experimental data and conclusions at a meeting of the Brunn Society for the Study of Natural Science and publishing his observations in 1866, his work remained largely ignored by a scientific community which at the time did not appreciate the importance of quantitative science. Unrecognised as a scientist but respected as a priest and teacher, Mendel became Abbot of a monastery in 1868.

MENDEL'S EXPERIMENTAL DESIGN

Although Mendel's genius lay in his ability to interpret results in the abstract terms of genes, his achievements also owed much to his experimental design and quantitative approach.

Mendel also shrewdly chose the pea plant as an experimental subject. It was easy to grow and many varieties were freely available which offered a wide range of traits to study. The position of the female stigma relative to the male pollen-producing anthers also made it an excellent subject by facilitating controlled pollination (see Fig. 1.1). In a simple, if laborious, process, the anther could be removed from one plant to prevent self-pollination and used as a source of pollen for cross-pollination. Furthermore, the flower petals also effectively protect the reproductive organs from air-borne sources of pollen. They could therefore be easily protected from pollinating insects by covering the flower with a small bag, thereby ensuring only controlled experimental pollinations occurred.

Before beginning his cross-breeding experiments, Mendel also spent many years establishing true breeding pea plant strains. From these he selected seven strains which exhibited clearly defined and contrasting pairs of visible traits. These are now referred to in

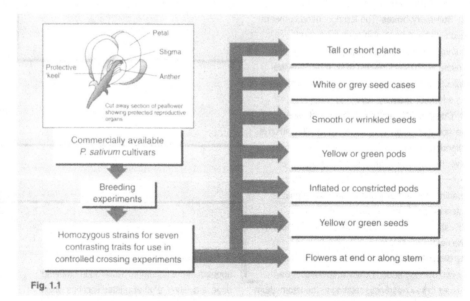

Fig. 1.1

terms of gene alleles with organisms carrying the same pair of alleles termed homozygotes and those with one copy of each allele termed heterozygotes.

Having established true breeding (homozygous) strains, Mendel initially studied the appearance of traits in first generation offspring (referred to as the first filial or F1 generation)

produced by crossing two parental strains. Importantly, he also obtained second filial (F2) generations by (i) crossing F1 offspring, (ii) self-pollination of an F1 offspring, or (iii) crossing F1 progeny and parental plants. Finally, Mendel formulated his concepts by statistical analysis of the quantitative data from many such experiments.

Segregation and dominance

The first conclusions derived by Mendel were based upon the early observations of the inheritance of just one pair of traits (termed monohybrid inheritance).

The most striking features were the appearance of only one type of F1 progeny and the appearance of both types of progeny (in the ratio of 3:1) amongst the F2 generation produced by self-pollination of the F1 products (see Fig. 1.2). Mendel correctly recognised that each individual pea plant must carry two copies (alleles) of the gene controlling a single inherited trait, one inherited from each parent. Furthermore, only one allele must be present and transmitted by the sex cells (or gametes) produced by each parent. These are then combined in the offspring, i.e. they independently segregate.

To account for the disappearance of one trait in the F1 generation and its re-emergence in a 1:3 ratio in the F2 generation, Mendel further proposed that each allele varied in its strength to dictate the physical outcome (or phenotype). In other words, one allele was dominant, 'masking' the expression of the other recessive allele. Consequently, a recessive trait (e.g. wrinkled seed) only appears in homozygous individuals carrying two copies of the recessive allele, whereas the dominant trait (e.g. smooth seed) appears in heterozygous individuals (carrying both alleles) *and* homozygous individuals carrying two copies of the dominant gene.

Independent assortment

Mendel also performed dihybrid crosses using plants with two pairs of contrasting traits. For example, crossing of plants producing smooth yellow seeds with plants producing wrinkled green seeds gave F1 progeny all producing smooth yellow seeds, since genes for smooth seeds and yellow seeds were dominant (see Fig. 1.3).

However, four types of progeny were produced in the F2 generation in ratios of 9:3:3:1. In order to explain this numerically consistent result Mendel formulated the principle of independent transmission of pairs of different traits to the gametes. Gametes produced by the F1 progeny would therefore contain any of four combinations of genes which could combine to give four types

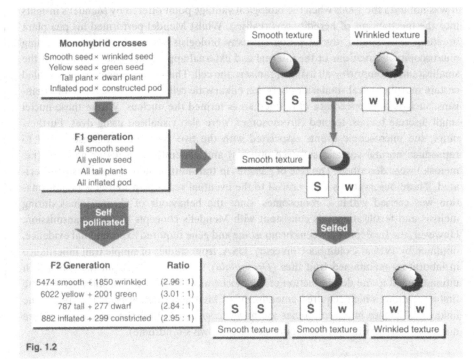

Fig. 1.2

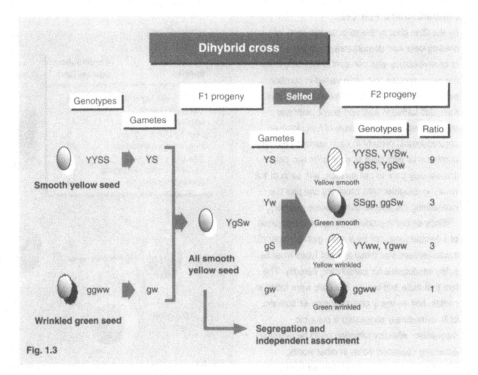

Fig. 1.3

of F2 progeny following self-pollination. Fortuitously, Mendel selected traits coded by alleles on different chromosomes.

OAs: 1. RAs: 1. SFR: 1–4, 36–9, 62–8.

11

2. GENES ARE LOCATED TO CHROMOSOMES

The behaviour of chromosomes during cell division and their number per cell correlated with, and extended, Mendel's principles thereby suggesting that genes were located within chromosomes.

It was not until the 1900s when the biological vantage point offered by Mendel's insights into the mechanisms of heredity was realised. Whilst Mendel performed his pea plant breeding experiments, the attention of many biologists was focused upon recording microscopic observations of the internal and external appearance (morphology) of the smallest units comprising all living organisms, the cell. This cataloguing process revealed certain morphological similarities between eukaryotic cells from a wide range of organisms, such as the appearance of defined areas termed the nucleus. Within these nuclei small discrete bodies, termed chromosomes, were also visualised using dyes. Furthermore, the microscopic events associated with the two types of cell divison used to reproduce normal somatic cells (i.e. mitosis) and generate sex cells or gametes (i.e. meiosis) were described. The role of gametes in transmitting heredity was also appreciated. These observations were crucial to the eventual realisation that genetic information was carried within chromosomes, since the behaviour of chromosomes during meiosis and fertilisation was consistent with Mendel's concepts of gene transmission. However, the final link between chromosome and gene required experimental evidence, obtained by 1915 at Columbia University, USA, from studies of simple trait inheritance in laboratory-maintained fruit flies (*Drosophila*). These results began the trail which ultimately led to the determination of the biochemical basis of genes, and provided the first exceptions which required amendment of Mendel's original principles (e.g. gene linkage). Changes in chromosomes are now recognised to underlie, and are therefore diagnostic for, several disease syndromes (e.g. Down's syndrome).

NON-CHROMOSOMAL DNA

Staining of eukaryotic cells demonstrates that the majority of DNA is located within the nucleus as chromosomes. Extrachromosomal DNA is however also located in cellular organelles (i.e. mitochondria and chloroplasts). Such organellar DNA molecules are generally circular and double-stranded, encoding specific proteins and RNAs required for organellar transcription and translation. The sizes and amount of genetic information encoded by mitochondrial genomes varies across species. Chloroplast genomes contain a larger number of genes and are less variable. Analysis of the relatively conserved mitochondrial and chloroplast DNAs may however provide insights into their evolutionary origins, and early human ancestry.

OAs: 2–6, 255, 256.
RAs: 2, 16, 117.
SFR: 5–7, 9, 22, 62–8.

CHROMOSOMES PER CELL

By the 20th century the microscopic study of dividing cells had demonstrated that the number of chromosomes was constant within cells from the same species, but often varied in number between species (Fig. 2.1). Within a given cell they also varied in size and shape, with two copies of each type (i.e. pairs of homologous chromosomes) present in each somatic cell. The number was also seen to double (to two pairs) immediately prior to cell division, with each of the resulting daughter cells receiving one pair (i.e. maintaining the normal chromosome number).

Since sexual reproduction required fertilisation of a female (egg) cell by a male gamete (sperm), it was realised that these two cell types must be solely responsible for transmitting heredity. The fact that male and female sex cells were found to contain half as many chromosomes as somatic cells immediately suggested a plausible mechanism whereby offspring could be produced exhibiting combined traits, in other words, inherited from chromosomes derived from both parents according to Mendel's principles of gene segregation and independent assortment. The relatively small amount of cytoplasm of sperm compared to its much larger nucleus, coupled

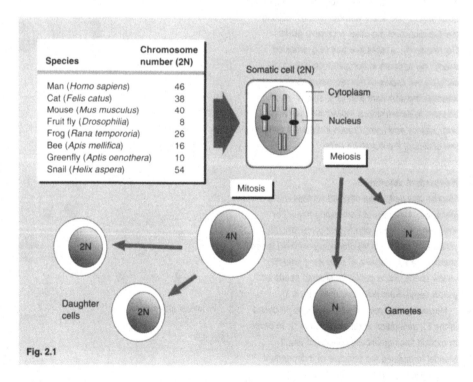

Species	Chromosome number (2N)
Man (*Homo sapiens*)	46
Cat (*Felis catus*)	38
Mouse (*Mus musculus*)	40
Fruit fly (*Drosophilia*)	8
Frog (*Rana tempororia*)	26
Bee (*Apis mellifica*)	16
Greenfly (*Aptis oenothera*)	10
Snail (*Helix aspera*)	54

Fig. 2.1

with the visible behaviour of chromosomes during mitosis and meiosis, also provided strong evidence that genes were located within the cell nucleus, and most probably associated with the chromosomes.

CHROMOSOMES DETERMINE SEX

The initial breakthrough for scientists attempting to pinpoint the cellular location of genes came in 1905 from work at Columbia University, USA, by Nettie Stevens and Edmund Wilson. They proposed that sex traits were located on a single pair of chromosomes. From their careful microscopic analysis of the appearance of chromosomes in mammalian cells, they discovered the existence of what are now known as the sex chromosomes. Whilst most chromosomes within a cell existed as homologous pairs (now termed the autosomes), they noted that cells from females contained two copies of an X chromosome, whilst male cells contained one copy of the X chromosome and one copy of a smaller, distinctly shaped Y chromosome. During meiosis all gametes received one of each of a pair of autosomes.

Female gametes also recieved one X chromosome whereas male gametes could receive either an X or a Y chromosome (see upper half of Fig. 2.2). Consequently, fertilised egg cells contain a full complement of autosomes but either two X chromosomes, or an X and Y chromosome; depending upon which of the two possible male gametes achieved fertilisation. The fertilised egg would therefore develop into either a female (XX) or male (XY).

Although the X/Y chromosome combination dictates the sex of almost all animals, there are several exceptions (e.g. hermaphroditic animals and most flowering plants lack sex chromosomes). There are also many variations of the XY mechanism (e.g. male birds and butterflies are XX and females are XY).

GENE LINKAGE AND CROSSING OVER

The first evidence that traits other than sex were determined by genes located within chromosomes was also derived from experiments performed at Columbia University. Between 1910 and 1915 T.H. Morgan and colleagues focused their research upon the inheritance of simple traits in the red-eyed fruit fly (*Drosophila*). Their work firstly indicated that the mutant gene for white eye colour in *Drosophila* was always transmitted with the X chromosome, i.e. was sex linked, (see lower half of Fig. 2.2). They also mapped several other mutant genes both to the X

chromosome and the remaining autosomes. Furthermore, it was evident that many genes were linked together on the same chromosome and thus inherited together by the offspring.

Another key observation was that the degree of linkage occasionally varied, with linked traits somehow separating during meiosis. The explanation for this again came from observing chromosome behaviour. Accordingly, Morgan had observed that during the coming together of homologous chromosomes (termed synapsis) many pairs formed intimate links. He correctly surmised that during synapsis parts of each chromosome were physically exchanged in a process now termed crossing-over or recombination. The degree of recombination which occurs is directly proportional to the degree of physical linkage (i.e. distance) between genes. Consequently, the measurement of recombination frequency was used as an early means of mapping the chromosomal location of several genes, including several human disease genes (e.g. for cystic fibrosis). This approach was also

used to demonstrate that genes were linearly arranged along the chromosomes.

Why study *Drosophila*?

Drosophila provided geneticists with almost ideal early experimental subjects. Firstly, laboratory populations of *Drosophila* could be easily and cheaply maintained on simple media. New generations of hundreds of offspring could also be obtained within a few days of controlled matings. Such large numbers of offspring also meant that rare genetically stable variants (or mutants), either arising spontaneously or induced (e.g. by X-rays), could be detected and propagated. *Drosophila* also contain only four pairs of chromosomes which could be readily visualised in some cells (e.g. larval salivary gland cells) by light microscopy. Their early adoption as targets for gene analysis means we presently know more about their genome, particularly in relation to development, than those of most other organisms.

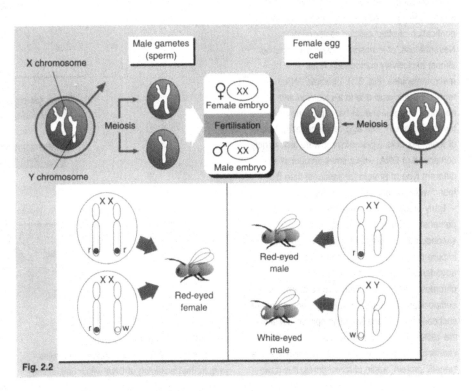

Fig. 2.2

3. GENES ARE COMPOSED OF DNA

Molecular genetics has developed from the demonstration that genes are composed of DNA, and several recombinant techniques ultimately derive from the nature of the early experiments used.

Most aspects of modern genetics can be traced back to a small number of significant experimental breakthroughs which ultimately combined to establish the molecular basis of inheritance. The establishment of the location and biochemical nature of inherited factors (i.e. genes) were two such important discoveries arising from early research. Thus, by 1910 it was widely accepted that the physical characteristics comprising the phenotype were determined by the combination of genes comprising the genotype. A decade later it was also established that genes were most probably located within nuclei, and probably associated with the chromosomes observed by microscopic analysis of cells during the metaphase stage of division. Eventually, many lines of circumstantial evidence combined to enable the design and performance of experiments that definitively established which of the many complex cellular macromolecules contained within the nucleus actually performed the functions of genes. Although it would be many years before it was determined how genetic information is encoded and transmitted by DNA, the initial demonstration that genes were composed of DNA was a turning point in the development of modern molecular genetics. Equally significant, the process of discovery involved many scientific disciplines (e.g. chemistry, microbiology) and technologies (e.g. microscopy) which in themselves provided new avenues for intellectual and technical development. For example, the ability of DNA to transform the character of a cell is now widely employed in many molecular genetic or recombinant DNA technologies.

CHROMOSOMES AND DNA VISUALISED

In 1869, the Swiss chemist Johann Frederieck Miesher was able to extract from human pus cells and cell nuclei a material which he termed 'nuclein'. Now identified as deoxyribonucleic acid or DNA, nuclein differed from protein because it contained phosphorous and was not degraded by proteolytic enzymes (e.g. pepsin). The term chromosome (meaning 'coloured body') was first used in 1888 to describe the densely stained structures observed within the nuclei of dividing eukaryotic cells (i.e. during metaphase). In the 1920's, Robert Feulgen developed a DNA-specific purple dye (named after him) which stained cells for the presence of DNA. This was used to demonstrate that DNA was located within the chromosomes of both somatic and sex cells. Although still used, many additional methods and stains are now available for visualising DNA.

'CIRCUMSTANTIAL' EVIDENCE

Many lines of 'circumstantial' evidence suggested that the molecular vehicles of heredity (i.e. genes) resided within small bodies termed chromosomes within the cell nucleus. Initially, available methods for chromosome extraction made it difficult to obtain highly purified preparations (i.e. without co-purification of other cellular components). Nevertheless, chromosomes were shown to be almost exclusively composed of two macromolecules (Fig. 3.1): (i) acidic DNA (now termed nucleic acid due to it's location and negative charge), and (ii) basic proteins (now termed histones). The egg cell-penetrating heads of male sex cells (spermatozoa) were also largely composed of DNA, with a small amount of a different type of protein (protamines) than that found within somatic cell chromosomes.

Early analyses of isolated chromosomes further demonstrated that the amount of DNA far exceeded the amount of protein even in relatively impure preparations. Despite the difference in the abundance of DNA and protein within chromosomes, the relatively simple biochemical composition of nucleic acids compared to proteins succeeded in dividing scientific opinion well into the 1930s. The comparative lack of nucleotide variation also gave credance to the existence of a 'genetic protein' within chromosomes. It is now

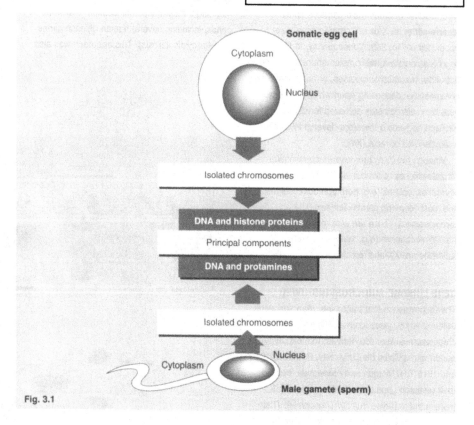

Fig. 3.1

realised that the proteins co-purified with chromosomal DNA actually perform vital roles (e.g. in the packaging of DNA within cell nuclei).

OAs: 7–9, 281–4.
RAs: 2, 16, 117, 129–31, 172–5.
SFR: 8, 9, 27, 29, 30, 37–9, 99–102.

DEFINITIVE EXPERIMENTAL PROOF

Experimental evidence linking DNA with inheritance was initially, if unintentionally, provided in 1928 by the microbiologist Fred Griffith whilst studying the pathogenicity of the pnuemonia-causing bacterium *Diplococcus pnuemoniae*. He noted that injection with mixtures of heat-killed pathogenic (smooth) bacteria and living non-pathogenic (rough) bacteria could induce pneumonia in experimental animals (Fig. 3.2). This resulted from the transformation of a small percentage of the non-pathogenic bacteria into virulent pathogenic bacteria. Pathogenicity was realised to be related to the presence of a smooth bacterial outer capsule. Thus, the transformation of non-pathogenic rough cells was accompanied by the formation of a new smooth polysaccharide-rich outer capsule.

Almost definitive evidence that the 'transforming factor' was actually bacterial DNA was not provided until 1944, and represented over a decade of studies by Oswald Avery and colleagues at the Rockefeller Institute, USA. They first demonstrated that transformation required breaking or rupturing of the bacterial cell walls to release the transforming factor (Fig. 3.3). Biochemical separation of the components of bacterial lysates demonstrated that the actively transforming fractions were predominantly composed of DNA. Furthermore, transforming activity was lost if the fractions were pre-treated with DNA degrading enzymes (DNases) but retained if pre-treated with protein (proteases) or RNA degrading enzymes (RNases).

Final proof of DNA's role as the compositor of genes came in 1952 with the elucidation of the mechanisms whereby certain viruses (bacteriophages or simply phages) replicate within living bacteria. In a series of experiments by Alfred Hershey and Martha Chase at Cold Spring Harbor, USA, phage protein and DNA were differentially labelled with radioisotopes (Fig. 3.4). These experiments demonstrated that infection of bacteria by phages involved insertion of the phage DNA but not its outer protein coat. Phage infection provided the early means of inserting DNA into host bacteria, and is still widely used for introducing (or cloning) DNA into bacterial cells in order to rapidly propagate and express recombinant DNA molecules.

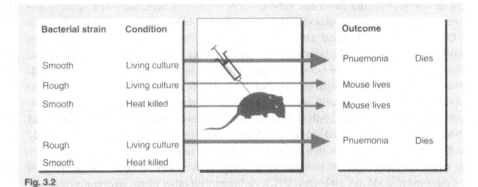

Fig. 3.2

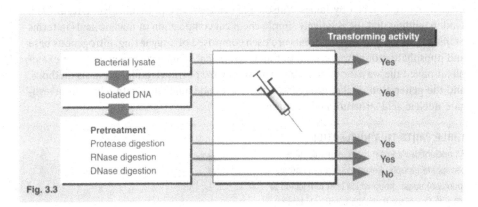

Fig. 3.3

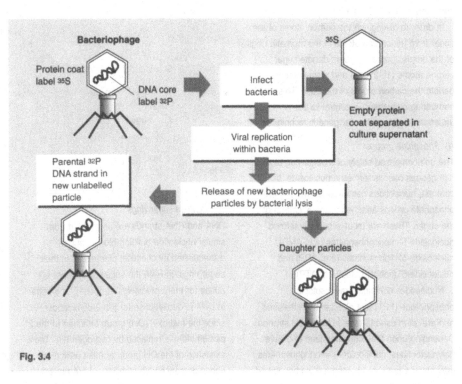

Fig. 3.4

4. THE CHEMICAL BUILDING BLOCKS OF NUCLEIC ACIDS

Nucleic acids are linear polymers constructed from a small number of chemically similar nucleotide building blocks composed of phosphate groups, a pentose sugar ring, and a nitrogenous base.

By the early 1920s the molecular vehicles of heredity had been narrowed down to the two principal macromolecules within chomosomes: proteins and nucleic acid. During the 1920s much of the chemical nature of nucleic acids was determined, principally from work in Hamilton Todd's laboratory in London, UK, and by Phoebus Levine at the Rockefeller Institute in the USA. They identified that in common with other macromolecules such as proteins which are composed of varying combinations of 20 amino acid building blocks, nucleic acid was composed of regularly repeating subunits termed nucleotides. Furthermore, only a limited variety of nucleotides existed in nature, and these were somehow linked together to form two types of polynucleotides remarkably similar in chemical composition, i.e. DNA and RNA. In the 1930s the simple chemistry by which these nucleotides are linked together as DNA or RNA strands was determined by Alexander Todd in Cambridge, UK. Combined with other discoveries concerning the physicochemical nature of DNA, the way was opened for the deduction of the double helix structure of DNA by James Watson and Francis Crick in Cambridge, UK, in 1953. Understanding that the relatively simple chemical composition of nucleic acids in terms of linear strands of nucleotide units are each comprised of a sugar ring, nitrogenous base and phosphate groups was crucial to the development of molecular genetics. For example, it paved the way for the development of *in vitro* polynucleotide synthesis methods, and the generation of chemically modified nucleotide analogues with which to investigate nucleic acid structure and function.

NUCLEOTIDE ANALOGUES

Chemical modification of phosphate groups, bases or the sugar-phosphate backbone of nucleotides can be used to generate a wide range of what are termed nucleotide analogues. Their incorporation into nucleic acid chains during their extension by DNA polymerases is central to many *in vitro* methods for gene detection (e.g. in gene probe labelling), isolation (e.g. by affinity purification), manipulation (e.g. mutagenesis), and analysis (e.g. nucleotide sequencing). For example, di-deoxynucleotides in which the 3' OH groups are replaced by hydrogen atoms are widely used in nucleotide sequencing methods since they specifically prevent chain extension.

OAs: 10–15, 297, 334, 448, 527, 531.
RAs: 2, 210, 211.
SFR: 10–13, 17–22, 143–5.

THREE PARTS TO A NUCLEOTIDE

All nucleotides contain three elements: (i) a phosphate group(s) linked to (ii) a five carbon (pentose) sugar group which is in turn joined to (iii) a flat aromatic molecule commonly termed a nitrogenous base (Fig. 4.1).

In order to distinguish the carbon atoms of the sugar from the carbon atoms in the aromatic rings of the bases, 'prime' numbers denote sugar carbon atoms (1', 2' etc.) and plain numbers denote the carbon atoms of the base. This numbering convention is central to the understanding of molecular genetic techniques.

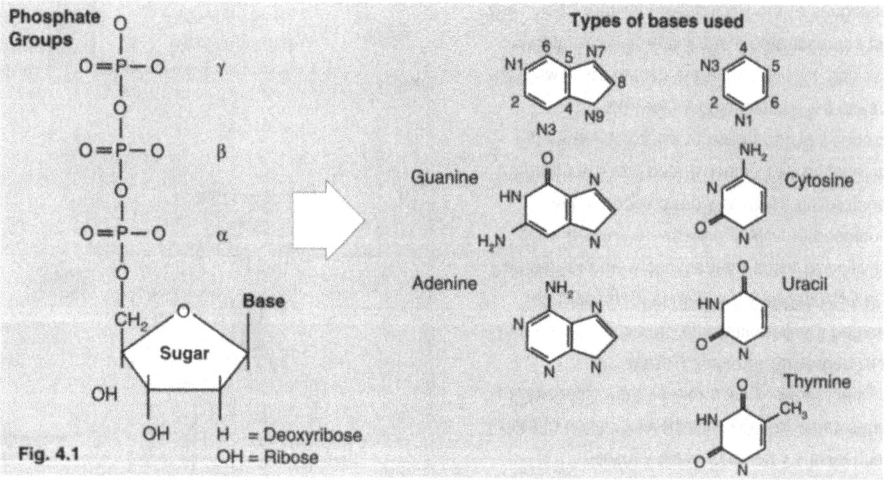

(i) Phosphate groups
The unit comprised solely of a sugar ring and a nitrogenous base is termed a nucleoside. In contrast, nucleotides contain one, two or three phosphate groups attached to the 5' carbon of the sugar. These are most accurately termed nucleoside-5'-monophosphates (e.g. AMP), nucleoside-5'-diphosphates (e.g. ADP) and nucleoside-5'-triphosphates (e.g. ATP).

Nucleotides containing radioisotopic phosphorous (^{32}P) have also been synthesised and are often used to label nucleic acid strands. Depending upon the chemistry used to cleave polynucleotides, nucleoside-3'-monophosphates may also be produced having a 3' carbon-linked phosphate group.

Fig. 4.1

(ii) Pentose sugar rings
DNA and RNA strands are assembled from similar nucleotide building blocks, but are distinguished by chemical differences in their sugar rings. In RNA the sugar component is ribose (or more precisely β-D-ribose). The sugar in DNA is deoxyribose (or β-D-2-deoxyribose) since the hydroxyl (OH) group attached to the 2' carbon atom is replaced by hydrogen (H). The existence of the OH group at this position in ribose decreases the stability of RNA under conditions such as high pH.

(iii) Nitrogenous bases
Two types of naturally occurring nitrogenous bases exist termed purines and pyrimidines, consisting of one or two (linked) aromatic rings respectively. DNA and RNA are both built up from two purine-containing nucleotides and two pyrimidine-containing molecules. Both DNA and RNA are constructed from the same purines, adenine and guanine (abbreviated as A and G), and the pyrimidine cytosine (C). The pyrimidine thymine (T) is limited to DNA, being replaced in RNA by the pyrimidine uracil (U).

FORMING POLYNUCLEOTIDE STRANDS

Nucleotides used to construct polynucleotide chains are termed nucleotide triphosphates since they contain three phosphate groups linked to the 5′ carbon of the pentose sugar (their relative positions designated α, β and γ). The bonds between phosphate groups are energy rich and represent an energy store used by many cellular processes. The linking or polymerisation of nucleotides into new strands involves an enzyme-mediated reaction between the 5′ end of one with the 3′ OH group of another (Fig. 4.2). This leads to the formation of a phosphodiester bond and the accompanying release of the α and β phosphates as pyrophosphate (PPi).

Polynucleotide chains therefore possess a characteristic sugar–phosphate backbone from which the protruding bases are available for hydrogen bonding as base pairs. Consequently, directionality can be ascribed to polynucleotides based upon the presence of a free 3′ OH group at one end and a free 5′ phosphate group at the other. This is critical in understanding and manipulating the natural functions of nucleic acids. For example, *in vivo* replication and *in vitro* synthesis of DNA or RNA by enzymes occurs by addition of new nucleotides to the 3′ end of the strands during extension (i.e. added in a 5′ to 3′ direction). Contrastingly, *in vitro* chemical synthesis occurs in a 3′ to 5′ direction. Altering 5′ and 3′ reactive groups can therefore profoundly affect nucleotide polymerisation. Indeed, this forms the basis of several methods for nucleic acid analysis or manipulation.

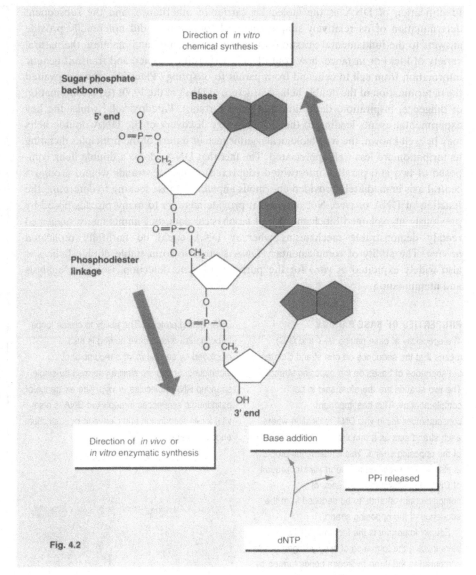

Direction of *in vitro* chemical synthesis

Sugar phosphate backbone

Bases

5′ end

Phosphodiester linkage

OH
3′ end

Direction of *in vivo* or *in vitro* enzymatic synthesis

Base addition

PPi released

dNTP

Fig. 4.2

Abbreviations and Conventions

Several abbreviations are commonly used for different types of nucleotides, their analogues, and their combination into polynucleotides. The term deoxynucleotide triphosphate is often abbreviated to dNTP (each base denoted dATP, dTTP, etc.). Ribonucleotides are denoted by the prefix 'r', and dideoxynucleotides are denoted by a 'dd' prefix (e.g. ddATP). The linear sequence of polynucleotides are represented as single letters denoting individual bases and/or including phosphate linkages. Nucleic acid sequences are also usually written in a 5′ to 3′ direction although in a duplex, the upper sequence (coding strand) runs in a 5′ to 3′ direction, and the lower (non-

coding) sequence runs in a 3′ to 5′ sequence.

Rare Nucleotides

A small proportion (varying between organisms) of cytosine and adenine residues have methyl groups added (by DNA methylases) to carbon 5 or 6 of the base rings, and are termed 5-methyl cytosine and 6-methyl adenine respectively. In eukaryotes, 5-methyl cytosine often occurs at a dinucleotide GC pair. Undermethylation of certain genes may be a key event leading to their expression, whilst methylated GC dinucleotides may be features of genetically silent sequences. Although methyl cytosine is not as prevalent in

prokaryotes, the methylation of adenines is an important part of their defence mechanism against viral infection, whereby non-methylated (viral) DNA is cleaved by restriction endonucleases. Several additional nucleotides have also been described which have subtle structural and functional roles. For example, transfer RNA (tRNA) molecules utilised during *in vivo* translation may contain pseudouridine and dihydrouridine, and several bases may have one or two methyl groups added. Some tRNA molecules may also utilise the base inosine within their anticodons, allowing for the degeneracy in the genetic code.

5. FORMATION OF THE DNA DOUBLE HELIX

The specific association of complementary DNA strands into a double helix (or duplex) is fundamental to life, and is exploited in numerous recombinant DNA techniques and approaches.

Identification of DNA as the molecular carrier of inheritance, and the subsequent determination of its relatively simple chemical composition, did not readily provide answers to the fundamental questions of how it could dictate and maintain the natural variety of life. For instance, how could DNA be faithfully copied and transmit genetic information from cell to cell, and from parent to offspring? Plausible answers awaited the determination of the double helical structure of DNA in the 1950s (by a combination of diligence, inspiration, deduction and good fortune). Paradoxically, whilst the key experimental events leading to the revolutionary discovery of the DNA double helix may be well known, the true biological significance or nature of the principles dictating its formation are less well appreciated. The fact that DNA adopts a double helix composed of two anti-parallel, intertwined (duplexed) chains or strands wound around a central axis immediately provided enormous impetus to those seeking to determine the function of DNA *in vivo*. Not only did it provide answers to many puzzles posed by previously unexplained biochemical and biophysical data, but immediately suggested readily demonstrable mechanisms whereby DNA could be faithfully replicated *in vivo*. The ability of complementary DNA strands to form stable double helices is also widely exploited *in vitro* for the purposes of gene detection, isolation, analysis and manipulation.

PROPERTIES OF BASE PAIRING

The specificity of base pairing (A-T and G-C) means that the sequence on one strand dictates the sequence of bases on the opposing strand. The two strands are therefore said to be complementary. This has important consequences for *in vivo* DNA replication where each strand acts as a template for the generation of the opposing strand. This fundamental property is also widely exploited for the *in vitro* replication of DNA, and allows the sequence of complementary strands to be deduced from the sequence of the opposing strand.

Equally important is the fact that A-T base pairs involve the formation of two hydrogen bonds compared to the three hydrogen bonds formed by a G-C base pair (Fig. 5.1a). Duplexes rich in G-C base pairs therefore require relatively more energy to separate than those with equal or greater proportions of A-T base pairs. *In vitro*, such differences between sequences can be detected experimentally by determining the amount of energy (i.e. heat) required to disrupt or denature the duplexes. Base pairing can also occur between nearby complementary, inverted repeat (or palindromic) sequences within a single strand of nucleic acid (Fig. 5.1b). It has been proposed that such transient base pairing between inverted repeats may occur in DNA duplexes *in vivo* forming semi-stable cruciform (or stem) loops. These may permit the interaction of

DNA-binding proteins. The ability to create loops through intra-strand base pairing is also employed by several *in vitro* recombinant techniques. Loops are similarly formed by single stranded RNA molecules *in vivo*. The existence of palindromic sequences in duplexed DNA is also vital for its recognition and cleavage by restriction endonucleases.

EARLY DATA EXPLAINED

By 1951 Edwin Chargaff at Columbia University, USA, had established that DNA from different organisms was composed of differing amounts of the four nitrogenous bases. The amount of the purine adenine was however roughly equivalent to the amount of the pyrimidine thymine. Similarly, the amount of the purine guanine was roughly equal to that of the pyrimidine cytosine. Known as Chargaff's rules, the significance of this relationship was only realised following the demonstration by X-ray diffraction studies performed by Maurice Wilkins and Rosalind Franklin in London, UK. They discovered that DNA adopts a regular and precise helical structure 20 Å in diameter. These data provided the clues necessary for the, now familiar, double helix to be deduced by James Watson and Francis Crick in Cambridge, UK, in 1953. Accordingly, a double helix structure fitted the experimentally measured diameter of crystallised DNA fibres. Hence, hydrogen bonding between bases generates base pair units occupying equal areas along the helix. Furthermore, because of the base pair symmetry, they can be accommodated into the double helix in any permutation.

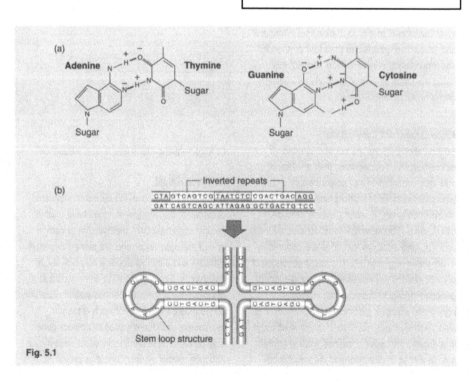

Fig. 5.1

THE DOUBLE HELIX

The double helix is derived from the anti-parallel winding of two complementary strands around a central axis. This leaves the DNA sugar-phosphate backbones on the solvent-exposed outer surface and the purine and pyrimidine bases facing into the centre of the helix. The energy required to hold the double helix together comes partially from the formation of hydrogen bonds between bases. Perhaps more importantly, it also derives from base stacking which brings the atoms into favourable Van der Waal's contact allowing the overlapping of π orbitals. Exposure of the polar and charged groups on the sugar-phosphate backbone to the solvent phase and shielding of the largely non-polar bases also significantly stabilises the double helix structure. Because the sugar-phosphate backbones of two DNA strands extend further from the helix axis than the bases, the double helix contains two grooves of different size termed the major and minor grooves (Fig. 5.2). These are important structural features determining the recognition by and interaction with DNA-binding proteins, including restriction endonucleases.

Right or left handed

The first three-dimensional model of the DNA double helix was constructed in a right-handed form with the two strands turning to the right up the helix. This was calculated to be the most stable structure. Although almost all DNA exists as a right-handed double helix (termed the B-form), it has been possible to synthesise DNA molecules *in vitro* which form a left-handed double helix (termed the Z-form). This has been particularly successful using repeating GCGC sequences, especially if the structure is stabilised by the methylation of C residues. Significant amounts of Z-form DNA have yet to be found *in vivo*.

Nevertheless, it is proposed that regions of Z-form DNA may be present in localised regions of DNA to permit relaxation and unwinding of portions of tertiary DNA structure. This would then allow the binding of regulatory proteins responsible for dictating gene expression (e.g. transcription factors).

OAs: 16–25, 393–402.
RAs: 2, 191, 192, 194.
SFR: 14–21, 119, 120, 122–4.

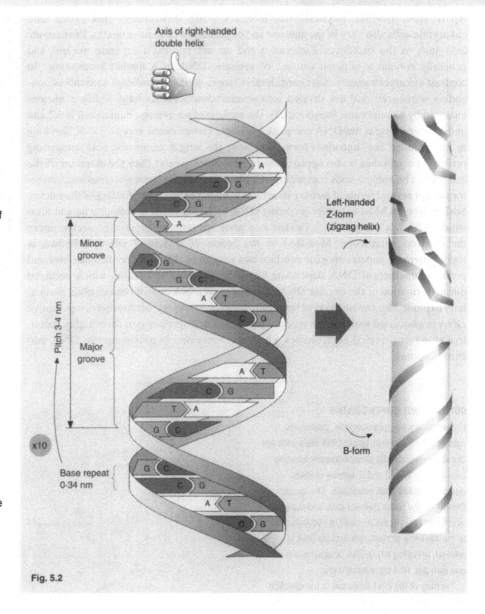

Axis of right-handed double helix

Pitch 3-4 nm

Minor groove

Major groove

x10

Base repeat 0·34 nm

Left-handed Z-form (zigzag helix)

B-form

Fig. 5.2

In vitro exploitation

Formation of the double helix requires the stable and specific association (or annealing) of two DNA strands into a duplex. In comparison to the strength of the chemical bonds holding each sugar-phosphate backbone together, the hydrogen bonds holding the DNA duplex together are relatively weak. *In vitro*, duplexes can therefore be broken into two separate intact DNA strands by simply adding heat energy, a process termed denaturation. Upon cooling the strands will re-form the original duplex, termed reannealing. This ability to break and re-form duplexes *in vitro* is widely exploited in many recombinant DNA technologies.

Confusingly perhaps, the *in vitro* process of duplex formation is also referred to as hybridisation since this artificial process may be manipulated to dictate the specificity of annealing. Thus, conditions may be chosen that allow formation of stable heteroduplexes, i.e. between strands of imperfectly matched sequence, or between RNA and DNA strands.

6. PACKAGING OF DNA WITHIN CELLS

DNA within cells adopts several forms of ordered tertiary structures beginning with the formation of coiled and supercoiled helical DNA under the control of enzymes known as topoisomerases.

Apart from obvious physical (i.e. morphological) differences, prokaryotic and eukaryotic cells also vary in the amount and organisation of their genomes. Prokaryotic cells such as the bacterium *Escherichia coli* do not possess a separate nucleus and generally contain a minimal amount of genomic DNA in a single chromosome. In contrast eukaryotic genomes are considerably larger, contain significant amounts of non-coding sequences, and are divided into several chromosomes held within a nucleus enclosed by a membrane. Paradoxically, the nucleus of an average human cell is 0.2 mm in diameter whereas the DNA comprising a single chromosome may be 1.7–8.5 cm long in its extended (i.e. uncoiled) form. Similarly, the length of nucleic acid comprising prokaryotic genomes is also approximately 1000 times greater than the diameter of the cell itself. The nucleic acids comprising eukaryotic and prokaryotic genomes must therefore adopt several forms of tertiary structure involving twisting and folding of the nucleic acid molecules. Most naturally occurring DNA molecules exist in a double helical form which, with few exceptions, is twisted into what are termed 'supercoils' visible under the electron microscope. Mediated by the action of a class of enzymes known as topoisomerases, supercoiling can produce two structures. These are termed negative and positive superhelical DNA depending upon the direction of twisting which occurred during formation of the circular DNA. The ability of DNA to form supercoils is particularly exploited for the isolation of bacterial plasmids. In eukaryotes however, supercoiled DNA is associated with various types of proteins (nucleoproteins) to form highly repetitive structures termed nucleosomes which are important in packaging the DNA into chromosomes.

THE KEYS TO DNA RELAXATION

The processes of gene transcription and replication require that localised regions of the highly compacted DNA (termed heterochromatin) must be relaxed and unwound (referred to as euchromatin). This is achieved by the interactions of histone and non-histone nucleoproteins (e.g. high mobility group proteins or HMGPs). These are in turn regulated by a series of *in vivo* reactions such as methylation, acetylation and phosphorylation which alter their affinity for DNA. For example, histones are highly phosphorylated just before mitosis when maximum compaction of the DNA occurs. Although the precise mechanisms remain to be defined, both types of nucleoprotein are involved and may control gene expression by binding hormones and other regulatory molecules. Nucleoproteins therefore, represent potential candidates for therapeutically modulating the expression of specific genes.

COILING AND SUPERCOILING

Unlike eukaryotic chromosomes, prokaryotic, mitochondrial and chloroplast DNA molecules are circular, generated by phosphodiester bonding between the free 3' and 5' termini of linear double-stranded DNA molecules. This circular DNA contains twists (termed coils and supercoils) which form during the process of circularisation. In the absence of twists the circular DNA is relaxed, adopting the B-DNA double helix (i.e. one turn per 10.4 base pairs (bp)).

Twisting of the DNA molecule in the direction which unwinds the double helix results in circular DNA containing a deficiency of turns (Fig. 6.1). It therefore adopts a strained structure referred to as negative (or right-handed) superhelical DNA. All naturally occurring DNA adopts a negative superhelical structure with one negative twist approximately every 20 turns. Twisting in the opposite direction (which adds turns) generates positive (or left-handed) superhelical DNA. The strain caused by the loss of turns in negative superhelical DNA is relieved by the disruption of hydrogen bonds or formation of supercoils, i.e. by twisting in the opposite direction. Under *in vivo* conditions both forms of negative superhelical

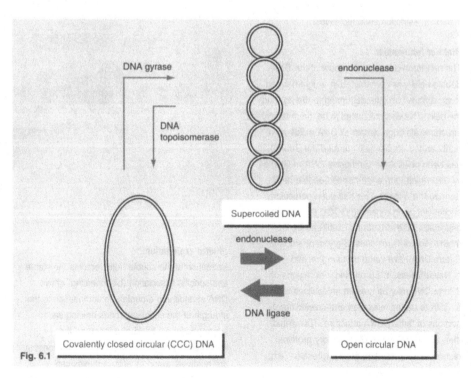

Fig. 6.1

DNA may co-exist. A DNA molecule may therefore contain regions of twisted structure and regions of disrupted hydrogen bonds (termed bubbles).

Positive superhelical DNA can however only compensate for the torsional strain by forming positive superhelices.

INFLUENCE ON DNA STRUCTURE

Supercoiling has a profound effect upon both the structure and function of DNA. To an extent structural changes occur by the introduction and removal of supercoils. Supercoil formation requires the input of energy and consequently imparts energy into the DNA structure. The storage of energy in the form of torsional stress within local DNA regions influences the equilibrium between the various DNA structures and also allows strand separation to occur. Accordingly, negative supercoiling is thought to promote DNA processes such as replication, recombination and transcription since the torsional stress imposed provides the energy required for localised unwinding and DNA strand separation. The introduction of approximately one negative turn for every 200 base pairs in negative supercoiled DNA may explain the proposed structural change from the B (right-handed) form to the Z (left-handed) form of DNA double helices. This change is however also influenced by the presence of GC dinucleotides, which in turn become more readily accessible for methylation.

TYPE I AND TYPE II TOPOISOMERASES

In addition to local environmental conditions, variations in DNA structure are under the control of specific enzymes termed topoisomerases. These alter the degree of DNA supercoiling by catalysing the breakage and rejoining of DNA strands (Fig. 6.2). Two types of topoisomerase have been described: (i) type I enzymes which break only one strand, and (ii) type II (e.g. gyrases) which break both strands. The action of topoisomerase type I results in relaxation of the supercoiled structure whereas topoisomerase type II introduces or removes two negative supercoils per reaction.

Their action essentially involves three steps: (i) the breakage of one or both strands, (ii) the passage of a segment of DNA through the break, and (iii) resealing of the break. The precise mechanism for the action of *E. coli* topoisomerase II (gyrase) reveals that the 5′ end of the cleaved DNA remains covalently attached to the enzyme via a tyrosine residue hydroxyl group. The 3′ hydroxyl at the other end of the DNA chain then attacks the DNA intermediate

and restores the circular form of the DNA. This reaction also involves hydrolysis of ATP to release sufficient energy to restore the enzyme's active conformation. Type II topoisomerases are therefore also ATPases.

Linking number of DNA

The coiling and supercoiling topology of DNA may be defined mathematically, and related to what is termed the linking number (L). This is defined as the number of times one strand of DNA winds around the other strand in a right-handed direction. DNA structures which differ in linking number are termed topological isomers or topoisomers. Two additional terms have also been introduced: (i) the writhing number or degree of supercoiling (W), and (ii) the twisting number (T) or number of Watson–Crick turns. The relationship between these parameters is $L = T + W$. Consequently, a change in the linking number usually leads to differing degrees of supercoiling (W) rather than altering the number of Watson–Crick turns. Changing the supercoiling (W) may lead to negative or positive supercoiling.

OAS: 26–30, 36–8. **RAs:** 3–5, 16, 17, 216, 217. **SFR:** 22, 24, 27, 28, 30, 31, 36–9.

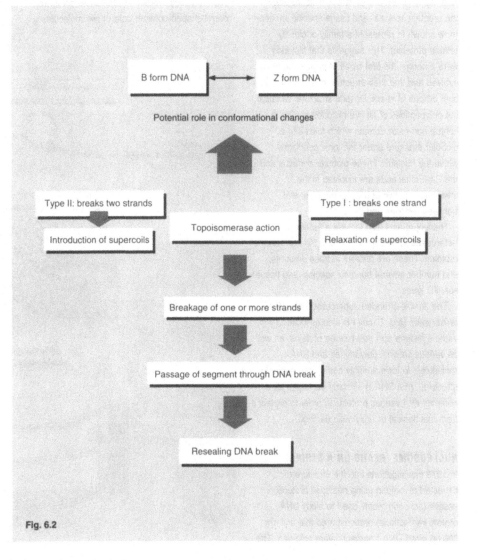

Fig. 6.2

DNA within eukaryotic cell nuclei is associated with histone proteins (i.e. as chromatin) and can adopt several levels of packaging by the formation of nucleosomes and their linkage into a solenoid structure.

NUCLEOPROTEIN TYPES AND FORMS

The DNA in eukaryotic cells, termed chromatin by early cytologists, is associated with various proteins. In fact these comprise approximately half of the chromatin mass. The most intimately associated proteins are termed nucleoproteins, of which the most abundant class is the highly basic histone proteins.

Five different types of histone polypeptide, denoted H1, H2A, H2B, H3 and H4, have been described. Their structure varies in the degree of similarity (or conservation) between different species and tissues. For example, the histones H4 and H3 are highly conserved, with only two of the 102 amino acids differing between the H4 histone from cows and peas. H2A and H2B are less highly conserved whilst H1 histones show the greatest species- and tissue-specific variation (now known to represent a family of closely related proteins). This suggests that histones were amongst the first types of proteins to have evolved and that their structure has changed little over billions of years. Despite structural variation, the polypeptides of all five histones contain a central non-polar domain which folds into a globular structure under the ionic conditions within the nucleus. These globular domains and the C-terminal ends are involved in the interactions between histone proteins, and between histones and DNA.

Nucleoproteins also include a further heterogeneous group of proteins, the non-histone proteins. These are present in trace amounts, and number several hundred species- and tissue-specific types.

The double-stranded supercoiled DNA of prokaryotes (e.g. *E. coli*) is also associated with various histone and non-histone proteins, as well as various cations, polyamines and RNA molecules, to form a highly compacted nucleoid structure. Viral DNA is similarly packaged as a complex with various proteins in order to protect it from mechanical or enzymatic damage.

NUCLEOSOME 'BEADS ON A STRING'

In 1973 investigations into the structure of extracted chromatin using nuclease protection assays (now commonly used to study DNA–protein interactions) demonstrated that uniform 200 bp sized DNA fragments were obtained. This provided the first conclusive evidence that DNA was intimately associated with protein and suggested that the nucleoproteins within chromatin were interacting with DNA in a regular manner. Electron microscopy also revealed that preparations of isolated nuclear DNA consisted of linear arrays of spherical structures termed nucleosomes, giving the appearance of beads on a string (Fig. 6.3a). By 1975 Robert Kornberg was able to demonstrate that each nucleosome contained equal amounts of each of the histones, except histone H1.

Definitive proof that the acidic and negatively charged DNA remained on the outside of the nucleosome complex was subsequently provided by neutron scattering studies. They revealed that each regularly repeated nucleosome comprises a barrel-shaped octamer core of two molecules each of histones H2A, H2B, H3 and H4 (Fig. 6.3b). These are organised into tetramers comprising $(H3)_2$ $(H4)_2$, with a H2A-H2B dimer on each face of the barrel. It is now known that approximately 146 bp of negatively supecoiled DNA are wound twice around each nucleosome core, although this figure varies from 160 to 240 across different eukaryotic species. This effectively reduces the DNA strand length by a factor of about 7, which alone is insufficient to package the DNA within a cell nucleus.

The lysine-rich histone H1 is bound to the outer surface of the nucleosome. This seals the DNA turns and aids the linking of nucleosomes.

OAs: 31–7, 42, 91. **RAs:** 6–8, 10. **SFR:** 5, 22–4, 26, 27, 30, 31, 37–40.

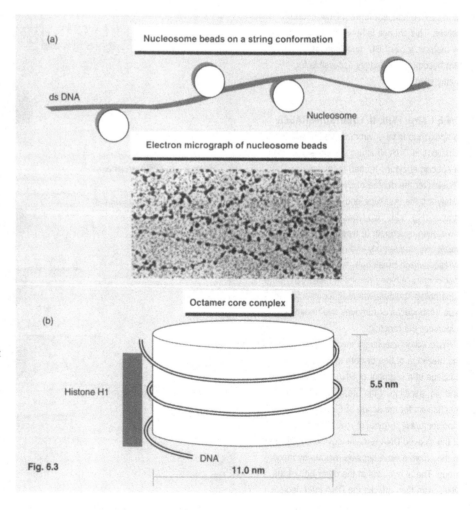

(a) Nucleosome beads on a string conformation

ds DNA

Nucleosome

Electron micrograph of nucleosome beads

(b) Octamer core complex

Histone H1

5.5 nm

DNA

11.0 nm

Fig. 6.3

Linking nucleosomes together

Nucleosomes are linked together into polynucleosome chains (giving the original 'beads-on-a-string' observed by electron microscopy) by stretches of linker DNA varying in length from 20 to 80 bp. This gives a minimum nucleosome repeat frequency of 165 bp. This repeat frequency may be related to DNA sequence, since it varies according to the transcriptional activity of the cell (Fig. 6.4a). Accordingly, transcriptionally inactive chicken erythrocytes have a repeat frequency of 212 bp whereas transcriptionally active yeast cells have a repeat frequency of 165 bp. The location of histone H1 on the linker DNA, and the phosphorylation/ dephosphorylation of H1 accompanying chromatin condensation and decondensation during replication, therefore suggests that it plays a key role in the linking of nucleosomes.

THE 30 nm CHROMATIN FIBRE

Polynucleosomes must however adopt a further level of packaging in order for the DNA to be condensed sufficiently to be accommodated within the eukaryotic nucleus as chromatin. Electron microscopy studies by Aaron Klug and colleagues at the MRC Laboratories, Cambridge, UK, in the late 1970s established that chromosomes were composed of a chromatin fibre 30 nm in diameter. At low ionic strength this dissociated to form 10 nm wide fibres known as nucleofilaments. They therefore proposed that the 30 nm chromatin fibre was composed of nucleosomes packaged together to form a solenoid structure (Fig. 6.4b). This is composed of 6 nucleosomes per turn with histone H1 located inside the 30 nm chromatin fibre (i.e. one H1 on the inner face of each radiating nucleosome). Despite controversy based on conflicting data provided by protease or antibody accessibility studies, the solenoid model has now been confirmed using more precise neutron scattering techniques.

PACKAGING INTO CHROMOSOMES

The 30 nm chromatin fibre is the normal form of DNA within the nucleus of resting eukaryotic cells. This reduces the area occupied by the bead-on-a-string structure by about seven times.

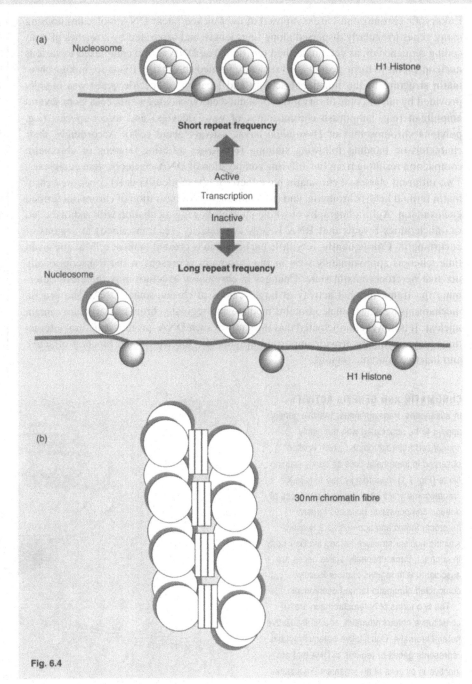

Fig. 6.4

During cell division (i.e. at metaphase) chromosomes become visible as the solenoid structure is further condensed to form supercoiled loops radiating from a non-histone protein core referred to as the scaffold. Approximately 85 kbp of DNA is contained within each loop giving a final structure of about 750 nm in width. This represents the highest order of packaging displayed by chromosomal DNA. The chromatin structure of a eukaryotic chromosome also resembles the nucleoid structure adopted by genomic DNA within bacteria. For example, the *E. coli* nucleoid is now known to be formed from 50–100 loops, each containing about 100 000 bp (100 kbp) of supercoiled DNA, attached to a central scaffold rich in proteins and RNA.

7. CHROMATIN STRUCTURE AND THE FUNCTIONAL ACTIVITY OF GENES

Transcriptionally active regions of chromosomes are characterised by a relaxed chromatin structure (termed euchromatin) whilst inactive regions adopt a more condensed structure (termed heterochromatin)

Eukaryotic chromosomes are composed of unbranched linear DNA molecules housing many genes irregularly dispersed along their length and separated by stretches of non-coding sequence of, as yet, unascribed function. The DNA is also complexed to various nucleoproteins to form a compacted structure termed chromatin. Evidence linking chromatin structure to the functional activity and location of specific genes was initially provided by microscopic observations of stained chromosomes within cells from several amphibian (e.g. lampbrush chromosomes of newt oöcytes) and insect species (e.g. polytene chromosomes of *Drosophila* larvae salivary gland cells). Accordingly, their characteristic banding following staining represents differing degrees of chromatin compaction resulting from the differing composition of DNA–nucleoprotein complexes. Two different classes of chromatin have therefore been identified: (i) condensed chromatin termed heterochromatin, and (ii) a more open configuration of chromatin termed euchromatin. Autoradiography of whole nuclei following incubation with radiolabelled ribonucleotides reveals that RNA is only synthesised (i.e. transcribed) at regions of euchromatin. Consequently, very little euchromatin is present in most cells at any given time whereas approximately 90% of the chromatin is present in the transcriptionally inactive heterochromatin state. Changes in chromatin structure may therefore determine the transcriptional activity of large regions of chromosomal DNA. The precise mechanisms by which nucleoproteins mediate and regulate chromatin structure remain unclear. It is however anticipated that the study of such DNA–protein structural interactions may eventually provide important insights into the nature and function of active and inactive genomic regions.

POLYTENE CHROMOSOMES
Chromosomal structure is difficult to discern in interphase cells due to the extended nature of the chromatin. This restricts karyotype analysis to metaphase cells. However, chromosomal structure and specific gene regions may be studied during interphase by staining and banding techniques using the secretory cells of some insects (e.g. from the salivary gland of *Drosophila* larvae) due to a process termed polyteny. This produces polyploid cells in which the chromosomal DNA is replicated a thousand or more times without mitosis or homologous chromosome separation (i.e. no cytokinesis); producing what are termed polytene chromosomes representing many homologous chromosomes arranged in register. Their existence not only aided by early physical localisation of some genes but also facilitated by the study of the genetic processes controlling embryo and larval development.

CHROMATIN AND GENETIC ACTIVITY
In eukaryotes, transcriptionally inactive genes appear to be associated with the highly condensed heterochromatin which is often observed in interphase cells as darkly staining fibres (Fig. 7.1). Accordingly, the female X chromosome inactivated during the process of dosage compensation becomes largely heterochromatic and appears as a densely staining nuclear structure termed the Barr body. In contrast, transcriptionally active genes are associated with regions of more loosely compacted chromatin termed euchromatin.

The two forms of heterochromatin are: (i) constitutive heterochromatin, and (ii) facultative heterochromatin. Constitutive heterochromatin represents genes or regions of DNA that are inactive in all cells of an organism. Facultative heterochromatin represents genes or regions which are inactive in some but not all cells. Constitutive heterochromatin is located either in bands throughout a chromosome or in only one arm. During interphase, regions of constitutive heterochromatin may aggregate to form chromocentres, varying between cell type and stage of development. Much of the constitutive heterochromatin contains simple repeated DNA sequences or satellite DNA, some of which may play a role in meiotic chromosome pairing. Facultative heterochromatin may represent stable differences in genetic activity within different cells. Consequently, embryonic cells possess very little compared to specialized cells.

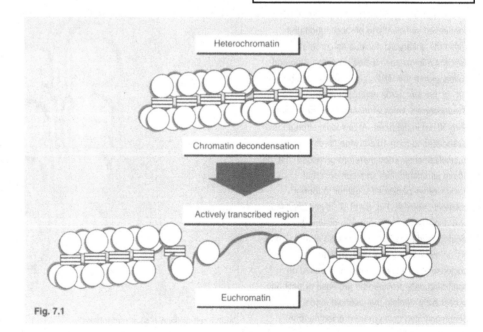

Fig. 7.1

OAs: 39–43, 105, 191–3. RAs: 9–11, 15. SFR: 22, 25, 27, 28, 36–9, 41.

MAPPING ACTIVE *DROSOPHILA* GENES

Staining of polytene chromosomes produces alternating light and dark bands which represent sets of 1024 homologous looped domains arranged in a register. The five thousand or so visible bands vary in thickness but are consistent between genetically indentical individuals allowing a physical map to be produced. Since approximately five thousand essential proteins are produced in *Drosophila* it has been suggested that each individual band represents a unique genetic region, or possibly an individual gene. Analysis of polytene chromosome banding patterns between mutant flies has therefore allowed the order of at least some genes to be partially established. This often provides a starting point for fine mapping using techniques such as chromosome walking.

Further evidence that individual bands in *Drosophila* polytene chromosomes represent a separate gene is provided by the observation of chromosome 'puffs' (Fig. 7.2(a)). Appearing and vanishing in a temporal pattern, these are now known to be sites of intense RNA synthesis (i.e. actively transcribed). They arise by localised chromatin decondensation to form loops of euchromatin which extend outside the longitudinally arranged multiple heterochromatin fibres.

'LAMPBRUSH' CHROMOSOMES

Visual evidence for the chromosomal location of transcriptionally active genes has also been provided by the analysis of 'lampbrush' chromosomes within some amphibian oöcytes. These arise in early oöcyte differentiation and may be maintained for months or even years, and result from the pairing of homologous chromosomes to produce a structure containing a total of four chromatids (Fig. 7.2b). Throughout this time they periodically expand and become transcriptionally active, producing both RNA and then protein. This active gene expression is accompanied by the formation of long chromatin loops covered with newly transcribed RNA. It is this that gives them a 'lampbrush'-like appearance visible under light microscopy.

Each lampbrush chromosome is thought to contain approximately 10 000 different chromatin loops with the intervening DNA condensed into structures termed chromomeres. Transcriptional activity varies from one loop to another although as transcription increases, the chromosomes gradually become larger. Shorter RNA transcripts tend to accumulate at one end of the chromosome giving the structure a tapered appearance. Some loops are transcribed discontinuously so that their diameter varies along the length of the chromosome as one transcription unit ends and another one begins.

In situ hybridisation studies have confirmed that these loops are analogous to chromosome puffs, i.e. they reflect a transcriptional unit.

(a)

Polytene chromosome banding patterns

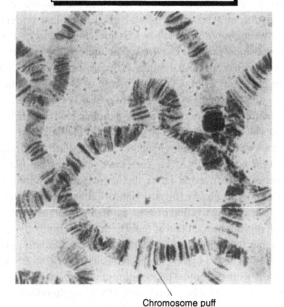

Chromosome puff

(b)

Lampbrush chromosome

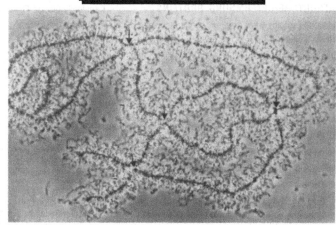

Fig. 7.2

Lampbrush chromosomes do however exhibit some unusual features such as the production of long transcripts from highly repeated sequences not normally transcribed. Furthermore, in *Xenopus* frogs most of the RNA required for development of the oöcyte is already produced by the time the lampbrush chromosomes become fully active. Although studies of lampbrush chromosomes confirm that extended looped domains in chromatin represent functional units, their complete function remains to be determined.

25

8. TYPES AND FUNCTIONS OF DNA–PROTEIN INTERACTIONS

An enormous diversity of long-lasting and transient DNA–protein interactions, involving many types of proteins, may be defined which are vital in determining the structure and function of DNA.

The co-isolation of DNA with histone proteins in early eukaryotic chromosome preparations initially confused the issue of which macromolecule was the vehicle of inheritance. Significantly however, it also provided the first evidence for the intimate interaction between proteins and DNA. Considerable research has subsequently confirmed the functional importance of DNA–protein interactions and has revealed the existence of an enormous range of DNA-binding proteins. For example, histone proteins play a vital role in maintaining the tertiary structure of DNA molecules which allows them to be packaged within cells. A host of different proteins have also been found to be either permanently or transiently bound to genomic DNA, performing essentially overlapping roles in determining both its structure and functional activity. Accordingly, the binding of topoisomerase enzymes mediates the localised unwinding of the normally highly condensed nuclear DNA (i.e. chromatin) to allow the binding of other enzymes (e.g. polymerases) involved in DNA replication, transcription or repair. Additional DNA-binding proteins also perform regulatory functions which determine the temporal and cell type-specific activation of gene expression. For example, transcription factors interact with promoter or enhancer sequences to determine the activation of transcription by influencing the binding of RNA polymerases. Similarly, the transcription of bacterial operons is regulated by several mechanisms employing DNA-binding proteins to induce or repress transcription. The isolation and characterisation of DNA-binding proteins, and analysis of their interactions with DNA, are therefore vital to the understanding and exploitation of genetic mechanisms both *in vitro* and *in vivo*.

MODELS FOR DRUG DEVELOPMENT

Molecular biological techniques are enhancing the ease and ability with which structural and functional regulatory sequences comprising individual genes may be detected. This is further being enhanced by the availability of precise three-dimensional models of the interaction of DNA with DNA-binding proteins. These are often based upon atomic coordinates provided by X-ray crystallography of DNA–protein complexes, or computer-based simulations. Such information increases our fundamental understanding of normal and pathological gene regulation *in vivo*. It is also opening up new avenues for exploitation, for example, the development of new drug compounds capable of modulating the transcription of specific genes by interacting with structures on either the DNA-binding proteins or their corresponding regulatory sequences.

STRUCTURE-DETERMINING PROTEINS

Histone proteins were amongst the first classes of DNA-binding proteins to be defined based upon their intimate association with the highly condensed DNA (i.e. chromatin) found within eukaryotic chromosomes. Several forms of histones are intimately involved in determining the level of DNA tertiary structure and packaging of DNA within cells. For example, they play a vital role in the condensation of the DNA molecules into easily separated chromosomes during cell division.

A range of additional proteins, many of which are enzymes, have since been defined whose transient binding and activation determines the transition of DNA from highly condensed to open structures (Table 8.1). These include enzymes which induce localised structural relaxation or unwinding of the DNA (e.g. the topoisomerase enzymes), and those involved in maintaining open conformations (e.g. the single stranded DNA binding proteins or SSBs). Changes in DNA structure ultimately determine its accessibility to further DNA-binding proteins required for functional activities including transcription, replication and repair. Several enzymes also transiently bind to DNA and thereby modify its primary structure (e.g. endonucleases and methylases).

OAs: 44–56, 122, 156–62, 204, 225–35, 296.
RAs: 12–15, 53, 77–81, 94–8.
SFR: 27, 28, 37–9, 72, 74–8, 90, 104, 110.

Table 8.1

Type of process: protein	Characteristics	Function
DNA packaging: histone	11–15 Kdal monomer H2A H2B H3 H4	Nucleosome formation core octamer
DNA packaging: histone	23 Kdal monomer H1	Nucleosome linking
DNA: conformation: high mobility group	25 Kdal monomer HMG 12	Non-specific binding affecting nucleosome displacement
DNA: conformation: high mobility group	10 Kdal monomer	Nucleosome associated factors
DNA conformation: *E.coli* in single strand binding	19 Kdal tetramer	Prevent reassociation during replication of *E.coli* DNA
DNA conformation: rep protein	68 Kdal monomer	Involved in DNA unwinding in replication of *E.coli* DNA
DNA conformation DNA B (helicase)	50 Kdal hexamer	Involved in DNA unwinding in replication of *E.coli* DNA
DNA conformation: DNA topoisomerase I and II	100 Kdal monomer (I) 375 Kdal tetramer (II)	Relaxes negative supercoils (I) induces negative supercoils (II)
DNA recombination: RecA protein	40 Kdal monomer	Involved in recombination and repair activities in *E.coli*
Chromosome structure: non-histone proteins	Heterogeneous	Involved in maintenance of DNA structure in nucleus

FUNCTION-DETERMINING PROTEINS

The transient binding of proteins to DNA is a key element in determining the transcription, replication, recombination and repair of DNA (Table 8.2). Accordingly, several enzymes are required for DNA repair which recognise and bind to damaged regions or chemically modified nucleotides. This is followed by the interaction of the DNA with other enzymes which excise and replace the damaged DNA, culminating with the sealing of the resulting gap in the sugar–phosphate backbone.

Protein–DNA interactions are also required during DNA replication to unwind both the supercoiled DNA and the double helix, thereby making single strands available to the DNA-synthesising enzymes. These in turn must transiently and specifically bind to single-stranded DNA (although their activity is dependent upon the availability of free 3′ OH groups provided by RNA primers hybridised to the single-stranded DNA). Similarly, the process of transcription requires access to single-stranded DNA in order to allow the binding of specific RNA-synthesising enzymes (i.e. DNA-dependent RNA polymerases).

By far the most varied and complex types of DNA–protein interactions are those involved in the regulation of gene transcription. Unfortunately, the study of such DNA-binding proteins and their interactions with DNA sequences has been more difficult than the detection of the interaction sites (i.e. promoters and enhancers). Nevertheless, three main classes of regulatory DNA-binding proteins have been described which differentially exert their effects upon transcription by virtue of their characteristic structures.

For example, many eukaryotic transcription factors involved in the regulation of ubiquitous or developmental gene expression contain a helix-turn-helix structure which allows them to bind to specific DNA sites (Fig. 8.1). Similar structures are also present in bacterial repressor proteins which modulate operon functions. Others interact with specific DNA sites (e.g. the GC box) by virtue of structures termed zinc fingers composed of a zinc atom complexed with various polypeptides. The last class contain structures termed leucine zippers generated by the distribution of leucine residues lying on the surface of an α-helix of the polypeptide. These leucines do not directly attach to the DNA but form sites for subunit dimerisation thereby correctly orientating the resultant dimeric protein for DNA binding.

Table 8.2

Protein	Structure	Function
Restriction endonucleases: type II	Dimeric bacterial restriction/ modification enzyme	Cleaves DNA at specific palindromic target sequences
DNA polymerase: types I, II and III	Multi-subunit structure (400 Kdal)	Involved in DNA replication
RNA polymerase: types I, II and III	Multi-subunit structure (400 Kdal)	Involved in transcription and formation of mRNA
Restriction methylase: type II	Dimeric bacterial restriction/ modification enzyme	Methylates DNA at A or C preventing digestion
Transcription factor: e.g. Sp1, GAL4, TFIIIA	Contains zinc finger DNA binding motif	Involved in transcriptional activation by promoter binding
Transcription factor: e.g. OCT-1, ANT-C	Contains helix-turn-helix DNA binding motif	Involved in transcriptional activation by promoter binding
Transcriptional factor: e.g. AP1, Fos CREB	Contains leucine zipper DNA binding motif	Involved in transcriptional activation by promoter binding
DNA cleavage enzyme: DNaseI	Non-specific dimeric endonuclease	May nick DNA to convert supercoiled form to relaxed form
RecA protein	40 Kdal monomer	Involved in recombination and repair activities in *E.coli*

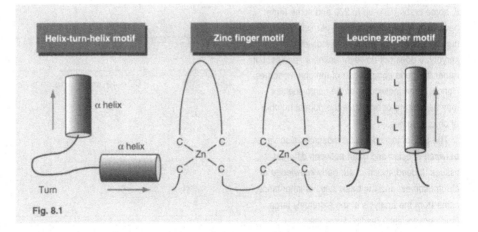

Fig. 8.1

27

9. ORGANISATION OF GENOMES INTO MULTIPLE CHROMOSOMES

In contrast to prokaryote genomes, eukaryotic cell genomes are divided into several separate DNA molecules organised into pairs of homologous chromosomes of characteristic size and shape.

Prokaryotic DNA is compactly organised into adjoining (i.e. contiguous) or overlapping genes allowing their entire genome to be contained within a single molecule, often inaccurately referred to as a chromosome. An important exception is the existence of additional independently replicating small circular DNA molecules within some bacteria (referred to as a mini-chromosome or plasmid). In contrast, eukaryotic genomes are much larger and more complex in organisation. They are consequently accommodated within several separate DNA molecules, each visible by light microscopy during the metaphase stage of cell division as individual chromosomes. Beginning in the latter part of the 19th century, cytologists derived susbstantial information regarding the basic structure and behaviour of eukaryotic chromosomes which played a vital role in the development of modern molecular biology. For example, the identification of DNA as the molecular vehicle of inheritance was aided by the discovery that each mitotic cell of a given species contained a constant number of pairs of characteristically shaped (i.e. homologous) chromosomes. Furthermore, their number was exactly halved in sex cells or gametes. Analysis of the chromosomal composition (or karyotype) of a cell (i.e. karyotype analysis) also established many primary differences between the genomes of organisms of different species. Amongst the most important were differences in the karyotypes of individuals of the same species. These were often associated with the development of several genetic diseases or syndromes and characterised by alterations in the structure and/or number of chromosomes. Consequently, karyotype analysis was, and still remains, an essential component of clinical cytogenetic approaches for the pre-natal diagnosis of and screening for several human diseases (e.g. the muscular dystrophies).

BACTERIAL 'CHROMOSOMES'

Like many prokaryotes, bacteria such as *E. coli* accommodate almost their entire genome (99–100%) into a single DNA molecule. This cannot be visualised by light microscopy since it is not packaged into a separate nucleus. The advent of electron microscopes has however allowed the definition of a discrete area within bacterial cells, termed the nucleoid, in which this single DNA molecule is supercoiled around several types of proteins. In addition to the nucleoid, some bacteria also harbour an additional 'plasmid' DNA molecule containing extra genes. These often confer specialised phenotypes upon the bacterial cell (e.g. antibiotic resistance or the ability to metabolise unusual compounds). Such plasmids are often exploited as cloning vehicles or vectors in recombinant DNA techniques.

OAs: 2, 36–8, 56, 69, 106–8, 255, 257, 1000. RAs: 16, 17, 34–6, 117, 124. SFR: 29–33, 43–5, 51–7.

CHROMOSOME NUMBER AND SIZE

The karyotype of each eukaryotic cell from a given species has a characteristic number of chromosomes (Table 9.1a). Most animal and plant cells contain 10–50 chromosomes, although a few exceptions are known. For example, human cells contain 46 chromosomes, whereas the roundworm has only 2, *Drosophila* flies have only 8, some crabs have up to 200 and some ferns have over 1000 chromosomes per cell. The number of chromosomes in an individual karyotype does not however define a species, but rather the gene composition of the chromosomes. Hence, some individuals of the same species may have more or less than the normal number of chromosomes.

The size of individual chromosomes also varies between species and often between different tissues. Indeed, much of our early knowledge of chromosomes, and the basic rules of inheritance, came from the analysis of the extremely large (and consequently readily discernible) chromosomes found in the salivary gland cells of *Drosophila*.

The majority of chromosomes in a cell (termed the autosomes) are also present in pairs of equal size and appearance (termed homologous chromosomes). Accordingly, clinical cytogeneticists are able to classify human chromosomes into groups of homologous chromosomes according to their relative size and appearance (Table 9.1b). The exceptions are the pair of sex chromosomes which differ in size and shape.

Table 9.1

(a) Organism	Chromosome No.	Genome size	DNA/chromosome
Maize	10	15 000 000 kb	150 000 kb
Human	23	3 000 000 kb	130 000 kb
Drosophilia	4	165 000 kb	41 250 kb
S. cerevisiae	16	20 000 kb	1250 kb
E.coli	1	4000 kb	4000 kb
Phage M13	1	6.4 kb	6.4 kb

(b) Size	Chromosomes	(group)	Appearance of chromosome
Largest	1–3	A	Metacentric (central centromere):
	4, 5	B	Chromosomes 1, 3, 16, 19, 20
	6–12, X	C	Sub-metracentric (centromere between the
	13–15	D	centre and end of chromosome):
	16–18	E	Chromosomes 2, 4–12, 17, 18, X
	19, 20	F	Acrocentric (centromere close to one end):
Smallest	21, 22, Y	G	Chromosomes 13–15, 21, 22, Y

KARYOTYPE ANALYSIS

The characteristics of individual chromosomes comprising a eukaryotic cell karyotype may be determined by microscopic examination of mitotically dividing cultured cells metabolically arrested at metaphase (e.g. by incubation with colchicine). Such cells are swollen in hypotonic solution to spread out the individual chromosomes, squashed onto slides and then appropriately stained. Generally, a photograph is taken and the homologous pairs of chromosomes arranged in size groups (Fig. 9.1).

Each type of homologous chromosome can be identified by its characteristic morphology such as length, position of the centromere and existence of protrusions (or knobs). In addition, most commonly used stains are not taken up evenly across the arms of an individual chromosome giving each a characteristic pattern of light and dark bands for each homologous pair. Each band is designated by a number according to its relative position on the long (q) or short (p) arm of the chromosome. This allows the detection of both numerical and structural chromosome abnormalities (e.g. linked with disease manifestation). In many cases it has also been extremely useful for the mapping of specific genes. For example, chromosome banding combined with *in situ* hybridisation studies using fluorescent probes (e.g. derived from cosmids, centromeric repeat sequences or whole chromosome libraries) has enabled several disease genes associated with subtle chromosome abnormalities (e.g. in cystic fibrosis) to be localised to individual bands or chromosomal regions.

Variations in ploidy status

The number of complete sets of chromosomes per cell is referred to in terms of 'ploidy status'. Gametes contain only one set of chromosomes and are therefore referred to as haploid. Normal somatic cells bear two complete sets of homologous chromosomes and are therefore described as diploid. Animal cells are generally diploid whereas many plant cells are polyploid, containing multiple sets of chromosomes. For example, cells from several wheat varieties are tetraploid, containing four complete sets of chromosomes. Polyploidy is lethal in humans and many animals.

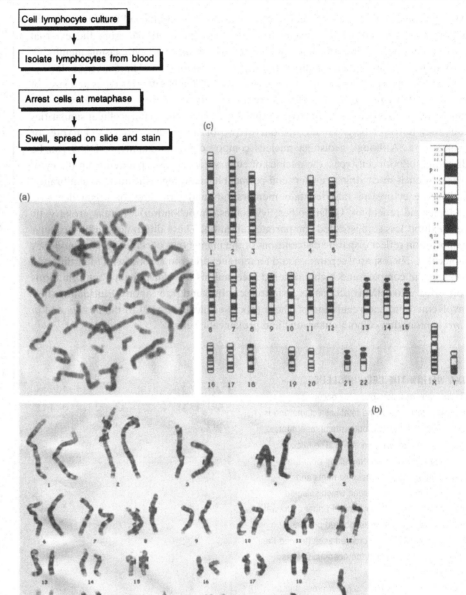

Cell lymphocyte culture

↓

Isolate lymphocytes from blood

↓

Arrest cells at metaphase

↓

Swell, spread on slide and stain

↓

(a)

(b)

(c)

Fig. 9.1

CHROMOSOME ABERRATIONS

Large mutations involving changes in whole chromosomes (numerical abnormalities) or the appearance of chromosomes (structural abnormalities) are termed chromosome aberrations. These are relatively common in animals, and may prevent foetal survival or lead to disease manifestation. Any increase or decrease in the number of individual chromosomes is termed aneuploidy. Normal individuals are termed disomic, whereas individuals lacking one chromosome of a pair or with an extra chromosome are termed monosomic and trisomic respectively. Structural alterations invariably arise from chromosome breakage and may take several forms. These may occur due to random events (e.g. during recombination) or result from an increased susceptibility to breakage at specific points.

10. DISTRIBUTION OF NUCLEIC ACIDS WITHIN EUKARYOTIC CELLS

Many types of DNA and RNA molecules are found within the complex structure of eukaryotic cells, located in different cellular compartments according to their individual biological functions.

All organisms, with the exception of viruses, are composed of small units termed cells within which the complex biochemistries of life occur protected from the external environment by a semi-permeable lipid bilayer membrane. This cytoplasmic membrane is composed of various combinations of protein, carbohydrate and lipid macromolecules which combine to allow selective uptake or export of many different types and sizes of molecules. Microscopic studies of cells from many tissues and organisms reveal a vast range of cell types bearing distinctive morphologies. Based upon intracellular similarities cells may however be classified into one of two broad groups, termed eukaryotes or prokaryotes. Although a distinct nucleus composed predominantly of DNA can be discerned in both cell types, the nucleus of eukaryotic cells is separated from the cytoplasm by enclosure within a further semi-permeable membrane (the nuclear membrane). Eukaryote cytoplasms also contain membrane-bound structures or organelles (e.g. endoplasmic reticulum, Golgi bodies, lysosomes, mitochondria) whereas prokaryotic cells exhibit less sophisticated compartmentalisation. These differences in intracellular composition reflect disparate distributions of the many types of cellular DNA and RNA molecules. Their spatial separation and temporal expression accords with their differing roles in the complex and highly ordered mechanisms by which genetic information is converted into cellular actions (e.g. via transcription and translation). Elucidating their involvement in, and regulation of, these processes has been vital to the *in vivo* and *in vitro* exploitation of both eukaryotic and prokaryotic cells.

ORGANELLE-BASED DNA

DNA is also found outside cell nuclei within mitochondria and plant cell chloroplasts. The genome of these small structures determines all of the tRNAs they utilize for translation, some of the rRNAs and many of the major components of the energy-producing oxidative phosphorylation system. Many of the proteins within these organelles are however nuclear encoded, and are transported across their membranes from their sites of synthesis in the cytoplasm. Analysis of DNA within these organelles has suggested that they may have originated from prokaryotes which infected or became entrapped within primordial eukaryotic cells and subsequently lost the ability to survive independently from the eukaryotic host during evolution. Indeed, identification of sequence variations within human mitochondrial DNA has recently been used to construct a human evolutionary tree and the tentative identification of an ancestral human female, termed 'mitochondrial Eve'.

DNA WITHIN THE CELL NUCLEUS

The nucleus of eukaryotic cells occupies approximately 10% of the total cell volume and is formed by an inner and outer membrane, termed the nuclear envelope (Fig. 10.1). The nuclear envelope is fixed within the cell by what are termed intermediate protein filaments and contains small pores through which selected macromolecules enter or leave. Further structural proteins, termed matrix-associated proteins (MAP), provide a highly ordered shell (termed the nuclear lamina) in which the genomic DNA is arranged.

In living cells this DNA is never present in a free extended form but is packaged into chromatin by its association with many different proteins. Only during mitosis does it become more tightly packaged into chromosomes. The DNA is also attached at certain sites within the nucleus to the scaffold provided by the lamina protein complex. These attachment sites play important roles during DNA replication. They are also involved in the formation of mitotic spindles utilised for the organisation of homologous chromosomes in the central plane during cell division. These structures are also important in the separation of homologous chromosomes into the resultant daughter cells.

OAs: 57–9, 69, 80, 150, 151, 168–70, 187. **RAs:** 16, 27, 28, 102, 143, 144. **SFR:** 22, 29–35, 37–9, 72, 86, 103.

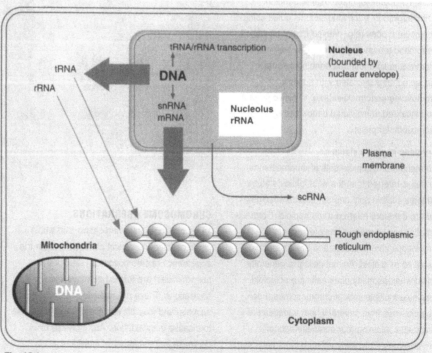

Fig. 10.1

RNA WITHIN THE CELL NUCLEUS

RNA also exists within the nucleus as small nuclear RNA (snRNA) (Fig. 10.2). This is involved in the complex post-transcriptional events that occur following the production of a nuclear RNA (hnRNA) or primary transcript. Consequently, snRNA molecules involved in post-transcriptional processing are transcribed and retained within the nucleus. These RNA species are complexed with small ribonuclear proteins (snRNPs) and associated into a structure termed a spliceosome. This structure sediments at 60S and is approximately the same size as a ribosome. Within the cell it may be found clustered in what are termed splicing islands where it may be involved in the transport of a processed mRNA from the nucleus to the cytoplasm.

Following processing the primary hnRNA transcript forms a messenger RNA (mRNA) which is transported from the nucleus to the sites of protein synthesis in the cytoplasm, i.e. the ribosomes. The continuous flow of mRNA from the nucleus to the cytoplasm is under complex control that includes specific RNA-binding proteins and small RNA species. Other small RNA species termed small cytoplasmic RNA (scRNA) molecules are also found in the cytoplasm; their function is less well defined.

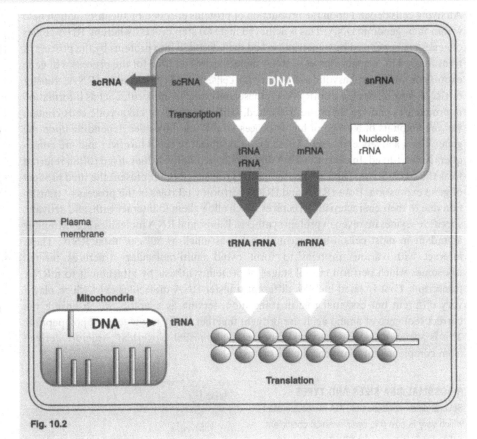

Fig. 10.2

THE NUCLEOLUS

Early cytologists described a dense structure within the nucleus of cells which they termed the nucleolus. This is now known to be the site of ribosome synthesis. Subsequently, electron microscopy has revealed distinct areas of ribosome protein and ribosomal RNA (rRNA) transcription, maturation and binding within the nucleolus. The nucleolus is therefore not a static structure but increases in size during periods of active protein synthesis. It also undergoes changes during cell division, appearing to redistribute at metaphase across the condensed chromosomes. Similarly, mini-nucleoli reappear during telophase at the tips of chromosomes in regions containing multiple copies of rRNA genes.

RNA WITHIN THE CYTOPLASM

Ribosomal RNA also exists within the cytoplasm complexed with ribosomal proteins to form free ribosomes. These are utilised for the synthesis of the proteins required to maintain cellular structure and biochemistry, and the synthesis of proteins for export (e.g. hormones).

Identical ribosomes are also found at the endoplasmic reticulum (ER), a continuous internal membrane that exists throughout the cytoplasm enclosing an internal space termed the ER lumen. Rough ER may be clearly defined under the electron microscope due to the spotted appearance created by the attachment of ribosomes complexed to mRNA molecules containing export signals.

Transfer RNA

Since protein synthesis occurs at ribosomes located in the cytoplasm, the adaptor RNA molecules (termed transfer RNAs or tRNAs) which specifically carry the amino acid building blocks to the ribosomes are also located within the cytoplasm. There is a constant turnover of tRNA molecules and therefore a constant flow of newly transcribed tRNA molecules from the nucleus to the cytoplasm. Recent reports suggest that the processing of tRNA molecules may occur near the nuclear envelope and that some of the base modifications occurring within tRNAs are performed in the cytoplasm, rather than in the nucleus (similar to other RNA species).

11. TYPES OF RNA MOLECULES

Cells contain three major types of RNA molecule termed messenger RNA (mRNA), ribosomal (rRNA), and transfer (tRNA), and several minor forms, each of which perform specific roles in gene expression.

All living cells depend upon the production of proteins encoded by the information held within their genomic DNA. This is achieved in a two step process whereby RNA copies of genes are generated by transcription and then decoded into proteins by the process of translation. Three major classes of RNA molecule are required for the processes of gene expression within prokaryotic and eukaryotic cells, termed ribosomal or rRNA, transfer or tRNA and messenger or mRNA. Of these, only mRNA molecules act as information intermediates and are therefore translated. Different types of eukaryotic cells contain varying amounts of a range of highly processed mRNA molecules depending upon the genes being actively expressed. These are also subject to rapid turnover and are consequently short-lived. In contrast rRNA and tRNA are relatively long-lived (often referred to as the stable RNAs) and are not themselves translated but represent the final product of gene expression. Both rRNA and tRNA perform vital roles in the process of translation due to their characteristic structures which allow them to interact with, and activate, specific enzymes involved in protein synthesis. Ribosomal RNA molecules are extremely abundant in most cells, often accounting for as much as 80% of total RNA. These interact with various proteins to form ovoid multi-molecular structures termed ribosomes which perform crucial stages of protein synthesis by attachment to mRNA transcripts. First isolated in 1959, different transfer RNA molecules exist which play a very different but essential role in translation, serving as adaptors which ensure the correct sequence of amino acids are brought together in the newly formed polypeptide. This is achieved by virtue of their conserved and unusual 'cloverleaf' structure derived from complementary intrachain base pairing.

RNA STABILITY IN CELL EXTRACTS

Messenger RNA molecules have short half-lives (between 1 and 6 hours) due to their rapid degradation *in vivo* by cytoplasmic ribonucleases. Unless stringent attempts are made to inactivate or inhibit ribonucleases, mRNA molecules will not survive long within *in vitro* cell lysates or cytoplasmic extracts. In contrast, cell extracts containing intact and viable rRNA and tRNA molecules are relatively easy to prepare by virtue of their relative abundance. Ribosomes may also be isolated intact thereby facilitating *in vitro* study of their structure and functions in translation. Prokaryotic and eukaryotic cell lysates may also be obtained devoid of mRNA which allow the *in vitro* translation of introduced mRNA transcripts.

OAs: 60–8, 90, 92, 167–70, 172–5, 188, 512. **RAs:** 18, 19, 75, 76, 102, 103. **SFR:** 36–9, 74, 86, 87.

RIBOSOMAL RNA SIZES AND TYPES

Several different ribosomal RNA molecules exist which vary in size (i.e. sedimentation coefficient, defined in Svedberg units (S)). These vary in nucleotide composition between prokaryotes and eukaryotes (see Table 11.1), and between different eukaryotic species. They are also associated with various polypeptides to form the two structural subunits comprising active ribosomes. Ribosomes also vary in size, from 2.5 million daltons in prokaryotes to 4.2 million daltons in eukaryotes, according to the size and composition of the subunits. Thus, the 70S *E. coli* ribosomes are composed of a large subunit containing two rRNA molecules of 23S and 5S associated with 34 polypeptides, and a small subunit with a single 16S rRNA molecule associated with 21 polypeptides. In comparison the large subunit of a typical mammalian 80S ribosome contains three rRNA molecules (of 28S, 5.8S and 5S) with 49 polypeptides, whereas the small subunit contains one rRNA molecule of 18S associated with 33 polypeptides. Nucleotide sequencing has however revealed that the 5.8S rRNA of the mammalian ribosome large subunit is also present in *E. coli* as an integral part of the 23S rRNA molecule.

Table 11.1

Ribosome	Prokaryotes	Eukaryotes (Mammalian)
Sedimentation coefficient (S)	70S	80S
Molecular mass (daltons)	2 520 000	4 220 000
Large subunit composition		
S value	50S	60S
Molecular mass (daltons)	1 590 000	2 820 000
Number of RNA molecules	2	3
RNA S values (nucleotides)	23S (2904)	28S (4718)
	5S (120)	5S (120)
Number of polypeptides	34	49
Small subunit composition		
S value	30S	40S
Molecular mass (daltons)	930 000	1 400 000
Number of RNA molecules	1	1
RNA S value (nucleotides)	16S (1541)	18S (1874)
Number of polypeptides	21	33

FEATURES OF TRANSFER RNA

Transfer RNAs are less abundant (about 15% of total RNA) and smaller than rRNAs, ranging in size from 65 to 110 nucleotides. Multiple copies of at least 56 different types of tRNA per cell have been described. Each tRNA is specific for a single amino acid, although due to the 'wobble' in the genetic code more than one type of tRNA may bind to the same amino acid (termed iso-acceptors). With a few exceptions (e.g. many of the mitochondrial tRNAs), most tRNAs adopt the same structure (Fig. 11.1). Referred to as the tRNA cloverleaf, this allows them to act as molecular adaptors which ensure the correct order of amino acids are incorporated into the sequence of a newly synthesised polypeptide. tRNA molecules are transcribed from clusters of multiple tRNA genes arranged into transcription units. They subsequently undergo a range of enzyme-catalysed processing and chemical modification reactions. Such nucleotide modifications affect both tRNA structure and stability, and may affect their interaction with ribosomes, enzymes and various proteins.

CLOVER LEAF STRUCTURE

The cloverleaf structure shown in Fig. 11.1 is formed by complementary base pairing into a stem and four loops (denoted I–IV). The two functionally active regions are termed the acceptor arm (to which amino acids are enzymatically attached) and the anticodon loop (which recognises mRNA codons). The acceptor arm forms the cloverleaf stem and contains 7 base pairs (bp) with a vital 5′ CCA-OH-3′ terminal trinucleotide critical for recognition by the amino acid 'charging' enzymes, termed the aminoacyl tRNA synthetases. The anticodon arm is formed from a 5 bp stem and 7 nucleotide loops containing certain conserved and variant nucleotides. The 3rd, 4th and 5th nucleotides of the loop represent the anticodon. This is always flanked on the 5′ by a pryrimidine, and usually a modified purine on the 3′. Named after the characteristic nucleotide sequence thymine-pseudouracil-cytosine-purine found at the same position in the 7 bp loop of virtually all tRNAs, the TψC arm also has a 5 bp stem with a CG base pair next to the loop. The D or DHU arm (so named because it contains the unusual

pyrimidine dihydrouracil) has a stem composed of 3 or 4 bp, usually with CG bp at both ends, and between 7 and 10 nucleotides per loop. Most of the tRNA size differences arise from variation in the number of nucleotides (between 3 and 21) within the variable loop (loop III). Nevertheless, the majority (80%) of tRNAs have small loops of 4–5 nucleotides.

Although the same structure is formed in most tRNAs, only certain nucleotides are invariant. The vast majority are variable or semi-invariant (e.g. always a purine or always a pyrimidine). Many of the conserved invariant and semi-invariant nucleotides are involved in the formation of the more compact 'L-shaped' tertiary structure of tRNAs. This ensures that the acceptor arm and anticodon loop are located at opposing ends of the molecule.

Minor forms of eukaryotic RNA

Unlike prokaryotes, eukaryotic cells also contain two additional types of RNA molecules termed small cytoplasmic (sc) and small nuclear (sn) RNAs. For example, several types of short uracil-rich RNA molecules, termed U-RNAs (a form of snRNA), are found in most vertebrate cell nuclei. Some of these (e.g. U1, U2, U4, U5 and U6) combine with various proteins to form small nuclear ribonucleoproteins which are involved with intron splicing during mRNA maturation. Some (e.g. U3 and U11) perform roles in pre-mRNA processing such as the addition of a poly(A+) tail, whereas others have not as yet been ascribed a particular function. Similarly, several other types of RNA molecules (e.g. 7SK RNA and *Alu*RNA) have been identified but no functions ascribed. A further type of RNA termed 7SL RNA has however been shown to be involved in the transport of newly synthesized proteins within the cell, although how this occurs has not been determined.

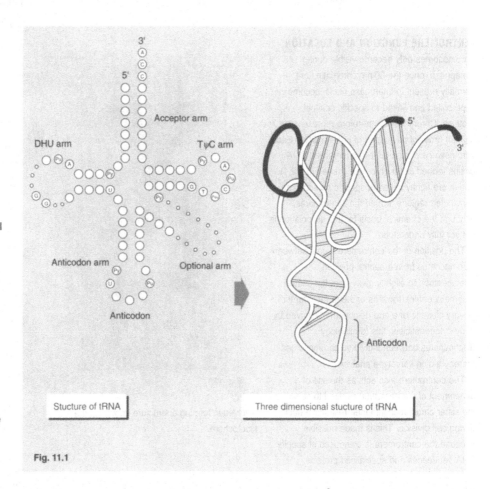

| Stucture of tRNA | Three dimensional stucture of tRNA |

Fig. 11.1

12. THE ANATOMY OF EUKARYOTIC CHROMOSOMES

Eukaryotic chromosomes comprise two DNA molecules (or chromatids), with protected ends (telomeres) linked together by a centromere containing a kinetochore structure used for microtubule attachment.

The development of DNA-specific dyes and the improvement of light microscopes in the late 19th century allowed detailed observation of the behaviour and structure of chromosomes within eukaryotic cell nuclei. Cytologists must have been initially puzzled by the transient appearance and disappearance of chromosomes, now known to result from changes in the packaging of chromatin. However, their observations provided the basic information from which many central concepts of genetics and molecular biology were developed, including the identification of DNA as the genetic material. The use of high resolution microscopy techniques means that we can now be extremely precise in terms of describing the morphology of chromosomes. Accordingly, we now have a clearer understanding of chromosome function in relation to the processes of gene replication and transmission central to the formation of natural genetic variation. For example, it is now known that chromosomes only become visible by light microscopy once DNA replication has occurred. Furthermore, chromosomes are actually composed of two DNA molecules joined together by a structure termed a centromere which is important in the separation of homologous chromosomes into daughter cell nuclei. The position of the centromere and the length of the arms (termed sister chromatids) also vary, giving each chromosome comprising the cell karyotype (i.e. chromosomal repertoire) a characteristic morphology. The ends of the chromosome, termed telomeres, also contain specialised sequences which play important roles in protecting the integrity of chromosomes. This knowledge has been directly exploited in many ways, including the development of eukaryotic cloning vectors (e.g. yeast artificial chromosomes).

CHROMOSOME BANDING

In addition to morphological differences, individual chromosomes may be distinguished by their banding pattern following staining with DNA-specific dyes. This occurs because each chromatid takes up dye in a non-uniform manner producing a characteristic series of light and dark bands (arising from differences in chromatin condensation). Banding can be generated using five different chromosome staining techniques, although how this is achieved is only known for replication staining which detects regions undergoing active DNA replication. G-band staining is perhaps the most useful technique and is conjectured to highlight AT-rich regions deficient in genes. If true, areas not producing G-bands may therefore represent regions containing gene clusters. Chromosomal banding has been extremely useful for the primary chromosomal localisation of some genes and is routinely used in karyotype analysis.

CENTROMERE FUNCTION AND LOCATION

Chromosomes only become visible during metaphase, once the 30 nm chromatin fibre normally present in interphase nuclei becomes supercoiled and linked to specific scaffold proteins. Chromosomes therefore represent the highest order of packaging of nuclear DNA. Each chromosome comprises two supercoiled DNA chains termed sister chromatids (see Fig. 12.1). These are tightly linked at specific non-staining, constricted regions termed the centromeres, although the chemical basis for this association is not yet fully understood.

The position of the centromere varies between chromosomes from a central position (metacentric), to slightly away from the centre (submetacentric), towards one end (acrocentric), or very close to one end (telocentric). Derived by early cytogeneticists, this terminology discriminates between individual chromosomes thereby aiding karyotype analysis.

The centromere also acts as the site of attachment of the microtubules used to separate the sister chromatids into their respective nuclei during cell division. This is made possible because the centromere is composed of specific DNA sequences and specialised proteins involved in forming a structure known as the kinetochore.

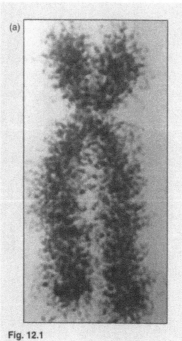

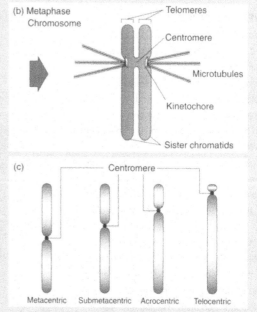

Fig. 12.1

OAs: 69–72, 255, 257, 258, 622–4.
RAs: 20–4, 117, 150.
SFR: 29–33, 40–5, 88–90.

CENTROMERIC STRUCTURES

Most of our knowledge of centromere–kinetochore structures is derived from yeast chromosomes. These are smaller and therefore easier to study than other eukaryotic chromosomes. The centromeres from the 11 yeast chromosomes so far studied in detail are well-conserved structures composed of approximately 220 bp of DNA which does not form into nucleosomes (Fig. 12.2). The central 125 bp serves as the point of formation of the kinetochore complex and is comprised of an AT-rich sequence bounded by two conserved regions (plus a few nucleotides on either side).

The kinetochore of yeast is not however typical of all eukaryotes since it is always present and permanently linked to a microtubule. In contrast, the kinetochores of higher order eukaryotes are more complex, and mature kinetochore structures are only present on metaphase chromosomes. Consequently much less is known about their composition and architecture. It is currently known that mammalian centromere–kinetochore DNA contains highly repetitive sequences to which at least five specific proteins bind (in an as yet unknown manner). For example, human centromeric DNA contains multiple 171 bp long repeats (termed alpha DNA), each of which binds a single molecule of a centromere-specific protein (termed CENP-B) (Fig. 12.3).

TELOMERE STRUCTURE AND FUNCTION

The telomeres at the ends of chromosomes are highly specialised structures which fulfil three important roles: (i) they provide protection from degradation by nucleases, (ii) they prevent joining together of chromosomes, and (iii) they prevent the problem of progressive shortening of the linear DNA comprising chromosomes. The latter would normally occur during the process of DNA replication by virtue of loss of the extreme 5′ end comprising the initiating RNA primer.

The first two functions are accomplished by the presence of the specialised telomeric DNA sequences which in all eukaryotes consist of multiple copies of short repeat sequences. In humans the repeat sequence of 5′-AGGGTT-3′ may be present over one thousand times. The overall structure is different for each telomere, and varies amongst different eukaryotes. The

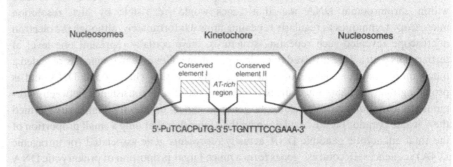

Fig. 12.2

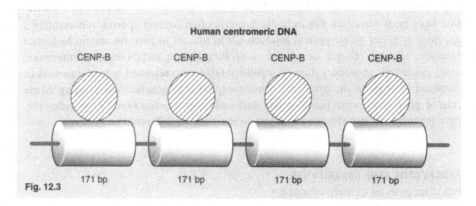

Fig. 12.3

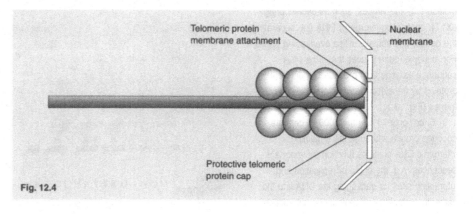

Fig. 12.4

extreme ends of the telomeres are however characterised by having one G-rich strand which contributes the 3′ end and overhangs the end of the C-rich strand by 12–16 nucleotides. This allows the association of telomere-specific proteins rather than the nucleosome-forming histones. These telomere-specific proteins are speculated to form a protective cap, or to protect the chromosome ends by attaching the telomere to other nuclear structures (e.g. the inner

surface of the nuclear membrane) (Fig. 12.4).

The mechanism whereby telomeres prevent gradual shortening of chromosomes remains to be determined. However, specific enzymes termed telomerases, each comprising a protein subunit and an RNA molecule complementary to the G-rich strand of an individual telomere, may play a vital role by continually extending the G-rich strand to counterbalance any shortening that may occur during replication.

13. ORGANISATION OF GENES WITHIN CHROMOSOMES

Prokaryote genomes contain little non-gene-associated (extragenic) DNA and are compactly arranged, whereas eukaryote genomes contain large amounts of extragenic DNA including several forms of repetitive sequence.

The naive hope of early geneticists following the demonstration that genes resided within chromosomal DNA was that genes would be visible by high resolution microscopy techniques as regularly repeating units. Unfortunately, although the electron microscope revealed such repeating structures, these actually represent one level of tertiary DNA structure involved in nuclear packaging. Genome mapping has revealed a much more complex picture of gene organisation within chromsomal DNA and is providing many insights into our evolutionary past. The inconsistency between enormous size variation amongst eukaryotic genomes and an organism's complexity (termed the C-value paradox) is now explained by the realisation that only a small proportion of the total eukaryotic genomic DNA actually represents gene-associated (or intragenic DNA) sequences. In contrast, genes form a much larger proportion of prokaryotic DNA and some viral genomes actually contain overlapping genes. Although most genes are randomly scattered throughout the chromosomal DNA separated by irregular lengths of apparently non-functional extragenic DNA, several exceptions to this random organisation have been described. For example, bacteria often contain operons representing a grouping of genes whose protein products act in concert to perform certain biological functions (e.g. the *E. coli lac* operon controlling lactose metabolism). Furthermore, many eukaryotic sequences (termed repetitive DNA) are repeated in tandem arrays or scattered throughout the genome; a feature exploited in genetic fingerprinting. Many related gene sequences (i.e. gene families) including non-functional gene relics (i.e. pseudogenes) are also clustered into loci on the same or different chromosomes.

THE C-VALUE PARADOX

Eukaryotes exhibit an enormous variation in gene size and haploid chromosome number (referred to as the C-value). Thus, the haploid genome of the yeast *Saccharomyces cerevisiae* is about 20 000 kb divided into 16 chromosomes, the 3 million kb of the human genome is divided into 23 chromosomes and the 90 million kb of the salamander genome is divided into just 12 chromosomes. Although a limited correlation exists between an organism's C-value and its genome complexity, this is not a strict relationship. For example, the haploid genome of the newt *Triturus cristatus* is much larger than is necessary to encode all the required proteins, and is seven times larger than the more complex human genome. This situation is referred to as the C-value paradox. The answer partly lies in the way genes are organised within the genome and the way in which eukaryotic genes are organised.

PROKARYOTIC GENE ORGANISATION

Prokaryotic genes are generally arranged in a compact manner within a single DNA or RNA molecule. Some viruses, such as bacteriophage φX174, contain no intergenic DNA (i.e. separating the genes) and actually utilise overlapping genes. For example, gene E uses the same DNA sequence as gene D to produce two different proteins by translation from different reading frames (Fig. 13.1).

In *E. coli* only 75% of the 4000 kb genome encodes protein, although the remaining intergenic DNA contains functionally important sequences (e.g. the origin of replication and interaction sites for packaging the DNA into the bacterial nucleoid). The majority (about 75%) of the estimated 2800 genes are also present as single copies randomly scattered throughout the 'chromosome'. However, the remaining 25% are organised into various operons (i.e. units of several genes simultaneously expressed for a specific biochemical purpose).

Many prokaryotes actually contain identical genes arranged within different sized genomes (e.g. from 1200 kb in *Haemophilus influenzae* to 30 000 kb in *Bacillus megaterium*). Although some of the size difference may be due to the presence

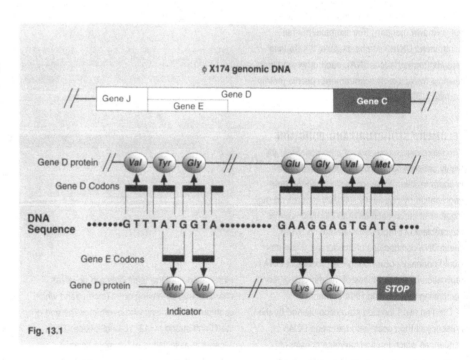

Fig. 13.1

of additional genes (e.g. for antibiotic resistance), it largely reflects disparities in the amount of intergenic DNA. Some bacteria also contain an additional circular 'plasmid' DNA molecules.

OAs: 73–88, 326–30, 739–41.
RAs: 33, 155, 164, 262.
SFR: 46, 47, 97–102, 148, 149, 167.

EUKARYOTE GENE ORGANISATION

By the 1970s two broad categories of eukaryotic DNA were defined whose differing amounts partly answered the C-value paradox (i) intragenic DNA, associated with a functional gene unit (i.e. introns, exons, promoters) or non-functional genes (e.g. pseudogenes), and (ii) extragenic DNA, not involved with a gene or associated sequences. The amount of extragenic DNA varies (e.g. 70% in the human genome) and also contains multiple copies of certain sequences termed repetitive DNA (Fig. 13.2).

Repetitive DNA

Repetitive DNA varies in size from a few base pairs to several kilobases and comprises multiple copies of a single gene element. Two categories of repetitive DNA have been defined: (i) highly repetitive DNA consisting of several hundreds or millions of repeats, and (ii) moderately repetitive DNA consisting of tens of or a few hundred repeats. These may either be clustered in tandem arrays or widely dispersed.

In the human genome, dispersed repetitive DNA is usually found as short or long interspersed repetitive nuclear elements, termed SINEs and LINEs respectively. SINEs are scattered throughout the genome and vary in length from 130 to 300 bp and are repeated some 7000 times. An important class of SINE is the *Alu* family (so named because they usually contain the recognition sequence for the restriction endonuclease *Alu*I). Their function is still unknown but they tend to resemble the sequence of 7SL RNA known to be involved in protein secretion. LINEs are usually found as truncated elements, believed to be propagated by transposition events. LINE-1 is a type of non-viral retroelement (i.e. a form of transposon) present as 60 000 copies and retaining the ability to move around the genome. Clustered repetitive or satellite DNA contains repeat sequences arranged into long tandem arrays. Three main types have been defined: (i) classical satellite DNA ranging between 100 and 5000 kb in length and found at chromosome centromeres, (ii) minisatellite DNA such as the telomeric repeat clusters, ranging between 100 bp and 20 kb, and (iii) microsatellite DNA consisting of dinucleotide (e.g. CA/GT) or mononucleotide (AT) repeats.

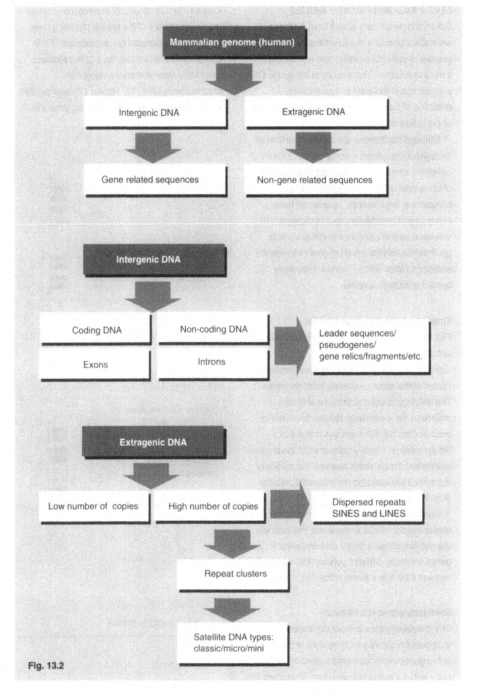

Fig. 13.2

The detection of these differing numbers of minisatellite DNA repeats forms the basis of genetic fingerprinting.

Inverted repeat sequences capable of forming stable hairpin structures have also been identified in eukaryotes. Averaging 200 bp, they range in size from a few base pairs to over a thousand. Although their function is poorly understood, they appear to be extremely common with about 2 million inverted repeats present per human genome.

13. ORGANISATION OF GENES WITHIN CHROMOSOMES/Contd

Eukaryotic genomes also contain two types of gene families composed of clustered or dispersed genes of identical or related sequence, and often including non-functional gene relics or pseudogenes.

GENES AND MULTIGENE FAMILIES

Eukaryotic genes vary in size from 100 bp to several kb, primarily due to differences in the amount of coding (i.e. exon) and non-coding (i.e. intron) sequences. The amount of intragenic DNA per genome is therefore not an accurate reflection of the total number or molecular mass of the proteins encoded.

Although most genes are randomly scattered throughout the genome, some genes of identical or related sequence are grouped together into clusters referred to as multigene families. In comparison to prokaryotic operons which are composed of coordinately expressed genes of unrelated sequence, genes within eukaryotic gene families share a high degree of sequence homology (often >90%). They are therefore termed homologous genes.

Simple and complex gene families

Two forms of multigene family may be discerned, termed simple and complex.

Simple multigene families represent multiple copies of the same or virtually identical genes. This strategy is used to produce sufficient molecules for a particular cellular function. For example, the 5S rRNA gene is present in a simple multigene family composed of thousands of identical copies which ensures that sufficient ribosomes are available for efficient translation (Fig. 13.3).

Complex multigene families represent genes of similar but not identical sequence. For example, the globin multigene family contains several genes encoding different polypeptides varying in sequence by a few amino acids (Fig. 13.4).

Gene family genomic locations

The individual genes comprising a specific multigene family may be organised in three ways within genomes: (i) clustered together at a single point within a single chromosome, (ii) scattered at different positions on the same chromosome or scattered between more than one chromosome, or (iii) a combination of clustered and scattered (see Fig. 13.5).

The human genome contains examples of all three arrangements. Accordingly, the 5S rRNA gene family is composed of 2000 tandemly arrayed copies located in a single position on the long arm (designated q) of chromosome 1, and the five members of the growth hormone gene family are all clustered on chromosome 17. In comparison, the five members of the aldolase gene family are scattered amongst five chromosomes (3, 9, 10, 16 and 17), and the 280 or more transcription units comprising the rRNA gene family are dispersed in clusters of 50–70 copies in five locations on the short arms (designated p) of chromosomes 13, 14, 15, 21 and 22. Furthermore, the immunoglobulin gene family is composed of non-identical gene sequences scattered across numerous chromosomes.

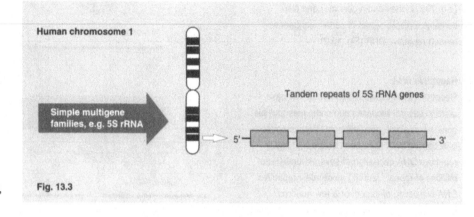

Human chromosome 1

Simple multigene families, e.g. 5S rRNA

Tandem repeats of 5S rRNA genes

Fig. 13.3

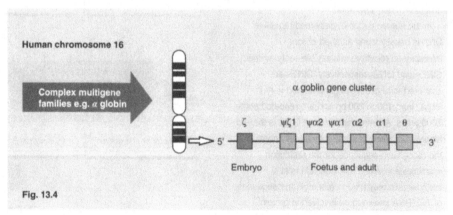

Human chromosome 16

Complex multigene families e.g. α globin

α goblin gene cluster

ζ   ψζ1   ψα2   ψα1   α2   α1   θ

Embryo   Foetus and adult

Fig. 13.4

Gene family locations	
CLUSTERED	5S rRNA gene: chromosome 1q
SCATTERED	Aldolase genes: chromosomes 3, 5, 10, 16, 17
COMBINATION	rRNA genes: chromosomes 13, 14, 15, 21, 22

Fig. 13.5

Evolution of multigene families

Several genetic processes have been proposed to account for the existence of different types and arrangements of multigene families. The fact that the members of many multigene families are clustered to allow coordinated expression suggests they have arisen by gene duplication from an ancestral gene (Fig. 13.6). In addition, random large-scale rearrangement events (e.g. during meiotic recombination) may be responsible for the scattering or 'shuffling' of individual or multiple genes throughout the genome in the form of multigene clusters. Additional random mutation events may also have given rise to sequence divergence. Consequently, such random evolutionary processes have led to divergence of some gene family members to produce new proteins with differing biological roles (e.g. the α- and β-globins differentially expressed during development). They may also be responsible for the formation of non-functional gene relics, termed pseudogenes.

Gene relics or pseudogenes

Functionally inactive genes produced by one or two mutations which introduce a stop codon or change the reading frame of the coding sequence are termed 'conventional pseudogenes' (Fig. 13.7). Numerous conventional pseudogenes have been isolated scattered throughout the entire genome and are especially found at multigene family loci. For example, several pseudogenes have been found in the apolipoprotein, β-globin, and immunoglobulin gene loci. This initial loss of functional activity is often closely followed by the relatively rapid introduction of additional mutations which eventually make it difficult to recognise the original ancestral gene sequence.

A further class of pseudogene termed 'processed pseudogenes' has also been described. These represent the mRNA form of the gene rather than the original DNA-encoded gene sequence, and consequently lack both intron and promoter sequences. The mechanism(s) by which processed pseudogenes have arisen remain to be determined. It may however have involved some form of reverse transcriptase event converting the processed mRNA into DNA which is then inserted back into the genome.

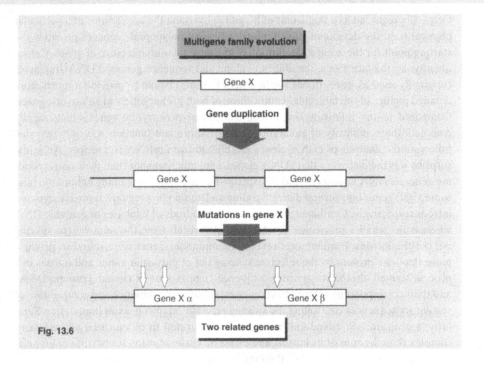

Fig. 13.6

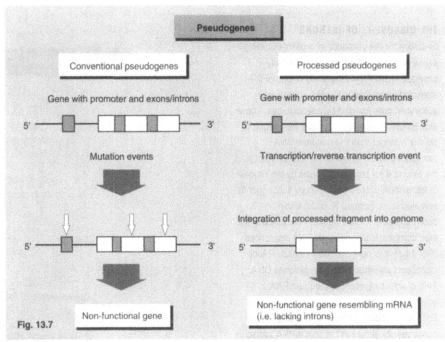

Fig. 13.7

Many eukaryotic genomes also contain gene relics representing truncations of either the 5' or 3' end of an ancestral gene. These may have arisen due to deletional or unequal recombination events.

OAs: 89–96, 895, 896, 1069, 1075.
RAs: 25–8, 389, 390.
SFR: 46, 47, 216, 217, 223.

14. THE MOLECULAR ANATOMY OF EUKARYOTIC GENES

Most eukaryotic genes are divided into separate coding (exon) and non-coding (intron) regions, making them much longer than their corresponding mRNA transcripts which have undergone intron removal.

Originally regarded as a single unit of heredity governing the appearance of a particular physical trait, the development of the one gene–one polypeptide concept provided the starting point for the accurate definition of the molecular architecture of genes. Consequently, by the late 1970s, the ability to clone and sequence genomic DNA fragments (originally used as gene probes for Southern blotting) began to provide a much more detailed picture of the molecular composition of both prokaryotic and eukaryotic genes. Compared to the relatively simple construction of prokaryotic genes which rapidly emerged, the complexity of eukaryotic gene structure and function was only revealed following the analysis of cDNA clones established from mRNA transcripts. An initial surprise was the discovery that cDNA clones were much smaller than their corresponding genomic DNA sequences. This is explained by the division of many eukaryotic (and some viral) genes into several discrete coding and non-coding regions, termed exons and introns respectively. Combined, these may span hundreds of kilobases of genomic DNA whereas the smaller messenger RNAs (mRNAs) result from the removal (or splicing out) of the introns. Furthermore, several discontinuous genes were found to produce more than one protein by the selective splicing out of particular exons and introns in a process termed alternative splicing. Sequence comparisons of cloned genomic DNAs and their corresponding cDNAs have since established the presence and composition of regulatory sequences controlling the splicing reactions at intron–exon boundaries. Similarly, a gene transcriptional unit has been demonstrated to be much larger and more complex than the sum of its introns and exons by virtue of many transcriptional regulatory sequences located within or flanking the unit.

INTRONS IN PROKARYOTES

Discontinuous genes (i.e. containing introns) are very common in higher eukaryotic organisms and in many types of viruses. Initial analysis of genes in the most common bacterial species (i.e. eubacteria) led to the belief that bacteria did not contain any discontinuous genes. However, recent studies have established that some bacterial genes such as the ribosomal RNA genes of archaebacteria (often inhabiting lake sediments or environments of extreme heat and salinity more common to the remote evolutionary past) also contain introns. Two classes of archaebacteria introns have so far been defined which bear no structural or functional resemblance to eukaryotic gene introns.

OAs: 97, 98, 873, 886, 895, 896, 1065. **RAs:** 29–32, 269–271. **SFR:** 46, 47, 49, 214, 216, 217.

THE DISCOVERY OF INTRONS

Evidence for the existence of protein-encoding exons and non-coding introns was intially provided in 1977 from the study of pre-mRNA transcripts produced within the nucleus of eukaryotic cells infected by adenoviruses. These viral mRNA transcript precursors were found to be much longer than the mature mRNA transported to the cytoplasm where they act as the template for protein synthesis by the process of translation. Electron microscopy subsequently revealed loops (termed R-loops) within heteroduplexes formed between the mRNAs and their complementary genomic DNA sequences (Fig. 14.1). It is now realised that each R-loop represents an intron within the genomic DNA. This is actually present in the pre-mRNA transcript but is removed by a process termed RNA splicing during mRNA processing, and is consequently absent in the final mRNA transcript.

This phenomenon was also rapidly demonstrated for the SV40 virus, and subsequently many eukaryotic mRNAs were demonstrated to be initially produced as much longer pre-mRNA molecules. It was originally believed that this size difference resulted from the removal of terminal sequences rather than

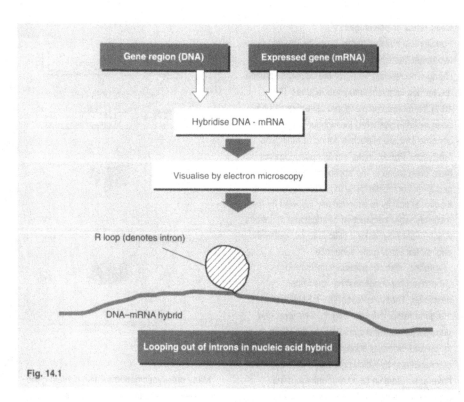

Fig. 14.1

internal sequences. However, electron microscopy and S1 nuclease mapping studies of the β-globin, ovalbumin and immunoglobulin genes soon provided conclusive proof for the existence of internally located introns separating exons within eukaryotic chromosomal genes.

INTRON NUMBER AND LENGTH

Eukaryotic genes may contain no introns or have greater than one hundred (see Table 14.1). These vary in length from 31 nucleotides in one of the SV40 genes to 210 000 nucleotides in the 2.3 Mb human dystrophin gene. Introns are generally much larger than exons and may be evenly or unevenly distributed within a gene. For example, the 250 kb human cystic fibrosis transmembrane conductance gene (CFTR) contains 24 exons scattered throughout the gene. The exons represent a total of only 2.4% of the entire gene and are separated by 23 introns which vary in length from 2 to 35 kb.

The human apolipoprotein (a) gene is similarly composed of numerous exons and introns. Each apolipoprotein (a) molecule is composed of various numbers of Kringle IV protein domains. These are each encoded by approximately 4.2 kb of DNA comprising two exons (totalling 342 bp) separated by an intron of consistent size but variable sequence. Each domain-encoding unit is also separated by an intron of variable size and sequence.

Why have introns and exons?

At first appearance, it makes no sense for a cell to waste energy on the production of long pre-mRNA transcripts, and complex strategies for the subsequent removal of introns (often representing up to 75% of the original transcript). A plausible hypothesis which suggests a positive benefit for this costly arrangement is that individual exons encode functionally discrete protein domains (such as those found in several enzymes, haemoglobins and immunoglobulins). This may allow new proteins to be rapidly produced during evolution by the recombination or shuffling of exons into new combinations (Fig. 14.2).

Additionally, their unusually long length may serve to maintain exon integrity during genetic recombination or crossing over. Consequently, imprecise joining within protein-encoding exons will produce an abnormal protein (e.g. by altering the reading frame) whereas imprecise joining of introns may have no effect upon the mature mRNA or translated protein.

The existence of introns also allows some genes to generate several forms of a protein in a tissue-specific manner by alternative splicing. This process was first demonstrated to occur in a number of eukaryotic cell-specific viruses. However, current evidence suggests that this process also occurs in eukaryotes with the aid of specialised cell- or tissue-specific protein splicing factors.

Table 14.1

Gene	Length	Intron no.	Chromosome
Histone H4	0-4 kb	0	1
Insulin	1-4 kb	2	11
Albumin	18 kb	13	4
Apolipoprotein B	43 kb	17	11
CFTR	250 kb	26	7
Dystrophin	2300 kb	100+	X

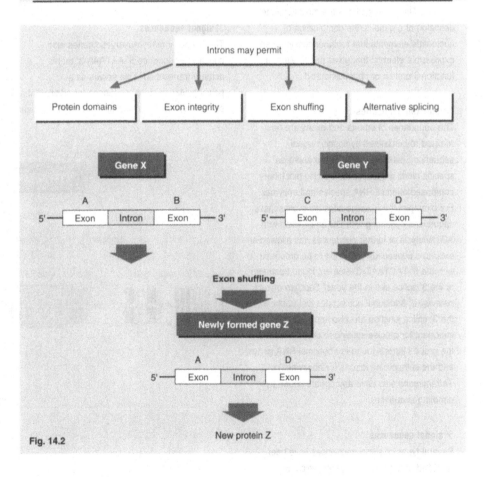

Fig. 14.2

41

Eukaryotic genes contain several types of signals and regulatory sequences which dictate the level and efficiency of transcription, and pre-mRNA processing, transport, stability and translation.

DEFINITION OF A EUKARYOTIC GENE

Prokaryotic genes are usually randomly scattered or clustered into operons composed of contiguous genes under the control of a single upstream promoter sequence. This minimises the energy required for the control of gene expression. Furthermore, prokaryotic mRNAs undergo minimal post-transcriptional processing. In contrast, each eukaryotic gene is usually associated with its own promoter or sets of promoters and various additional signal sequences (Fig. 14.3). These may be located upstream, downstream or within the gene, and are responsible for regulating transcription or the processing of pre-mRNAs (e.g. splicing and polyadenylation). These regulatory sequences are however not expressed within the final protein product. Therefore, perhaps the most accurate definition of a gene is 'the combination of nucleotide elements that together constitute an expressible element that gives rise to either a functional protein or ribonucleic acid'.

Intron-exon boundaries

The boundaries of introns and exons are now realised to be defined by specific signal sequences (see Fig. 14.3). These serve as specific binding sites for the splicing machinery composed of small RNA species and enzymes. For example, the 5′ splice sites have consensus sequences beginning with GU and ending with AG. Analysis of further sequences has allowed an extended consensus sequence to be proposed with the 5′-TACTAAC-3′ element found upstream of the 3′ splice site in the yeast *Saccharomyces cerevisiae*. Additional nucleotides just upstream of the 3′ splice junction are also known to be important for precise splicing to occur. Similarly, the group I introns found in ribosomal RNA genes and the self-splicing introns (or ribozymes) of *Tetrahymena* also have appropriate boundary-denoting sequences.

5′ signal sequences

Several types of signal sequences have been identified at the 5′ end of genes which are normally absent in the final protein. They generally contain leader sequences varying in length from 20 to 600 nucleotides located at their 5′ ends which may act as transport or secretion signals. A further 5′ sequence also dictates the binding of ribosomes to ensure translation begins at the correct start position (i.e. to generate an open reading frame). In prokaryotes this ribosome binding site (RBS) is known as the Shine–Dalgarno sequence and has the consensus sequence 5′-AGGAGGU-3′, complementary to a sequence within the 16S rRNA component of prokaryote ribosomes. Although also present at the 5′ end of eukaryote genes, eukaryote RBSs are much more difficult to define. Protein-encoding genes usually have a 5′ ATG codon which acts as a translation initiation signal, as well as coding for methionine. Consequently, methionine is the first residue in many eukaryotic proteins, and formylmethionine is the first residue in prokaryotic proteins.

3′ signal sequences

The 3′ end of many eukaryotic genes also contains a sequence 5′-AATAAA-3′ that is actually transcribed. This serves as a polyadenylation signal sequence for addition of a poly(A+) tail to the 3′ end of a pre-mRNA nuclear transcript during transcript processing. Although transcription may proceed past this point to the termination signal codon (TAA, TAG or TGA), this sequence is now realised to be a signal for both poly(A) polymerase and an endonuclease.

Instability signals in genes

In comparison to the relatively unstable and rapidly degraded prokaryotic mRNA transcripts, eukaryotic mRNAs contain a number of additional signal sequences which serve to control mRNA stability. For example, histone protein-encoding gene transcripts contain a short stem-loop structure which replaces the poly(A+) tail. This DNA-encoded region confers instability upon the mRNA transcripts in order to restrict the synthesis of histones to the S phase of the cell cycle. Instability-conferring sequences composed of a 50 nucleotide AU-rich sequence have also been defined at the 3′ untranslated regions of other eukaryotic mRNAs.

OAs: 99–105, 196–208. RAs: 29–32, 83–6, 259. SFR: 46, 47, 50, 74–8.

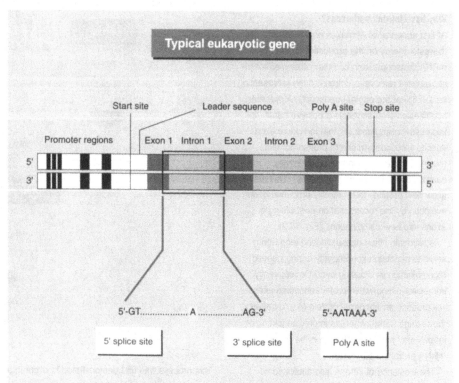

Typical eukaryotic gene

Start site | Leader sequence | Poly A site | Stop site

Promoter regions | Exon 1 | Intron 1 | Exon 2 | Intron 2 | Exon 3

5′ — 3′

3′ — 5′

5′-GT..............A ....AG-3′ | 5′-AATAAA-3′

5′ splice site | 3′ splice site | Poly A site

Fig. 14.3

EUKARYOTIC PROMOTER SEQUENCES

Eukaryotic genes may be divided into three classes (denoted I, II and III) depending on which type of RNA polymerase is used for transcription. Comparison of upstream flanking sequences has revealed a number of important control sequences which determine the binding and efficient transcription of a gene by its polymerase (Fig. 14.4). The most common sequence is termed the TATA box and is situated approximately 30 bases upstream (denoted −30) of the start site of class II genes. Several other conserved upstream sequences, including the GC box and CAAT box, have also been described at nucleotide positions −80 to −150.

Promoter sequences controlling the transcription of class I genes (e.g. rRNA genes) differ from those of class II genes since they appear to span the transcription start point (−45 to +20). This region contains an equivalent sequence to the TATA sequence with further elements known to be upstream. An 18 bp termination sequence has also been defined for the rRNA genes such that termination occurs approximately six hundred bases after the end of the gene. The promoters for class III genes (e.g. 5S rRNA, small rRNA and tRNA genes) are known as internal control regions and are also unusual in that they are located within the gene. Another unusual promoter element first found in the virus SV40 is known to be able to dramatically increase the expression of the particular gene with which it is associated. Termed enhancer sequences, these are distinct from other promoters since they are able to exert their effect even when located thousands of bases upstream or downstream of the gene in either orientation.

PROKARYOTIC GENE SIGNALS

Prokaryotic cells utilise two consensus promoter sequences (i.e. conserved sequences) similar to the eukaryotic −35 sequence but located at −10 and at −35. Sometimes called the Pribnow box the −10 element has the consensus sequence 5′-TATAAT-3′, whereas the −35 element has the consensus sequence 5′-TTGACA-3′. Many prokaryotic genes also possess leader and trailing sequences containing ribosome binding sites and termination signals respectively. The termination site also deviates from eukaryotic

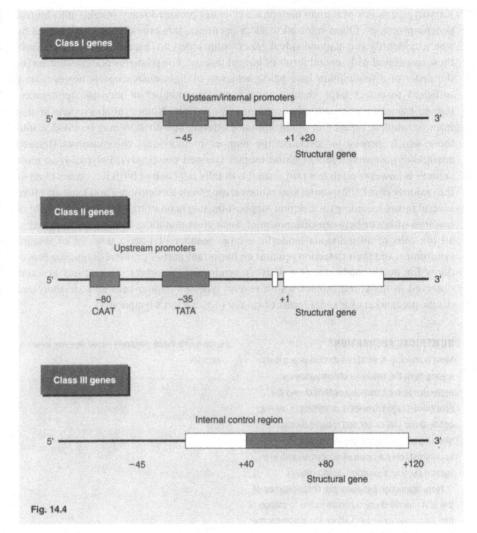

Fig. 14.4

gene signals by containing complementary palindromic sequences allowing the formation of a cruciform or stem-loop in double stranded DNA or single stranded RNA.

Other gene-associated features

Much of the DNA in higher eukaryotic cells is known to contain methylated cytosine residues, particularly within CG dinucleotides (denoted CpG). This element, sometimes called the CpG or HTF island (acronym for 'Hpall tiny fragments'), is a classic feature usually found associated upstream of transcriptionally active genes. In particular, these CpG islands surround the promoters of housekeeping genes which encode enzymes involved in essential metabolic

reactions such as glycolysis.

Different patterns of methylation are therefore believed to affect the expression of some genes such that undermethylated regions are usually active. Although undermethylation usually correlates with actively transcribed eukaryotic genes, there are several exceptions including non-methylated *Drosophila* DNA. The degree of methylation may be assessed by cleavage of the DNA with the restriction enzyme *Hpall* which recognises the methylated sequence CmCGG. Indeed, in many cases the isolation of a DNA fragment containing a CpG island may be the first point in the localisation of a gene (e.g. the cystic fibrosis gene).

15. CHROMOSOME ABERRATIONS AND LINKS TO HUMAN DISEASE

Large mutations causing numerical or structural aberrations within human chromosomes are correlated with several disease manifestations and are used both to localise and diagnostically detect disease genes

Karyotype analysis of human metaphase cells has provided many insights into human genetic processes. Often referred to as cytogenetics, this provided an early means by which to identify and map individual genes within individual chromosomes, particularly those associated with several forms of human disease. The relatively poor resolution (of the order of a few million base pairs) achieved by light microscopy is however only sufficient to detect large changes in chromosome number or physical appearance, termed numerical or structural aberrations respectively. Numerical aberrations arising from incomplete separation of chromosomes during cell division may be divided into those which increase or decrease the number of individual chromosomes (termed aneuploidy), or whole sets of chromosomes (termed polyploidy). Polyploidy in most animals is however relatively rare since it is usually fatal before birth if it occurs in all or the majority of cells. Structural aberrations arising from chromosome breakage may take several forms, including the deletion, duplication, inversion or transfer (translocation) of material within or between chromosomes. Such gross mutations in chromosomes within all the cells of an individual underlie a large number of human diseases or disease syndromes, and their detection remains an important part of prenatal diagnostic procedures for many conditions (e.g. Down's syndrome). Structural aberrations are also observed in the chromosomes within several types of tumour cells and are therefore diagnostic markers for some forms of cancer (e.g. Burkitt's lymphoma).

'INVISIBLE' SINGLE GENE DISORDERS

Cytogenetics can only detect aberrations involving greater than 4 million base pairs of DNA. Although greater resolution can be achieved using fluorescence *in situ* hybridisation, the detectable changes usually involve more than one gene. Most of the 5000+ identified human genetic disorders which segregate in a Mendelian fashion involve subtle mutations within a single gene. Many of these are common enough, and the gene mapped and identified along with disease markers (e.g. restriction fragment length polymorphisms), to make diagnostic detection feasible. However, this requires more sensitive techniques (e.g. Southern analysis, and the polymerase chain reaction).

OAs: 106–8, 885, 1000–2, 1004.
RAs: 34–6, 361. SFR: 29, 43, 51–7.

NUMERICAL ABERRATIONS

Most numerical aberrations involve aneuploidy, arising from the failure of chromosomes to separate (termed non-disjunction) during the anaphase stage of meiosis or mitosis. This may occur at the first or second meiotic division affecting all cells of an organism (see Fig. 15.1). In contrast, only a clone of abnormal cells are formed by non-disjunction during mitosis.

Non-disjunction between sex chromosomes at the first meiotic division results in the formation of half the gametes lacking either sex chromosome, and half with both. Non-disjunction during the second meiotic division produces one gamete with no sex chromosome, two with one sex chromosome and one with two copies of a single sex chromosome (i.e. XX or YY). Sex chromosome monosomy appears to be lethal, resulting in prenatal death (i.e. miscarriage), whereas trisomy is the most common form of aneuploidy. Sex chromosome abnormalities appear to be well tolerated, probably due to the phenomenon of sex chromosome dosage compensation whereby one of the two X chromosomes in females is inactivated. In comparison, autosomal abnormalities often have devastating consequences. Accordingly, monosomy of autosomes is incompatible with life, and trisomy of autosomes is very rare (see Table 15.1. Consequently few autosomal trisomies are

known and these generally result in early infant mortality.

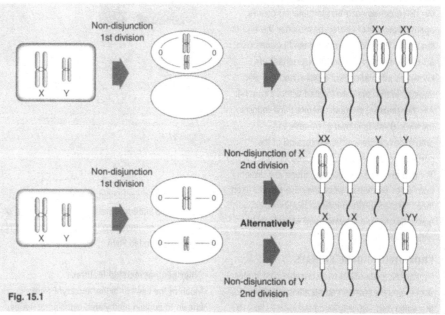

Fig. 15.1

Table 15.1

Karyotype	Common name	Clinical consequences
Trisomy 13 (47, XY,+13)	Patua syndrome	Multiple defects, death by age 1–3 months
Trisomy 18 (47, XY,+18)	Edward's syndrome	Multiple defects, death by age 1 year
Trisomy 21 (47, XY,+21)	Down's syndrome	Mental retardation, cardiac deformities, etc.
X0 (45, X) (female)	Turner's syndrome	Gonadal dysgenesis, mental retardation, etc.
XXY (47, XXY) (male)	Klinefelter syndrome	Slowly degenerating testes, enlarged breasts

STRUCTURAL ABERRATIONS

Structural aberrations may be formed *de novo* by chromosome breakage, or inherited by the production of unbalanced chromosomes in carriers. Six main classes of structural chromosomal aberration are detectable by cytogenetics: (i) translocation of material between chromosomes (e.g. by unequal crossing over), (ii) deletion of a chromosomal region, (iii) duplication of a chromosomal region, (iv) inversion of a chromosomal region, (v) formation of an isochromosome, arising from breakage of the centromere perpendicular to the normal division axis, and (vi) the formation of centric fragments. Each of these may dramatically affect the function of one or more genes by altering transcriptional activity or the nature of the protein(s) produced.

Although relatively few patients with inherited diseases have associated cytogenetic defects, several genetic diseases may be characterised by structural abnormalities (see Table 15.2). Such cytogenetic defects have therefore been used diagnostically. They have also been useful in the localisation of several disease genes prior to their cloning, isolation and identification.

ONCOGENE CONNECTIONS

Analysis of structural rearrangements has been particularly important in the study and cloning of some cancer-causing genes (i.e. oncogenes). For example, several B cell tumours arise from translocations joining the normally unexpressed c-*myc* proto-oncogene (located on the long arm of chromosome 8) to highly expressed immunoglobulin genes. Accordingly, in the majority of cells from individuals with Burkitt's lymphoma the c-*myc* oncogene is brought directly under the control of an immunoglobulin enhancer sequence located on the long arm of chromosome 14. This results in unregulated expression of the c-*myc* gene (Fig. 15.2). Similarly, in chronic myelogenous lymphoma structural rearrangements result in the formation of a chracteristically shaped chromosome termed the Philadelphia chromosome. This involves the translocation of the tip of chromosome 9 to chromosome 22. Consequently, the major portion of the oncogene c-*abl* is consistently joined to a genetic region known as *bcr* resulting in unregulated expression of a bcr-abl fusion protein (a protein kinase) important in the uncontrolled growth of leukaemic cells.

Table 15.2

Genetic disease (manifestations)	Structural aberration(s)
Cri-du-chat syndrome (cat-like cry of infant, with congenital heart disease and severe mental retardation)	Deletion of part of the smallarm of chromosome 5 (karyotype designated 46, XY, del(5Xp25))
Isochromosome Xq	Formation of an X Isochromosome containing only long arms, plus an Xp isochromosome or two acentric pieces (effective Xq trisomy and Xp monosomy)
Retinoblastoma (eye tumour formation in children)	Deletion of part of chromosome 13 (band q14.11)
Prader–Willi syndrome (neurological disorder)	Deletions and duplications of the long arm of chromosome 15
Neurofibromatosis (neutral tumour formation	Translocations involving the long arm of chromosome 17(band q11.2)
Lowe syndrome (congenital cataracts, mental retardation, kidney defects)	Translocations with breakpoints in the long arm of the chromosome (band q25)
DiGeorge syndrome (thymus, parathyroid and congenital heart defects)	Rearrangements within the long arm of chromosome 22 (band q11)
Wilm's tumour (kidney tumour formation in children)	Deletion of part of the short arm of chromosome 11 (band p13)
Fragile X syndrome (mental retardation, altered speech, large testes, etc	Breakage of the tip of the chromosome (although usually identified by the lack of chromatin condensation at *FMR-1* gene site)
Burkitt's lymphoma (B lymphocyte tumour)	Reciprocal translocation of part of the long arm of chromosome 8 (band q24) and most often chromosome 14 (band q32)

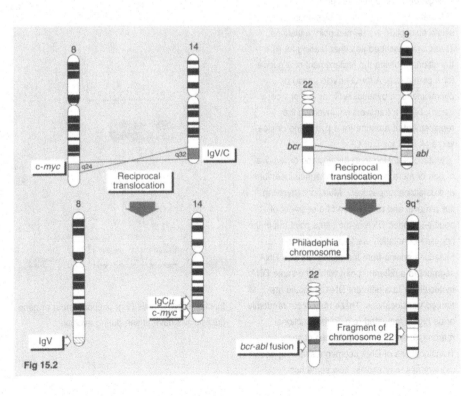

Fig 15.2

16. TYPES OF MUTATIONS AND THEIR EFFECTS

Spontaneous or induced mutations affecting one nucleotide or a large DNA segment may either be silent (i.e. have no effect) or produce a profound change in the structure and/or function of a gene(s).

Much of our current knowledge of the structure and function of genes has been derived from the analysis of alterations (i.e. mutations) in their nucleotide sequence. Indeed, the early steps towards the molecular definition of genes largely depended upon the analysis of alterations in an organism's phenotype arising from naturally occurring (i.e. spontaneous) or artificially induced mutations. The subsequent elucidation of the biochemical composition of genes has led to a greater understanding of how different types of mutations arise, and the cellular mechanisms used to correct them (i.e. DNA repair mechanisms). Thus, the proof-reading activity of DNA polymerases and DNA repair mechanisms serve to reduce the amount of mutations which are stably incorporated into the genome. Nevertheless, several mutations evade correction and are therefore able to be transmitted throughout subsequent generations. These range from the stable incorporation of small-scale mutations arising from errors in DNA replication (e.g. mismatched bases) to large-scale mutations involving the random or ordered movement (or transposition) of large segments of DNA within or between individual chromosomes by the process of genetic recombination. Mutations may be silent (i.e. have no detectable effect) or have a pronounced and detectable effect. The latter may either be beneficial to the organism and/or its progeny (i.e. by increasing genetic variation) or profoundly deleterious (e.g. lethal or leading to disease manifestation). Many recombinant DNA methods have therefore been developed in order to either detect or introduce mutations. Such methods have been widely employed for clinical diagnosis, gene mapping and functional analyses, and the production of new genetically engineered proteins.

ARTIFICIALLY INDUCING MUTATIONS

Mutations affecting one or more nucleotides may be randomly induced within DNA molecules *in vivo* by several types of chemical or physical agents, termed mutagens. These may then be stably incorporated during DNA replication. For example, ethidium bromide can be used to intercalate between the two DNA strands of a double helix thereby distorting its structure leading to incorrect base pairing. Physical agents such as heat, or exposure to ultraviolet or gamma radiation may also be used to chemically alter the nucleotides (e.g. forming thymine dimers). Many alternative approaches have now been developed to introduce mutations into DNA *in vitro* (i.e. by *in vitro* mutagenesis). Based on the use of base analogues (e.g. dITP and 5-bromo-uracil) or synthetic oligonucleotides, these allow the random or site-specific introduction of single or multiple nucleotide changes.

CLASSES OF MUTATION

Several classes of mutations affecting the nucleotide sequence of a DNA molecule may be defined (see Fig. 16.1). Each involve a change in a single or multiple nucleotide(s).

Mutations leading to the replacement of a single nucleotide are termed point mutations. These are described as either leading to: (i) a transition, involving the replacement of a purine for a purine (e.g. A for G or vice versa) or pyrimidine for a pyrimidine (T for a C or vice versa), or (ii) a transversion, involving the replacement of a purine for a pyrimidine or vice versa (A or G for T or C).

Mutations arising from the addition or removal of one or more nucleotides are termed insertions and deletions respectively. Mutations involving the removal and re-insertion of a segment of double-stranded DNA at the same point but in the opposite orientation are termed inversions. Mutations arising from the movement of a DNA segment to a different point within the same DNA molecule, or to a different DNA molecule are termed translocations. These may occur randomly or be highly regulated (e.g. in the functional rearrangement of immunoglobulin genes). Translocations of DNA segments spanning one or more genes may explain how some non-functional gene relics (e.g. pseudogenes) or gene duplications have arisen during evolution.

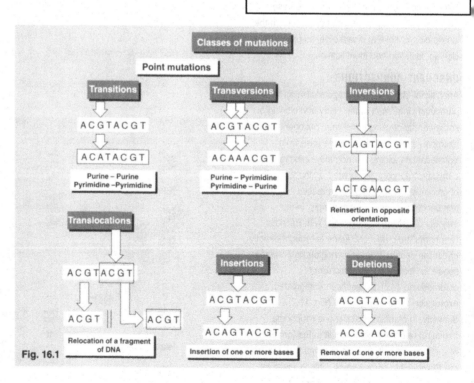

Fig. 16.1

OAs: 109–12, 115, 276, 1000–9, 1071, 1072.
RAs: 37, 38, 41, 44, 129–31.
SFR: 58–61, 94, 95, 197, 201.

EFFECTS OF MUTATIONS

Mutations may either be (i) silent, having no discernible effect on the cell or organism, or (ii) result in structural and/or functional changes within the gene and/or its protein product.

Silent mutations

All types of mutations, but especially those occurring in the extensive non-coding intronic or extragenic DNA sequences found in eukaryotic genomes, may be silent (Fig. 16.2). Some point mutations within coding regions may also be silent due to the degeneracy or 'wobble' of the genetic code. This is especially the case for those affecting the third base of a codon since this will often not produce a change in the order of amino acids encoded by the gene. Such mutations are therefore undetectable at the protein level but are detectable at the nucleotide level by several methods. For example, they may be detected based upon the creation or destruction of a restriction enzyme recognition site leading to a restriction fragment length polymorphism (RFLP).

Some point mutations (termed missense mutations) resulting in a change in the encoded amino acid sequence are also functionally silent since they do not affect the activity or expression of the encoded protein. These may be detected at both the nucleotide level and the protein level (e.g. as antigenic variants). Depending upon the amino acid involved and its location, missense mutations may however affect the structure and/or function of the encoded protein.

Functional mutations

All types of mutations occurring within both coding and non-coding DNA sequences can affect the functioning of a gene and its encoded protein. For example, point mutations within the coding portion of a gene may create a premature stop codon (i.e. a nonsense mutation) leading to the expression of a truncated polypeptide and mutant phenotype (Fig. 16.3). Alternatively, point mutations within the non-coding regions may alter a functional regulatory sequence, such as a promoter or initiation signal, leading to increased or decreased gene expression. Insertions or deletions within coding regions and involving multiples of three nucleotides may also lead to the production of incorrect length polypeptides. However, deletions or insertions not involving

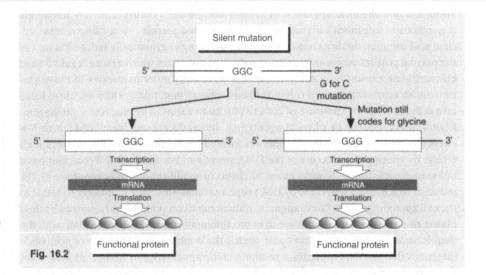

Fig. 16.2

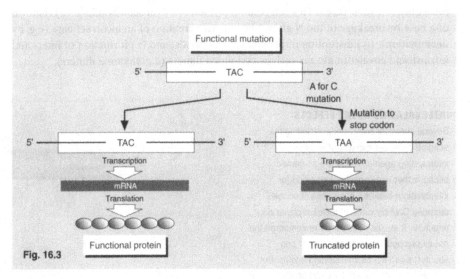

Fig. 16.3

multiples of three nucleotides will result in a frameshift mutation. This effectively scrambles the code preventing expression of a functional protein. Both insertions and deletions may also affect the expression of a given gene if they affect gene-associated regulatory sequences.

Similar effects are most frequently observed when inversions or translocations occur. However, translocations involving whole genes may not affect the function of the encoded protein but may significantly alter its level of expression (e.g. by bringing it under the control of a new transcriptional regulatory sequence).

MUTATION DETECTION

Although the appearance of mutant phenotypes was originally used to detect mutations, the identification of most forms of mutations (particularly silent mutations) requires alternative approaches. Translocations involving large segments of DNA may be detected by visual examination of chromosomes. However, more sensitive and sophisticated techniques are required to detect the more common forms of small-scale mutations. These include nucleotide sequencing, Southern blotting, oligonucleotide hybridisation assays, RFLP analysis and the polymerase chain reaction.

17. FORMS OF CHEMICALLY ALTERED DNA

Various chemical and physical mutagens can modify the chemical nature and properties of nucleotides thereby generating point mutations, deletions, insertions, frameshifts, and strand breaks.

Structural and chemical alteration of cellular nucleic acids occurs relatively frequently throughout an organism's lifetime with differing consequences. Accordingly, some nucleic acid chemical modifications occur naturally, being enzymatically induced as part of the normal post-transcriptional or post-replicative processes which ensure regulated and efficient gene expression. Less common, but potentially more hazardous to the organism, are structural alterations (often termed chemical mutations) which are introduced into cellular DNA by exposure of cells to particular external agents, termed mutagens. Chemical compounds or physical agents (e.g. heat and various forms of high or low energy radiation) are the most common forms of such mutagens. These exert their effects by altering the structure of the DNA strands and/or the chemical composition of individual nucleotides. A wide range of chemical modifications may consequently be induced, which, if not corrected by DNA repair mechanisms, may ultimately be lethal to the cell and/or organism. For example, certain organic compounds characterized by their planar ring structures (e.g. acridine dyes or ethidium bromide) can intercalate with the double stranded DNA to disrupt and stretch the double helix. This allows nucleotide insertions during DNA replication resulting in the appearance of frameshift mutations. Other compounds (e.g. nitrous acid, hydroxylamine and alkylating agents) directly alter nucleotides or their bases. This can result in: (i) altered hydrogen bonding, (ii) removal of a base by breakage of the N-glycosidic link, (iii) creation of an incorrect base (e.g. by deamination), (iv) disruption of phosphodiester bonds, and (v) formation of inter- and intra-strand covalent linkages such as cyclobutyl dimers (e.g. thymine dimers).

NATURAL NUCLEOTIDE MODIFICATION

In addition to chemical mutation of DNA, the structure of various parts of a nucleotide may also undergo chemical modification as part of the normal process of *in vivo* gene expression. This particularly occurs during the post-transcriptional modification of ribonucleotides within ribosomal, messenger and transfer RNA molecules. Indeed, the generation or incorporation of chemically modified ribonucleotides is critical in determining their secondary and tertiary structures, and is often required for functional activity. For example, the 5′ end of eukaryotic messenger RNAs contains a post-transcriptionally added cap structure composed of methylated cytosine and guanine ribonucleotides. Transfer RNA molecules also contain several different unusual ribonucleotides formed by post-transcriptional modification, e.g. 7-methylguanosine, dihydrouridine, queosine and pseudouridine (ψ).

INTERCALATING AGENT EFFECTS

Several chemical compounds, including the acridine dyes, are capable of acting as intercalating agents. These are so named because they are capable of inserting (or intercalating) between base pairs in double-stranded DNA by virtue of their flat planar ring structure. It should however be remembered that not all compounds containing planar ring structures can act as intercalating agents, but only those of the correct size and geometry. For example, the ethidium bromide (EtBr) is an intercalating agent because its four planar rings have overall dimensions very similar to those of a purine–pyrimidine base pair (Fig. 17.1).

Insertion of an intercalating agent causes stretching or elongation of the double helix by the enforced separation of neighbouring base pairs. This generally results in the insertion of a new base adjacent to the site of intercalation during the replication of DNA by DNA polymerase. This causes a frameshift mutation and most frequently results in a phenotypic change by disrupting the nature of an encoded protein. It is this property which led to the early experimental use of EtBr as a mutagen, and the continued use of intercalating agents in mutation research. Despite its

mutagenicity and carcinogenicity, the intercalation of EtBr into DNA duplexes is still routinely exploited for the visualisation of nucleic acids following gel electrophoresis.

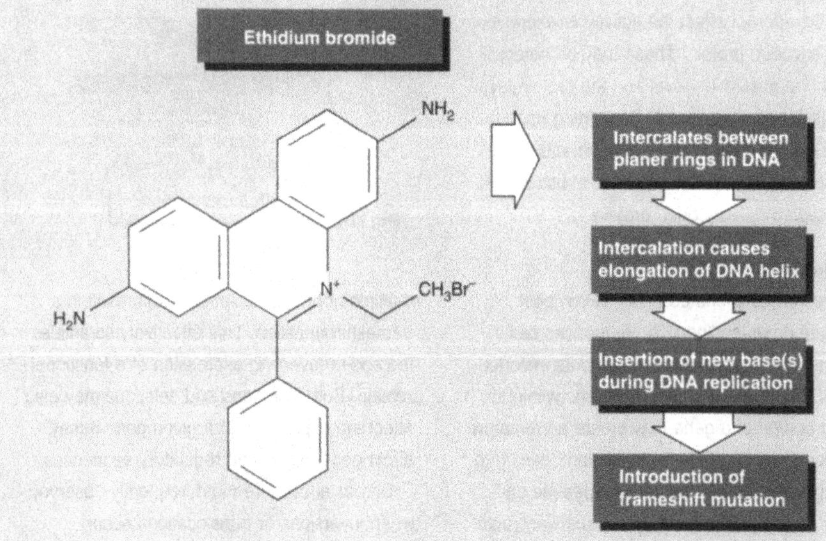

Fig. 17.1

OAs: 109–13, 115, 188, 191, 771, 776, 543. RAs: 19, 37, 38, 44, 76–8, 272, 273. SFR: 10, 11, 13, 36, 58–61, 197.

4. Chemical building blocks of nucleic acids.
5. Formation of the DNA double helix.
16. Types of mutations and their effects.
18. DNA repair mechanisms.
22. Nature of the genetic code.
23. Transcription.
24. Post-transcriptional processing of mRNA transcripts.
25. tRNA and rRNA processing.
26. Mechanisms regulating gene expression.
37. *In vivo* DNA replication.
46. Visualising nucleic acids.
47. Electrophoresis of nucleic acids.
60. Types and applications of nucleotide analogues.
79. *In vitro* mutagenesis.

TYPES OF CHEMICAL DAMAGE

Various chemical compounds can alter the composition of bases, often affecting their behaviour within nucleic acids (Table 17.1). For example, deamination by nitrous acid can convert guanine to xanthine without altering the base pairing (i.e. xanthine still base pairs with cytosine). However, comparison deamination by nitrous acid also converts cytosine to uracil, and adenine to hypoxanthine. These conversions cannot subsequently hydrogen bond to the opposing base on the other strand of the DNA double helix, resulting in the formation of a point mutation in subsequent DNA copies. Alkylating agents (e.g. dimethyl sulphate) can similarly have several effects. These include disruption of phosphodiester bonds leading to DNA strand fragmentation, alteration of hydrogen bonding properties, base methylation, and the formation of covalent inter-strand linkages.

Table 17.1

Mutagen	Example of mutagen	DNA damage
Base analogues	Bromouracil (BU) Aminopurine (AP)	Base pair errors (tautomerism)
Alkylating agents	Ethylmethane sulphonate Sulphur mustard	Base pair error (G/T modification) Depurination/deletion/distortion
Acridines	Proflavin	Intercalation into double helix
Hydroxylamine	Hydroxylamine	Base pair error (C modification)

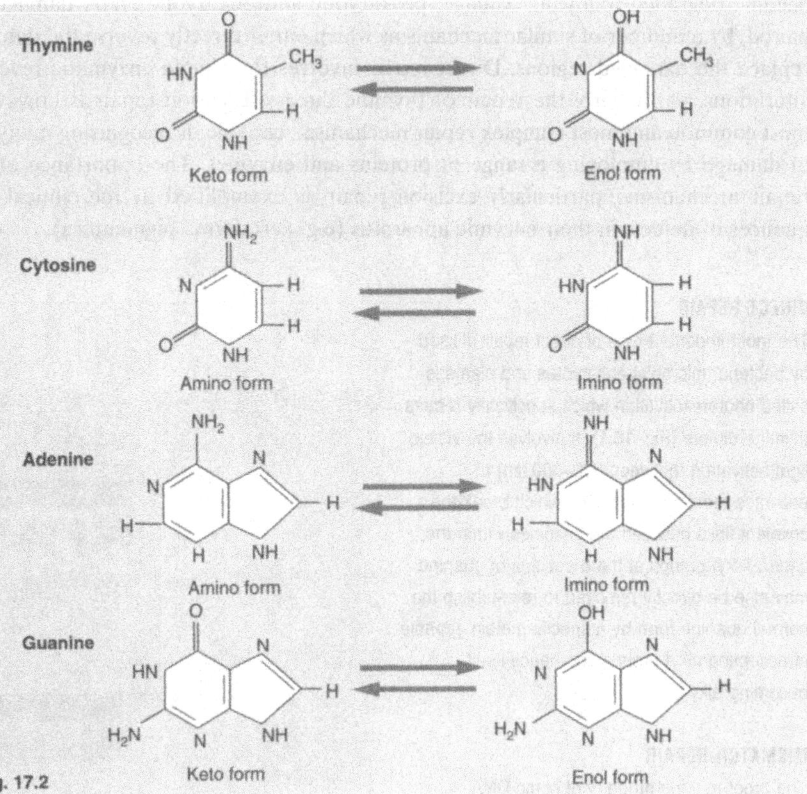

Fig. 17.2

Effects of radiation

High and low energy forms of radiation including ultraviolet (UV) light at about 260 nm wavelength, gamma (γ) and X-rays are also effective mutagens. Both UV and X-rays can induce mutations by causing what is termed a tautomeric shift. This arises because each of the purine and pyrimidine bases of DNA and RNA may exist in two slightly different structural forms termed structural isomers or tautomers. Thus, guanine and thymine can exist in keto or enol forms, whilst adenine and cytosine can exist in amino or imino forms (Fig. 17.2).

Under normal physiological conditions the equilibrium is shifted almost entirely towards the keto and amino forms, however absorption of radiation energy by the bases causes the equilibrium to shift towards the enol and imino forms. The altered molecular configurations of these minor forms imbues them with different hydrogen bonding properties leading to atypical base pairing. For example, the enol form of thymine base pairs with guanine (rather than adenine), and the imino form of adenine base pairs with cytosine (rather than thymine). This increases the frequency with which point mutations are incorporated during DNA replication. The base analogue 5-bromo-uracil (5bU) used for *in vitro* mutagenesis similarly performs this function by undergoing a tautomeric shift once incorporated into DNA.

X-rays and γ-rays may also lead to the opening of the heterocyclic rings of bases and the breakage of phosphodiester bonds leading to DNA strand breaks. A range of oxidative modifications can also occur in the presence of oxygen. UV irradiation can additionally cause the photochemical dimerisation of adjacent pyrimidines, particularly thymines, although C-C and T-C dimers can also be formed. Such local compaction of the stacked bases can result in the creation of deletion mutations.

Effects of heat

Heat is a common form of physical mutagen in the environment, leading to the creation of apurinic sites by depurination. This breaks the N-glycosidic link holding the base within the polynucleotide. If uncorrected, this results in the introduction of both point or deletion mutations. It is estimated that up to 10 000 purine bases may actually be removed from the nucleotides of the genomic DNA within each human cell every single day.

18. DNA REPAIR MECHANISMS

Excision–repair mechanisms are the most common and complex type of enzymatic repair mechanisms operating to directly or indirectly remove the many types of damaged DNA which arise *in vivo*.

DNA is extremely stable compared to proteins and RNA molecules which are continually 'turned over' (i.e. rapidly degraded and replaced by new molecules). Indeed, this fact provided an early clue to the role of DNA as the hereditary material. Genomic DNA is not however exempt from gradual changes in both conformation and structure. Accordingly, the entire genome or restricted portions may conformationally alter as it undergoes condensation and relaxation to allow vital functions including *in vivo* DNA replication or gene expression via transcription. Several types of structural changes also arise from chemical or physical damage to individual bases or phosphodiester bonds. Although these occur relatively infrequently and generally affect only a few nucleotides they can have profound effects upon DNA function. Errors such as mismatches may also be introduced during DNA replication despite the self-correcting mechanisms (i.e. proof-reading) of DNA polymerases. If undetected and not repaired, such alterations become stably incorporated into the genome. In contrast to other macromolecules which become degraded following damage, prokaryotic and eukaryotic DNA damage is repaired, by a number of similar mechanisms which either directly reverse the damage or replace the damaged regions. Direct repair involves the simple enzymatic reversal of alterations, particularly the repair of thymine dimers. Excision repair is however the most common and most complex repair mechanism, capable of recognising many forms of damage by employing a range of proteins and enzymes. The importance of DNA repair mechanisms, particularly excision repair, is exemplified by the clinical consequences of defects in their enzymic apparatus (e.g. xeroderma pigmentosa).

TYPES OF CHEMICAL MUTATIONS
Several forms of chemically modified nucleotides may be generated within DNA both *in vivo* and *in vitro* by certain chemical or physical mutagens. *In vivo* these modified nucleotides are referred to as chemical mutations (or DNA lesions) whereas *in vitro* modified nucleotides are referred to as nucleotide analogues. Chemical mutations commonly arise due to alkylation and oxidative deamination. This results in altered base pairing such as mispairing between guanine and thymine, or adenine and cytosine, and the formation of pyrimidine dimers. UV or high energy radiation may also efficiently generate chemical mutations by inducing a shift from the normal keto or amino forms of bases to the rarer enol or imino forms. This may create dimers between adjacent pyrimidines (e.g. thymine dimers or thymine–cytosine dimers), openings within the heterocyclic rings of bases, or the disruption of phosphodiester bonds.

DIRECT REPAIR
The most important type of direct repair utilised by bacteria, microbial eukaryotes and plants is called photoreactivation which specifically repairs thymine dimers (Fig. 18.1). It involves the visible light activation (between 300–400 nm) of enzymes termed photolyases which break the covalent links between the thymines within the dimer. Alkyl groups at the 6 position of guanine may also be directly removed to re-establish the normal guanine form by a specifc protein capable of accepting alkyl groups and hence itself becoming alkylated.

MISMATCH REPAIR
The proof-reading property of some DNA polymerases during DNA replication, brought about by their 5'–3' exonuclease activity, is not a completely reliable process. Mismatched bases may therefore still be incorporated within daughter strands. These errors may, however, be corrected by several types of enzymes which combine to recognise the error, modify the base(s) to signal its presence, and then repair it. Since the error is generated specifically in the daughter strand, this repair system must be able to discriminate between parental and daughter DNA strands. In *E. coli* this is achieved by virtue of the specific methylation of nucleotides in the parental DNA strands (e.g. adenines). In contrast, newly formed daughter strands are not methylated for some time following the completion of DNA synthesis.

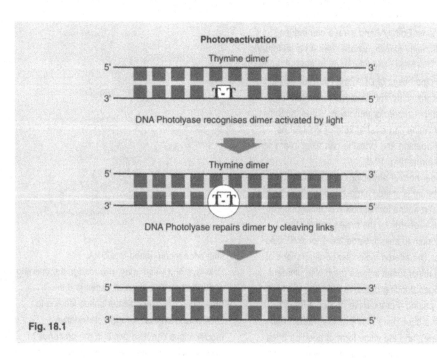

Photoreactivation

Thymine dimer

5' ——— 3'
3' —— T-T —— 5'

DNA Photolyase recognises dimer activated by light

Thymine dimer

5' ——— 3'
3' —— (T-T) —— 5'

DNA Photolyase repairs dimer by cleaving links

5' ——— 3'
3' ——— 5'

Fig. 18.1

OAs: 116–29, 333, 341.
RAs: 37–45, 178–81, 184, 185.
SFR: 36–9, 58, 59, 61, 125, 182, 183.

EXCISION REPAIR MECHANISMS

Excision repair mechanisms vary in the type of enzymatic systems used according to the nature of the DNA lesion. They are characterised by the following four distinct and sequential steps, as shown in Fig. 18.2.

(1) Incision: This is individual to each type of damage and is the rate controlling step since it involves recognition of the specific type of damage to be repaired. It involves one of several enzymes which mark the site. This is achieved by the introduction of one or two single-stranded nicks adjacent to the damaged nucleotide, or cleavage of the altered base to produce an apurinic or apyrimidinic site.

(2) Excision: This step is catalysed by one of several nucleases and involves the removal of the damaged nucleotide (and often some of its neighbouring nucleotides).

(3) Resynthesis: This step involves the resynthesis of the excised nucleotides by DNA polymerase I. The exonucleolytic activity of DNA polymerase I may also move the nick along the damaged strand for some distance.

(4) Ligation: The final step is the joining of the gapped ends by DNA ligase.

INCISION AND EXCISION MECHANISMS

The recognition and excision of bulky lesions such as pyrimidine dimers utilises a multi-enzyme complex. For example, repair of pyrimidine dimers in *E. coli* involves the protein products of four genes, denoted *UvrA*, *UvrB*, *UvrC* and *UvrD*. Accordingly, the local distortion in the DNA is first recognised by a tetrameric endonuclease complex composed of two UvrA and two UvrB proteins (see Fig. 18.3). Initially UvrA binds a short distance from the damaged site and, in an ATP-dependent manner, unwinds the DNA to allow the UvrB proteins to associate. Again at the expense of ATP, the newly formed UvrAB complex then migrates along the DNA until it reaches the damaged site. Incisions are then made by the UvrC protein either side of the damage, often spanning the dimer by as many as 12 bases. The excised oligonucleotide is then released by the unwinding of the DNA mediated by the helicase activity of the UvrD protein. The resultant gap is then filled by the action of DNA polymerase I and DNA ligase.

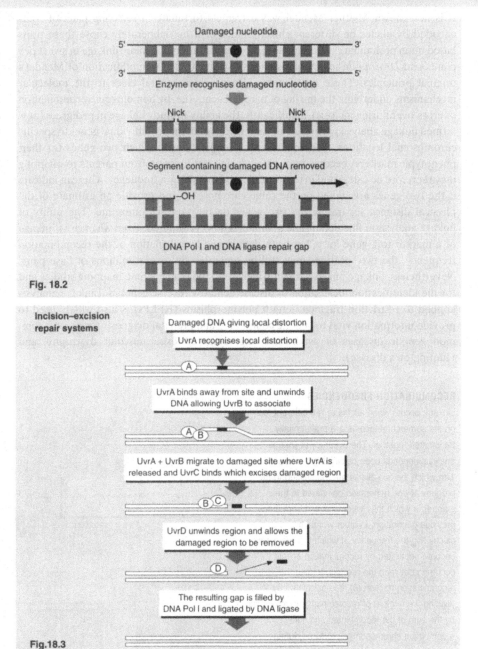

Fig. 18.2

Fig.18.3

Single base modifications may also be marked and removed by apurinic–apyrimidinic (AP) endonuclease. This enzyme catalyses depurination (i.e. the removal of the purine base) and the breaking of the phosphodiester backbone at depurinised sites. Deaminated bases may also be removed by enzymes known as DNA glycosylases (e.g. uracil DNA glycosylase), and the sugar-phosphate remnant removed by AP endonuclease. Thymine dimers may also be removed by the cooperation between a specific N-glycosylase and AP endonuclease. Thus, the N-glycosylase breaks the glycosidic bond that connects one of the thymines to the polynucleotide, and the AP endonuclease nicks the chain either side of the nucleotide sequence containing the thymine dimer.

19. LINKAGE ANALYSIS

Linkage analysis is a statistical method whereby the frequency of recombination events occurring between two polymorphic alleles can be calculated to establish their relative chromosomal locations.

It is not known whether Mendel was extremely fortunate in selecting pairs of traits encoded by alleles on different chromosomes or if he deliberately chose these traits based upon preliminary experiments. The demonstration of genetic linkage in sweet pea plants and *Drosophila* sex chromosomes subsequently led to modification of Mendel's original principles. These experiments also provided several clues to the molecular mechanisms underlying the mixing of parental genes (i.e. by homologous recombination or cross-over during meiosis). Significantly, the ability to study linkage in pedigrees (now termed linkage analysis) provided an early means of mapping individual genes to specific chromosomal locations. Consequently, the frequency with which two genes (or their phenotypic markers) become segregated during transmission from parents to offspring may therefore be calculated in terms of their recombination frequency. This can indicate if the two genes are linked on the same chromosome and provide an estimate of the physical distance separating the two genes on the same chromosome. The utility of linkage analysis in fine detail gene mapping is however limited. Even when close linkage of a marker to a gene locus has been determined by calculation of the recombination frequency, the two locations may still be separated by several millions of base pairs. Nevertheless, linkage analysis is still widely used today in genome mapping studies and for the identification/localisation of disease genes by reverse genetics. Linkage analysis applied to restriction fragment length polymorphisms (RFLPs) is also widely used to provide information vital to genetic counselling and prenatal diagnosis for several common genetic diseases or syndromes (e.g. cystic fibrosis, muscular dystrophy, and Huntington's disease).

REVERSE GENETICS

The term reverse genetics is often used to describe the process by which a gene (e.g. a disease-associated gene) is mapped based upon an initial clinical phenotype. For example, the clinical phenotype may be associated with a particular RFLP which has been located close to the disease locus, as determined by the calculation of the recombination frequency. Having obtained such a chromosomal marker, it may be used as a means to generate a probe for chromosome walking and/or jumping until the disease gene locus is reached. The identification of the gene associated with cystic fibrosis is a classical example of the process of reverse genetics in action.

OAs: 130–3, 859, 863, 866–8, 883, 890–2, 1138. **RAs:** 304, 306.
SFR: 46, 62–8, 160–3, 172–81.

RECOMBINATION FREQUENCIES

Linkage occurs when alleles at two different loci on the same chromosome are preferentially transmitted together. The accuracy of linkage analysis depends upon calculating the frequency with which the two alleles are transmitted to the progeny of one generation compared to the manner in which they were inherited from the previous generation; in other words, by calculating the frequency at which the two alleles become separated by recombination events occurring between the two loci, termed the recombination fraction (θ), it is calculated by dividing the number of recombinant chromosomes by the sum of the recombinant and non-recombinant chromosomes produced during meiosis (see Fig. 19.1). Loci which are physically very close will always be transmitted together giving a recombination fraction of 0. In comparison, unlinked loci give a θ value of 0.5 since the alleles are transmitted independently. Loci are therefore generally regarded as being linked if a θ value of less than 0.5 is obtained.

The recombination frequency can also provide direct information regarding the physical distance between the two loci. Thus, a recombination frequency of 1% (i.e. the two loci are separated

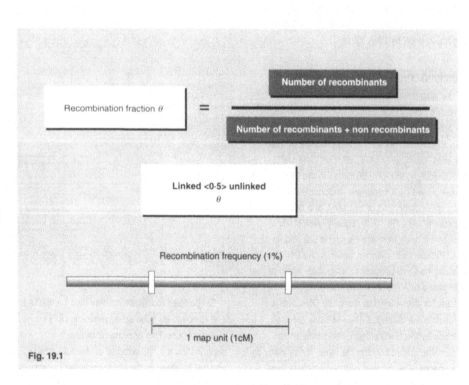

Fig. 19.1

in 1% of gametes) represents separation of the two loci by one map unit or centiMorgan (cM), representing 10^6 base pairs (1000 kb) in humans.

The accuracy of linkage analysis has been further increased by the development of a probability-based technique, termed the lod score method.

CALCULATING LOD SCORES

Developed by Newton Morton, the calculation of lod (log of the odds) scores tests the significance of any proposed linkage between two loci given their recombination frequency. The lod score (designated Z) calculates the likelihood of individual genotypes within a pedigree resulting from an experimentally derived recombination frequency by comparing it with the probability of obtaining those genotypes if the loci were unlinked. Using this method, a range of lod scores can be obtained at a range of θ values for any set of genotypes in a pedigree.

Linkage is statistically supported if the lod score is equal to or greater than 3.0. This indicates that there is a 1000 times higher probability that the genotypes have arisen by linkage than non-linkage. If the range of possible lod scores and θ values are examined, the θ value giving the highest lod score is most likely to be the most accurate. Consequently, this θ value gives the best approximation to the physical distance separating the two loci.

Calculating lod scores for one pair of loci in a relatively small family is not particularly difficult. However, applying such statistical analysis to several large families with several different loci represents a formidable challenge. Several computer programs and packages have thankfully been developed to assist this process.

Requirements for gene mapping

The successful application of genetic linkage techniques to the localisation of a gene (e.g. associated with a disease) depends upon the satisfaction of several important factors.

Since linkage analysis is a statistical method, a sufficient number of large families with affected members for which pedigrees can be established must be available. This allows any possible genetic heterogeneity between different families with the same apparent clinical or biochemical disease manifestations to be taken into account. Similarly, the disease (or phenotype) must be clearly and unambiguously defined to avoid complications. The mode of inheritance and penetration of the disease or phenotype must also be determined and accounted for in any calculation. Another prerequisite is the availability of at least two distinct forms of disease or phenotypic marker.

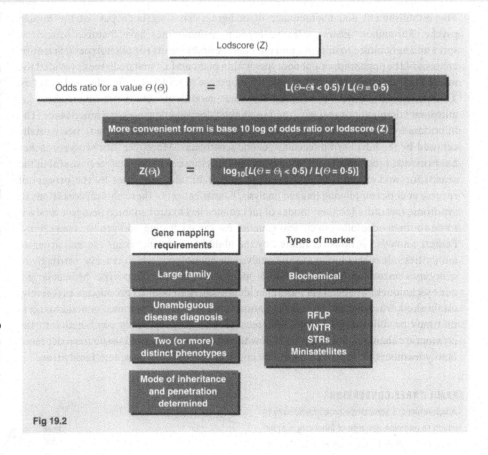

Fig 19.2

Types of marker

Several types of markers have been used in linkage analysis, including proteins implicated in disease pathogenesis by clinical/biochemical studies. The most important criterion for any marker is that it must be polymorphic, being present in the population in two or more forms. RFLPs have served as a particularly useful type of marker for linkage analysis. Often however, RFLPs may not be very informative and are limited to the detection of a relatively small proportion of DNA sequence variations. RFLPs are therefore becoming replaced by more informative, highly polymorphic DNA markers such as VNTRs (variable number of tandem repeats) and STRs (short tandem repeats) or DNA minisatellites. For example, the human genome can already be efficiently scanned using the 200+ STR loci which have been identified (out of the 130 000 estimated to be present in human DNA, and therefore available for exploitation as markers).

LINKAGE DISEQUILIBRIUM

In 1908, G.H. Hardy and W. Weinberg independently developed a mathematical rule relating allelic and genotypic frequencies in a theoretical population under idealised conditions (i.e. composed of a large number of diploid, sexually reproducing individuals not subject to selection, migration, mutation or mating restrictions). Under such idealised conditions, the frequency of genotypes and alleles is dependent upon the probability of selection of alleles from the gene pool into gametes (termed Hardy–Weinberg equilibrium). Alleles located at different loci on the same chromosome and which adhere to this completely random distribution into gametes are said to be in linkage equilibrium. Alleles at two different loci which are not randomly distributed to gametes are however said to be in linkage disequilibrium. This may be mathematically related to the recombination distance between the loci.

20. PEDIGREE ANALYSIS AND MODES OF INHERITANCE

Autosomal and sex-linked patterns of dominant or recessive inheritance may be established from analysis of pedigrees, thereby aiding prenatal diagnosis, carrier testing, or gene localisation by linkage analysis.

The establishment and maintenance of pedigrees has long been part of the human psyche. Throughout history selective breeding programmes have featured in society, sport and agriculture to improve and maintain desirable traits (or phenotypes). Scientific analysis of the appearance of phenotypes within plant and animal pedigrees provided the initial breakthroughs required for the development of molecular and classical genetics. Thus, pedigree analysis established the basic modes of eukaryotic inheritance, i.e. autosomal dominant or recessive, and sex-linked dominant or recessive inheritance. The importance of human pedigree study for human genetic study has not been totally eclipsed by the advent of molecular genetic techniques. Moreover, the two approaches have provided reciprocal benefits. For example, pedigree analysis has been useful in the search for, and chromosomal localisation of, specific disease genes by the process of reverse genetics employing linkage analysis. Classification of these clinical conditions or syndromes according to their modes of inheritance has in turn enabled pedigree analysis to be routinely exploited in clinical genetics for carrier testing and genetic counselling. Pedigree analysis is however limited by the ability to construct complete and accurate family trees. It is also limited to the analysis of phenotypes which correlate strongly to genotypes since some genotypes do not 'penetrate' into the phenotype. Molecular genetic techniques have been very useful in identifying both familial promiscuities (thereby establishing true paternal/maternal relationships) and low penetrance genetic markers that may be followed in human pedigrees. Genetic fingerprinting combined with the polymerase chain reaction may also allow the analysis of preserved tissue from deceased family members thereby facilitating the construction of more complete family trees.

UNMASKING INFIDELITY

The precision of pedigree analysis and its use in tracking disease genes depends upon establishing authentic familial relationships. Unfortunately, sexual promiscuity is remarkably common amongst humans (and several other species) but rarely admitted when attempting to construct family trees. Even when revealed during questioning, it is not always possible to determine if such infidelities have resulted in the transmission of 'foreign' genes to siblings (unless it explains the unexpected appearance of a trait otherwise absent in the pedigree). The advent of genetic fingerprinting techniques and their application to paternity testing has greatly aided the accuracy of family tree construction by unmasking any genetic products arising from infidelity. Perhaps unsurprisingly, such knowledge can pose investigators with ethical problems (e.g. who should be informed).

FAMILY TREE CONVENTIONS

A convenient, if sometimes problematic, way in which to examine patterns of inheritance within pedigrees is to establish a family tree. Certain conventional symbols are used to identify the various components of a family tree (see Fig. 20.1). Accordingly, circles are used to denote females, squares are used to represent males, and diamonds used to indicate cases where the sex of the individual is unknown. Filled symbols are used to indicate that an individual expresses the trait under study (i.e. an affected individual) whereas open symbols are used for unaffected individuals. Direct horizontal lines indicate marriage and vertical lines link the progeny produced by that union. Brothers and sisters from a single union (often referred to as siblings or sibs) are joined by horizontal lines above their symbols. The order of their births is also indicated by numbers beneath the symbols. Occasionally, data may not be available for all siblings although the number of siblings may be known. This is indicated by a number within the symbol rather than below the symbol. Identical or non-identical twins are distinguished by joining of the symbols with converging diagonal lines which indirectly or directly join the horizontal line respectively.

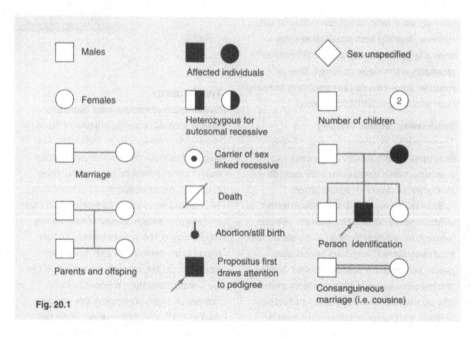

Fig. 20.1

The value of pedigree analysis

Examination of a pedigree may readily indicate whether a particular trait is inherited in a pattern consistent with known modes of inheritance. This may not however be clear, particularly if the trait is variably expressed due to poor penetrance.

Nevertheless, over 5000 gene loci have been identified which do follow conventional patterns of inheritance. Several of these are common enough to make their study at a diagnostic level both feasible and advantageous.

MODES OF INHERITANCE

Under ideal circumstances it may be possible to discern four basic patterns of Mendelian inheritance from the analysis of pedigrees.

(1) Autosomal dominant conditions: such as polydactyly (the appearance of extra digits on hands or feet), are indicated if: (i) the trait appears in all generations (unless penetrance is reduced); (ii) approximately 50% of offspring produced by union between an affected and a normal parent are affected, thereby indicating that the trait appears only in heterozygotes; (iii) the trait appears with equal frequency amongst the sexes (see Fig. 20.2).

(2) Sex-linked dominant conditions: are also indicated by the trait appearing in all generations, but (i) approximately half the children of an affected female are affected whilst all the daughters but none of the sons of an affected father are also affected; (ii) affected females come from affected mothers or fathers.

(3) Autosomal recessive conditions: such as cystic fibrosis, are indicated if: (i) the trait often skips a generation such that most affected individuals have normal parents; (ii) all children are affected if both parents are affected; (iii) the children of a normal parent and an affected parent are all normal. If however, the normal individual is a heterozygous 'carrier', approximately half the children should be affected; (iv) the trait appears with equal frequency amongst the sexes (see Fig. 20.3).

(4) Sex-linked recessive conditions: such as haemophilia and Duchenne's muscular dystrophy, are indicated if: (i) most affected individuals are male, produced from affected or carrier (i.e. heterozygous) mothers (identified by having affected brothers, fathers or maternal uncles); (ii) affected females are produced from affected fathers married to affected or carrier mothers; (iii) all sons of affected females are affected and approximately half the sons of carrier females are affected (see Fig. 20.4).

Traits may follow dominant or recessive inheritance patterns involving the transmission of genes on autosomes or sex chromosomes. Each inheritance pattern exhibits characteristic features useful in their detection.

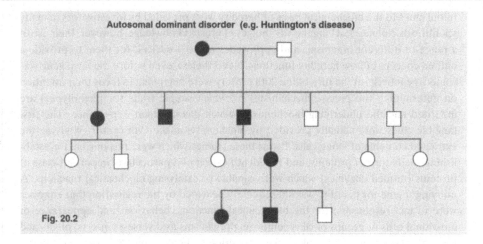

Autosomal dominant disorder (e.g. Huntington's disease)

Fig. 20.2

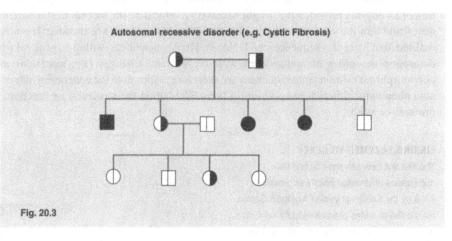

Autosomal recessive disorder (e.g. Cystic Fibrosis)

Fig. 20.3

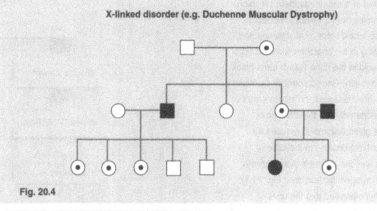

X-linked disorder (e.g. Duchenne Muscular Dystrophy)

Fig. 20.4

OAs: 130–3, 857–61, 864–6, 892.
RAs: 304, 306–8, 309, 311–3.
SFR: 46, 62–8, 160–3, 172–81, 229.

21. GENES DICTATE THE NATURE OF PROTEINS

Determining the biochemical basis of inherited traits established the link between genes and proteins resulting in the 'one gene–one polypeptide' rule, now understood in terms of the molecular composition of genes.

Initial clues to the biochemical basis of heredity were provided by investigators in many scientific disciplines. Lacking the advantages of present knowledge, however, their use of a range of different organisms and approaches made it difficult for them to provide a unified concept of how heredity functions. Nevertheless, even before the term gene was coined, geneticists at the turn of the 20th century were beginning to focus their attention on determining the precise mechanisms by which simple traits (or phenotypes) are inherited and the underlying biochemistry which dictates their appearance. The first tangible clues were actually provided by biochemists in the 19th century studying the extracted contents of living cells. It was these studies which were responsible for establishing the nature of proteins, and the identification of a particularly important class of proteins (termed enzymes) which were capable of catalysing biochemical reactions. A unifying theme for heredity was subsequently provided by the realisation that enzymes were in fact responsible for the biochemical functions, behaviour and appearance of individual cells or groups of cells comprising organisms as diverse as insects, plants and humans. Subsequently, the link between inheritable diseases and the absence or alteration of an enzyme provided the insight necessary to elucidate the biochemical nature of genes, and how the variety of inheritable traits observed in nature are encoded. It is now realised that it is the sequences of DNA or RNA nucleotides within a gene which determine the order of amino acids within a protein. Changes (e.g. mutations or polymorphisms) within gene sequences are therefore responsible for generating inheritable phenotypic differences by virtue of their effect upon the structural or functional proteins of a cell.

WHICH CAME FIRST?

The order of nucleotides within the DNA comprising a gene determines the amino acid sequence and synthesis of a protein. However, the construction or replication of DNA requires both an existing DNA molecule as a template and protein enzymes (termed polymerases) to catalyse the addition (or polymerisation) of the nucleotide building blocks. Scientists and philosophers are therefore faced with the 'chicken and egg' conundrum of which came first: genes or proteins. Attempts to synthesise self-replicating amino or nucleic acids from simple chemical mixtures (i.e. representing a 'primordial soup') subject to extreme heat, pressure and electrical discharges remain inconclusive. However, certain RNA molecules (e.g. ribozymes) have been shown to possess catalytic acitivity suggesting that RNA may actually have been the evolutionary precursor of life.

LINKING ENZYMES TO GENES

The first link between proteins and the appearance of inherited traits was provided in 1908 by the English physician Archibold Garrod. He studied diseases characterised by the failure of cells to perform known biochemical reactions. For example, alkaptonuria is a rare genetic disease inherited in a simple Mendelian fashion. This disease arises from a failure to correctly break down the amino acids phenylalanine and tyrosine, resulting in the characteristic build up of homogentisic acid in the urine (which turns black upon exposure to air). Garrod correctly surmised that such metabolic diseases, or 'inborn errors of metabolism', stemmed from the absence or deficiency of a given enzyme(s) required for normal cellular biochemistry. In the case of alkaptonuria it was the absence of an enzyme that oxidises homogentisic acid (see Fig. 21.1). He therefore hypothesised that the units of inheritance (i.e. genes) dictated the production of proteins. Therefore, the efficiency of an organism's enzymes depended upon their level of production which is in turn dictated by the nature of the organism's genes. Thus, a single gene must control the nature and production of each enzyme or protein.

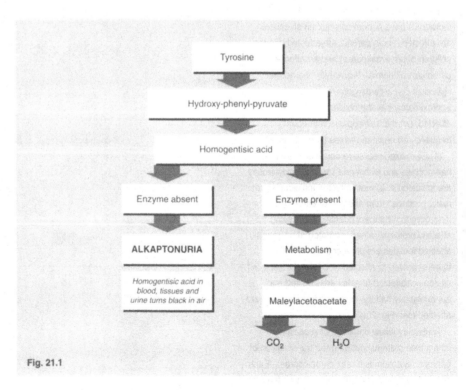

Fig. 21.1

Experimental verification of his hypothesis was however significantly delayed by the rarity of such genetic diseases for analysis.

OAs: 134–8, 147. RAs: 46–9, 270, 271. SFR: 18, 36–9, 69–71.

PROVING 'ONE GENE–ONE ENZYME'

It took almost 30 years for sufficient knowledge of the biochemistry of cellular metabolism to be generated before the definitive link between genes and proteins could be established. It also required a new experimental approach; one provided by George Beadle and Edward Tatum working at Stanford University, USA. Rather than attempting to understand the biochemistry of previously identified traits (e.g. *Drosophila* eye colour), they decided to look for mutations causing defects in the already well-understood biochemical reactions used to produce amino acids and vitamins.

Their investigations utilised the bread mold *Neurospora* for essentially three reasons.

(1) Wild type *Neurospora* can synthesise their own essential amino acids and consequently readily grow on minimal media containing sugar, salts and the vitamin biotin. However, mutants unable to synthesise a given amino acid can also be grown on media to which the amino acid has been added.

(2) They grow as a haploid organism (i.e. they contain only one copy of each chromosome) allowing each mutation to be clearly identified without being masked by a normal allele.

(3) They produce haploid spores (termed conidia) which can produce haploid progeny, or fuse and undergo meiotic division to form haploid sexual spores. This allowed sexual crosses to be performed for genetic analyses.

Beadle and Tatum began by exposing wild type *Neurospora* to X-rays or ultraviolet radiation (to induce mutations) followed by growing the conidia on complete medium (Fig. 21.2). This allowed them to isolate a large range of mutants lacking the ability to synthesise a specific amino acid, or vitamin. The nature of the metabolic defect could then be deduced from the pattern of growth of the mutant strain on selective minimal media (i.e. supplemented by a single amino acid or vitamin). They then used sexual crossing to establish that each mutant strain contained a single mutation in a single gene. From numerous such experiments involving many metabolic pathways, they established that each mutation within a gene only affected one enzyme. This correspondence between a gene and an enzyme

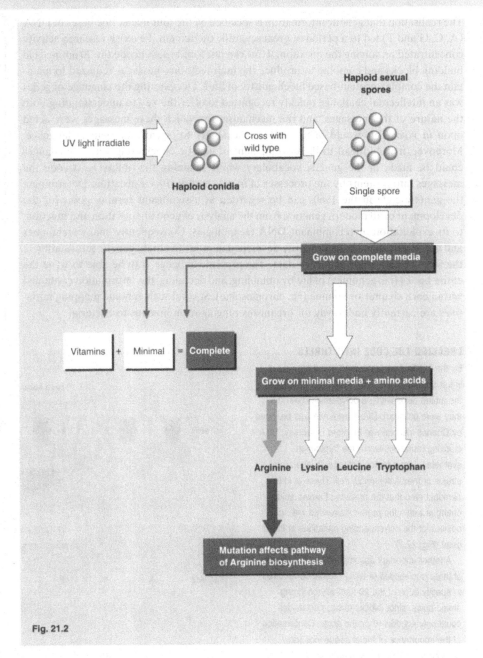

Fig. 21.2

led them to formulate the 'one gene–one enzyme' hypothesis.

Within the following decade biochemical and genetic evidence from the analysis of non-enzyme proteins quickly led to the more generalised 'one gene–one protein' rule. This was subsequently refined to 'one gene–one polypeptide' based upon studies by Linus Pauling at CalTech, USA. He demonstrated that the

haemoglobin molecule is composed of more than one polypeptide subunit and that chemically altered forms of haemoglobin cause sickle-cell anaemia. Following the identification of DNA as the genetic material, the validity of this concept has been well established and the molecular composition and organisation of many genes precisely described.

22. THE NATURE OF THE GENETIC CODE

The order of the individual amino acids comprising a protein is determined by groups of three nucleotides (termed codons), read in an open reading frame from the 5′ end of the mRNA intermediate.

The realisation that genetic information is specified by the four nucleotide bases of DNA (A, C, G and T) led to a period of great scientific excitement. Feverish research activity concentrated on solving the question, 'How can just four bases encode the 20 amino acid building blocks and combine to produce the multitudinous messages required to maintain the complex protein-based biochemistry of life?' Deciphering the language of genes was an intellectual challenge quickly recognised to offer the key to understanding both the nature of the messages, and the mechanisms by which these messages were acted upon *in vivo*. This would in turn afford the ability to 'speak' the genetic language. Moreover, it would lead to the development of genetic engineering whereby changes could be made in the genetic vocabulary whilst retaining the ability to decode the messages into proteins by the processes of *in vivo* and *in vitro* translation. Determining the genetic code in the 1960s can be regarded as a significant turning point for the development of all modern genetics, from the analysis of gene organisation and function, to its exploitation by recombinant DNA technologies. Consequently, many techniques and approaches have been primarily developed in order to obtain genetic information in the form of nucleotide sequence data. The present challenge is to be able to write the entire books (i.e. genomes) of life by obtaining, and decoding, the information contained within each chapter or volume (i.e. chromosome). Several such genome mapping initiatives are currently under way for organisms ranging from humans to bacteria.

THE CODE IS DECODED FROM RNA

Proteins are not directly synthesised from genomic DNA. Specific RNA copies, termed messenger RNA (mRNA) molecules, are used as intermediate nucleic acids to convey the information from the genetic material (DNA) in the nucleus to the sites of protein synthesis in the cytoplasm. Consequently, codons are written using the four bases of RNA and not DNA, i.e. substituting uracil (U) for thymine (T). The overall process of decoding, termed translation, also involves a further group of ribonucleic acid molecules termed transfer RNAs (tRNAs). This important family of molecules recognises a codon on the mRNA molecule via interaction with an anticodon sequence. They also carry the amino acid specified by that codon to allow the formation of the desired polypeptide.

BREAKING THE CODE INTO THREES

By the 1950s it was assumed that there was a linear correlation between nucleotide bases and the amino acids within a polypeptide chain (i.e. they were colinear). Direct evidence was provided by Charles Yanofsky at Stanford University, USA, studying mutations within the tryptophan synthetase gene of non-pathogenic laboratory strains of the bacterium *E. coli*. These studies demonstrated that the position of amino acid changes within the protein correlated with the position of the corresponding mutations in the gene (Fig. 22.1).

Another important assumption was that blocks of three nucleotides (a triplet) would be required to specify each of the 20 naturally occurring amino acids, since blocks of two nucleotides could only encode 16 amino acids. Confirmation of the importance of triplet sequences (now referred to as codons) was provided by Sydney Brenner and Francis Crick at Cambridge, UK, in 1961. Studying extensive mutations created within the bacteriophage virus T4, they observed that mutants containing one or two base additions or deletions led to the production of non-functional proteins. In contrast, totally active proteins could often be produced if three bases were deleted or inserted. Nevertheless, the fact that 64 different triplets were possible still perplexed molecular biologists of the time.

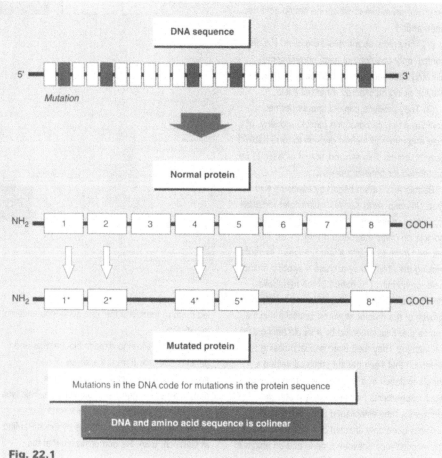

Fig. 22.1

The first codon assignments

Determining which triplets (or codons) of the possible 64 encoded which amino acid began with an historic set of experiments performed by Marshall Nirenberg and Heinrich Matthaei at the National Institute of Health, Bethesda, USA in 1961. They first synthesised a polyribonucleotide composed of uracils (i.e. UUUU . . .) which they added to a cell-free *in vitro* translation system. This produced a synthetic polypeptide composed only of phenylalanine indicating that the codon UUU encoded phenylalanine (see Fig. 22.2). Similar experiments revealed that CCC coded for proline and AAA coded for lysine. A few additional codons were tentatively assigned based on the formation of random copolymer sequences (mixtures of two, three or four bases). However, amino acid assignments of the remaining codons required the ability to generate defined co-polymers of repeating sequence (e.g. GUGGUG or AAGAAG), a task achieved by H. Gobind Khorana. Using this technology all the codon assignments were completed by 1966.

Reading frames

The completion of codon assignments revealed that the code apparently lacked internal punctuations analogous to commas. Consequently, since each amino acid is specified by a codon composed of three bases, a given sequence may be read in any of three ways or reading frames (Fig. 22.3). The true positions of the codons relative to a protein's amino acids are termed the open reading frame (ORF) which is in turn defined by a specific start codon. Misreading of the frame can occur due to several forms of mutation and results in the production of a non-sense, truncated and/or abnormally functioning protein. Such mutations therefore have profound and potentially lethal consequences upon the cell or organism (e.g. muscular dystrophy).

Direction of the code

The direction in which the genetic code is read is a direct consequence of the way in which mRNA is transcribed from DNA. Indeed, only the sequence of the positive (or sense) strand of DNA is decoded by translation into polypeptides. Accordingly, it is the negative or non-sense strand of DNA which serves as the template for the production of messenger RNA. The mRNA is consequently a direct copy of the sense DNA strand. Synthesis of protein also begins from the 5′ end of the mRNA molecule. The amino acids comprising a polypeptide are therefore written, and assembled, from the NH (amino or N) terminal to the COOH (carboxy or C) terminal end, just as the nucleotide sequence of DNA/RNA is written down and assembled in a 5′ to 3′ direction.

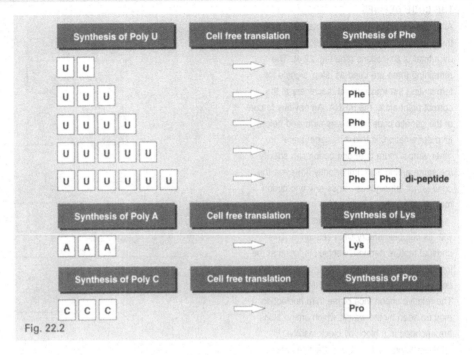

Fig. 22.2

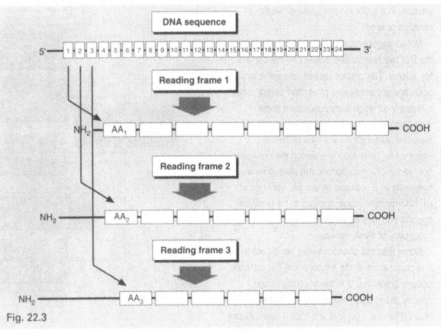

Fig. 22.3

OAs: 138–46, 239–41. RAs: 47–52, 101, 104, 257, 266. SFR: 18–20, 36–9, 70, 71, 120, 124, 222.

The genetic code is degenerate in that each of the 20 amino acids is encoded by two or more codons, with 61 of the 64 possible codons specifying amino acids and three specifying translational stop signals.

THE CODE DEFINED

All 20 amino acids (normally abbreviated to a three letter or one letter designation) are encoded by a total of 61 codons (see Fig. 22.4). The remaining three are used as 'stop' signals for terminating the translational machinery at the correct point along the mRNA. An obvious feature of the genetic code which was surmised before its determination, is that it is 'degenerate'. In other words, more than one codon can specify the same amino acid. Importantly however, the code is not 'ambiguous', since any one codon may only specify a single amino acid.

The third base of a codon does not always pair with its complementary DNA sequence. This unusual feature, termed 'wobble', is found at the interaction between the codon and the anticodon deciphering unit of the transfer RNA molecule. The relative importance of the third nucleotide may be seen by the way in which amino acids are encoded in a block by block fashion. In general, when the first two nucleotides are identical in a block the third may either be cytosine or uracil.

Not all degeneracy is based on the similarity of the first two nucleotides, typified by the codons for leucine. The precise reason why some amino acids are encoded more often than others is still a mystery although it may represent some evolutionary process to combat the effects of mutation. Although most amino acids are specified by more than one codon their usage is not random. Certain codons are utilized more frequently in *E. coli* and others are not used at all. Non-random usage appears to be organism specific and may be associated with the availability of tRNA molecules.

Some features of codons may be related to the chemical nature of the amino acids. For instance, codons containing U in the second position specify amino acids with a hydrophobic side chain (Phe, Leu, Ile, Met and Val). Those utilising A in this position specify amino acids with a polar or fully charged atom (Tyr, His, Gln, Asn, Lys, Asp and Glu).

Stopping and starting the code

Three codons (UAA, UAG and UGA) do not encode an amino acid and have therefore been termed non-sense codons. These codons are translational stop signals that specify the end of a particular message. Interestingly, it appears that UAA is the preferred stop codon, although this is not exclusively used *in vivo*. The codon UAG is found to encode not only the start of a reading frame but also encodes the amino acid methionine. Consequently, newly synthesised polypeptide chains start with this amino acid, although it may be subsequently removed by post-translational processing and therefore absent in the mature functional protein. The ability to start and stop the code is fundamental to life and was the first clue to the nature of gene organisation within genomic DNA.

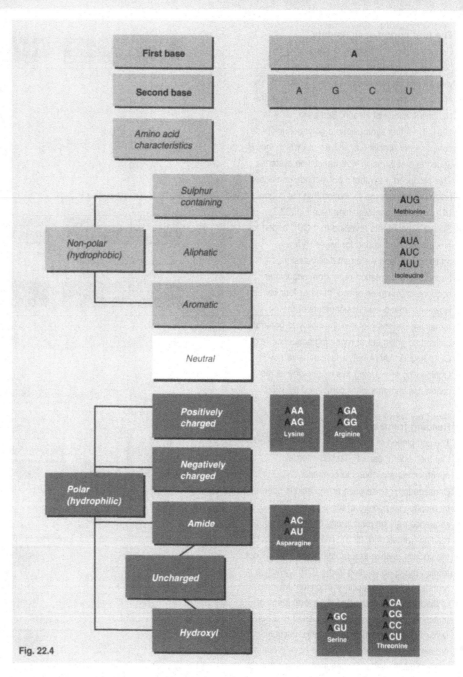

Fig. 22.4

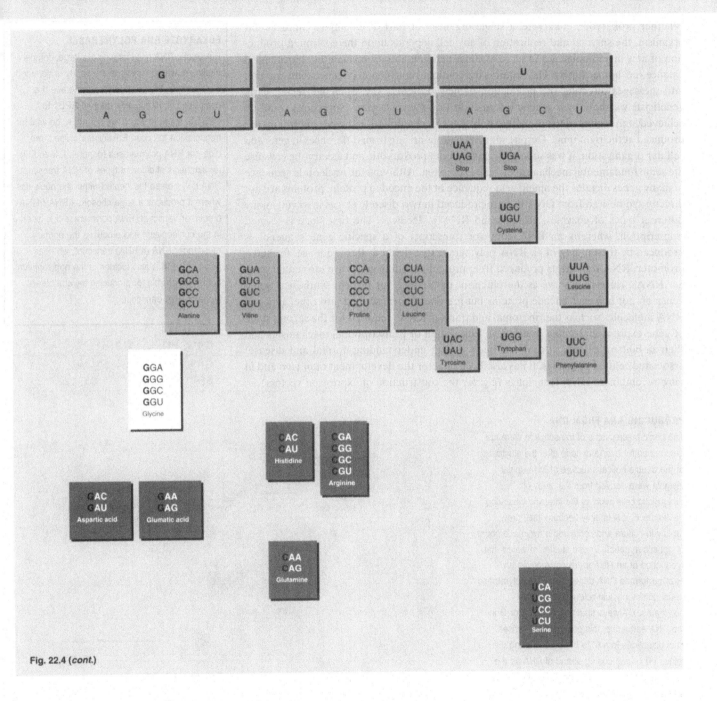

Fig. 22.4 (*cont.*)

Exceptions to a universal code

Initial studies of the genetic codes utilised by several different organisms suggested that the code, and its translational machinery, were universal and highly conserved. The discovery that subcellular organelles such as mitochondria and chloroplasts had an altered code was totally unexpected and initially confounded geneticists. These organelles have been found to (i) utilise UGA as a codon for tryptophan and not the usual stop codon, (ii) utilise four compared to the usual three codons as translational stops, and (iii) use four codons for methionine rather than the one utilised in the mammalian code. It is speculated that these organellar genomes are streamlined versions of the universal code and restricted by the availability of their own transfer RNA molecules, possibly as a result of genetic drift.

23. TRANSCRIPTION: FORMING GENETIC MESSAGES

The flow of genetic information from genes to sites of cellular action requires the creation of messenger RNA transcripts copied from specific gene sequences by DNA-dependent RNA polymerase-mediated transcription.

Whether prokaryote, eukaryote, a single organism or part of a complex multicellular organism, the survival and replication of any cell depends upon the continued production of new macromolecules to serve as construction materials or maintain biochemical balance (i.e. homeostasis). This requires the continual conversion of genetic information into messages directing the production of new proteins. The challenge facing early geneticists was how such genetic information flow (termed gene expression) could be achieved, and regulated to ensure that only the correct amounts and types of proteins are produced at the right time. Despite many variations arising from differences in gene and cellular organisation, it was soon realised that both prokaryotic and eukaryotic cells use the same fundamental mechanism of gene expression. Although the nucleotide sequence of many genes dictates the amino acid sequence of the encoded protein, proteins are not directly synthesised from DNA but are produced in two discrete stages involving many different types of enzymes, proteins and RNA molecules. The first stage is termed transcription, whereby an RNA copy (or transcript) of a specific gene sequence is produced by the binding of an RNA polymerase to a defined start sequence, termed a promoter. RNA transcripts produced from protein-encoding genes are termed messenger RNAs, since they serve as the blueprint or message for protein synthesis. Some genes do not however encode proteins but are directly transcribed into other types of RNA molecules such as the ribosomal and transfer RNAs required for the second stage of gene expression, termed translation. Elucidation of transcriptional mechanisms and their in-built regulatory circuits has been vital in understanding normal and disease-associated cellular processes. It has also been vital for the development of *in vivo* and *in vitro* recombinant DNA techniques (e.g. for the construction of expression vectors).

EUKARYOTE RNA POLYMERASES
Eukaryotic RNA polymerases are large, complex multimers whose functions are poorly understood. For example, the yeast RNA polymerase II is composed of 10 subunits (termed RPB1 to RPB10), some of which appear not to be vital for transcription to occur. Eukaryotes utilise three different RNA polymerases (denoted I, II and III) for synthesis of different types of RNA transcript. RNA polymerase I is located within the nucleolus where it produces a large ribosomal RNA (rRNA) precursor transcript. RNA polymerase II is located in the nucleoplasm and produces the primary messenger RNA (mRNA) transcript, whereas polymerase III is also located in the nucleoplasm but is responsible for producing several other types of RNA transcript.

OAS: 147–65, 512, 513.
RAs: 53–9, 62, 82, 85, 101,
SFR: 36–9, 50, 72–4, 85, 120, 125.

PRODUCING RNA FROM DNA
The basic mechanisms of transcription were not determined for over a decade after the elucidation of the double helical structure of DNA. Initial insights were derived from the study of prokaryotic cells such as the intestine inhabiting bacterium *E. coli* (mainly because they grew rapidly in culture and contained a minimal amount of genetic material). These studies revealed that production of an RNA transcript requires an enzyme termed DNA-dependent RNA polymerase (most commonly just referred to as RNA polymerase). Always starting from an A or G in the DNA sequence, this enzyme polymerises ribonucleotides in a 5′ to 3′ direction using the minus (–) or non-coding strand of DNA as the template (see Fig. 23.1). The transcribed RNA sequence is therefore an exact copy of the sequence of the plus (+) or coding DNA strand, with the exception that the T in the DNA is replaced by U in the RNA.

Prokaryotes use only one DNA-dependent RNA polymerase for transcription, with each *E. coli* RNA polymerase composed of five or four polypeptide subunits (termed holoenzyme ($\alpha_2\beta\beta'\sigma$) and core enzyme ($\alpha_2\beta\beta'$) respectively).

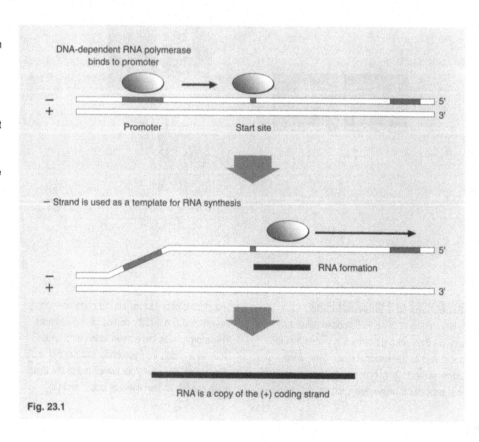

DNA-dependent RNA polymerase binds to promoter

Promoter   Start site

– Strand is used as a template for RNA synthesis

RNA formation

RNA is a copy of the (+) coding strand

Fig. 23.1

The three stages of transcription

Transcription may be divided into three phases (Fig. 23.2).

(1) Initiation: In *E. coli*, this requires the binding of the holoenzyme form of RNA polymerase to a defined region (termed a promoter sequence) within the double-stranded DNA located upstream of the sequence to be transcribed. Initiation of eukaryotic transcription is however more complex, involving three different RNA polymerases. Each of these recognises different promoter sequences, often located at different sites upstream or even, in the case of RNA polymerase III, within the gene to be transcribed. Several of these sequences may also be involved in regulating gene expression by helping or preventing the initiation of transcription. RNA polymerase II-initiated transcription also requires the co-ordinated binding of a set of additional proteins termed transcription factors.

(2) Transcript elongation: this is very similar in prokaryotes and eukaryotes, except that in *E. coli* the RNA polymerase is converted into its core enzyme form following the addition of about 8 ribonucleotides. This extends the growing RNA chain by moving along the template in a zipper-like fashion, opening the duplex ahead of it and closing up again behind. Such a locally unwound region formed during transcription is termed a transcription 'bubble'. Unwinding can however only occur over a short distance (between 12 and 17 bases) since it causes overwinding of adjacent regions leading to intolerable stresses on the structure of the DNA duplex. Several additional molecules (e.g. *trans*-acting factors and topoisomerase enzymes) are known to be involved in this localised unwinding of the DNA. Interestingly, but as yet unexplained, the rate of elongation varies as the transcript is formed as it appears that the RNA polymerase often slows down, pauses and then reaccelerates.

(3) Transcription termination: the RNA chain stops growing in both prokaryotes and eukaryotes when the RNA polymerase reaches defined sequences, termed terminator sequences. *E. coli* terminators generally have a GC-rich, often palindromic, sequence followed by an AT-rich sequence. A specific protein called the rho (ρ) protein may also be bound to some terminators (so-called *rho-dependent terminators*). The

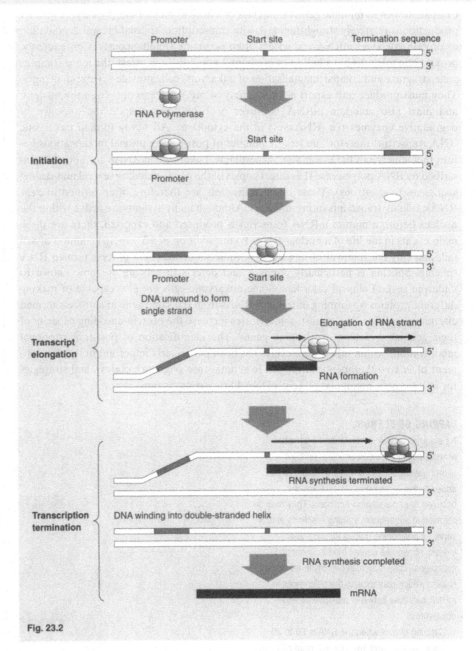

Fig. 23.2

palindromic nature of such terminator sequences is thought to cause the formation of stem-loop structures which, in ways which as yet remain poorly understood, are vital to the termination process. Eukaryotic transcription termination sequences and termination processes vary according to the gene being transcribed and the type of RNA polymerase used. It is however clear that eukaryotic transcription termination does not involve the formation of stem-loop structures.

Transcript processing

Transcripts may undergo several forms of post-transcriptional processing (e.g. editing and base or sequence modification) before a mature transcript is finally produced. These modifications have a significant impact upon the structure, stability and function of the transcript (e.g. ensuring intracellular transport and accurate initiation or termination of the translation of protein-encoding sequences).

24. POST-TRANSCRIPTIONAL PROCESSING OF MESSENGER RNA

The first stages required for the production of a mature eukaryotic mRNA transcript involve the modification of the 5' and 3' ends by the addition of a 'cap' and a poly(A) tail respectively.

Consistent with its minimal genome and lack of a nuclear membrane, prokaryote mRNA production is relatively straightforward, with transcription, translation and degradation of protein-encoding mRNA transcripts often occurring simultaneously. Consequently, prokaryotic mRNAs have half-lives of about 2 minutes. In contrast, the more complex gene structure and compartmentalisation of eukaryotic cells provide several challenges. They must produce and export a wide variety of mRNA transcripts from the nucleus, and must also maintain mRNA half-lives of 1–24 hours despite the presence of degradative enzymes (i.e. RNAases) in the cytoplasm. All newly formed eukaryotic RNA transcripts therefore undergo a number of post-transcriptional modifications before they emerge in the cytoplasm. Accordingly, eukaryotic mRNAs are initially transcribed by RNA polymerase II as direct copies of the gene, including several non-coding sequences (e.g. introns). These initial transcripts are therefore often termed nuclear RNAs, fidelity transcripts or pre-mRNAs. Although many systems are active within the nucleus before a mature mRNA transcript is produced and exported, there are three main events in the life of a nuclear RNA: (i) the addition of a 5' cap, (ii) addition of a 3' tail, and (iii) the removal of unwanted intronic sequences by a process termed RNA splicing. Splicing is particularly complex and defects in splicing are now known to underpin several clinical disorders. Some eukaryotic cells are also capable of making different proteins by forming different mRNAs from the same gene in a process termed alternative splicing. In contrast, some viruses increase the protein-encoding capacity of their genome by using overlapping genes. The identification of post-transcriptional modifications within mRNA molecules has been particularly important in the development of *in vivo* therapeutic strategies (e.g. anti-sense oligonucleotides), and strategies for *in vitro* gene manipulation (e.g. cDNA library construction).

COMPLEXITY OF NUCLEAR EVENTS

Many complex transcriptional processes occur simultaneously within the small enclosed nucleus. For example, given that only one or a few copies of each protein-encoding gene is present per genome, the production of sufficient protein requires the formation of many mRNA transcripts from the same gene. Consequently, more than one RNA polymerase is actively engaged in transcribing a given gene at any one time. This creates a characteristic transcription pattern comprising many nuclear RNAs at various stages of maturation. Since many genes are simultaneously transcribed, the nucleus also contains many different types of mRNA molecules (termed heterogeneous nuclear RNA or hnRNA). Initial attempts to elucidate transcriptional processes by analysing hnRNA isolated from eukaryotic nuclei were therefore severely hampered. Analysis was further complicated by the movement of newly transcribed RNAs coordinated by several nuclear accessory proteins and associated with the nuclear scaffold.

CAPPING OF 5' ENDS

As eukaryotic RNA is transcribed it becomes associated with small ribonuclear proteins allowing what is termed a cap structure to be added to the 5' end (Fig. 24.1). Caps are believed to serve several functions. They may be involved in subsequent splicing reactions and/or serve as positioning guides for translation. Indeed, ribosomes cannot bind to uncapped messages. They may also determine transcript stability which may explain the differences in mRNA half-lives between prokaryotes and eukaryotes.

Capping occurs when the RNA is 20 to 30 nucleotides long and involves the initial addition of a guanine nucleotide to the first RNA base by a reaction catalysed by the enzyme guanalyl transferase. The cap is unusual in that it is attached backwards involving the formation of a 5' to 5' linkage of the first RNA base. Under all other *in vivo* circumstances nucleic acids are synthesised using 5' to 3' linkages. Following addition, the N7 atom of the guanine is also methylated to generate 7-methylguanine (often referred to as the cap 0 structure). The two adjacent ribose units may also be methylated at the C2 position, giving cap 1 if only the first ribose is methylated and cap 2 if both are methylated.

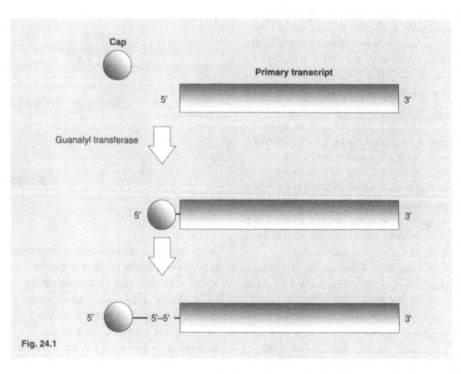

Fig. 24.1

ADDITION OF A 3' POLY(A) TAIL

The 3' end of prokaryotic mRNA has now been well characterised, and shown to form a hairpin loop followed by a series of uracil residues. With a few exceptions (e.g. those coding for histone proteins involved in the tertiary folding of DNA to allow compact packaging within cell nuclei), the 3' end of most eukaryotic mRNAs is more complex. Each contains a long stretch of up to 250 adenylate residues, termed a poly(A) tail. These are not encoded in the DNA sequence of the gene, but are added by a non-template-mediated process in essentially two stages (Fig. 24.2).

(1) Almost simultaneously as the 5' cap is added to the newly formed nuclear RNA transcript, a specific endonuclease recognises and cleaves the 3' end about 10–30 nucleotides downstream of a specific signal sequence. This is known as the polyadenylation signal and has the consensus sequence 5'-AAUAAA-3'. Recent evidence does however suggest that additional sequences or the presence of precisely positioned nucleotides are also important.

(2) The intermediate 3' end is then bound to a further enzyme called poly(A) polymerase which catalyses the transfer of adenylate residues to the 3' terminus. Subsequently a number of 78 kDa poly(A) tail-binding proteins become associated with the nuclear RNA.

Function of poly(A) tails

The precise role of the poly(A) tail in eukaryotic mRNAs has been the subject of speculation ever since its discovery. It was initially thought to provide mRNAs with much needed stability by protecting them from enzymes which possess the potential to degrade the 3' end. Although this has not been ruled out, the existence of a number of eukaryotic mRNA transcripts devoid of poly(A) tails, but which are nevertheless effective templates for cytoplasmic protein synthesis, suggests this may not be their sole function. It is therefore also possible that the poly(A) tail structure may act as a transport sequence essential for nuclear–cytoplasmic transfer of at least some mature mRNAs.

OAS: 166–70, 185, 186, 364, 439–45, 515. **RAs:** 60–3, 186.
SFR: 36–9, 120.

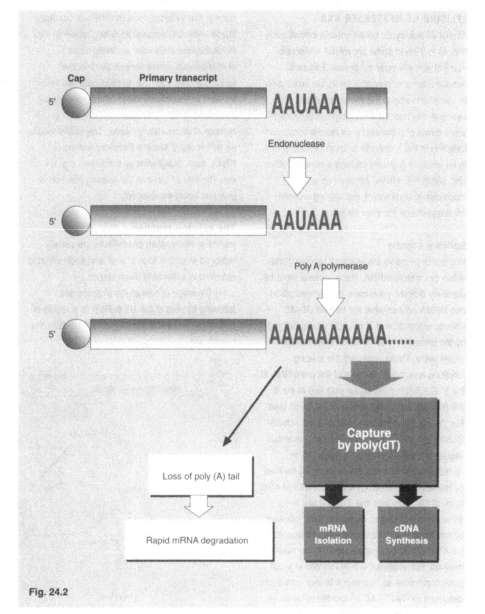

Fig. 24.2

Purification of messenger RNA

The presence of a poly(A) tail on most eukaryotic mRNAs has been widely exploited for their *in vitro* isolation. The most commonly employed approach is to isolate total cytoplasmic RNA (e.g. by controlled cell lysis) and then isolate poly(A) tail-bearing transcripts by means of hybridisation with synthetic poly(U) or poly(dT) oligonucleotides immobilised to a suitable matrix (e.g. cellulose or paramagnetic beads). RNA transcripts devoid of a poly(A) tail (including partially degraded mRNAs) do not bind and are washed away. Bound poly(A)+ mRNA is then eluted by altering the buffer conditions (i.e. decreasing the salt concentration) and thus breaking the inter-strand base pairing. This approach forms the basis of many commercial kits for the isolation of poly(A) mRNA in high yields and purity within a few hours or minutes. The poly(A) tail may also be used for the production of complementary DNA (e.g. for cDNA library construction).

The final stage in the production of a mature eukaryotic mRNA transcript involves a series of complex splicing reactions involving the removal of non-coding intronic sequences and, in some cases, RNA editing.

SPLICING OF MESSENGER RNA

Almost all eukaryotic genes contain introns such that up to 75% of some pre-mRNA transcripts may not actually code for protein. Failure to remove introns or precisely rejoin the exons prior to translation affects the protein produced. For example, the truncated haemoglobin produced in some forms of β-thalassaemia results from mutation in the splice site junction (from a G to A in an intron of β-globin) causing a reading frame shift within the mRNA. Although as yet incompletely understood, the splicing process involves several complex stages.

Splicing signals

In order to preserve the intended reading frame within the mature mRNA, intron removal must be precisely defined. All nuclear genes transcribed into mRNA contain what are termed GT–AG introns, whose 5′ and 3′ boundaries are defined by the presence of GT and AG dinucleotides respectively. These represent the splicing junctions specifying cleavage of the pre-mRNA at the 5′ GU (termed the donor site) and at the 3′ AG (termed the acceptor site) of the intron (see Fig. 24.3a). Recognition sequences are actually longer than this implies, with several consensus sequences now identified. For example, in vertebrates the 5′ splice site is generally defined by the sequence 5′-AGGTAAGT-3′ (AGGUAAGU in the RNA). Similarly, the 3′-splice site is generally composed of six pyrimidines, followed by any base, a C, and then the AG dinucleotide (i.e. Py6XCAG). Internal sequences may also be involved. For example, GT–AG introns in yeast (*Saccharomyces* sp.) contain a branch consensus sequence (5′-TACTAAC-3′) located between 18 and 140 bp upstream of the 3′ splice site. Similar branch sequences have also been identified in mammals.

The splicing machinery

The splicing reaction is performed by a large nuclear-located structure of 50-60S, termed a spliceosome. It contains a number of small

OAs: 103, 171–84, 541, 542. **RAs:** 61–74, 79, 259, 408. **SFR:** 36–9, 74, 120.

nuclear ribonucleoproteins (snRNPs or 'snurps'). These were first discovered during research into the autoimmune disorder systemic lupus erythematosus, characterised by defective splicing arising from the reaction between splicing proteins and autoantibodies. Snurps are usually composed of 100–300 bases of RNA and a number of ribonuclear proteins. The RNAs tend to be rich in uracil and are therefore termed U-RNAs, each designated by a number, e.g. U1, U2 etc. The role of these in the splicing reaction is only just being established.

The splicing reaction

Introns in mammalian pre-mRNAs are usually removed stepwise from 5′ to 3′ and each intron is removed in essentially three stages.

(1) Cleavage occurs at the 5′ splice site following binding of the U1-snRNP to a region of homology, and whilst the U2-snRNP binds to the branch site.

(2) The resulting free 5′ G residue of the GU donor site then appears to interact with an A residue some 60 base pairs from the 3′ splice site (i.e within the branch site). In yeast this is the A residue indicated in bold in the TACT**A**AC sequence. This interaction appears to be mediated by U2-snRNP and results in the joining of the A and G residues in an unusual 2′ to 5′ linkage. This results in the formation of a lariat or branched circle.

(3) The reaction then proceeds with cleavage at the 3′ splice site, removal of the lariat intron (which is then debranched and degraded) and ligation of the two exons (Fig. 24.3b). U5-snRNP is known to bind to this site and may participate in the ligation of the right and left exon sequences. Once the stepwise removal of introns is completed the spliceosome dissociates into its various components and the mature mRNA is released (see Fig. 24.4).

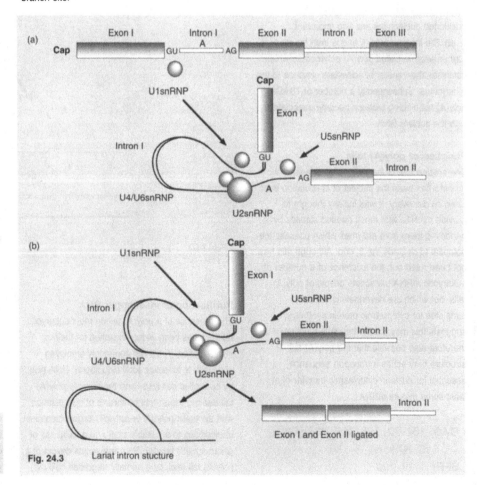

Fig. 24.3   Lariat intron stucture

Why bother with introns?

The purpose of introns remains largely unknown. One possible clue has recently been found in the fungus *Physarum*, where some mitochondrial introns potentially encode proteins required for the efficient removal of the intron from which they originated. It is also suggested that the possession of introns allows a greater diversity of proteins to be formed more rapidly during evolution by the selective recombination of exons. Such an argument is made more plausible by the alternative splicing of the same mRNA transcript in some cells.

Alternative splicing

An important event in some genes is the ability to use different combinations of exons and introns to produce two distinct proteins. Termed alternative splicing, it is frequently used by bacteriophages to maximise the potential to encode proteins from the restricted size of their genomes. Alternative splicing is also found in mammalian cells and was first described for the tissue-specific production of calcitonin and calcitonin gene-related protein (CGRP) (see Fig. 24.5). Extensive alternative splicing of transcripts from the same tropomyosin gene is also used to produce a wide variety of tissue-specific tropomyosin molecules. The strategy of alternative splicing therefore avoids the need for evolution to produce a multigene family in order to generate several different forms of tissue-specific proteins.

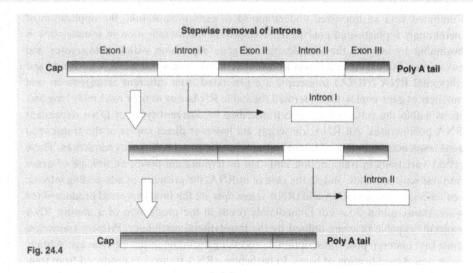

Fig. 24.4

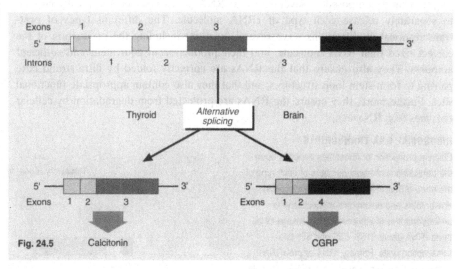

Fig. 24.5

MESSENGER RNA EDITING

A rare form of mRNA processing termed RNA editing was recently discovered and is the focus of much scientific debate. First discovered in the parasitic protozoan *Trypanosoma brucei*, it involves the addition of new nucleotides, or the removal or replacement of existing nucleotides. The degree of editing can be very extensive. In *Trypanosoma* many new U nucleotides are added, possibly accounting for their ability to rapidly change surface protein composition in order to evade immune recognition.

In human apolipoprotein (b), a single C to U change results in tissue-specific expression of two forms of protein. Thus, mRNA editing occurring in intestinal cells produces a 2153 amino acid molecule. In contrast editing does not occur in other cells of the body, resulting in production of a 4563 amino acid polypeptide.

Ribozymes: self-splicing RNAs

Although the cutting and joining of GT–AG introns must involve several enzymes, RNAs may themselves possess enzymatic activity. For example, an unusual self-splicing intron (termed a ribozyme) has been discovered within the protozoan *Tetrahymena thermophila*. Under specialised conditions, this ribozyme can perform the splicing reaction without the need of any protein enzyme or splicing complex. It begins by folding the intron using intra-strand base pairing. The splice sites are then brought together by the 3′ OH group of a G residue reacting with the 5′ end of a downstream intron. The spliced exons are then joined and the intron is converted to a circle via the attack of an internal phosphate by the 3′ end. This results in the generation of a circular intron and a small 15 bp oligonucleotide.

Ribozymes have since been found in fungal mitochondria and bacteriophage T4. Several new forms of ribozymes have also been discovered in the mitochondria of yeast and some fungi which involve formation of a similar lariat intermediate to that involved in mRNA splicing. Another type is also known to exist in the genomes of some plant viruses where a folded RNA (termed a hammerhead) self-cleaves during formation of a single genome from large precursor RNAs. Ribozymes have recently become the focus of considerable attention due to the possibility of engineering them for use in human gene therapy.

25. TRANSFER AND RIBOSOMAL RNA PROCESSING/MODIFICATION

Ribosomal and transfer RNA transcripts undergo several types of processing, including editing and several forms of base modification, before a mature, functional transcript is formed.

Combined with an improved understanding of gene organisation, the application of increasingly sophisticated analytical techniques to the investigation of transcription is beginning to identify the complexities of gene expression within prokaryotes and eukaryotes. For example, it is now known that in eukaryotes transfer RNA (tRNA) and ribosomal RNA (rRNA) transcripts are generated from different arrangements and numbers of gene copies. Often termed the stable RNAs due to their extremely long life-spans within the cell, they are also transcribed by different types of DNA-dependent RNA polymerases. All RNA transcripts are however direct copies of the transcribed gene sequences including additional terminal and internal non-coding sequences. These reflect variations in transcription initiation or termination positions, linkage of genes into transcription units, and, in the case of hnRNA, the presence of non-coding intronic sequences. Although rRNA and tRNA transcripts are the final expressed product of the gene, transcription does not immediately result in the production of a mature RNA molecule capable of being utilised by the translational machinery. Primary transcripts must first undergo post-transcriptional processing including sequence cleavage, editing, or chemical modification of bases. In particular, rRNA transcripts produced from transcription units composed of several linked rRNA genes must be specifically cleaved to separately release each type of rRNA molecule. The different types of post-transcriptional modifications serve several functions including the production of the correct sized final RNA molecule, and their production in equivalent and sufficient amounts. They also ensure that the RNAs are correctly folded by intra-strand base pairing to form stem-loop structures, and that they also contain appropriate functional sites. Furthermore, they ensure the RNAs are protected from degradation by cellular enzymes (e.g. RNases).

RIBOSOMAL RNA 'S' VALUES

Several ribosomal RNA molecules contribute to the subunits used in the construction of ribosomes. The size of ribosomes, their subunits and each type of rRNA molecule is constant within a given cell. They may however vary between prokaryotic and eukaryotic cells, and amongst different eukaryotes. Their sizes were initially determined by their sedimentation velocity during ultracentrifugation. The rate of sedimentation reflects both the size and shape of the molecule and is expressed as a sedimentation coefficient or S value where S = Svedberg units (after the inventor of the first ultracentrifuge). Consequently, each rRNA gene is classified according to the S value of the processed transcript which, like tRNA transcripts, is the final expressed product.

OAs: 60, 90, 92, 187–90, 520–3.
RAs: 18, 19, 67–9, 75, 76, 274, 275.
SFR: 11–13, 36–9, 50, 74, 87, 163.

RIBOSOMAL RNA TRANSCRIPTS

Efficient production of ribosomes depends upon the transcription of equal numbers of each type of the three or four rRNA molecules produced in prokaryotes and eukaryotes respectively. In prokaryotes this is achieved by the linkage of all three rRNA genes (16S, 23S and 5S) into transcription units. Primary rRNA or pre-rRNA transcripts therefore contain each rRNA molecule separated by short spacer regions (Fig. 25.1). These are subsequently cleaved in two stages by specific ribonucleases to yield mature transcripts. Eukaryote rRNA transcripts are similarly produced, with the exception that only the 28S, 18S and 5.8S genes are linked within a transcription unit whilst the 5S gene is located elsewhere and transcribed independently.

The large demand for ribosomes is met by the existence of multiple copies of rRNA transcription units. In eukaryotes these may number up to 5000 copies arranged in closely linked multigene families. Alternatively, some eukaryotes achieve the same level of rRNA production by gene amplification, whereby extra DNA copies are replicated within the nucleus and remain loosely attached to the chromosomal copy.

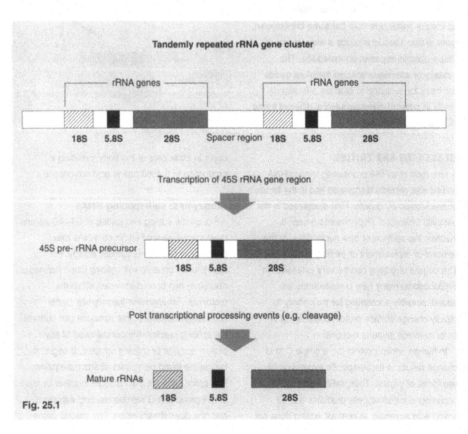

Fig. 25.1

EDITING OF tRNA TRANSCRIPTS

Both eukaryotic and prokaryotic tRNA transcripts are initially transcribed as larger pro-molecules (by about 20%) containing additional 5′ and 3′ sequences. In *E. coli* there are several separate transcription units containing up to seven tRNA molecules linked together in a cluster. Due to the massive demand for tRNA molecules in eukaryotic translation, eukaryote tRNA molecules are similarly clustered, and are also present as multiple copies.

In prokaryotes the terminal additions and spacer sequences linking tRNA molecules are cleaved by combinations of specific enzymes such as ribonuclease P (cleaving 5′ sequences) and ribonuclease D (cleaving 3′ sequences) (see Fig. 25.2). Interestingly, ribonuclease D is an example of an RNA enzyme or ribozyme. It is composed of a small protein and a 377 ribonucleotide RNA molecule with at least some intrinsic enzymatic activity. Although similar mechanisms employing similar ribonucleases are thought to operate in eukaryotes, the events involved are as yet poorly characterised.

Following completion of transcription, tRNA transcripts also undergo sequence alteration of the 3′ end. This ensures the presence of the 3′ trinucleotide CCA crucial for amino acid attachment. The CCA sequence is not present in eukaryotic DNA but is added by the enzyme tRNA nucleotidyl transferase. Although prokaryotic genes often contain the 3′ CCA sequence, its presence in tRNAs is also ensured by subsequent post-transcriptional processing using the same enzyme.

CHEMICAL MODIFICATIONS

Up to 50 types of ribonucleotide chemical modifications have now been described within the transcribed tRNA genes (Table 25.1). These occur at positions involved in stem-loop formation within the typical tRNA clover leaf structure. Each is mediated by a different tRNA-modifying enzyme. The most common types of modification are: (i) methylation of the base or sugar, (ii) base rearrangement, (iii) double bond saturation, (iv) deamination, (v) sulphur substitution, and (vi) addition of more complex groups. The purpose of some of these modifications, such as those occurring within the anticodon loop, have not yet been determined.

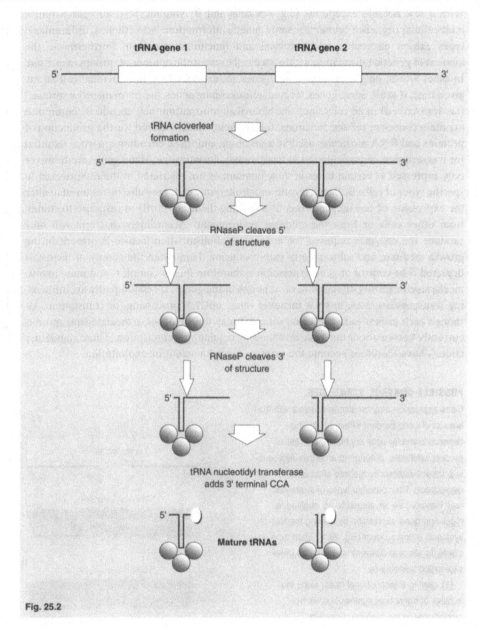

Fig. 25.2

Table. 25.1

Bases	Methylation	Deamination	Sulphur addition	Ring addition	Bond saturation
Uridine	Ribothymidine	Pseudouridine	4-thiouridine		Dihydrouridine
Cytidine	5′/3′-methylcytidine				
Adenosine	Methyladenosine				
Guanosine	Methylguanosine			Queosine Wyosine	

26. MECHANISMS REGULATING GENE EXPRESSION

The temporal, cell-specific and developmental regulation of gene expression may involve one or more mechanisms which control the rates of transcription, mRNA turnover and processing, or translation.

With a few notable exceptions (e.g. sex cells and B lymphocytes), all cells within a multicellular organism contain the same genetic information. Nevertheless, different cell types exhibit enormous morphological and functional variation. Furthermore, the amount of genetic information within each cell exceeds the amount of protein expressed. In other words, only a fraction of the genes are expressed by an individual cell at any given time, if at all. Some genes, termed housekeeping genes, are permanently expressed (i.e. transcribed) in all cells since the biological information they encode is required to maintain common, cellular functions; for example, those required for the production of proteins and RNA molecules used in translation, and those encoding enzymes required for transcription or participation in basic metabolic pathways. Many genes are however only expressed at certain times in development, or are restricted in their expression to specific types of cells. Both eukaryotic and prokaryotic cells may also be required to alter the expression of certain genes (i.e. by switching them on or off) in response to stimuli from other cells or from the external environment. Accordingly, bacteria will only produce the enzymes required for lactose metabolism when lactose is present in the growth medium, and subsequently stop producing them when the supply of lactose is depleted. The control of gene expression is therefore highly complex, and may involve mechanisms exerting effects at several points in the gene expression pathway, influencing transcription rates, mRNA turnover rates, mRNA processing, or translation. Although each prokaryotic and eukaryotic cell may utilise all these mechanisms, more is currently known about the basic mechanisms regulating transcription. These regulatory circuits have therefore become the primary focus for scientific exploitation.

HOUSEKEEPING GENES

Many different types of genes may be classified as housekeeping genes based upon their ubiquitous expression in all living cells. Classical examples include those encoding rRNAs, the ribosomal proteins required for the continuous production of the ribosomes vital for maintaining translation, and the enzyme RNA polymerase required for transcription. Their ubiquitous expression makes them ideal genes to use as transcriptional markers. For example, they may be used as internal standards in Northern blotting, ribonuclease protection assays or reverse transcription polymerase chain reaction (RT-PCR). Such markers can be used to indicate the presence or absence of a specific temporally expressed gene transcript, and improve the accuracy of their quantitation.

OAs: 155, 185, 186, 189–96,209, 229, 231–4. RAs: 59, 63, 77–82, 87, 89–92, 110. SFR: 36–9, 50, 74–84.

POSSIBLE CONTROL STRATEGIES

Gene expression may be simply equated with the amount of gene product within a cell. This depends primarily upon controlling the rate of product synthesis, although to a certain degree it is counter-balanced by the rate of product degradation. Four possible types of strategies may therefore be envisaged to be involved in regulating gene expression by altering the rate of synthesis of the product (Fig. 26.1). Each type exerts its effect at different stages in the gene expression pathway by:

(1) altering transcriptional rates, since the number of transcripts synthesised will be proportional to the amount of product;

(2) altering mRNA turnover rates (i.e. transcript stability), since this influences the availability of transcripts for translation and hence product synthesis;

(3) altering the rate of mRNA processing, since translation can generally only occur from mRNAs which have undergone post-transcriptional modifications (e.g. capping, splicing and polyadenylation);

(4) altering the rate of translation (e.g. by altering the rate of attachment or translocation of ribosomes).

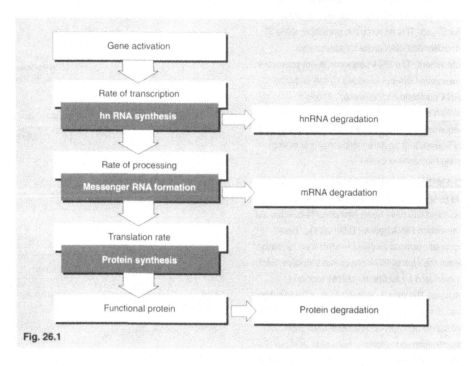

Fig. 26.1

Current evidence supports the involvement of all four strategies, possibly in different combinations for any given gene. However, mechanisms regulating transcriptional rates appear to be most commonly utilised, and as a result are certainly the best characterised.

TRANSCRIPTIONAL REGULATION

Many of the transcriptional regulatory mechanisms of prokaryotes and eukaryotes have been defined over the last 30–40 years. The foundations were laid in the early 1960s, by analysing the regulation of key enzymes involved in the metabolism of simple substrates (e.g. lactose) by *E. coli* mutants. This research established the existence of defined units, termed operons, in which the transcription of the enzyme-encoding genes is linked. Furthermore, transcription is controlled (i.e. switched on or off) by the binding of regulatory proteins (e.g. inducers or repressors) to defined DNA control sequences located upstream of the linked structural genes.

Similar mechanisms have since been characterised in eukaryotes. These involve the interaction of regulatory DNA-binding proteins (e.g. transcription factors) with transcriptional regulatory DNA sequences (such as promoters and enhancers) to influence binding of the transcribing enzyme RNA polymerase II (see Fig. 26.2). The picture is however far more complex in terms of the nature of the proteins and the DNA regulatory sequences. Nevertheless, the basic mechanisms of transcriptional regulation governing temporal, cell-specific, and developmental gene expression appear to have been elucidated. This knowledge has more than academic importance. It affords many opportunities for direct exploitation by allowing the artificial modulation of gene expression *in vitro* and *in vivo* (e.g. in the development of some forms of gene-based therapies).

mRNA STABILITY AND PROCESSING

Several lines of independent evidence have indicated that mRNA processing events and mRNA stability may also regulate gene expression *in vivo* (Fig. 26.3). Although the factors determining mRNA stability are as yet poorly understood it is suggested that the inclusion of AU-rich tandem repeats may confer instability on the transcript. Similarly, the nature of sequences upstream of the polyadenylation sequence may also be important. Thus, levels of mRNA are reduced if these sequences are removed or varied. It is also possible that the 5′

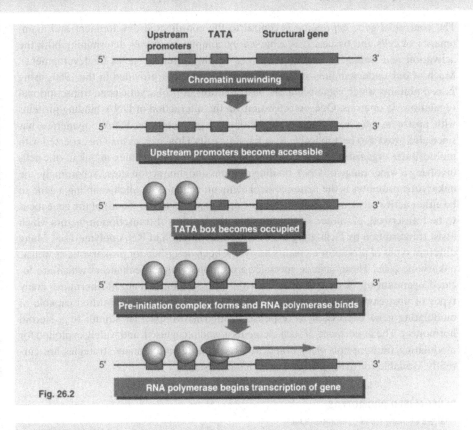

Fig. 26.2

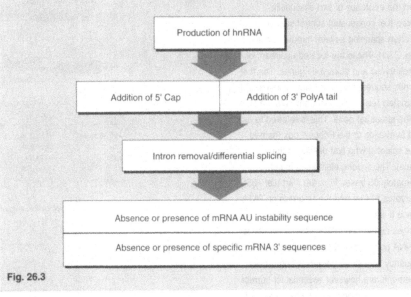

Fig. 26.3

cap of mRNA may be involved in determining the nuclear and cytoplasmic stability of a given transcript. The expression of some genes within individual prokaryotic and eukaryotic cells may also be regulated by the process of differential splicing of some transcripts. Consequently, the same transcript may be used to generate different types of gene products in different amounts in different cells (e.g. cell-specific forms of tropomyosin).

71

27. TRANSCRIPTIONAL REGULATORY SEQUENCES

A range of regulatory sequences, including promoters, enhancers and transient response elements, serve to modulate gene transcription by influencing the binding and activation of RNA polymerases.

The control of gene expression is central to the coordinated development and maintenance of cells and tissues. It is achieved by complex processes determining both the activation and timing of transcription (e.g. during different stages of development). Much of our understanding of these mechanisms was first provided in the 1960s using *E. coli* mutants which established the basic features of prokaryotic gene transcriptional regulation via operons. Operon activation via the interaction of DNA-binding proteins with upstream sites to influence the binding of promoters by RNA polymerase has since also proved to be applicable to eukaryotic cells. However, as may be expected with multicellular organisms, the situation is generally far more complex in eukaryotic cells involving a wider range of DNA-binding proteins and interaction sites. Accordingly, the eukaryotic promoter is the major control element, acting as a switch enabling a gene to be either active or silent. Generally located upstream (i.e. to the 5′ end) of the gene locus to be transcribed, promoter sequences allow the binding of transcription factors which assist transcription by facilitating the binding and initiation of RNA polymerases. Many different types of promoter elements have now been described for prokaryotic as well as eukaryotic cells. These include specialised promoters termed enhancers which are located upstream of, downstream of, or within the structural gene. Furthermore, many types of upstream 'transient response' elements have also been identified capable of modulating gene expression in response to internal or external stimuli (e.g. steroid hormones). The importance of such elements is well recognised, and widely exploited for modulating transcription *in vitro* and *in vivo*. Consequently, many strategies are currently available for their detection and characterisation.

MAPPING REGULATORY SEQUENCES

Many methods have been developed for the detection and fine mapping of regulatory sequences such as promoters and enhancers. These methods vary in their sophistication and discriminatory power. For example, the primary location of regulatory sequences may be provided by methods such as gel mobility shift (or gel retardation) assays based upon the detection of DNA–protein complexes. Regulatory sequences may also be located based upon the identification of consensus sequences, or their ability to drive transcription following cloning into reporter gene vectors. Fine mapping is generally achieved by *in vitro* mutagenesis strategies based upon the creation and analysis of site-directed or deletion mutants. They also provide data regarding the nature of their functional interactions with DNA-binding proteins. Such knowledge permits the development of new drugs and therapeutic strategies based upon modulating transcription.

GENERALISED PROMOTERS

Analysis of many types of promoters has revealed the existence of two essentially conserved (i.e. consensus) control sequence motifs, often spanning several hundred bases (see Fig. 27.1). These are located upstream of most prokaryotic and eukaryotic genes. The first prokaryotic sequence is an AT-rich region found approximately ten bases upstream from the start site of the structural gene. This is referred to as the −10 sequence or the Pribnow box (named after the scientist who first described the sequence). The second element is found approximately 35 bases from the start (i.e. −35). In eukaryotes it is termed the Hogness or TATA box, since it is usually composed of eight nucleotides centred on TATA. This is the site to which RNA polymerase II binds and is subsequently activated by transcription factors. Both elements are however essential for correct transcription, as demonstrated by the effects of naturally occurring or experimentally induced mutations within promoter sequences (see Fig. 27.1c). Such experiments have also shown that the substitution of A for T within the TATA box within individual gene promoters markedly affects transcriptional efficiency. In addition to the

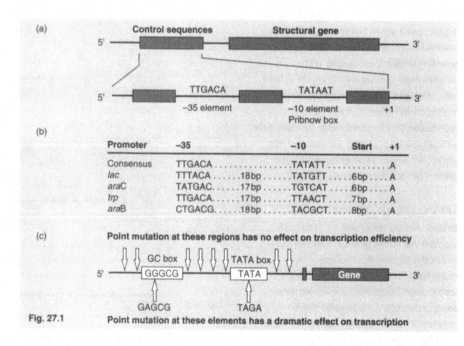

Fig. 27.1

generalised promoters, the transcription of many eukaryotic genes may be finely controlled by a variety of upstream elements. These are often specific for a single gene or a restricted number of genes.

OAs: 160–3, 184, 194–208, 225, 230–2, 516–17. **RAs:** 53, 57–9, 80, 82–8, 207, 272. **SFR:** 36–9, 50, 72–6, 80.

EUKARYOTIC 5' PROMOTERS

Sometimes described as upstream activating sequences (UASs), these elements are thought to be important in determining the level of transcriptional activation. For example, a region known as the CAAT box with the consensus sequence 5'-GGNCAATCT-3' (where N represents any nucleotide) may be found at –80. Another relatively common motif termed the GC box (with the consensus sequence 5'-GGGCGG-3') is also found between –40 and –110. This is generally associated with constitutively (continuously) expressed genes rather than developmentally regulated genes. Both of these elements seem to be effective when present on the non-coding DNA strand, in comparison to the TATA element which appears on the coding DNA strand (see Fig. 27.2).

Many gene-specific promoters have also been described, such as the heat shock element (HSE) associated with a gene involved in the expression of proteins (e.g. hsp70) during the heat shock response. Lying 15 bp upstream of the TATA sequence, it is present in multiple copies which serve as binding sites for the heat shock transcription factor (HSTF) which mediates the expression of hsp70.

Some genes are also associated with so-called 'transient response' elements which increase (upregulate) transcription in response to internal or external factors. Accordingly, the transcription of the low density lipoprotein receptor gene may be transiently upregulated by the interaction of steroids with upstream steroid hormone response elements (HREs). Similarly, the human metallothionein gene may be transiently upregulated via metal response or glucocorticoid response elements activated in response to the presence of heavy metals or glucocorticoid hormones respectively (see Fig. 27.3).

ENHANCER ELEMENTS

Enhancers are a particular type of promoter with unusual properties. Thus, they are able to exert their effects when located on either strand or in either orientation, and may alter transcription rates even when located several kilobases from their associated gene. The first enhancers were discovered in the 5.2 kb virus SV40, where they were able to modulate transcription when moved from the normal position to anywhere in the

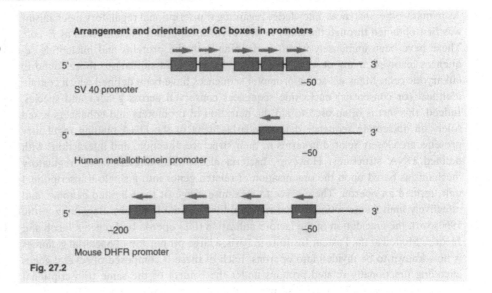

Arrangement and orientation of GC boxes in promoters

SV 40 promoter

Human metallothionein promoter

Mouse DHFR promoter

Fig. 27.2

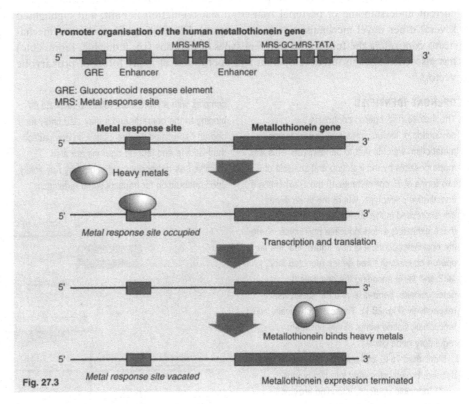

Promoter organisation of the human metallothionein gene

GRE   Enhancer   Enhancer
MRS-MRS   MRS-GC-MRS-TATA

GRE: Glucocorticoid response element
MRS: Metal response site

Metal response site   Metallothionein gene

Heavy metals

Metal response site occupied

Transcription and translation

Metallothionein binds heavy metals

Metal response site vacated   Metallothionein expression terminated

Fig. 27.3

circular genome. Sequencing of the SV40 enhancer revealed that it is composed of a repeated 72 bp region and a number of small modular elements related to transcription factor binding sites. However, their precise mode of action has yet to be determined.

Enhancers have since been demonstrated to

be common features within the genomic DNA of many eukaryotic cells where they are regulated by external or tissue-specific signals. One of the most studied tissue-specific enhancer sequences is the immunoglobulin heavy chain enhancer which is able to increase immunoglobulin expression several thousand-fold.

28. OPERONS AND PROKARYOTIC CONTROL OF GENE EXPRESSION

The expression of prokaryotic genes within operons is regulated by several mechanisms, including the repression, attenuation and/or induction of transcription by the presence of product/substrate.

As in many other instances, knowledge regarding transcriptional regulatory mechanisms was first obtained through the study of prokaryotes, especially bacteria such as *E. coli*. These have been immensely useful in defining both the proteins and nucleotide sequences involved, many of which are identical (or at least similar) to those found in eukaryotic cells. Many *cis*-acting promoter sequences have been defined which contain identical (or consensus) nucleotide sequences conserved across genera and species. Indeed, this fact is often used to aid the detection of promoters and enhancers based solely on nucleotide sequence data. Similarly, many of the DNA-binding regulatory proteins are closely related in terms of their structure, function, and interactions with defined DNA structures. However, bacteria also utilise transcriptional regulatory mechanisms based upon the organisation of related genes into a single transcriptional unit, termed an operon. These serve to maximise usage of their limited genome, and effectively limit transcription to that required for essential functions. Beginning in the 1960s with the elucidation of the lactose utilization (*lac*) operon by Francois Jacob and Jacques Monod at the Pasteur Institute in Paris, a large proportion of bacterial genomes is now known to be divided into operons. Each of these is composed of several genes encoding functionally related proteins under the control of the same transcriptional regulatory sequences. Their analysis provided the foundations upon which much of our current understanding of bacterial transcriptional regulation is built, and highlighted several other novel mechanisms (e.g. attenuation). Elucidating the molecular mechanisms controlling the function of different types of operons (i.e. inducible, repressible) has also been vital to the development and application of many forms of prokaryotic vectors.

OPERONS IDENTIFIED

The lactose (*lac*) operon controlling the production of the enzymes required for lactose metabolism was the first to be identified. This was made possible by the isolation and analysis of two forms of *E. coli* mutants: (i) those exhibiting a constitutive phenotype, where the three genes are expressed in the absence of lactose, and (ii) those exhibiting a non-inducible phenotype where the opposite occurs. It is now known that the *lac* operon consists of three genes (denoted *lacZ*, *lacY* and *lacA*) encoding the enzymes β-galactosidase, permease and transacetylase respectively (Fig. 28.1). These are physically and functionally linked with a single upstream regulatory gene (*lacI*).

More than 75 *E. coli* operons have since been defined, broadly classified into two groups.

(1) Inducible operons, encoding enzymes involved in metabolic pathways whose transcription is induced by substrate (e.g. the lactose, galactose and arabinose operons).

(2) Repressible operons, encoding biosynthetic enzymes whose transcription is regulated by the product of the biosynthetic pathway via product co-repression and/or attentuation. For example, tryptophan acts as a co-repressor, forming a complex with a repressor protein to facilitate its binding to the operator and it may also finely tune operon expression by attenuation. In comparison, the histidine and leucine operons are also classified as repressible, although they rely solely upon attenuation for transcriptional regulation.

OAs: 189, 203, 209–24, 609–12.
RAs: 88–93, 218, 278.
SFR: 36–9, 76–8, 81–4, 98, 101, 102.

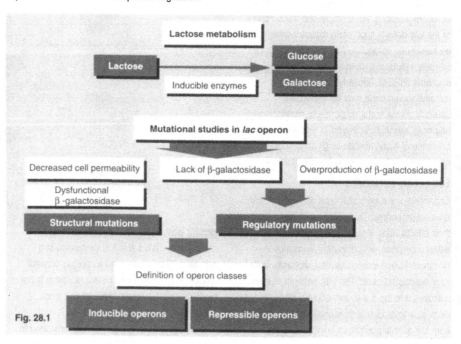

Fig. 28.1

REGULATION OF THE *lac* OPERON

In the absence of lactose there is no need to expend energy producing lactose-metabolising enzymes. Instead, a repressor protein is produced from the *lacI* gene which binds to a DNA sequence, termed the operator, located adjacent to the promoter and the structural genes (Fig. 28.2). The binding of this regulatory protein sterically blocks the attachment of RNA polymerase to the promoter, thereby preventing transcription of the *lac* operon. However, in the presence of lactose, allolactose or a synthetic inducer (e.g. IPTG), an inducer–repressor protein complex is formed. The binding of the inducer alters the structure of the repressor protein such that it can no longer bind to the operator sequence, thereby allowing transcription of the *lacZ*, *Y* and *A* genes.

The transcription of the *lac* operon is also influenced by the availability of glucose (a breakdown product of lactose) in a process termed catabolite repression. This involves a further regulatory protein (catabolite activator protein or CAP), and a further upstream CAP site to which cAMP produced by adenylate cyclase can bind. In fact, it is the attachment of CAP-cAMP which stimulates binding of RNA polymerase to the *lac* promoter. Consequently, in the presence of glucose and lactose, *lac* operon transcription is kept to a minimum by the fact that glucose inhibits adenylate cyclase and thus keeps the CAP site vacant. Once the glucose is depleted however, the CAP site is filled due to the rise in cAMP levels, allowing transcription and lactose metabolism to occur.

TRYPTOPHAN OPERON REGULATION

Transcription of the tryptophan (*trp*) operon is primarily regulated by product co-repression (i.e. via tryptophan) (see Fig. 28.3). In the absence of tryptophan, the repressor molecule cannot bind to the *trp* operator to prevent transcription. Binding can only occur in the presence of newly synthesised tryptophan (i.e. produced by transcription of the *trp* operon). This newly synthesised tryptophan, binds to the repressor enabling its interaction with the operator (i.e. tryptophan serves as a co-repressor).

Transcription of several operons including the *trp* operon is also finely controlled by a

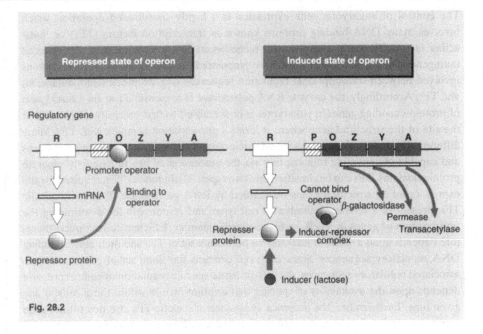

Fig. 28.2

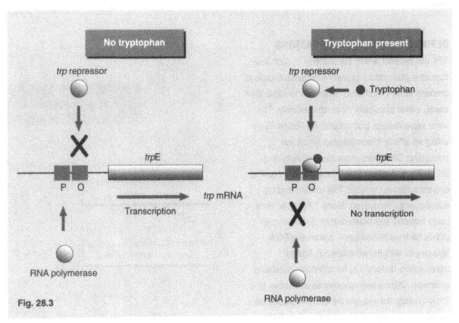

Fig. 28.3

mechanism termed attenuation. This relies upon the fact that in bacteria transcription and translation are closely linked, occurring almost simultaneously. Thus, one of two possible stem-loop structures can be formed between the promoter and the first 5′ operon gene (e.g. encoding tryptophan) during formation of the mRNA transcript: (i) a large loop which has no effect on transcription, and (ii) a small form of

loop which can act as a transcriptional terminator. The type of loop formed depends upon the relative availability of the repressor (e.g. tryptophan) which in turn influences the speed at which ribosomes translate individual transcripts. This complex system allows precise control of operon expression to meet the needs of the individual bacterium at any given time, dependent upon product availability.

29. TRANSCRIPTION FACTORS AND GENE EXPRESSION

Transcription factors are forms of DNA-binding proteins which regulate the developmental and temporal expression of genes by influencing the initiation of RNA polymerase II transcription.

The control of eukaryotic gene expression is a highly coordinated operation which involves many DNA-binding proteins known as transcription factors (TFs) or *trans-acting factors*. Recently, the powerful recombinant DNA techniques of site directed mutagenesis and DNA footprinting have pinpointed the precise molecular interactions involved between transcriptional regulation sequences (e.g. promoters and enhancers) and TFs. Accordingly, the enzyme RNA polymerase II responsible for the transcription of protein-encoding genes in eukaryotes is now realised to first precisely locate itself at the site of the structural gene where it forms a pre-initiation complex with TFs. Many different TFs have now been described, each recognising specific regulatory sequences and capable of activating transcription via the possession of two functionally separate protein domains. TFs can be classified into two types: (i) ubiquitous TFs, required for the expression of all structural genes transcribed by RNA polymerase II, and (ii) specific TFs, present in a restricted number of cell types and responsible for determining the transcription of genes in a cell- or tissue-specific manner. Efficient transcription therefore depends upon a series of interactions between a set of TFs and their corresponding DNA regulatory sequences. Since each cell contains the same set of genes and their associated regulatory sequences, the cell- or tissue-specific regulation of gene expression depends upon the availability of specific transcription factors within those cells at any given time. Furthermore, the presence or absence of specific TFs also determines gene expression during development, differentiation and cellular responses to environmental stimuli.

TFs AND DEVELOPMENTAL GENES

Studies of *Drosophila* fruit flies have indicated that each different developmental gene contains a characteristic 180 bp nucleotide sequence termed the homeo box. The homeo box has since been detected in many invertebrates and vertebrates, including humans. Homeo boxes encode for a 60 amino acid protein homeo domain which serves as the helix-turn-helix DNA-binding domain of transcription factors. Their significance in development has been powerfully demonstrated in *Drosophila* by the introduction of site-specific mutations which dramatically and bizarrely alter its final body plan (e.g. the growth of legs from the head instead of antennae). The similarity of homeo box sequences between species suggests that studies of genetically less complicated organisms such as *Drosophila* may provide insights relevant to human genetic processes.

DEFINING TRANSCRIPTION FACTORS

TFs are present within the nucleus in such low concentrations that it proved a formidable task to prepare them in sufficient quantities to allow their study, either physically or mechanistically. TFs were nevertheless first isolated by Robert Tjian using an affinity chromatography column containing DNA from promoter elements (Fig. 29.1). Thus, nuclear preparations applied to the columns allowed specific TFs to bind and be subsequently recovered. Many TFs have since been isolated and defined from genomic and cDNA libraries based upon comparing DNA sequences with those of already known transcription factors (i.e. by computer database analysis). Clones with sequences indicative of a DNA-binding domain are therefore regarded as representing potential transcription factors.

For example, although the proto-oncogene c-*myc* was found to be transcriptionally active in disorders such as Burkitt's lymphoma, many years of biochemical characterisation were unable to ascribe it any specific function. However, database analysis clearly indicated that it was a potential transcription factor, a prediction subsequently confirmed using molecular genetic techniques. Despite the diffculty of isolating and studying TFs, the general scheme of events

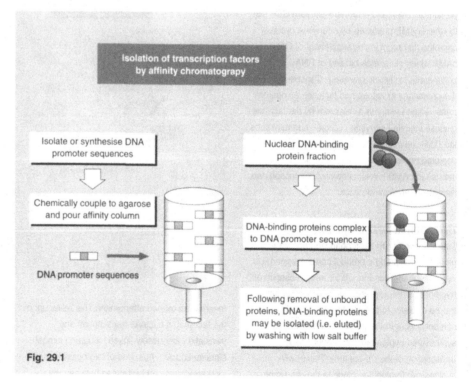

Isolation of transcription factors by affinity chromatograpy

Isolate or synthesise DNA promoter sequences

Chemically couple to agarose and pour affinity column

DNA promoter sequences

Nuclear DNA-binding protein fraction

DNA-binding proteins complex to DNA promoter sequences

Following removal of unbound proteins, DNA-binding proteins may be isolated (i.e. eluted) by washing with low salt buffer

Fig. 29.1

whereby TFs initiate RNA polymerase II-mediated transcription have now been largely defined.

OAs: 46–54, 160–2, 225–35, 312, 317, 320. **RAs:** 13–15, 57–9, 94–100. **SFR:** 36–9, 50, 72, 74, 75, 85.

Activation of RNA polymerase II

The precise modes of action of different TFs in initiating RNA polymerase II binding to the promoter have not been defined. However, it appears that ubiquitous TFs fulfil this function by enhancing the formation of stable multi-subunit complexes at the promoter located upstream of the transcription initiation site. Accordingly, the transcription factor TFIID is thought to initially bind to the –35 TATA box followed by the factors TFIIA and TFIIB (Fig. 29.2). The formation of this complex appears to signal the RNA polymerase II to bind, with the resultant complex further stabilised under the influence of the TFIID protein. The arrival of a further factor TFIIE appears to complete what is then termed the pre-initiation complex. This enables the RNA polymerase II to begin transcribing the gene.

In some cases the TFs bind to regulatory sequences many hundreds or thousands of bases away from the RNA polymerase initiation site. This requires the formation of a DNA loop in order to bring the distant regulatory sequences and the activating domains of bound TFs close enough to participate in the formation of the initiation complex (see Fig. 29.3).

The nature of TF domains

TFs have two functionally discrete domains, one for DNA binding and one participating in their interaction with the other components of the transcription initiation complex. TF cloning has now allowed comparisons to be made between the deduced amino acid sequences of TFs. These data suggest that the TF DNA-binding domains have evolved a limited number of structures. Thus, 80% of TFs utilise DNA-binding domains composed of one of three structural motifs: (i) helix-turn-helix motif, such as that found in many factors that regulate development of *Drosophila* (termed homeo domain proteins); (ii) zinc fingers in which a repeated Cys-His motif folds around a zinc atom (e.g. TFIIIA); and (iii) leucine zippers, in which leucine residues within separate TF domains interact to form homo- or heterodimers capable of DNA binding.

Another family, termed helix-loop-helix TFs, also share the dimerisation properties of leucine zippers but have a positively charged DNA-binding domain. Other structural variants have

also been discovered which indicate that many factors are modular in construction.

Three types of activating domains have also been defined: (i) those with an acidic region bearing a net negative charge, (ii) those with a high proportion of proline residues, and (iii) those

with a high proportion of glutamine residues. Little is yet known about the precise nature of the interactions between TFs, between TFs and RNA polymerase, or with other as yet unidentified components of the initiation complexes formed during transcriptional activation.

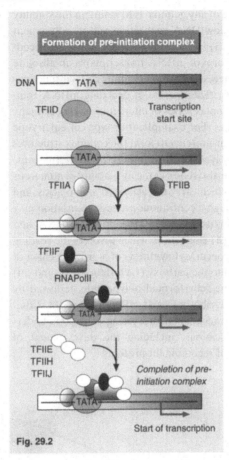

Fig. 29.2

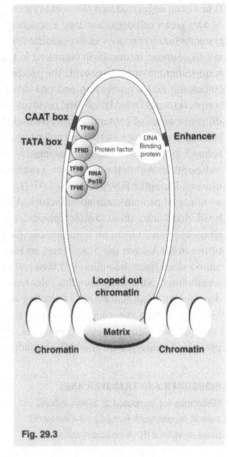

Fig. 29.3

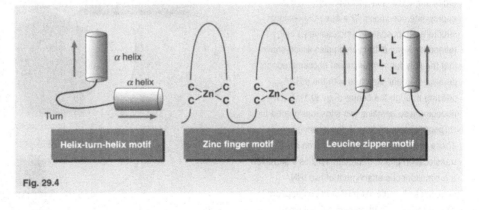

Fig. 29.4

30. *IN VIVO* TRANSLATION: DECODING GENETIC MESSAGES

Protein synthesis by translation begins with the formation of an initiation complex between the ribosome small subunit, mRNA, an amino acid-charged tRNA, and proteins termed initiation factors.

It has been suggested that the primary purpose of any genome is to maintain the viability of any given cell/organism long enough to allow DNA replication and its subsequent transmission to progeny cells/organisms. Whether or not this is the sole purpose, decoding of genomic information (supplied in the form of mRNA transcripts) is an absolute requirement in order to provide the proteins necessary for maintaining cellular functions (including DNA replication and cell division). The process of decoding mRNA transcripts is termed translation, and involves a complex series of interactions between many different types of proteins and RNA molecules. For example, four types of eukaryotic rRNAs are complexed with over 50 ribonuclear proteins to form the complex structures termed ribosomes which are intimately involved in the synthesis of new (nascent) polypeptides within the cytoplasm. Translation also requires the availability of numerous universal adaptor RNA molecules (tRNAs) which carry the activated amino acids, and a variety of protein translation factors. Although a continuous process, translation may be divided into three distinct phases: (i) initiation, where the mRNA, tRNA and ribosome join to form an initiation complex; (ii) elongation, which involves the reading of the mRNA from the 5′ to 3′ end and the concurrent synthesis of a growing chain of amino acids from the amino (N) terminal end to the carboxy (C) terminal end; and (iii) termination, which involves the release of the fully formed polypeptide, followed by dissociation, degradation and recycling of the ribosome and mRNA transcript. Combined with data regarding subsequent post-translational processing steps, this understanding has led to several practical applications, including the development of optimised strategies for the bulk production of recombinant proteins.

SUGAR, SLICE AND TRANSPORT SITES

Nascent polypeptides produced by translation often undergo several forms of post-translational modification within the cell before they attain their active conformations or reach the desired internal or external destinations. For example, many enzymes are only active following the removal of peptides from, or proteolytic cleavage of, inactive forms. Proteins also acquire N-linked carbohydrate residues transferred to asparagine residue side chains by a hydrophobic carrier (e.g. a dolichol pyrophosphate donor composed of a long lipid of 20 isoprene units containing a terminal phosphoryl group to which activated oligosaccharides are attached). These sugar residues are then trimmed within the endoplasmic reticulum (ER) and subsequently transferred from the ER to the Golgi complex for transport. Lysosomal enzymes also contain a conformational motif that signals the addition of mannose-6-phosphate which is recognised by receptors on the lysosome allowing their uptake.

RIBOSOMES AND TRANSFER RNAs

Ribosomes are composed of a large subunit and a small subunit, each formed by the interaction between various rRNA molecules and ribosomal proteins. They contain many base paired helical regions and can self-assemble *in vitro* under appropriate conditions. The three-dimensional structure of ribosomes is increasingly being refined by X-ray diffraction studies which show that the fully formed functional ribosome adopts a deformed 'donut' structure with the mRNA passing through the centre (Fig. 30.1). Each ribosome also contains two sites, designated the P (peptidyl) site and the A (aminoacyl) site. These are required for the elongation or translocation phase of protein synthesis and allow the simultaneous attachment of two tRNA molecules. A particular mRNA codon appears to dictate the precise three-dimensional structure or cavity for its associated tRNA–amino acid molecule.

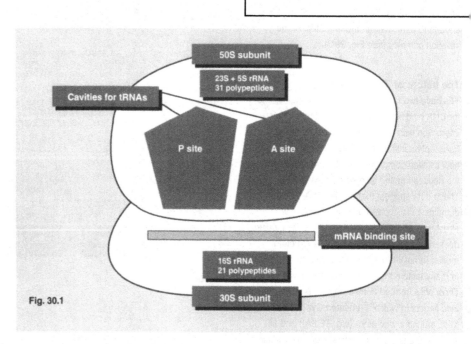

Fig. 30.1

Each cell contains many different tRNAs, each specific for one of the 20 naturally occurring amino acids (e.g. tRNA^Tyr specific for tyrosine). More than one tRNA may however recognise and bind to the same amino acid (by virtue of the degeneracy of the code). These amino acids are covalently linked to the acceptor arm of tRNAs by a process termed aminoacylation or charging. This process is catalysed by enzymes termed aminoacyl-tRNA synthetases, each specific for a single amino acid. Codons on the mRNA are in turn recognised by the triplet of nucleotides forming the anticodon loop of the tRNA clover leaf.

OAs: 61–8, 147–50, 188, 236–43.
RAs: 18, 19, 50, 51, 101–10.
SFR: 36–9, 86, 87.

TRANSLATION INITIATION

Translation in *E. coli* is initiated by the binding of the small ribosome subunit at a specific site, termed the ribosome binding site (RBS). This is located within the mRNA just upstream of the AUG initiation codon (coding for methionine) from which protein synthesis begins. The *E. coli* RBS is known to be a purine-rich sequence of approximately 10 nucleotides with a consensus sequence 5′-AGGAGGU-3′. It is also frequently termed the Shine–Dalgarno sequence after its discoverers (see Fig. 30.2a). Analysis of the 3′ end of the *E. coli* 16S rRNA molecule revealed remarkable complementarity to the Shine–Dalgarno sequence. Indeed, it is now known that the Shine–Dalgarno sequence base pairs with the 16S rRNA of the small ribosomal subunit.

Eukaryotes do not possess a defined ribosome binding site equivalent to the Shine–Dalgarno sequence (see Fig. 30.2b). Instead, the ribosome appears to bind to the extreme 5′ end of the capped mRNA and moves or 'scans' along the mRNA until the 18S rRNA defines the initiator AUG codon.

Initiation complex formation

Translation can only proceed once an initiation complex has been formed. This requires the binding of an initiating aminoacyl-tRNA (i.e. tRNA charged with an amino acid) to the small ribosome subunit bound to the mRNA. The initiator tRNA is always charged with methionine (i.e. tRNA^{Met}) because it bears the anticodon to the initiating AUG codon. In bacteria, but not eukaryotes, the methionine is further modified following charging of the tRNA to form N-formylmethionine, or fMet. This substitution of a formyl group (-COH) effectively blocks the amino terminal group such that it may only act as the start point for polypeptide formation by the process of elongation.

The formation of the initiation complex also requires various non-ribosomal proteins termed initiation factors. These vary in number and respective functions between prokaryotes and eukaryotes. Thus, three initiation factors (IF1, IF2 and IF3) are used in *E. coli* to facilitate ribosome binding and the provision of energy in the form of GTP (see Fig. 30.3). Eukaryotes use more initiation factors (designated by an 'e' prefix, i.e. eIF1, etc.) and use ATP as the energy source.

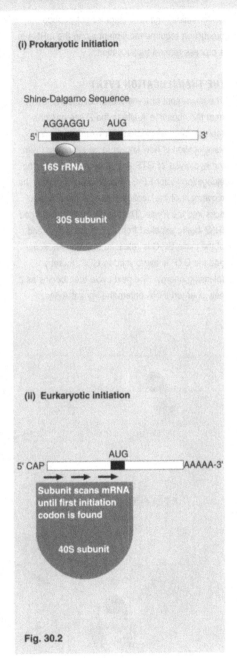

(i) Prokaryotic initiation

Shine-Dalgarno Sequence

AGGAGGU   AUG

16S rRNA

30S subunit

(ii) Eurkaryotic initiation

AUG

5′ CAP ━━━━━━━━━━ AAAAA-3′

Subunit scans mRNA until first initiation codon is found

40S subunit

Fig. 30.2

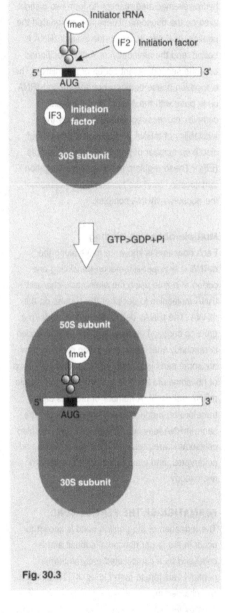

Prokaryotic initiation complex formation

Initiator tRNA

fmet

IF2 Initiation factor

AUG

IF3 Initiation factor

30S subunit

GTP>GDP+Pi

50S subunit

fmet

AUG

30S subunit

Fig. 30.3

Initiation factors

Ribosomes not actively engaged in translation are present within the cell as separate subunits. In prokaryotes these are maintained by binding of IF3 to the small (30S) subunit. This is believed to be achieved by IF3 distorting the 30S subunit and abolishing its affinity for the large (50S) subunit. However, when mRNA levels increase within the cell, IF2 becomes associated with GTP, the hydrolysis of which provides the energy for initiation complex formation. This interaction also allows the association of the factors IF1 and IF2-GTP to the 30S subunit. The initiator tRNA^{fMet} is then recognised by the factor IF2 residing on the ribosomal subunit. IF3 is subsequently released after formation of the initiation complex. This allows the binding of the 50S subunit, the release of IF1 and IF2, and the hydrolysis of the GTP.

Binding of the ribosome large subunit, tRNAs, and elongation factors, allows cycles of peptide bond formation as the ribosome moves along the mRNA until a stop codon terminates polypeptide growth.

POLYPEPTIDE ELONGATION

Once formed, the large ribosomal subunit can bind to the initiation complex by a GTP hydrolysis-mediated reaction to form two distinct sites on the ribosome. These are designated the peptidyl or P site, to which the initiator tRNA is bound, and the aminoacyl or A-site, positioned over the second vacant codon of the mRNA. The elongation phase begins when the second tRNA base pairs with the A site. This is a highly complex process regulated not only by the availability of mRNA and activated tRNAs, but also by a number of protein elongation factors (EFs). These regulate the stages of elongation, translocation, and the termination or release of the ribosome–mRNA complex.

Multiple ribosome binding

Each ribosome is known to move along the mRNA in a zipper-like motion translating one codon at a time using the amino acid-charged tRNA molecules to decipher the codons on the mRNA. The tRNAs deliver the amino acids to a growing carboxyl terminal end until a stop codon is reached. After a few elongation cycles a new ribosome can bind to the mRNA allowing a group of ribosomes to become associated with a single mRNA transcript. A number of proteins can therefore be produced simultaneously from the same mRNA template. These structures, termed polysomes, are present in both prokaryotes and eukaryotes, and may be visualised by electron microscopy.

FORMATION OF THE PEPTIDE BOND

The formation of the peptide bond is known to occur in the larger ribosomal subunit and is catalysed by a complicated enzyme termed peptidyl transferase (see Fig. 30.4). This joins the two charged amino acids held in the P- and A-sites by the tRNAs. The enzyme tRNA-deacylase then cleaves the link between the amino acid and the first tRNA present in the P site. This leaves a dipeptide (termed a peptidyl-tRNA) bound to the tRNA occupying the A-site. The precise role of the peptidyl transferase however remains elusive. It has been suggested that rather than being involved in the formation of the peptide bond itself, its main purpose is to align the amino acyl tRNAs onto the P and A sites in a favourable

conformation. The next stage of polypeptide production requires movement along the mRNA in a process termed translocation.

THE TRANSLOCATION EVENT

The movement or translocation of peptidyl-tRNA from the ribosome A-site to the P-site has only recently been elucidated. For example, *E. coli* translocation is now known to be associated with the hydrolysis of GTP and to be regulated by the elongation factor EF-G. Translocation involves the movement of the peptidyl-tRNA by three nucleo-tides into the P-site. This results in the uncharged tRNA being expelled from the P-site. At the end of the cycle EF-G is released from the ribosome and the GTP is hydrolysed to GDP, thereby releasing energy. The next cycle then begins as a new charged tRNA enters the vacant A-site.

EFs and GTP cycling

In *E. coli*, complementary aminoacyl-tRNAs are brought to the A site over a particular codon by the protein elongation factor EF-Tu which also contains a bound GTP molecule. This is hydrolysed to GDP, and the EF-Tu GDP then dissociates from the ribosome and subsequently becomes complexed with a second elongation factor (EF-Ts), which allows the release of the GDP. A new molecule of GTP is then able to bind the EF-Tu–EF-Ts complex which in turn promotes the release of the EF-Ts part of the structure. Finally, the cycle is repeated where EF-Tu GTP is able to accept another incoming aminoacyl-tRNA and deliver it to the A site on the ribosome. Similar events occur in eukaryotic cells during the elongation phase of translation.

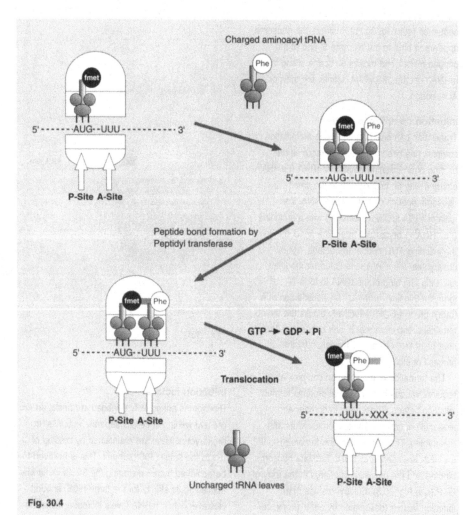

Fig. 30.4

CHAIN TERMINATION AND RELEASE

The termination of a growing polypeptide chain is determined by the presence of a termination codon (for which there is no corresponding tRNA) within the ribosome A-site (see Fig. 30.5). Termination also involves several proteins termed release factors (RFs).

Three release factors have been discovered in *E. coli*, RF1, RF2 and a further as yet uncharacterised RF3. It is these release factors which actually recognise the various stop codons. It is known that RF1 reads UAG and UAA codons, whereas RF2 reads UGA and UAA codons. The principal function of RF1 and RF2 is to bind to the A site of the ribosome and cleave the completed polypeptide from the last tRNA. This effectively transfers the growing polypeptide chain to a molecule of water rather than a new aminoacyl-tRNA. RF3 plays a supporting role in this process. Release factors which play similar roles have also been defined in eukaryotes (designated eRFs), with the difference that the process requires the hydrolysis of GTP to accelerate the reaction.

Release factors also cause the ejection of the polypeptide chain from the mRNA which subsequently gives rise to the dissociation of the large and small ribosomal subunits. These return to the cell pool to be available for use in further rounds of translation. The new (nascent) polypeptide may then undergo one or more forms of post-translational modification, followed by transport around, or secretion from, the cell.

Inhibition of translation

A number of compounds are known to interfere with the translation machinery (see Table 30.1). Some of the most potent inhibitors are antibiotics that specifically block prokaryotic protein synthesis at different stages. Indeed, it is this activity that makes antibiotics useful in combating bacterial infections. Compounds such as cyclohexamide are also available which block eukaryotic translation, e.g. by binding to the slightly larger 60S ribosomal subunit and inhibiting the translocation stage. These compounds can be used with *in vitro* cell cultures to arrest translation. This greatly facilitates the investigation of the various components and mechanisms utilised during the translational process. It also aids the analysis of mRNA transcripts (e.g. by increasing the representation of specific transcripts in the cytoplasm).

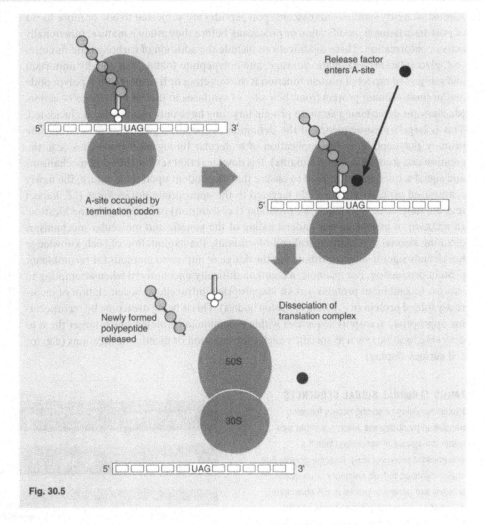

Release factor enters A-site

A-site occupied by termination codon

Newly formed polypeptide released

Dissociation of translation complex

Fig. 30.5

Table 30.1

Inhibitor	Effect (in prokaryotes)
Streptomycin	Causes misreading of mRNA and prevents initiator tRNA entering the P-site
Tetracycline	Binds 30S subunit and prevents aminoacyl/tRNA binding
Chloramphenicol	Peptidyl transferase activity inhibited
Erthromycin	Binds 50S subunit inhibiting translocation
Kirromycin	Binds to EF-Tu preventing peptide bond formation and its own release from the A-site
Thiostrepton	Binds to EF-G and prevents translocation

31. SEQUENCES INVOLVED IN CELLULAR PROTEIN TARGETING

The transport of newly synthesised proteins to the desired intracellular, membrane or extracellular locations depends upon the presence and recognition of several types of specific signal sequences.

Almost all newly synthesised (nascent) polypeptides are subjected to one or more forms of post-translational modification or processing before they attain a mature, functionally active conformation. These modifications include the addition of carbohydrate moieties (e.g. glycosylation), enzymatic cleavage, and polypeptide folding. An equally important and integrated aspect of protein function is the targeting or transport of the polypeptide and/or multi-subunit protein from their sites of synthesis to their required site of action. Mechanisms determining accurate protein targeting have only recently been elucidated. This is largely a consequence of the definition of specific signal sequences within the primary polypeptide by the application of molecular biological technologies (e.g. the creation and study of deletion mutants). It is now clear that several different mechanisms and signal sequences are utilised to ensure that, dependent upon their nature, the newly synthesised proteins are correctly targeted to the appropriate intracellular (i.e. import into the nucleus or organelles), extracellular (i.e. secretion) or intra-membrane location. In addition to increasing our understanding of the genetic and molecular mechanisms dictating normal and pathological cell functioning, the exploitation of such knowledge has already significantly contributed to the design of improved methods for recombinant protein production. For example, a common difficulty encountered when attempting to express recombinant proteins was an inappropriate intracellular accumulation of incorrectly folded protein (e.g. within inclusion bodies). This is being overcome by incorporating appropriate transport sequences within recombinant proteins which target them to desirable locations such as specific vesicles for secretion or membrane locations (e.g. for cell surface display).

FOLDING FOR FUNCTION

The production of many functionally active proteins requires that the nascent polypeptide produced by translation is folded into the correct tertiary structure. This important process occurs close to the site of protein synthesis and involves many as yet poorly understood mechanisms. The central dogma of protein folding originally stated that the primary structure of a polypeptide was sufficient to dictate its tertiary structure. However, it is now realised that several types of accessory protein molecules, termed molecular chaperones, are involved. Originally described and still best characterised in bacteria such as *E. coli*, they appear to act as catalysts which increase the rates of the final steps of the folding process. Molecular biology is both aiding and benefiting from the delineation of such intracellular folding mechanisms and the accessory proteins involved.

AMINO TERMINAL SIGNAL SEQUENCES

Protein targeting or sorting occurs following translation by ribosomes which are either free within the cytosol or associated with the endoplasmic reticulum (ER). Nascent polypeptide chains destined to form secretory or membrane proteins are generally formed at ER ribosomes. These then cross the ER membrane into the lumen by an ATP-driven process during which time they may also be glycosylated. From here they move to the Golgi complex before being transported to the desired destination. This process occurs via the recognition of specific amino acid signal sequences at the N-terminal end of the protein (first discovered by Sabatini and Blobel in 1970). Signal sequences of most secretory proteins contain several highly conserved features in both prokaryotic and eukaryotic cells (see Fig. 31.1). They range in length between 13 and 36 residues with basic residues preceding a hydrophobic core. These are followed by a cleavage site allowing the removal of signal sequences from the protein by an ER membrane-bound signal peptidase. Ovalbumin is however a classic exception, shown by deletion mutagenesis to contain an internal rather than an amino terminal signal sequence

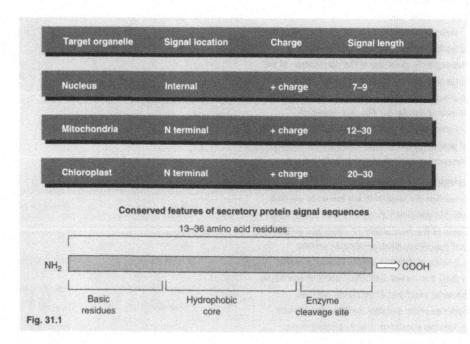

Target organelle	Signal location	Charge	Signal length
Nucleus	Internal	+ charge	7–9
Mitochondria	N terminal	+ charge	12–30
Chloroplast	N terminal	+ charge	20–30

Conserved features of secretory protein signal sequences

13–36 amino acid residues

NH_2 — COOH

Basic residues — Hydrophobic core — Enzyme cleavage site

Fig. 31.1

located between residues 22 and 41. Recombinant techniques have also demonstrated that proteins normally targeted to an intracellular site may be converted into secretory proteins purely by the addition of an amino terminal signal sequence.

OAs: 244–54, 595–600, 1036–9.
RAs: 111–16, 340–3, 379, 380.
SFR: 36–9, 93, 212, 213.

7. Chromatin structure and gene activity.
10. Distribution of nucleic acids in cells.
11. Types of RNA molecules.
23. Transcription.
30. *In vivo* translation.
37. DNA replication *in vivo*.
63. The nature of cloning vectors.
65. Bacteriophage vector development.
66. Plasmid vector development.
67. Yeast-derived plasmid vectors.
68. Phagemid vectors.
69. Animal and plant cell vectors (e.g. SV40).
79. *In vitro* mutagenesis methods.
92. Cellular receptors.
96. Protein engineering.

NUCLEAR IMPORT SIGNALS

All of the proteins and enzymes required for the processes of transcription and DNA replication which occur in the nucleus are synthesised in the cytosol. A mechanism is therefore required to direct these proteins to, and across, the nuclear membrane. Although the nuclear membrane contains pores of approximately 7 nm which allow small proteins such as histones to readily pass through, these block the passage of larger proteins such as polymerases.

Studies of the 92 kDa T antigen from the SV40 virus have shown that these large nuclear-destined proteins contain a nuclear localisation sequence of five consecutive positively charged amino acids (e.g. proline and lysine). This transit signal has been confirmed by recombinant DNA techniques which allowed a non-nuclear protein, pyruvate kinase, to be imported when the nuclear signal sequence was fused to the amino terminus. This heptapeptide is thought to direct a number of proteins to the nucleus, and, unusually, does not appear to be cleaved from the protein.

IMPORT INTO MITOCHONDRIA

Most mitochondrial proteins are encoded by nuclear DNA and are synthesised by free ribosomes in the cytosol. They therefore contain amino terminal sequences termed mitochondrial pre-sequences or transit sequences, which direct their transport into the the mitochondria. The mitochondrial entry sequence differs from the ER signal sequence, being rich in positively charged amino acids (e.g. serine and threonine). A mitochondrial protein remains in the outer membrane if this sequence is followed by a membrane anchoring sequence and a second positively charged sequence. It is also thought that proteins undergoing translocation across mitochondrial inner membranes become partly unfolded and are refolded once in their correct orientation. Such pre-sequences are also found in chloroplast proteins directing them to a number of sites depending on the nature of the particular pre-sequence.

TARGETING FOR DESTRUCTION

Cells have also developed a mechanism for detecting and degrading defective or oxidatively damaged proteins. This is an important part of metabolic regulation. The mechanism is based on a small highly conserved 8.5 kDa protein termed ubiquitin. Using an ATP-driven reaction, ubiquitin becomes covalently attached through its carboxy terminal glycine to the ε-amino group of lysine residues on certain proteins. This induces the activation of degradative measures which result in protein removal (by a mechanism as yet largely unknown). Targeting for destruction by ubiquitin appears to be related to the nature of the amino terminal residue. Highly destabilising residues such as arginine and aspartate have short half-lifes and favour ubiquitinisation, whereas methionine and serine do not. This process, and the amino acids that determine half-life and degradation, appear to have been highly conserved throughout evolution.

Signal sequence recognition

The adaptor linking signal sequences and ER-bound ribosomes is a 325 kD ribonucleoprotein complex, termed the signal recognition particle (SRP). It contains a 300 nucleotide 7SL RNA molecule bearing the repetitive *Alu* sequence (Fig. 31.2). This SRP participates directly in binding to a free ribosome. Furthermore, when translation of a mRNA transcript bearing a signal sequence begins, the SRP delivers the ribosome–mRNA complex to an SRP receptor docking protein (ribophorin) located at the ER (Fig. 31.3). The SRP is then cycled back into the cytosol and translation continues. Translocation across the ER membrane is thought to be a separate process from that involved in polypeptide elongation, although they usually occur simultaneously. This is supported by the fact that unfolded polypeptide chains serve as optimal substrates for translocation. An exception is the 18 kD yeast mating protein precursor which is known to be released from the ribosome and folded into a functional protein before it is translocated across the ER membrane.

Integral membrane proteins (e.g. cellular receptors) pose another problem. Their progress has to be arrested during translocation in order that a number of the component chains can be removed in the reverse direction. This is in part accomplished by the presence of a 'stop transfer' or membrane anchor sequence located on the nascent polypeptide.

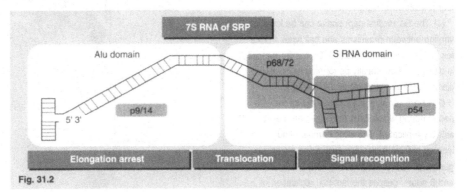

Fig. 31.2

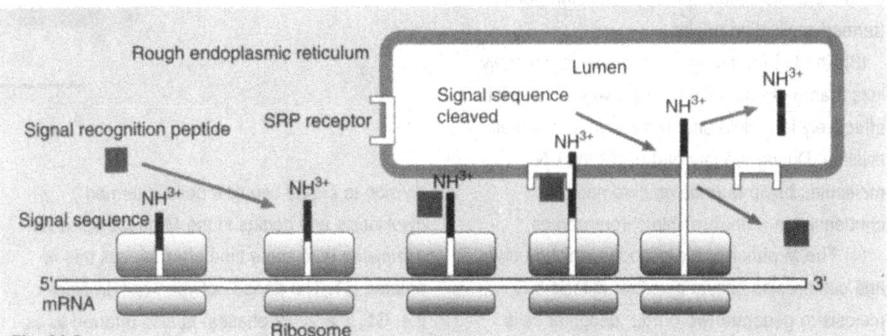

Fig. 31.3

32. EUKARYOTIC CELL DIVISION: MITOSIS AND MEIOSIS

The cell cycle may be divided into several phases during which DNA and cytoplasmic components are replicated followed by cell division, in the case of mitosis producing two diploid daughter cells.

All new cells are formed by the division of pre-existing cells in a series of events which control the transmission of genetic material from parent to daughter cells. With the exception of terminally differentiated cells (e.g. adult neuronal cells, erythrocytes and mature lymphocytes), most cells comprising a eukaryotic organism retain the capacity to undergo cell division. Although bacteria may undergo several forms of sexual replication involving gene transfer, the lack of a membrane-bound nucleus and single genomic DNA molecule makes prokaryotic cell division a relatively simple and rapid process. Accordingly, following DNA replication, the two resultant DNA molecules may be simply moved to opposing ends of the cell by the action of microtubules (termed spindles) and the cell divided across the middle. In contrast, the possession of a diploid genome composed of multiple chromosomal DNA molecules located within an inner nuclear membrane makes eukaryotic cell division a more complex process. Furthermore, eukaryotic cell division must be tightly controlled in order that it occurs at the right time and that the correct number and type of parental chromosomes are transmitted to daughter cells. The utilisation of sexual reproduction by eukaryotic organisms also requires two different forms of cell division, termed mitosis and meiosis. Vegetative reproduction by mitosis therefore generates diploid cells containing one copy of each parental chromosome. In contrast, sexual reproduction by meiosis involves the production of haploid sex cells or gametes (i.e. egg or sperm) containing copies of only one of each pair of parental homologous chromosomes. Errors occurring during cell division (e.g. due to incomplete separation or non-disjunction of homologous chromosomes), or the breakdown of cell cycle control mechanisms, can have profoundly deleterious or lethal consequences.

CYCLING OUT OF CONTROL

Cell cycling is normally tightly regulated and involves cellular cross-communication mediated by a range of soluble molecules. Some of these (e.g. steroid hormones) permeate the cell membrane and directly interact with DNA regulatory sequences (e.g. steroid regulatory elements). Others mediate their effect indirectly by binding to, and activating, cell surface receptors. This initiates the production of one of several types of second messengers (e.g. cyclic AMP) which subsequently leads to the activation of DNA regulatory sequences. Breakdown of the cellular cross-talk mechanisms by the production of constitutively activated or inactive receptors, or the constitutive activation of certain genes (e.g. oncogenes), leads to abnormal cell cycling and uncontrolled cell proliferation (e.g. as seen in various tumours).

EUKARYOTIC CELL CYCLES

Growing and dividing cells may be described in terms of the cell cycle. This can be divided into four phases (see Fig. 32.1).

(1) The G1 (or first gap) phase can be long, varying between organisms and cell types from a few hours to weeks or months. During G1 the cell is diploid and essentially performing its normal biochemical and physiological functions.

(2) The S (or synthesis) phase marks the period during which DNA molecules are being actively replicated. The large numbers and lengths of DNA molecules comprising eukaryote genomes means that this may take between 6 and 8 hours, despite the fact that replication of each molecule occurs from multiple start points (termed replication origins).

(3) The G2 (or second gap) phase is generally less than 3–4 hours long, during which the cell is effectively tetraploid and awaiting inevitable cell division. During the late part of G2 the DNA molecules begin to undergo chromatin condensation to form visible chromosomes.

(4) The M phase is the period in which nuclear and cell division occurs by either mitosis or meiosis to generate two or four daughter cells respectively. The actual process of cytoplasmic

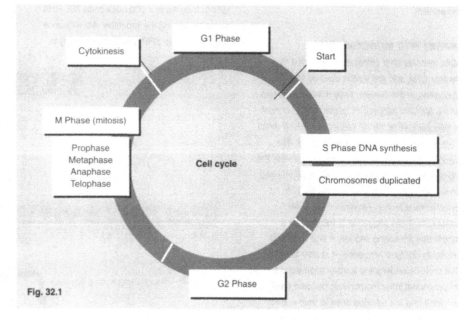

Fig. 32.1

division to create two new cells is termed cytokinesis and begins in the M phase but is not completed until some time after the cell has re-entered G1. The period between cell divisions (i.e. G1, S and G2 phases) is also referred to as interphase.

OAs: 106–8, 255–61, 276–9.
RAs: 16, 34–6, 117–26. SFR: 5, 22, 29–33, 36–45, 51–7, 88–90.

CELL DIVISION BY MITOSIS

By the end of interphase both the nuclear DNA (in the form of chromatin) and the centrioles have been replicated ready for cell division. The process of mitotic division may also be divided into four phases (Fig. 32.2).

(1) *Prophase.* Early prophase is marked by the disappearance of the nucleolus and the start of chromatin condensation. By late prophase the nuclear membrane has also disappeared and the chromosomes appear, each composed of two sister chromatids joined by a centromere. The two pairs of centrioles have by this time also migrated to opposite poles of the cell, and spindles have begun to form and attach to the kinetochore structures within the centromeres.

(2) *Metaphase.* Now visible by light microscopy, condensed chromosomes can be observed to line up along the middle or equatorial plane of the cell. The mitotic spindle is also fully formed. This comprises: (i) polar microtubules extending from the peri-centriolar material at each pole and overlapping at the equatorial plane; (ii) astral microtubules extending from the poles in all directions; and (iii) kinetochore microtubules. Similar spindle microtubules are also found in plant cells but are associated with a specialised region termed the microtubule-organising centre.

(3) *Anaphase.* Anaphase begins with the separation of sister chromatids, and finishes when they have been moved into separate poles by the spindle microtubules. This gives the chromatids a characteristic V-shaped appearance as the microtubules lead and the chromatid arms follow. This movement is achieved by 'pulling' as tubulin subunits are removed from the chromatid end of kinetochore microtubules, and/or by 'pushing' as the spindle microtubules lengthen and overlap.

(4) *Telophase.* This is the final phase of mitosis during which the chromosomal DNA is relaxed (i.e. the chromatin is decondensed), the spindle microtubules disappear, the nuclear envelope reforms separating two sets of chromosomes, and the nucleolus reappears. Cytokinesis also begins during telophase and ends during early interphase. In animal cells this is brought about by a ring of microtubules along the equatorial plane, whereas in plant cells it involves the formation of a discrete separating cell plate.

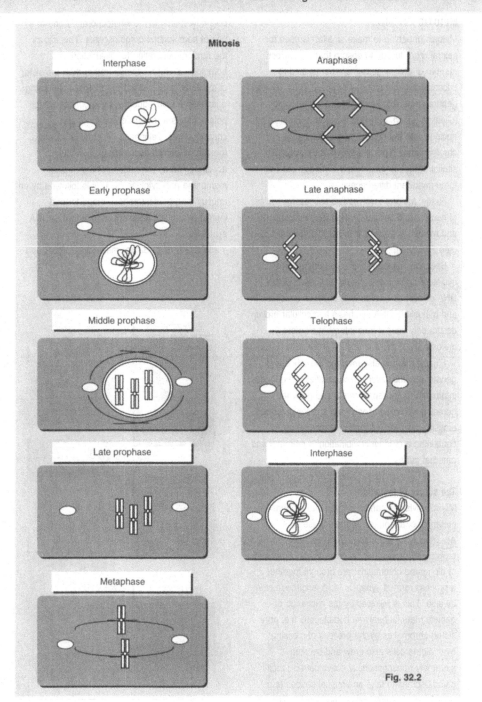

Fig. 32.2

CENTRIOLES

All animal cells, but not those of higher order plants, contain two centrioles arranged at right angles to one another. Each is composed of nine sets of three microtubules arranged into a hollow rod. Primarily involved in the formation of the mitotic spindle required for chromatid separation, they also serve vital functions in microtubule assembly in the basal bodies of the cilia and flagella found in some specialised cells (e.g. bronchial epithelial cells and sperm cells).

32. EUKARYOTIC CELL DIVISION: MITOSIS AND MEIOSIS/Contd

Meiosis employs similar stages to mitosis, but uses two cycles of nuclear and cell division, and crossing over between chromatids, to produce four genetically unique haploid daughter cells.

MEIOTIC DIVISION

Meiosis (meaning to make smaller) is used for sexual reproduction to ensure that the correct number of chromosomes are transmitted to successive generations of progeny. Thus, meiosis generates haploid sex cells or gametes which, following fusion, generate a zygote containing a nucleus with the normal diploid number of chromosomes. This is achieved by means of similar stages to those used in mitosis but with three important differences.

Firstly, meiosis involves two successive cycles of nuclear and cellular division (termed meiosis I and meiosis II) to yield four haploid daughter cells (see Fig. 32.3).

Secondly, although the cells undergo two cycles of division there is no replication of the DNA or cellular components between meiosis I and meiosis II. DNA synthesis and cellular growth occurs only once during interphase before meiosis I.

Finally, meiosis also involves the process of genetic recombination which shuffles the genetic information between maternal and paternal homologous chromosomes via a process termed crossing over. This results in each gamete containing a random combination of maternal and paternal genes.

The stages of meiosis

Meiotic divisions can be divided into prophase, metaphase, anaphase and telophase. Accordingly, in prophase I the previously replicated homologous chromosomes, composed of elongated chromatids, are brought together in a process termed synapsis to lie lengthwise side by side. This is followed by the exchange of genetic material between homologous (i.e. non-sister) chromatids by the process of crossing over. Some cells also grow and become extremely transcriptionally active causing their chromosomes to take on unusual shapes (e.g. appearing as lampbrush chromosomes). The nuclear envelope also disappears, animal cell centrioles migrate to opposing poles, and the spindle microtubules form.

In metaphase I the tetrads (i.e. the four chromatids) are still held together at the points of crossing over (termed chiasmata) and line up along the equatorial plane. Spindles from each

pole now attach to the kinetochores of one of a pair of homologous chromosomes. This allows the random separation (or disjunction) of homologous chromosomes (rather than the sister chromatids seen in mitosis) to either pole during anaphase I. Consequently, by telophase I each pole contains a mixture of maternal and paternal chromosomes. Telophase I is completed when the chromosomes decondense, generally accompanied by reformation of the nuclear membrane and cytokinesis. This is followed by an unusual and brief interphase stage termed interkinesis which does not involve further DNA synthesis.

Meiosis II (Fig. 32.3b) resembles mitosis and begins with a brief prophase II, followed by the lining up of chromosomes along the equatorial plane in metaphase II. Sister chromatids are then separated to opposite poles by means of the spindles attached to their kinetochores. Haploid cells containing one of each chromatid are finally formed during telophase II as the nuclear membrane is reformed and cytokinesis occurs.

OAs: 106–8, 2552–61, 263–72, 276–9. **RAs:** 16, 34–6, 117–26, 129, 131. **SFR:** 5, 22, 29–33, 36–45, 51–7, 88, 89, 94.

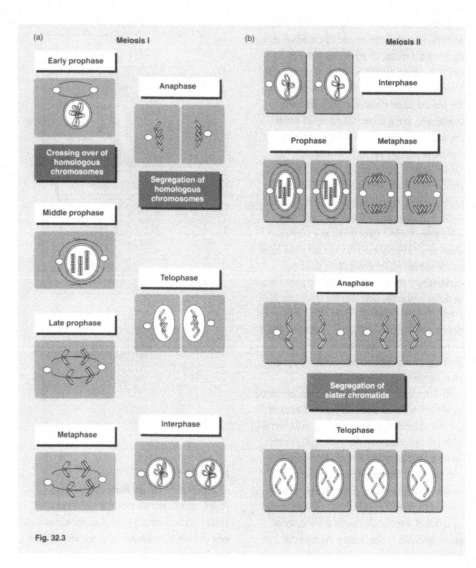

Fig. 32.3

6. Packaging of DNA within cells.
7. Chromatin and gene activity.
9. Organisation of genomes into multiple chromosomes.
12. Anatomy of eukaryote chromosomes.
19. Linkage analysis.
20. Pedigree analysis.
23. Transcription.
23. Regulation of gene expression.
33. Mechanisms controlling cell cycling.
34. Genetic recombination mechanisms.
37. *In vivo* DNA replication.
42. Enzymes in molecular biology.
46. Visualising nucleic acids.
92. Receptors and cellular signalling.
93. Oncogenes and the basis of cancer.
94. Cancer diagnosis and therapy.

Crossing over of chromosomes

Crossing over of chromosomes occurs during prophase I when the four chromatids become closely associated (i.e. as a tetrad). This forms a characteristic structure termed a synaptonemal complex. Genetic exchange between non-sister chromatids occurs by a process of generalised (or homologous) recombination involving regions of sequence with extensive homology. This involves the alignment of homologous DNA strands which are then cut by endonuclease action (Fig. 32.4). This is followed by strand displacement and rejoining by the action of DNA ligase. Until the homologous chromosomes are separated to opposing ends of the cell during anaphase I, the tetrads remain joined at the points at which crossing-over has occurred (termed chiasmata). These structures are often clearly visible in late prophase I cells using light or electron microscopy. Crossing over is however a random process resulting in the formation of many new combinations, thereby greatly increasing the genetic variation amongst offspring. The frequency of exchange of genes by crossing over is proportional to the distance between individual genes (i.e. the further they are apart the more chance that cross-over will occur between them). Measurement of this frequency therefore provides the basis of linkage analysis by which genes may be localised within eukaryotic chromosomes.

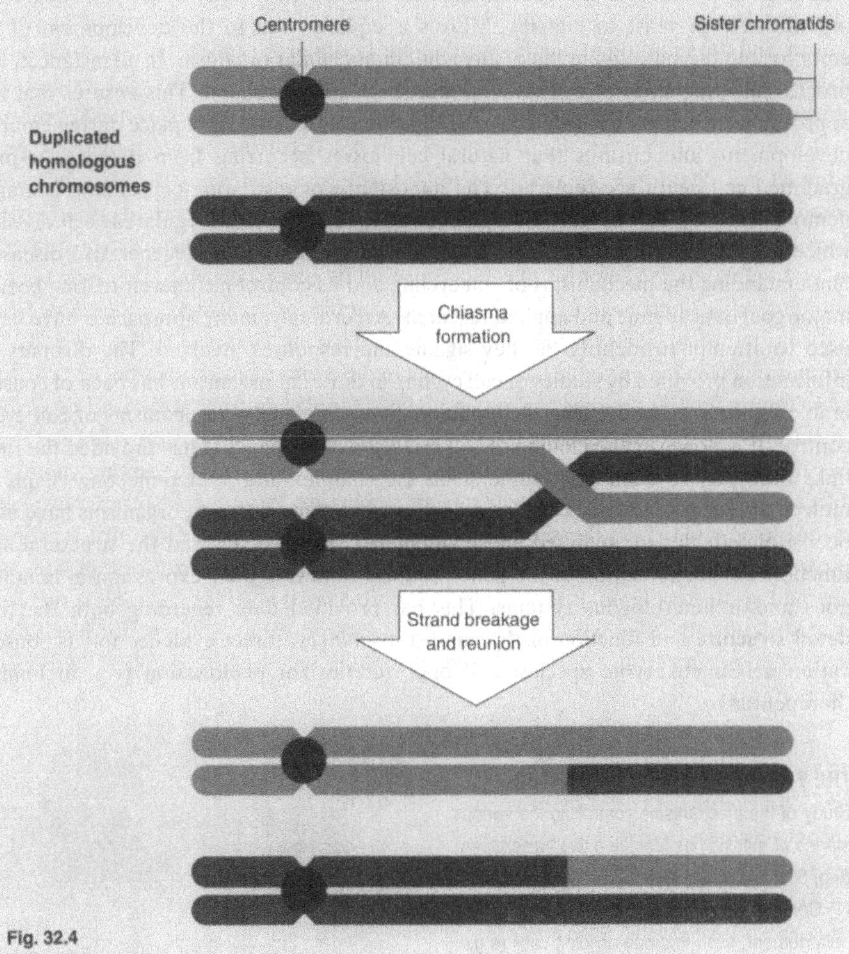

Fig. 32.4

CONTROLLING CELL CYCLING

Both the length of time taken for a cell to divide and the number of times a cell divides vary between species and cell types. For instance, some cells of the central nervous system apparently lose the capacity to divide within a few months following birth. In contrast, many cell types (e.g. skin cells) continue to divide throughout an organism's lifetime in order to replace those damaged. Other cells do not divide unless activated by an external signal.

Although the precise mechanisms controlling cell cycling are not fully established, it is clear that several crucial environmental factors may act upon in-built genetic programmes to influence cell division. For example, the length of time taken for a cell to divide, and the decision of a cell to pass the 'start' point at the end of G1 committing it to division, is partly dependent upon conditions of temperature and nutrient availability. Low temperatures and poor nutrient conditions will therefore lengthen the time taken to divide or may prevent cell division completely. The presence or absence of certain external signals may also influence different types of cells to divide, stop dividing or undergo programmed cell death (termed apoptosis) by activating or repressing in-built genetic programmes. For example, hormones like the plant cytokinins promote mitosis in normal cellular growth and wound repair.

Colchicine may be used to interrupt mitotic spindle function leading to the formation of polyploid cells. It may also be used to hold cells in metaphase to facilitate karyotype analysis. Indeed, drugs inhibiting cell division have formed the front-line approach to tumour chemotherapy based upon the fact that tumour cells are characterised by rapid growth and division.

Recent studies of two yeast species (*Saccharomyces cerevisiae* and *Schizosaccharomyces pombe*) have identified a number of genes that are involved in controlling the entry of cells into the division stage of the cell cycle. One basic genetic mechanism which appears to be used in all forms of eukaryotic cells involves a protein termed the maturation-promoting factor or MPF (encoded in *S. pombe* by the *cdc2* gene). MPF appears to activate other proteins in a non-uniform manner and has different effects at different stages during the cell cycle. Its own production appears to be controlled by the interaction of a protein, cyclin, with another protein generating an inactive form of MPF termed pre-MPF. This is then enzymatically converted to active MPF which initiates mitosis.

33. MOLECULAR MECHANISMS OF CELL CYCLE CONTROL

Temporal protein phosphorylation produced by the complex interaction between various protein kinases, proteins termed cyclins, and protein inhibitors, appears to play a vital role in cell cycle control.

Cell division by mitosis is a fundamental process almost identically used by all eukaryotic organisms from yeast to humans. Mitosis is equally vital to the development of an embryo, and the subsequent life of any adult multicellular organism. In all instances, the process and frequency of cell division must be highly regulated. This ensures that the appropriate structures are formed at the right time and in the right place during embryo development, and ensures that natural cell losses occurring from damage or programmed cell death are replaced. The importance of regulating cell cycling is amply demonstrated by the clinical consequences of abnormal (i.e. unregulated) cell division which accompanies many diseases, including cancers and neurodegenerative diseases. Understanding the mechanisms of cell cycling and its control has therefore long been a major goal of academic and applied research. Accordingly, many approaches have been used to attempt to identify the key signals and responses involved. The diversity of information provided by studies of cell cycling in different organisms has been of considerable importance in defining the commonality of molecular mechanisms of cell cycle control. It is however the cloning of cell cycle control genes that has provided the final links necessary to completely understand the complex and interconnecting chains of molecular events involved. Thus, many genes from many different organisms have now been isolated, cloned, analysed and manipulated. This has allowed the structural and functional characterisation of the gene products following their expression in homologous and/or heterologous systems. This has provided data regarding both the fine detail structure and function of the control machinery, direct evidence for its conservation across eukaryote species, and opportunities for exploitation (e.g. in human therapeutics).

PROGRAMMED CELL DEATH

First recognised over 20 years ago, the natural process of controlled cell deletion, termed apoptosis, appears to be vital in regulating animal cell populations by ensuring a balance between cell production and cell loss. Apoptosis differs in many ways from necrotic cell death (e.g. associated with cell replacement in response to injury). The principal difference is that apotosis follows an orderly and controlled programme, induced by remarkably weak stimuli. The molecular mechanisms of apoptosis are currently the focus of considerable research to define both the diversity and action of the stimuli involved. The association of abnormally controlled apoptosis with many major diseases (e.g. cancer, immune and neurodegenerative disorders, heart disease) and even ageing suggests this may provide data and insights useful for the development of several therapeutic intervention strategies and agents.

THE TWO BASIC TYPES OF CONTROL

Study of the mechanisms controlling the various stages of mitosis has identified two basic types (Fig. 33.1).

(1) Controls which regulate cell metabolism and enlargement, such that non-dividing cells (e.g. terminally differentiated cells) produce sufficient proteins for maintenance but do not increase in mass or size.

(2) Controls which coordinate the enlargement and division of growing cells, so that the daughter cells are accurate reflections of the original size and mass of the parental cells (i.e. the average size of the cells remains constant from generation to generation). This operates in most cells during the G1 phase of mitosis once they are fully (i.e. irreversibly) committed to division. Termed the commitment point, this is the main control point in the eukaryotic cell cycle.

Much of the information regarding cell cycle control was initially derived from recombinant DNA studies of genes isolated from yeasts, such as *Saccharomyces cerevisiae* (budding/baker's/brewer's yeast) and *Schizosaccharomyces pombe* (fission yeast). These studies were greatly enhanced by the isolation in the 1970s of

temperature sensitive (*ts*) cell division cycle (*cdc*) lethal mutants. This allowed arrest at different points in the cell cycle to give cells of uniform morphology.

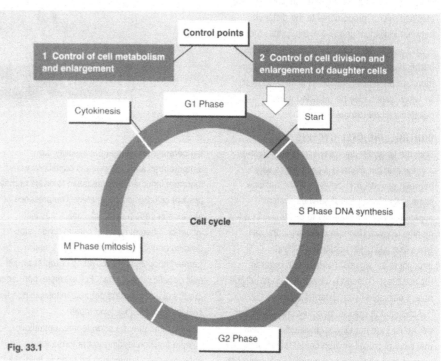

Fig. 33.1

OAs: 255, 260–73. **RAs:** 35, 36, 117–19, 123–8, 329. **SFR:** 88–93, 177–9.

cdc GENES AND PROTEIN KINASE

Over 40 different *cdc* genes have now been cloned, and their sequences and functions determined. An important discovery initially arose from the analysis of *cdc* genes arresting cells at the G1 commitment point, or at the boundary between the G2 and M phases of mitosis (Fig. 33.2). Comparison of the sequences of such genes in both *S. cerevisiae* and *S. pombe* showed they were remarkably similar. Computer database comparisons combined with enzyme assays revealed that they encoded defective forms of proteins with serine and threonine kinase activity (i.e. the ability to transfer a phosphate group from ATP to the hydroxyl group of serine and threonine residues). Since phosphorylation via protein kinases was already known to regulate the activity of other proteins, and be involved in cell signalling, the cloning of *cdc* genes revealed that protein phosphorylation is critical to cell cycle control. Phosphorylation is now known to be a general control mechanism in all eukaryotes. In fact, the equivalent human genes (e.g. *Cdc2Hs*) function perfectly following introduction into yeast *ts* mutants. Furthermore, the maturation-promoting factor (MPF) involved in inducing meiosis and mitosis in *Xenopus* frog oocytes also has protein kinase activity.

REGULATING PHOSPHORYLATION

The first clues regarding how cdc-mediated phosphorylation was involved in regulating the cell cycle came from western blotting studies using monoclonal antibody probes raised to yeast cdc proteins expressed in *E. coli*. These studies revealed that the proteins were present in constant amounts in cells at different stages throughout the cell cycle. However, isolation of the proteins at different stages revealed that their protein kinase activity varied. Thus, high levels of protein kinase activity were present in extracts from cells in mitosis, but little or no activity was present at other stages of the cell cycle. Further biochemical analysis of the cdc proteins revealed that their active form was complexed with other proteins.

A principal protein of this complex was first identified and cloned from sea urchin eggs. It was termed cyclin since it accumulates and disappears in a pattern matching the timing of cell division. Several classes of cyclins have since

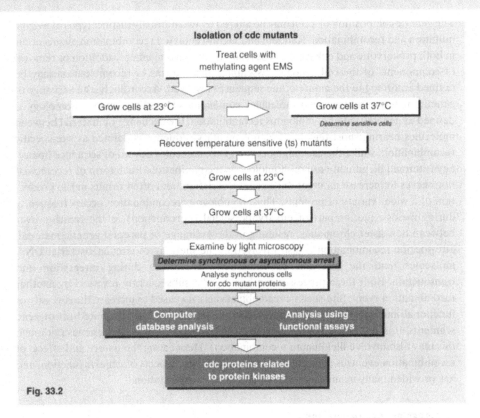

Fig. 33.2

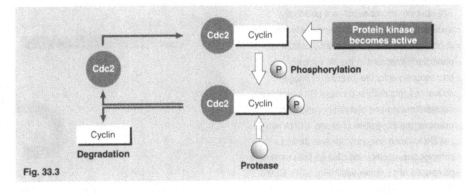

Fig. 33.3

been identified and cloned from many eukaryotic cells. Biochemical studies have subsequently confirmed that (in conjunction with other *cdc* genes) they play ubiquitous roles in regulating cdc protein kinase activity. Accordingly, activation of *cdc* gene activity depends upon the formation of cyclin–cdc protein complexes (Fig. 33.3). Consequently, *cdc* genes actually encode cyclin-dependent kinases or CDKs. It is now suggested that in their initial division phase cells contain inactive CDKs but no cyclin, but as the cell grows cyclins are synthesised which bind and activate

the CDKs. The resultant increase in phosphorylated proteins then pushes the cell into mitosis which can only end when the cyclins are destroyed, possibly by protease degradation of the phosphorylated cyclins. Recently however, several types of proteins have been identified which inhibit active CDK–cyclin complexes. These may also play an important role in preventing proliferation in quiescent or senescent cells. Furthermore, their loss or inactivation may partly explain the uncontrolled proliferation of tumour cells.

34. GENETIC RECOMBINATION MECHANISMS

Three recombinational mechanisms (termed homologous, site-specific and illegitimate) may be defined, involving genetic exchange between DNA molecules with extensive, limited, or no sequence homology respectively.

The genetic composition of a cell may be altered by two mutually distinct types of events: mutation and recombination. Random and highly regulated recombination events occur in both prokaryotic and eukaryotic cells resulting in rearrangement, addition or removal of components of the cell genome. Essentially three forms of recombination may be defined according to the nature of the sequences involved. Accordingly, the exchange of genetic material between two molecules containing extensive sequence homology is defined as generalised or homologous recombination, the exchange of material between molecules bearing short regions of complementary sequence is defined as site-specific recombination, whilst the exchange between two molecules devoid of sequence homology is termed illegitimate or non-homologous recombination. Each form of recombination serves to increase an organism's genetic variation, and often results in the production of a wider variety of proteins. Thus, homologous recombination occurs frequently during meiosis (i.e. as part of eukaryotic sexual reproduction) as the crossing over between non-sister chromatids, resulting in the exchange of parental genetic material. Site-specific recombination is also involved in the merging of circular bacterial DNA molecules, and the insertion of viral DNA molecules during integration and transduction. Both these processes can however be subsequently reversed by another recombination event. Site-specific recombination is also used to increase the diversity of functional immunoglobulin molecules which are produced from a limited set of gene segments. It is also utilised in the movement of transposons or jumping genes (although this may also involve illegitimate recombination). Delineating the nature and effects of recombination events is central to understanding many aspects of cellular function, and has provided many avenues for *in vivo* and *in vitro* exploitation.

EXPLOITING RECOMBINATION

Both the effects and mechanisms of genetic recombination are exploited by molecular genetic technologies. In the former case, calculating the frequency of recombination between two genes allows the degree of physical linkage to be established. In many situations this has greatly facilitated the localisation of genes by reverse genetics. Understanding the mechanisms of site-specific and homologous recombination has also led to the development of many approaches used for genetic engineering. Amongst the most recent and potentially important of these is the development of methods for the *in vivo* targeting of specific genes. These are central to the formation of transgenic organisms, and several forms of gene-based therapy.

OAs: 121–2, 274–81, 287, 794–5, 828, 1103–7. **RAs:** 129–31, 145, 138, 405. **SFR:** 94, 95, 97–9, 158, 159, 220.

SITE-SPECIFIC RECOMBINATION

Site-specific recombination is a potentially reversible event which occurs relatively frequently in both prokaryotic and eukaryotic cells. It is particularly important in the life cycles of bacteriophages for the insertion of phage sequences into circular bacterial DNA molecules, bacterial mating and plasmid genetics. For example, the integration of phage λ DNA into *E. coli* DNA during lysogeny involves distinct, homologous recombining sites on both molecules composed of 15 nucleotides (Fig. 34.1). Of the four phage integration sites identified to date, one has been demonstrated to be located in a region that is bound by a phage type I topoisomerase-like protein called integrase produced by the λ *int* gene. Following alignment of the two DNA molecules at the integration sites, integrase cleaves within the homology regions to produce 7 base pair staggered ends. It also catalyses the exchange of strands at the cut sites. Excision of the λ prophage during entry into the lytic cycle is highly regulated by the coordinated action of integrase and the *xis* gene product (excisionase) acting at the homologous regions now flanking the inserted prophage DNA.

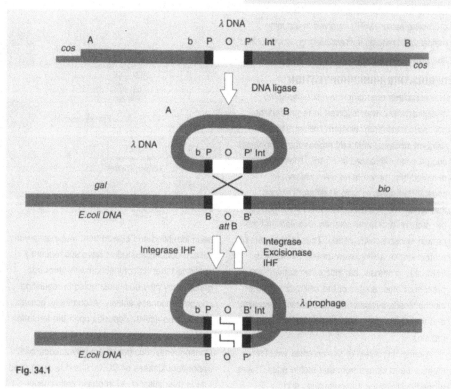

Fig. 34.1

Similar mechanisms are used during transpositional recombination controlled by the transposon-encoded transposase enzyme and an enzyme termed resolvase.

HOMOLOGOUS RECOMBINATION

Homologous or generalised recombination involves the highly precise strand breakage and reunion between two DNA molecules at potentially random sites by virtue of their extensive sequence homology. First identified in crossing over between homologous chromatids during meiosis, this does not generally lead to base additions or deletions, but rather a simple exchange of polynucleotides. The manner in which this cross-over exchange is achieved was initially suggested by Robin Holliday in 1964. It initially involves a process termed synapsis whereby the two DNA molecules are aligned so that homologous regions lie next to one another in register, forming what is termed a tetrad (Fig. 34.2). This is followed by breakage of single strands at equivalent positions on each molecule by the action of an endonuclease, and the formation of a heteroduplex by strand displacement cross-over. The nicked ends are then sealed and the cross-over point moves along the duplex in an ATP-mediated hydrolysis process termed branch migration. This heteroduplex structure is referred to as the Holliday intermediate (or chi) structure and is visible by electron microscopy.

This heteroduplex is only released following the formation of a second nick in each DNA molecule, and followed by a limited form of DNA repair. The position at which this occurs has very different genetic consequences depending upon whether the second cut occurs on the same DNA strands, or the previously uncut strands. In the former situation it results in the relatively minor exchange of small homologous regions and consequently little genetic change. The latter type of cutting however produces a more radical alteration in the genetic material by creating two new molecules with exchanged ends.

Many of the proteins and enzymes involved in generalised recombination have now been identified and some have been shown to be identical to those used by DNA repair mechanisms. The initial steps involve a RecA protein. This has ATPase and recombinase activity which allows strand displacement and invasion. The process also involves a three subunit protein RecBCD with helicase and exonuclease activity. This model must however

Fig. 34.2

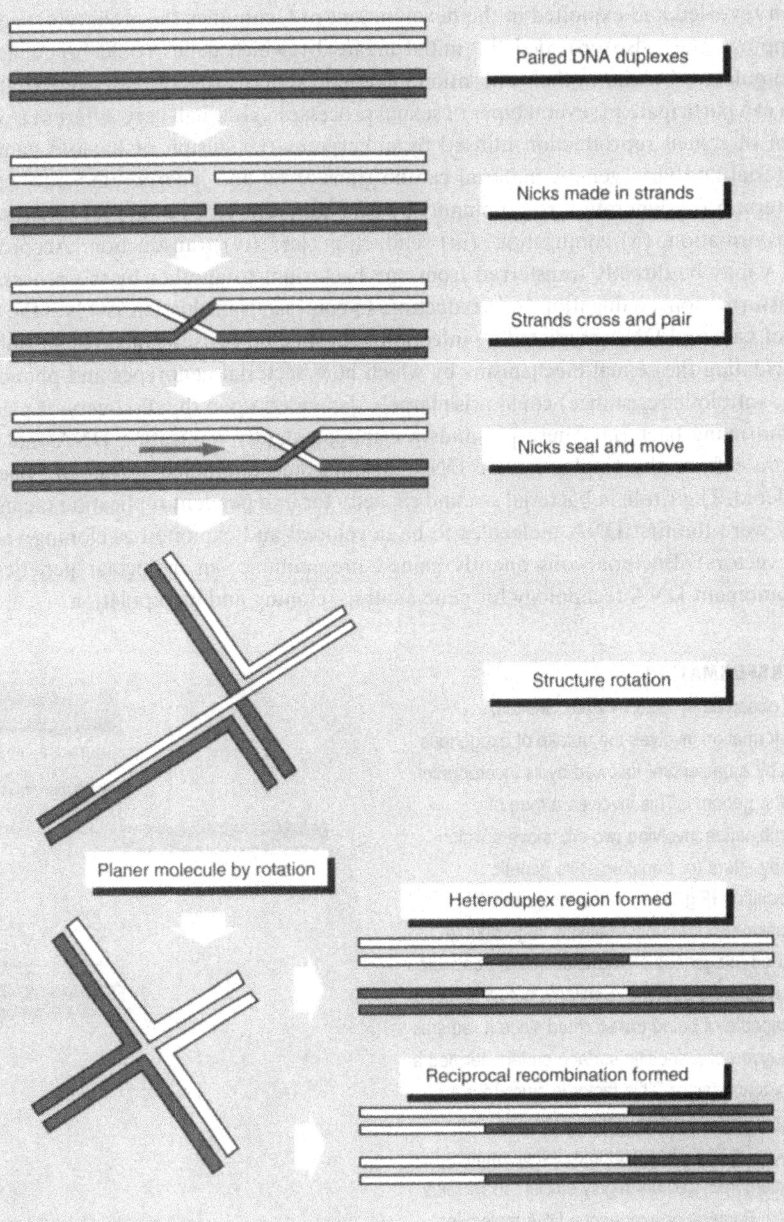

Paired DNA duplexes

Nicks made in strands

Strands cross and pair

Nicks seal and move

Structure rotation

Planer molecule by rotation

Heteroduplex region formed

Reciprocal recombination formed

be revised to account for differences in the exact position of crossing-over. This often occurs during the non-reciprocal incorporation of integrating foreign DNA into bacterial chromosomes where only a single strand of the foreign DNA invades.

This is mediated by RecA and RecBCD, followed by exonuclease degradation of unpaired segments of foreign and bacterial host DNA, and finally re-ligation of the helix.

35. GENE TRANSFER DURING BACTERIAL REPRODUCTION

Exchange of bacterial DNA may occur by the uptake of exogenous DNA (transformation), transfer of specialised F plasmids (conjugation and sexduction) or bacteriophage infection (transduction).

Reflected by their widespread experimental use, bacteria such as the intestine-inhabiting (entero-)bacterium *Escherichia coli* (*E.coli*) have long been favourite organisms for biochemical and genetic study. Many aspects of their molecular biology have therefore been revealed and exploited in the development of techniques for prokaryotic genome mapping. They also provided the initial means by which genes could be cloned and manipulated. Crucial to their utilisation in genetic analysis was the discovery that bacteria can participate in several types of sexual processes. Although very different from the form of sexual reproduction utilised by eukaryotes (i.e. fusion of haploid gametes), bacterial and viral genetic material can be transferred and incorporated into another bacterium (to generate a recombinant) by four different types of sexual processes: (i) transformation, (ii) conjugation, (iii) sexduction, and (iv) transduction. Accordingly, DNA may be directly transferred from one bacterium to another by the processes of transformation, conjugation and sexduction. In contrast, transduction involves the transfer of bacterial DNA mediated by infection with specific viruses called bacteriophages. Elucidating the sexual mechanisms by which new bacterial genotypes and phenotypes (e.g. antibiotic resistance) could arise largely depended upon the discovery of plasmids within many bacteria. These plasmids are independently replicating DNA molecules separate from the single circular DNA chromosome contained within the bacterial nucleoid. Their role in bacterial sex and capacity for independent replication meant that they were the first DNA molecules to be developed and exploited as cloning vehicles (or vectors). Bacteria consequently gained pre-eminence in molecular genetics and recombinant DNA technology for gene analysis, cloning and manipulation.

WHY AN EARLY FAVOURITE?

Several factors were responsible for *E. coli* rapidly becoming a favourite model organism for genetic analysis. Many laboratory strains were available which could be maintained and selectively propagated on simple and cheap media containing, or lacking, basic nutrients. Many generations could also be obtained in a few hours, increasing the chance of detecting spontaneously arising or artificially induced mutants (e.g. by exposure to chemical or physical mutagens). Many of their nutritional requirements were also discovered to be genetically controlled allowing mutants to be detected and selected using media in which a particular metabolite was omitted or added. Mutations could therefore be rapidly linked to a cellular or biochemical function, and since *E. coli* contains a relatively small haploid genome, the mutant gene could be readily isolated and investigated.

TRANSFORMATION

First observed in 1928 by Fred Griffiths, transformation involves the uptake of exogenous DNA by a bacterium, followed by its incorporation into the genome. This involves a form of recombination involving two cross-overs which thereby alters (or transforms) its genetic composition (Fig. 35.1). It has since been demonstrated that only relatively large double-stranded exogenous DNA molecules are capable of transforming bacteria. Similarly, not all bacteria are capable of being transformed since it requires possession of a specific surface protein, termed a competence factor. This factor is crucial for the energy-requiring binding and uptake of exogenous DNA. For example, transformation is inefficient in *E. coli* but highly efficient in *Bacillus subtilis*. Bacteria or exogenous DNA molecules capable of this function are described as 'competent'.

The ability of DNA to transform bacteria was first exploited by Avery and colleagues to definitively establish that DNA was indeed the genetic material. Transformation has since been extremely useful for mapping the genes of competent bacteria (e.g. *B. subtilis*) which inefficiently take up DNA by other means such

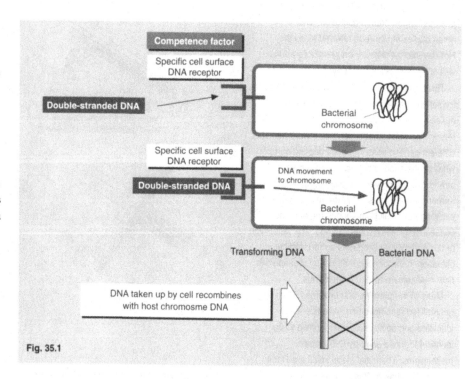

Fig. 35.1

as transduction. Indeed, transformation mapping methods were used to demonstrate that the *B. subtilis* chromosome (like all prokaryotes and some bacteriophages) is a circular molecule.

OAs: 274, 275, 281–4, 323–5, 619.
RAs: 132–4, 174, 175, 216, 217.
SFR: 36–9, 94, 95, 97–9, 101, 102.

CONJUGATION AND THE F PLASMID

Conjugation was first discovered in *E. coli* in 1946 by Joshua Lederberg and Edward Tatum. Their studies involved the use of multiple *E. coli* strains (termed auxotrophs) which lacked the ability to synthesise certain complex molecules such as amino acids or vitamins. In a series of experiments they were able to show that conjugation was a distinctly different form of genetic transfer. This required cell–cell contact and involved the one-way transfer of DNA from a 'male' F+ donor strain (where F represents fertility) to a 'female' F–recipient strain (see Fig. 35.2).

Transfer is now known to involve the linking of the two cell types via a tube-like structure called the sex pilus produced by F+ cells using genes encoded by a specialised plasmid. Termed the F plasmid, in *E. coli* it contains 30 (or more) genes including the transfer or *tra* genes. During conjugation a single-stranded copy of the F plasmid generated by rolling circle replication passes into the F– cell converting it to a F+ cell, whilst the original F+ cell retains a copy of the F plasmid. Integration of the F plasmid into a host cell can generate two forms of additional donor cells, termed Hfr (for high frequency recombination) and F′ (F-prime) cells. Subsequent conjugation of either type of donor cell can result in co-transfer of F plasmid and host genes, albeit by different mechanisms. The latter mechanism is termed sexduction.

SEXDUCTION

Also often referred to as F-duction, sexduction occurs when an F plasmid which has previously been integrated into a bacterial chromosome by recombination following conjugation is subsequently imprecisely excised (along with some host cell DNA) by looping out. This excised plasmid is then referred to as an F′ (F-prime) plasmid, and is capable of transferring to a new host cell by conjugation. Once inside, it can spontaneously integrate into the new host chromosome at the same sites by homologous recombination involving a single cross-over event. Although integration and transfer occur at high frequencies, sexduction is infrequently used in mapping. It is however useful in studying allele interactions in a normally haploid cell, since

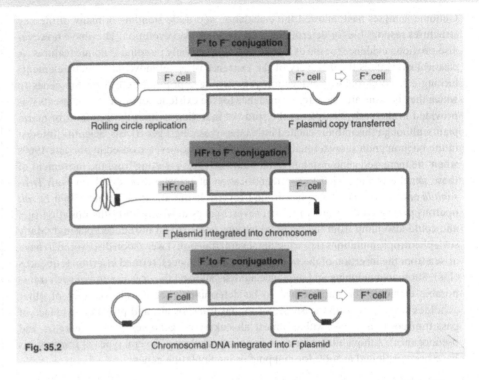

Fig. 35.2

(F⁺ to F⁻ conjugation)
F⁺ cell — F⁺ cell ⇨ F⁺ cell
Rolling circle replication   F plasmid copy transferred

(HFr to F⁻ conjugation)
HFr cell — F⁻ cell
F plasmid integrated into chromosome

(F′ to F⁻ conjugation)
F′ cell — F⁻ cell ⇨ F⁺ cell
Chromosomal DNA integrated into F plasmid

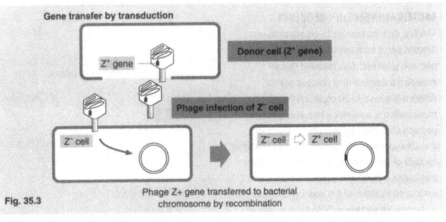

Gene transfer by transduction

Z⁺ gene   Donor cell (Z⁺ gene)

Phage infection of Z⁻ cell

Z⁻ cell   Z⁻ cell ⇨ Z⁺ cell

Fig. 35.3   Phage Z+ gene transferred to bacterial chromosome by recombination

sexduction results in the formation of a partial diploid (termed a merozygote) in which a gene is found both on the chromosome and the F′ plasmid.

TRANSDUCTION

The process of transduction involves the transfer of DNA between bacteria mediated by bacteriophage infection (Fig. 35.3). Accordingly, host bacterial DNA may be incorporated into a bacteriophage particle during its assembly in the infected host. This may occur either by faulty excision of the prophage (termed specialised or restricted transduction) or random inclusion of bacterial DNA into the phage head (termed generalised transduction). The original host bacterial DNA may then be transferred to another bacterium by subsequent infection with these new bacteriophage particles. This may then be recombined within the chromosome of the new host cell by genetic recombination. Generalised transduction has been extremely useful and widely adopted for the mapping of bacterial genes.

36. TRANSPOSABLE GENETIC ELEMENTS: TRANSPOSONS

Several forms of transposon have been found in both prokaryote and eukaryote genomes capable of moving DNA around the genome, either directly or via DNA and RNA intermediates.

Genome analyses have allowed the elucidation and understanding of many of the key structures responsible for determining gene function and evolution. They have however also provided evidence for the existence of several initially puzzling genomic features. A classical example is the discovery of the existence of novel moveable genetic elements lacking a fixed chromosomal location yet exerting a profound effect on the genes to which they became attached. Initial evidence for the existence of moveable elements was provided in the early 1950s from the study of mutations within the genomes of maize plants, although this initially elicited little interest amongst geneticists. Scientific interest in the phenomenon (now termed transposition) was however aroused by the late 1960s when the increased gene instability and mutation rates resulting from the movement of these genetic elements led to the identification of transposition events in both *Drosophila* and *E. coli*. For example, certain irreversible mutations occurring within *E. coli* operons prevented the transcription of several genes downstream of the mutation site, and could also jump from the bacterial 'chromosome' to an introduced plasmid. Many such plieotropic mutations (i.e. affecting several functions) were soon discovered to have arisen from the insertion of the same large DNA sequences, termed insertion sequences (ISs). Subsequent cloning and sequence analysis allowed the organisation of such transposable elements (or transposons) to be determined. Large-scale analyses of other genomes also revealed the existence of similar transposons, and provided evidence of past transposition events within almost all eukaryote genomes (e.g. gene relics and pseudogenes). Athough their function remains speculative, several types of transposons have been exploited as tools for mapping or manipulating genes.

EXPLOITING TRANSPOSONS

The unusual properties of transposons have been exploited in a variety of ways for gene mapping, gene manipulation, and the cloning of previously unknown genes. For example, transposons can be used as molecular tags in cases where no suitable hybridisation probe is available, a process termed transposon tagging. Mutant genes functionally inactivated by the insertion of a transposon may be similarly isolated with the advantage that the clone will contain additional host DNA adjacent to the inserted transposon. Transposons such as the *Drosophila* P element have also been used in a process termed enhancer trapping (a variation of transposon tagging) to specifically isolate enhancer-containing clones. Many transposons including the *Drosophila* P element have also become widely used as vehicles for delivering foreign DNA into cells or organisms. The P element is routinely used to insert foreign DNA into flies and has been invaluable in unravelling the complexity of the genetic mechanisms controlling their development.

BACTERIAL INSERTION SEQUENCES

The first data concerning IS organisation and function came from the analysis of mutated bacterial plasmids. This revealed that an individual bacterium may possess several different ISs in up to 20 copies. Although transposition is a relatively rare event, occurring perhaps only once per thousand cell divisions, ISs are capable of transferring between individual bacteria of the same or related species. When transposition does occur, mutations are introduced by virtue of the addition or deletion of a short target sequence (5–15 bp in length) termed a 'short target site duplication' (STSD).

Comparison of *E. coli* IS elements reveals that they vary in length from 750 to 2500 bp and are largely composed of a single gene encoding a transposase enzyme (an exception being IS1 which contains two genes). This enzyme is responsible for catalysing the transposition event (Fig. 36.1). The gene is flanked at either end by short inverted repeat sequences, varying in length between individual IS elements by between 9 and 41 bp. These are recognised by the transposase enzyme during the initial transposition step and therefore define the sequence to be transposed.

Complex or composite transposons of up to

Bacterial insertion sequence IS10

17bp IR   402 amino acid Transposase   23bp IR

E. coli chromosome   IS10 (1329bp)

Transposon Tn10

Transposase non-functional   Transposase IS10 functional

Tetracycline resistance

IS10L   IS10R

E. coli chromosome   Tn10 (9300bp)   *E. coli* chromosome

Fig. 36.1

10 kb have also been defined consisting of two IS elements located close together such that an internally located genomic DNA, often containing an antibiotic resistance gene, is transposed. Tn3 type transposons have also been described which encode a transposase but which lack flanking IS elements.

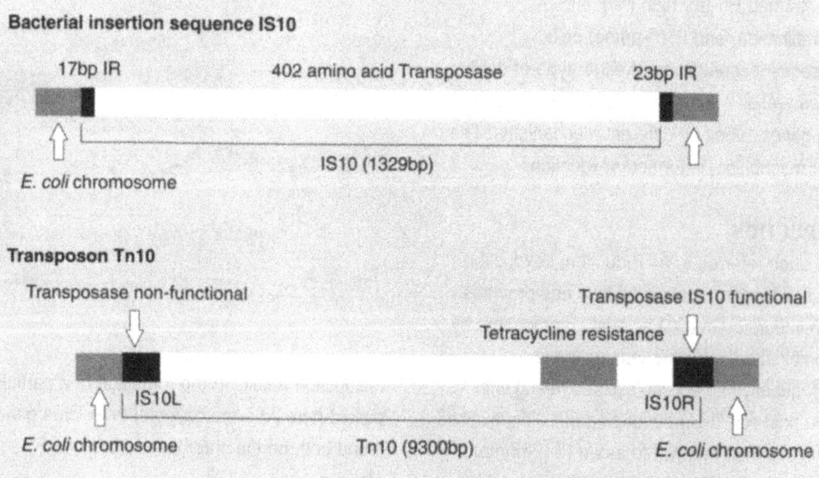

OAs: 82, 285–8, 290–2, 893, 1116.
RAs: 134–40, 195, 253, 386–8.
SFR: 36, 99–102, 106–9, 120, 158–9, 220.

MAJOR TRANSPOSON FAMILIES

Three broad classes of transposable elements have been identified based on their sequence similarities. The first family includes elements such as the *E. coli* IS1 and Tn10 elements, the P elements from *Drosophila*, and the *Ac* and *Ds* elements from maize which move directly as DNA. Interestingly, the bacterial Tn3 transposon is also a family member but moves in a two-step process involving a DNA intermediate.

The second broad family, including yeast *Ty* elements and *copia* elements from *Drosophila*, have a structure with directly duplicated long terminal repeats (LTRs). These encode a reverse transcriptase and allow the movement of the transposon via an RNA intermediate produced by a promoter in the LTR.

The final family, characterised by *L1* from the human genome and the *cin4* element from maize, also encode a reverse transcriptase and consequently also allow movement through an RNA intermediate. These contain a structure that is polyadenylated at the 3′ end of the RNA transcript, and often truncated at the 5′ end. Termed retrosposons because they move information from RNA to DNA against the normal direction of information flow, they may in fact be responsible for the generation of the processed pseudogenes observed in many animal genomes.

Use of DNA intermediates

Study of plasmid to plasmid transposition first revealed evidence that some transposons such as Tn3 do not themselves actually move. Instead, they utilise replication and recombination steps whereby the transposon sequence is copied into the target location. Accordingly, Tn3 transposition begins with the introduction of single stranded nicks in the flanking IR sequences and staggered nicks in the target DNA mediated by transposase (Fig. 36.2). An unstable association between the target DNA and free ends of the transposon is then formed (probably held together by the transposase) allowing DNA polymerase to synthesise new complementary strands. These are subsequently ligated to form a structure termed the cointegrate. This comprises the donor and recipient strands covalently joined, with a transposon located at each junction. The donor transposon is then regenerated by a recombinational process termed resolution, catalysed by an enzyme termed resolvase. The recipient therefore now contains the transposon and an additional short sequence produced by target site duplication (i.e. an STSD).

Use of RNA intermediates

Yeast *Ty* and *Drosophila copia* elements are distinct from bacterial transposons in both organisation and transposition mechanisms. These have a similar organisation to retroviral genomes, encode a reverse transcriptase capable of converting RNA to DNA, and are transcribed into RNA at a very high level. The determination of *Ty* element transcription rates has provided evidence that transposition involves the formation of DNA copies from a spliced donor RNA intermediate. Furthermore, newly transposed elements do not contain the intron sequences which have been artificially engineered into the donor *Ty* element.

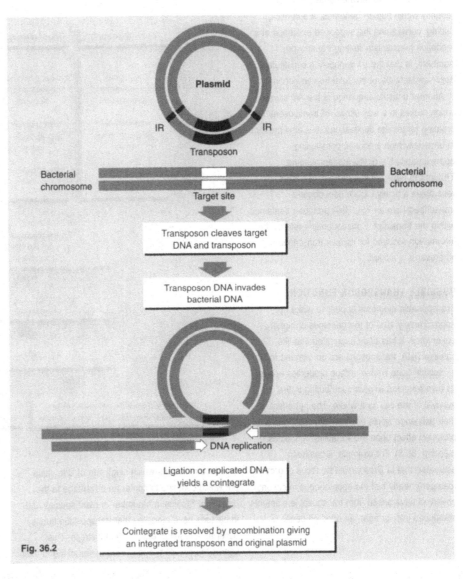

Fig. 36.2

95

The variety and prevalence of transposons, combined with the presence of large introns, suggests they may serve to increase genetic diversity by rearranging exons or altering transcriptional regulation.

BACTERIOPHAGE TRANSPOSONS

The DNA of several bacteriophages (e.g. Mu and D108) undergoes transposition as part of the normal infection cycles. Like many lysogenic phages (e.g. phage λ) their DNA is capable of integrating into the host DNA, but uniquely, they are also capable of moving around the host DNA. Although a contentious subject, their unusual behaviour suggests that transposable phages actually represent transposons which have acquired the genes necessary to form a phage coat and can thus exist outside the host cell.

UNUSUAL HUMAN TRANSPOSONS

High copy numbers of two transposable DNA elements appear to have emerged relatively recently within human genomes. A surprising finding, considering the supposed existence of a feedback mechanism limiting transposon numbers, is that the *L1* transposon constitutes approximately 4% of the total human genome.

Another unusual sequence is the *Alu* element which moves in a way similar to transposons by creating target site duplications. It is also present in unusually high amounts, constituting approximately 5% of the human genome. Thought to have originated from 7SL RNA, it resembles a highly mobile pseudogene transcribed from an 7SL RNA promoter contained within the transcript. Consequently it retains the information required for its own transcription wherever it is moved.

POSSIBLE TRANSPOSON FUNCTIONS

Transposable elements appear to make up approximately 10% of the genomes of higher eukaryotes. It has been suggested that like satellite DNA, transposons are an extreme form of 'selfish' gene whose unique properties ensure its own retention without contributing to the survival of the cell as a whole. The net effect of their existence would however appear to have a profound effect upon the variability of species (see Fig. 36.3). For example, it has been estimated that in *Drosophila* (the focus of greatest research) nearly half the spontaneous mutations observed have arisen from transposable elements introduced into, or near, an affected gene.

TRANSPOSITION AND BIODIVERSITY

A unique feature of transposable elements is the way in which they are relatively inactive at a single chromosomal region for long periods, after which a period of intense movement occurs. Such rapid changes involving the simultaneous activation of more than one transposon is termed a transposition burst. Transposition bursts were first discovered in maize genomes from plants subjected to repeated chromosomal breakage (a phenomenon termed hybrid dysgenesis). They were subsequently identified in *Drosophila* flies following crosses between certain strains. It is now believed that transposition bursts act to introduce new traits which singly are of little value but which combine to confer an advantage to the organism. Studies in a number of plant species and in bacteria have indicated that transposition bursts occur as a result of environmental stress. This serves to produce progeny cells better suited to their surroundings compared to parent cells (see Fig. 36.3). Indeed, the majority of antibiotic resistance transferred amongst bacteria and phages is thought to arise from transposition events involving complex transposons containing the antibiotic resistance genes.

Transposons are also important in determining the mating type of the yeast *Saccharomyces cerevisiae* with transpositions occurring at every cell division. Furthermore, certain parasites (e.g. *Trypanosoma*) employ infrequent transposition events to after the expression of antigenic surface glycoproteins in order to evade immune recognition.

OAs: 82, 289, 293–5, 690, 795, 893.
RAs: 134–42, 253, 386–8, 405.
SFR: 36, 99–102, 106–9, 120, 158–9, 220.

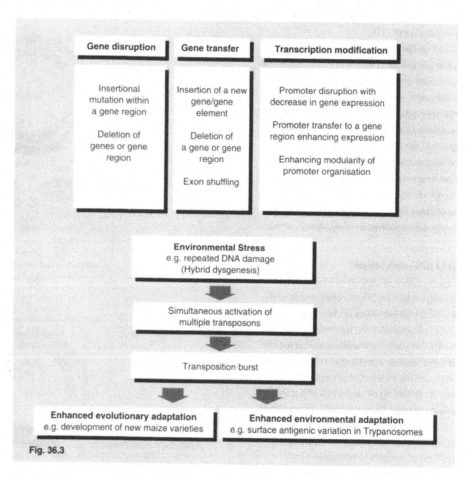

Fig. 36.3

EXON SHUFFLING

Transposons are thought to be involved in a particular form of molecular evolution termed exon shuffling (Fig. 36.4). Consequently, when two transposable elements recognised by the same site-specific transposase integrate close together (i.e. forming a complex transposon) the DNA between them may become a substrate for transposition by the transposase. This is thought to provide an effective pathway for the duplication of exons and their movement to a new chromosomal region. Indeed, the relatively small size of exons in comparison to introns would usually mean that a newly created exon would probably insert into a pre-existing intron. This concept is best confirmed by the presence of gene-like elements in a number of intron sequences, and the presence of repetitive elements and processed pseudogenes. These are all believed to represent evolutionary remnants of transposition events.

GENE REGULATORY FUNCTIONS

The rearrangements and mutations brought about by transposable elements often affect the regulation of nearby genes. This manifests itself in various ways during an organism's development (e.g. during morphogenesis). In some cases it may also confer some advantage upon the organism. Accordingly, insertion of new regulatory elements upstream of a structural gene may result in the movement of a DNA-binding region used for *trans*-activation into the vicinity of the gene (e.g. into the region of a proto-oncogene). Since *trans*-activation may be regulated at a distance (as exemplified by enhancer sequences), the transposition of genetic elements usually (but not exclusively) affects transcriptional regulation (Fig. 36.5). Transposition does not generally alter the nature of the encoded protein itself since the exons encoding the protein domains are relatively small.

The use of transposable elements in moving enhancer sequences around the genome may in fact allow optimised transcriptional regulation of important or emerging genes. It is also possible that the vast excess of non-coding DNA observed in most eukaryotic genomes confers a genetic advantage by increasing the opportunity for transposons to alter transcription rates without disrupting protein or RNA coding sequences.

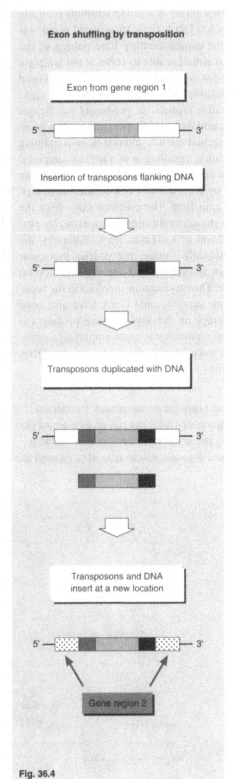

Exon shuffling by transposition

Exon from gene region 1

Insertion of transposons flanking DNA

Transposons duplicated with DNA

Transposons and DNA insert at a new location

Gene region 2

Fig. 36.4

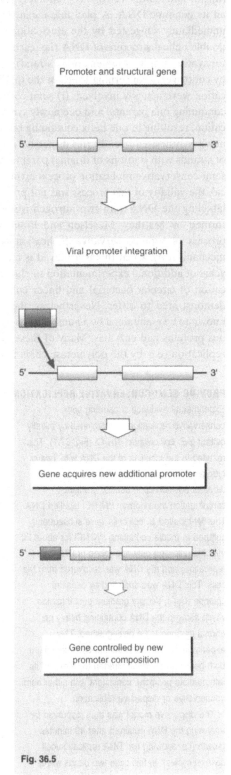

Promoter and structural gene

Viral promoter integration

Gene acquires new additional promoter

Gene controlled by new promoter composition

Fig. 36.5

97

37. *IN VIVO* DNA REPLICATION

New DNA strands are generated by semi-conservative replication of parental strands, from defined non-base-paired regions (replication origins and forks) created by the action of many proteins/enzymes.

Mitotic and meiotic cell division both rely upon the ability of a cell to faithfully replicate all its genomic DNA. A plausible means by which DNA replication could occur was immediately suggested by the elucidation of the complementary base pairing of the double helical structure of DNA (i.e. each DNA strand is able to serve as the template for synthesis of its complementary strand). This discovery was however rapidly followed by controversy concerning which of the three possible overall strategies for DNA replication were actually involved: (i) semi-conservative replication, producing two helices containing one parental and one newly synthesised DNA strand; (ii) conservative replication, resulting in one helix containing both parental strands and one helix containing newly synthesised strands; (iii) dispersive replication, resulting in two helices composed of strands with portions of original parental and newly synthesised DNA. Although the semi-conservative replication process favoured by Watson and Crick seemed most logical, the validity of this process was not proven until 1958. The evidence came from the labelling the DNA with two nitrogen isotopes during replication in experiments performed by Matthew Meselson and Franklin Stahl at CalTech, USA. Although the process of semi-conservative replication is elegantly simple, the precise molecular mechanism by which this is achieved is far from straightforward and required several years of additional experimentation to elucidate. The mechanisms involved in the replication of circular bacterial and linear eukaryotic chromosomal DNA have also been demonstrated to differ. Nevertheless, the accuracy or 'fidelity' of these processes is known to be maintained by a number of complex biochemical systems employing numerous proteins and enzymes. Many of these have since been exploited for *in vitro* DNA replication (e.g. by the polymerase chain reaction).

IN VITRO DNA REPLICATION

Many molecular biological or recombinant DNA techniques for gene isolation, analysis and manipulation utilise some form of limited *in vitro* DNA copying. This is reliant upon the use of many of the same enzymes employed *in vivo* although the principles are significantly simplified. For example, the polymerase chain reaction (PCR) can efficiently generate many millions of copies of a defined DNA sequence. This technique is dependent upon the use of a DNA polymerase to extend the 3' ends of synthetic priming oligonucleotides annealed to the target DNA (artificially maintained in a linear single stranded conformation by high temperature cycling). Similar primer-directed copying of DNA is used in dideoxy nucleotide sequencing. These methods are however limited to the copying of relatively short sequences, and do not generally benefit from the same *in vivo* error correction systems.

OAs: 27–9, 296–9. **RAs:** 3–5, 143, 144, 146, 148–50. **SFR:** 36–9, 101–5, 120.

PROVING SEMI-CONSERVATIVE REPLICATION

Experimental evidence supporting semi-conservative replication was provided by initially culturing *E. coli* cells in $^{15}NH_4Cl$ (Fig. 37.1). This resulted in the labelling of the DNA with 'heavy' nitrogen. This consequently migrated to a different position upon density gradient centrifugation than normal $^{14}NH_4Cl$-labelled DNA. The ^{15}N-labelled *E. coli* cells were subsequently cultured in media containing $^{14}NH_4Cl$ for about 20 minutes (allowing only one round of DNA replication) and the DNA was recovered from the cells. This DNA was analysed by caesium chloride (CsCl) density gradient centrifugation which allowed the DNA containing heavy or normal nitrogen to be distinguished. This experiment revealed that all the DNA contained both heavy and normal nitrogen (by virtue of its intermediate position), consistent with either semi-conservative or dispersive replication.

The dispersive model was then disproved by analysing the DNA obtained after 40 minutes growth (i.e. allowing two DNA replication/cell division cycles). In this case two bands were present, consistent with half the DNA containing only normal nitrogen and half the DNA containing

both heavy and normal nitrogen. This classical experimental result was now wholly in accord with the semi-conservative model of DNA replication, since dispersive replication would be expected to

yield only 'intermediate' DNA molecules containing both heavy and normal nitrogen.

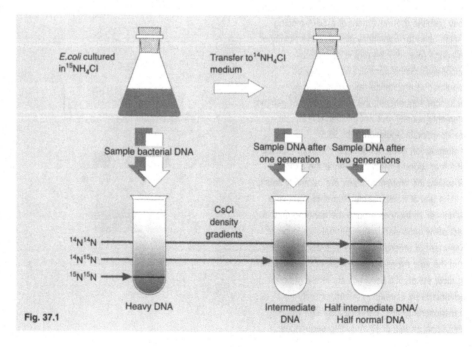

Fig. 37.1

INITIAL STEPS IN DNA REPLICATION

(1) *Replication origins.* Similar to the situation during DNA transcription (but unlike the *in vitro* DNA replication occurring in the PCR), only limited regions of the DNA of both prokaryotes and eukaryotes exist in a non-base-paired form at any given time during *in vivo* replication. Such non-paired regions are termed replication origins (Fig. 37.2). It is from these that new DNA strands are synthesised in both directions (i.e. bi-directionally) whilst un-zippering the DNA helix. The rate of DNA synthesis in eukaryotes is slower than that of bacteria such as *E. coli*, being between 2 and 4000 bp per minute. The large size of eukaryotic DNA molecules also means that replication proceeds from multiple replication origins per DNA molecule (one every 40 kb in yeast and one every 150 kb in mammals). These form what are referred to as replication bubbles. In contrast, replication of prokaryotic DNA involves the formation of only one replication origin per chromosome.

(2) *Replication forks.* The point at which the base pairs between parental strands are broken and nucleotides are added to the newly synthesised strand is termed the replication fork. The breakage of the base pairs between parental strands occurs via an ATP-mediated process involving the action of an enzyme called DNA helicase. It also migrates along the parental DNA molecule with the replication fork. In *E. coli* this process is also accompanied by the attachment of single-stranded binding proteins (SSBs) which prevent reassociation (reannealing) of the two parental strands separated by the DNA helicase.

(3) *Unwinding DNA helices.* Genomic DNA *in vivo* adopts a tertiary conformation composed of helices wound into supercoiled structures. Unwinding of the DNA during replication therefore requires the involvement of members of a further class of enzyme termed DNA topoisomerases. These enzymes mediate unwinding of the DNA in advance of the replication fork by forming transient breaks in the phosphate–sugar backbone of one (type I enzymes) or both (type II enzymes such as *E. coli* DNA gyrase) parental strands. These enzymes are also involved in the reverse processes, i.e. during the formation of DNA supercoils.

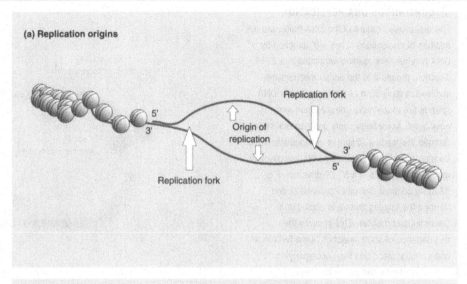

(a) **Replication origins**

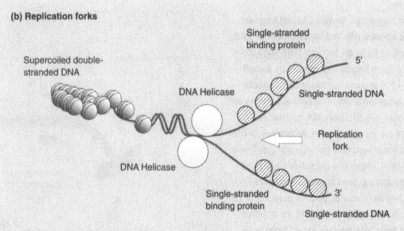

(b) **Replication forks**

(c) **Unwinding DNA helices**

Fig. 37.2

Only one parental (leading) strand is copied by continuous RNA-primed DNA synthesis whilst the other (lagging) strand is copied discontinuously, involving the linkage of small Okazaki fragments.

DISCONTINUOUS DNA REPLICATION

The anti-parallel nature of the DNA helix, and the addition of nucleotides in a 5′→3′ direction by DNA polymerases reading templates in a 3′→5′ direction, means that the actual mechanisms involved in replication of the two parental DNA strands are much more complex than first envisaged. Accordingly, only one parental strand (termed the leading strand) is continuously copied from an initial RNA 'primer' by DNA polymerase-mediated synthesis in a 5′→3′ direction (Fig. 37.3). In contrast, the other parental strand (termed the lagging strand) is copied in a discontinuous manner. This involves the maintenance of short stretches of the helix in an open configuration and the accompanying synthesis of new short complementary sequences.

These sequences, termed Okazaki fragments after the scientist who first isolated them in 1968, are generally between 100 and 1000 nucleotides in length. Each fragment is initiated by a short priming RNA sequence produced by a specific RNA polymerase enzyme. In both eukaryotes and prokaryotes, two different DNA polymerase enzymes are then involved in the process of DNA polynucleotide synthesis from the primers. These priming RNA sequences are subsequently degraded and replaced by a new portion of DNA polynucleotide (mediated by DNA polymerase I in *E. coli*). Following their synthesis, the adjacent Okazaki fragments are then joined together by the catalytic action of an enzyme termed DNA ligase, thereby completing the synthesis of the lagging strand.

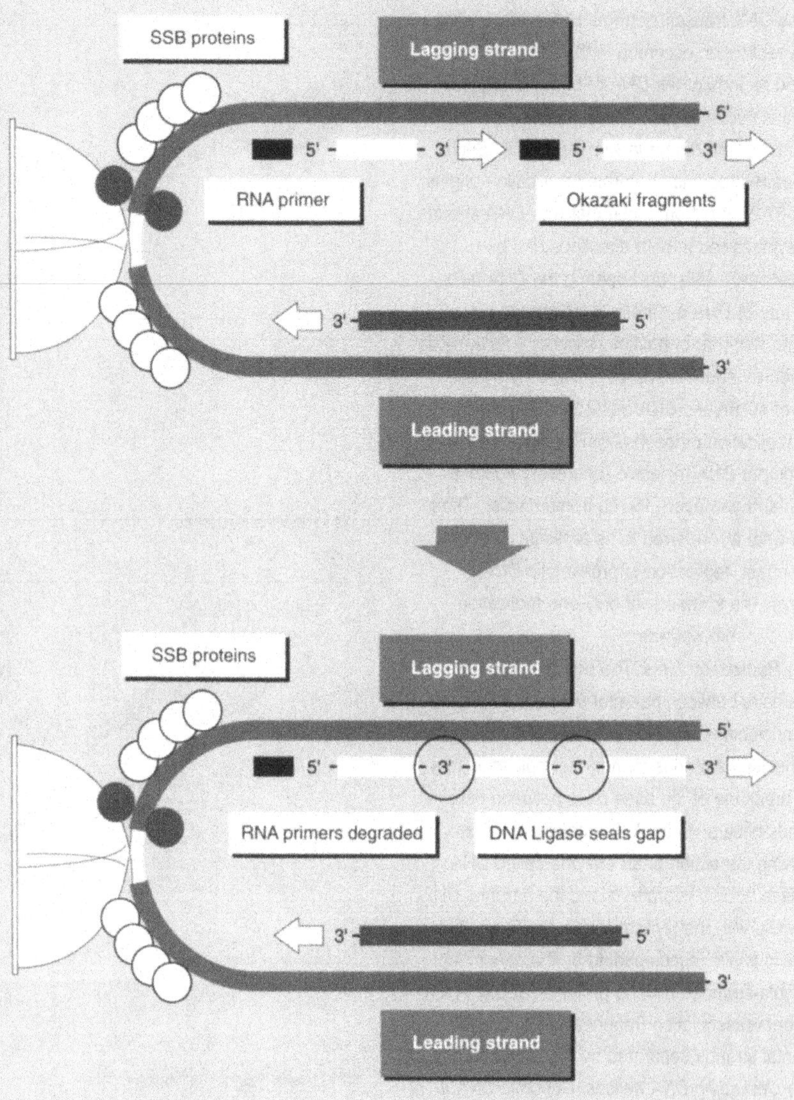

FORMING NEW POLYNUCLEOTIDES

New polynucleotides are formed from parental templates by DNA polymerase enzymes which catalyse the template-directed addition of new deoxyribonucleotides in a 5′→3′ direction. Several different types of DNA polymerases have been isolated and characterised from both prokaryotic and eukaryotic cells. Each have differing degrees of involvement in DNA replication and repair. *E. coli* DNA polymerase I and III are both involved in DNA replication (the latter serving as the primary replicating enzyme), and only two of the five mammalian DNA polymerases (termed α and δ) are utilised in replication. Both DNA polymerase I

and III possess 3′→5′ exonuclease activity (referred to as proof-reading) which allows them to remove and replace incorrectly incorporated nucleotides (e.g. mismatches). In contrast, only DNA polymerase I has the 5′→3′ exonuclease activity required to remove the RNA primers involved in the formation of Okazaki fragments during the synthesis of the lagging strand. Eukaryotic DNA polymerases α and δ do not

possess either 3′→5′ or 5′→3′ exonuclease activity and therefore are not able to remove the RNA primers or mismatched nucleotides from the lagging strand.

OAs: 27–9, 296–301, 336–8.
RAs: 143–50, 184, 214–18.
SFR: 36–9, 101–5, 116, 120.

Fig. 37.3

4. Building blocks of nucleic acids.
5. Formation of the DNA double helix.
6. Packaging of DNA within cells.
7. Chromatin and gene activity.
9. Organisation of genomes into multiple chromosomes.
10. Distribution of nucleic acids in cells.
12. Anatomy of eukaryote chromosomes.
16. Types of mutations and their effects.
18. DNA repair mechanisms.
23. Transcription.
41. Bacteriophage genetics.
42. Enzymes in molecular biology.

PRIMING DNA SYNTHESIS

As illustrated by the Okazaki fragments, DNA polymerases (unlike RNA polymerases) require a short polynucleotide priming sequence in order to begin synthesis. In other words, they can only add nucleotides to the free 3′ end of an existing polynucleotide. The initiation of DNA replication *in vivo* therefore requires the presence of a short (usually five nucleotide) RNA primer sequence annealed to the unwound parental DNA in order to provide a 3′ OH group.

Best characterised in *E. coli*, the process of RNA primer synthesis occurs at specific points along the DNA by the initial addition of at least six prepriming proteins (Fig. 37.4). These are then converted into a complex termed a primosome by the addition of the enzyme primase. On the leading strand, RNA synthesis occurs only at the replication fork. In contrast, the primosome continues to travel along the lagging strand making new RNA primers in a 3′→5′ direction at sites of defined sequence (e.g. TGGT and GGGT in the case of bacteriophage T7).

WAYS TO REPLICATE CIRCULAR DNA

The general processes of DNA replication are very similar between prokaryotes and eukaryotes. Several different strategies are however used to replicate linear (e.g. eukaryotic chromosomes) and circular (e.g. mitochondrial and bacterial) DNA molecules. Replication of a circular DNA molecule can occur by two means. The first strategy employed by many bacterial and phage molecules involves a single replication origin producing two replication forks travelling in opposite directions. This produces what is termed a θ intermediate (see Fig. 37.5).

The second method, also used in mitochondrial DNA replication, is referred to as the rolling circle method (see Fig. 37.6). This initially results in the production of repeated copies of the original circular DNA molecule in the form of linear DNA molecules termed concatamers. This method also involves the discontinuous replication of the lagging strand as small Okazaki fragments, but does not however require RNA priming. Instead, replication is initiated by the breaking of one of the strands by an endogenous endonuclease, thereby generating a free 5′ PO_4 and 3′ OH terminus. The newly synthesised leading strand is also not dissociated from the parental template. Instead, one parental strand is displaced to act as the template for lagging strand synthesis. The process of producing new circular molecules is completed by cutting the linear concatamers into appropriately sized fragments, and the subsequent joining of the free ends by the action of a DNA ligase.

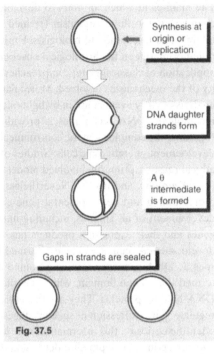

Fig. 37.4

Fig. 37.5

Fig. 37.6

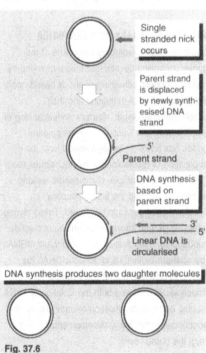

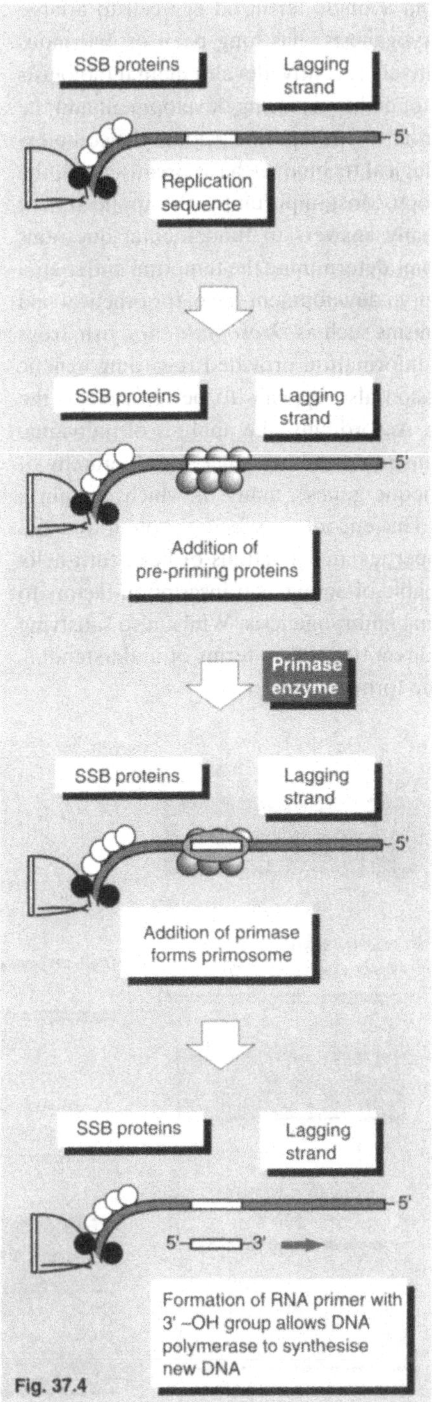

38. GENETIC CONTROL OF DEVELOPMENT

Genetic and molecular biological strategies have allowed the identification, isolation and characterisation of many genes determining embryo development, particularly in *Drosophila* flies.

The manner in which an embryo develops from a single fertilized egg cell to a large complex multicellular organism (termed embryogenesis) has long been of enormous interest and fascination to biologists. Until relatively recently, developmental biologists relied primarily on the histological observation of embryos during development and the application of classical genetic approaches in order to attempt to elucidate the complexity of the mechanisms involved. Molecular biological techniques have greatly aided the analysis of early events unseen by light microscopy. Most importantly, the application of recombinant DNA techniques is providing many answers to fundamental questions regarding the complex genetic control mechanisms determining the temporal and spatial development of embryonic cells. Studies of embryo development have, for practical and ethical reasons, primarily involved model organisms such as *Drosophila* flies, fish, frogs (e.g. *Xenopus*), and insects. Nevertheless, the information provided regarding genetic programming and developmental gene expression also appears to be relevant to the development of all animals, including humans. Accordingly, the analysis of individual genes and their expression products has demonstrated the existence of a hierarchy of developmental regulatory genes (termed homeotic genes), many of which contain a specific DNA sequence termed a homeo box. This encodes a specific protein domain, termed the homeo domain, which has the properties and functions of helix-turn-helix DNA binding proteins. They are therefore capable of acting as transcription factors to regulate gene expression at specific stages during embryogenesis. Whilst also satisfying scientific curiosity, this information may have direct benefits in terms of understanding and correcting developmental defects, and some forms of cancer.

ETHICS AND EMBRYO EXPERIMENTATION

Hitherto, developmental biologists have concentrated their efforts on the analysis of easily accessible and available non-mammalian sources of embryos which do not pose ethical problems. However, the increased availability of mammalian, and especially human, embryos for research purposes (resulting from improved *in vitro* fertilisation techniques) has aroused considerable concern. This revolves around the sanctity of human life and the central issue of when life begins; immediately upon conception or at a defined stage during fetal development? These concerns must be squarely faced and appropriate guidelines established for the use of human embryos. This decision-making process requires the involvement of all members of society since any research must ultimately provide a positive benefit to humankind, and not be seen to merely satisfy scientific curiosity.

MOLECULAR ANALYSIS STRATEGIES

The first insights regarding the nature of the genes controlling development were provided by classical genetic screening methods based upon the identification of mutations and their associated phenotypes. Current understanding of the structure and function of developmental genes has however largely derived from the application of recombinant DNA techniques (see Fig. 38.1). For example, chromosome walking techniques allowed the first *Drosophila* developmental genes to be cloned. These cloned genes were then characterised using nucleotide sequencing, hybridisation with embryonic mRNAs (by northern blotting) and genomic DNAs (by Southern blotting), and *in situ* hybridisation. These studies allowed both the isolation of cDNA clones, and the identification of particular gene sequences within many developmental genes (e.g. the homeo box).

The most effective recombinant DNA strategies have been those involving the manipulation of cloned genes (e.g. by *in vitro* mutagenesis) and the analysis of the effects following reinsertion of the modified gene into the embryo. Similarly, the modification and analysis of expressed cDNA clones has been extremely useful in defining the

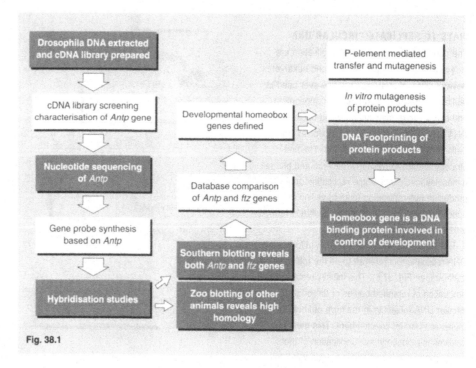

Fig. 38.1

functions of specific developmental genes. These strategies have been particularly extensively applied to the determination of the genetic control of development in *Drosophila*.

OAs: 295, 302–7, 795, 823, 828, 829. RAs: 151–9, 163, 360. SFR: 85, 110–15, 120, 158, 227.

DROSOPHILA DEVELOPMENT GENES

The *Drosophila* life cycle is complex. It involves metamorphosis from an early embryological state to a mature adult via three larval stages and a pupal stage, taking in all about 10 days. Over 50 developmental genes have now been identified which are involved in specifying the body plan of the adult fly. These may be divided into three main classes of genes: maternal-effect genes, segmentation genes, and homeotic genes (see Tables 38.1 and 38.2). It is the sequential or hierarchical expression of these genes during embryogenesis which determines the ultimate position and fate of cells within the larval imaginal discs, each pair of which are responsible for forming specific structures in the adult such as the head, wings, legs, thorax, and antennae.

Maternal-effect genes

Maternal-effect genes control the very earliest stages of development and determine the organisation of cells within the egg. As the name suggests, these genes are exclusively transcribed in the mother, and exert their effect following transmission of their mRNA to the egg whilst still in the mother's ovary. This stored mRNA forms concentration gradients which serve as signals to activate regulatory genes within individual cell nuclei. It is these signals which determine embryo polarity (i.e. the front, back, top and bottom of the fly). For example, mutations within the bicoid (*bcd*) gene results in an embryo with two tails and no head or thorax. This is a consequence of the failure of the mutant mRNAs to provide the correct signals, since normal maternal-effect gene mRNA injected into *bcd* mutant embryos results in the production of a normal fly.

Segmentation genes

Segmentation genes are transcribed in the zygote where they extend the pattern of development beyond that established by the maternal-effect genes. This leads to the segmentation of the embryo. Normally 15 discrete segments are formed, each destined to become a different part of the adult fly body.

At least 24 segmentation genes have now been described which fall into three categories, and which are transcribed in a defined sequence. The first genes to act are the 'gap' genes which

ensure that the correct number of segments are formed in the correct position (see Fig. 38.2). Mutations within these genes lead to the loss of one or more embryo segments (e.g. *knirps*). The remaining two classes, the 'pair-rule' and 'segment polarity' genes, then sequentially act on all segments. Consequently, mutations in 'pair-rule' genes delete every other segment (e.g. *fushi tarazu*), and mutations in 'segment polarity' genes produce segments lacking one part but duplicating the remainder as a mirror image (e.g. *engrailed*).

Homeotic genes

Homeotic genes were the first genes ever to be isolated by chromosome walking and jumping techniques. This was based upon the initial identification of chromosomal inversion mutants in which the positions of specific bands on polytene chromosomes were changed. These genes act to specify the fate of each segment in the final fly. They are located in two large gene complexes: the antennapedia complex (ANT-C) and the bithorax complex (BX-C). Mutations within homeotic genes often lead to the exchange of indentity of segments (e.g. the formation of legs in place of the antennae).

Table 38.1

Gene or locus	Active area of gene	Effect and function
Maternal effect genes	Maternal cells and tissues	Pattern formation initiated in regions of the embryo
Gap genes	Developing embryo	Determine segmentation and influences pair rule genes
Pair rule genes	Developing embryo	May affect segmentation and influence activity of polarity homeotic genes
Segment polarity genes	Developing embryo	Segmentation affected if gene is mutated, influences homeotic genes
Homeotic genes	Developing embryo	Determines segments and structures within segments

Table 38.2

Gap genes	Pair rule genes	Segmentation polarity genes
Giant (*gt*)	Barrel (*brr*)	Armadillo (*arm*)
Hunchback (*hb*)	Even-skipped (*eve*)	Cubitus interruptus (*ci*)
Knirps (*kni*)	Fushi tarazu (*ftz*)	Engrailed (*en*)
Kruppel (*kr*)	Hairy (*h*)	Fused (*fu*)
	Odd-paired (*opa*)	Hedgehog (*hh*)
	Odd-skipped (*odd*)	Paxh (*pat*)
	Paired (*prd*)	Wingless (*wg*)
	Runt (*run*)	

Fig. 38.2

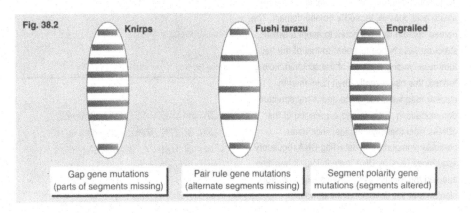

| Gap gene mutations (parts of segments missing) | Pair rule gene mutations (alternate segments missing) | Segment polarity gene mutations (segments altered) |

Embryo development in *Drosophila*, and a diverse range of organisms, is regulated by temporal expression of genes containing conserved sequences encoding transcription factor DNA-binding protein domains.

GENETIC CONTROL MECHANISMS

Recombinant DNA technologies have identified many *Drosophila* development genes. More importantly, they have clarified the understanding of the complex genetic control mechanisms operating during development. For example, the use of cloned developmental genes as probes has revealed that each gene is expressed at distinct times and locations. This cascade of gene activation occurs in the early stages of embryo development to ensure that cells progressively become more restricted in the way they develop. This increasingly subdivides the embryo into defined regions each of which is committed to forming the correct structure in the correct location within the adult fly (Fig. 38.3).

Comparing the nucleotide sequences of genomic and cDNA clones of developmental genes also suggests how expression of these genes is regulated in a complex and interactive manner. Consequently most, if not all, developmental genes contain sequences which encode for transcription factors which bind to DNA regulatory sequences to modulate transcription (i.e. as on/off switches). Accordingly, many segmentation genes have been demonstrated to encode DNA-binding proteins containing a 'zinc finger' binding domain. Furthermore, all homeotic genes, and some segmentation genes, contain a specific sequence termed the homeo box which encodes a DNA-binding domain (see Fig. 38.4).

HOMEO BOXES AND HOMEO DOMAINS

Homeo boxes were first defined in *Drosophila* homeotic genes as 180 bp sequences which were remarkably conserved across a diverse range of organisms (Table 38.3). Their deduced amino acid sequences revealed that they encode a 60 amino acid domain, termed a homeo domain. The homeo domain was predicted to adopt a similar structure (and hence function) to that of the helix-turn-helix binding domains of transcription factors. Indeed, this has recently been confirmed by nuclear magnetic resonance and X-ray structure determinations. Cloning and expression of the cDNAs from these genes has since been particularly important in detecting DNA regulatory sequences (e.g. by DNA footprinting). It has also enabled the identification of the specific control elements of the development pathway.

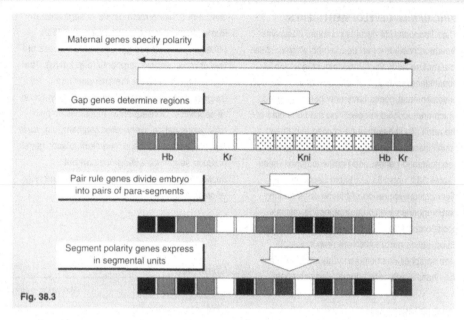

Fig. 38.3

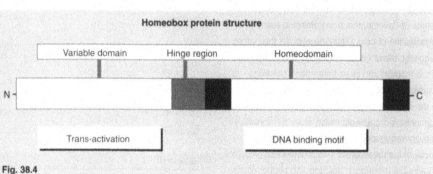

Fig. 38.4

Table 38.3

Homeodomain consensus sequence

Species	Consensus amino acid sequence of homeodomain
	RKRGRTTYTRYQTLELEKEFHFNRYLTRRRRIEIAHALCLTERQIKIWFQNRRMKWKKEN
Drosophila ftz	S T Q ⋯ I ⋯ D N S S ⋯ S DR
Drosophila Ubx	R Q ⋯ M Y ⋯ L I
Drosophila Antp	Q ⋯ T H
Mouse Hox 2.3	Q ⋯ Y ⋯ T
Frog MM3	N ⋯ V
Frog AC1	R QIYS ⋯ N ⋯ R

OAs: 44–54, 273, 302, 308–22, 828–9, 376. **RAs:** 12–14, 94–100, 126–8, 160–71, 360. **SFR:** 85, 110–15, 120, 158.

Vertebrate homeo box genes

As originally anticipated, the study of *Drosophila* developmental genes has provided information relevant to our understanding and further study of developmental processes in a diverse range of organisms, from sea urchins to humans. Cross-hybridisation experiments using cloned *Drosophila* genes as probes have identified remarkably conserved homeo box motifs in numerous mammals. These probes have also been used to retrieve developmental genes from genomic libraries thereby making them available for analysis and manipulation. Termed *HOX* genes in vertebrates, they exhibit remarkable similarities in terms of sequence, genomic organisation, and function. In fact, the order of *Drosophila* homeotic gene clusters perfectly matches that of the equivalent mouse *HOX* genes (see Fig. 38.8), although vertebrates have four rather than two homeotic gene complexes. Furthermore, the order of the genes along the chromosome reflects the order in which the genes are expressed to determine the segments they control (i.e. from head to tail, anterior to posterior). Similarly, *in situ* hybridisation has shown that the pattern of *HOX* gene expression within mouse embryos closely corresponds to the pattern of expression observed in *Drosophila* embryos.

The theory that homeotic genes represent generalised systems for controlling segmentation and segment identity has recently been reinforced by transgenic studies. Transgenic mice created by the insertion of *Drosophila*, human or other animal homeotic genes into embryos have developed quite normally, suggesting that the same signalling mechanism is involved. Furthermore, *HOX* genes have been introduced into *Drosophila* where they produce the same effects as the normal fly gene counterparts.

Are homeo box genes universal?

Homeo box-containing genes have also been found in the unsegmented roundworm *Caenorhabditis elegans*. Their differing sequence did however prevent their detection using cloned *Drosophila* genes as probes. The cloning of *C. elegans* homeo box genes is nevertheless expected to allow many homeo box genes to be detected in a much wider range of organisms.

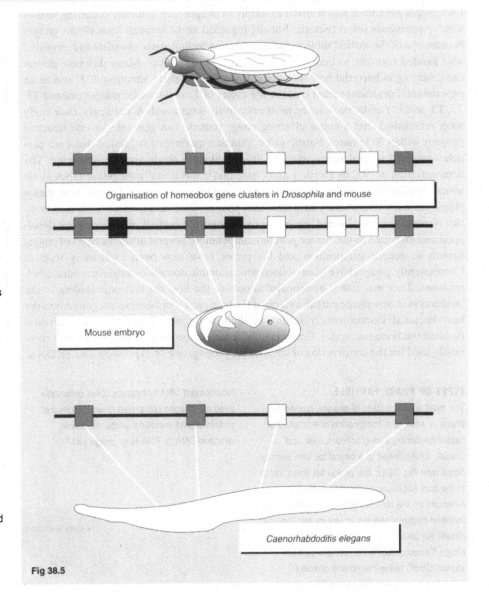

Organisation of homeobox gene clusters in *Drosophila* and mouse

Mouse embryo

Caenorhabdoditis elegans

Fig 38.5

Importantly, these studies are also shedding light on the genetic control of programmed cell death. This phenomenon is often observed during development in many organisms, and also frequently occurs in defined cell populations in adults (termed apoptosis). Study of developmental mutations in plants such as *Arabidopsis* is revealing that they also utilise homeotic-like genes encoding transcription factors. They also contain many sequences which correspond to those found within animal homeotic genes.

Homeo box gene evolution

The structural, organisational and functional similarity of homeotic genes in a wide range of organisms suggests that they may be very ancient genes which were crucial to the generation of multicellular organisms. Analysis of the differences between these 'master control' genes in different organisms (sometimes referred to as 'molecular fossils') may provide insights into evolutionary processes spanning millions of years.

39. THE NATURAL BIOLOGY OF BACTERIOPHAGES

Bacteriophages (or simply phages) are bacterial viruses which infect and replicate within bacteria by lysing the host cell to release new phage particles or by integrating their DNA into the host chromosome.

Bacteriophages (often also referred to simply as phages) are naturally occurring viruses which parasitically infect bacteria. Initially regarded as the smallest form of life, phages became one of the earliest subjects for investigation by biologists, chemists and physicists who banded together to become known as the 'phage group'. Many different phages exist, varying in bacterial host specificity. However, the early adoption of *E. coli* as an experimental organism meant that the life cycles of *E. coli*-specific phages (named T1, T2, T3, and λ) rapidly became the most extensively scrutinised. Accordingly, their study soon established that a single infecting phage particle can generate several hundred progeny within 20 minutes. Furthermore, genetic experiments suggested that each particle contained several genes arranged linearly along a single viral chromosome. The demonstration in 1952 by Hershey and Chase that it was actually the phage DNA alone which entered the bacterium during infection, and therefore induced production of new phage particles, immediately suggested that phages could be exploited as cloning vehicles (termed vectors). Indeed, phages were amongst the earliest organisms to be developed and exploited as vectors for genetic manipulation. Several different types of phage, varying in genetic composition and life cycles, have now been extensively studied. Consequently, phages have been isolated which contain double- or single-stranded DNA genomes. They may either be propagated outside the host chromosome leading to the production of new phage particles (termed the lytic cycle), or become integrated into the host bacterial chromosome resulting in their replication during bacterial division (termed the lysogenic cycle). Consequently, many phage-derived vectors are still commonly used for the construction of libraries for propagating or expressing foreign DNA.

A GLOWING EXAMPLE

The diversity of phages having varying host specificity is being inventively exploited to provide a sensitive and rapid means of detecting specific pathogenic bacteria via bioluminescence. Phages have been genetically engineered to contain a luciferase enzyme gene such that, following phage infection, live bacteria 'glow' as the expressed enzyme generates bioluminescence. Importantly, dead or dying bacteria do not 'glow'. Consequently only potentially harmful, proliferating bacteria are detected. The ability to sensitively and rapidly discriminate between individual bacterial strains potentially allows phage bioluminescent detection technology to be used for clinical diagnosis, and monitoring microbial contaminations in a wide variety of settings. It has also been suggested that luciferase-containing canine enterobacteria-specific phages could be created to make dog faeces glow in the dark and therefore be easily avoided!

OAs: 245, 323–5, 587–9, 592–5.
RAs: 172–4, 214, 215, 332–5.
SFR: 36, 78, 84, 116–17, 120–2.

TYPES OF PHAGE PARTICLE

The most common class of phages, typified by phage λ, exist as a combination of a polyhedral capsid containing a long coiled nucleic acid (usually DNA), linked to a helical tail with terminal fibres (see Fig. 39.1). The phage tail fibres attach to the host bacterium via specific surface receptors on the host cell membrane. These vary between bacterial species or strains and therefore dictate the phage's host specificity. For example, phage T4 recognises an *E. coli* porin protein termed OmpC (outer-membrane protein C). Following attachment, the phage DNA passes from the capsid into the tail which contracts and punches a hole in the bacterial cell wall. This causes the DNA to be injected into the bacterial cytoplasm leaving the phage coat outside. The DNA then replicates in a lytic or lysogenic infection cycle.

A less common class of phage is the single-stranded circular DNA-containing filamentous coliphages, typified by M13, f1 and fd. These are specific for enteric bacteria containing projections known as F pili. These act as the attachment sites for the coliphages via their gene 3 proteins. The unusual life cycle of filamentous coliphages makes them particularly suitable for some recombinant DNA techniques. Less commonly used are phages composed of an icosohedral protein capsid enclosing single- or double-stranded DNA or RNA (e.g. phage φX174).

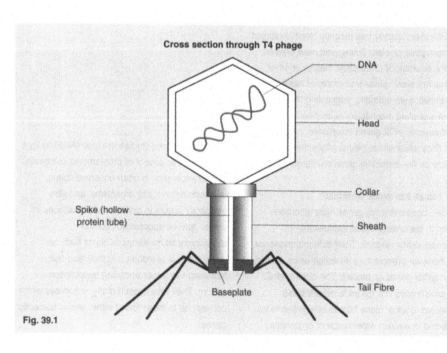

Cross section through T4 phage

DNA — Head — Collar — Sheath — Tail Fibre — Spike (hollow protein tube) — Baseplate

Fig. 39.1

LYTIC LIFE CYCLE

Many phages are termed virulent or lytic because their replication within a susceptible host results in lysis (rupturing) of the bacterium (Fig. 39.2). Once they infect a bacterium by attachment and injection of their DNA, they 'hijack' the host cell's metabolic machinery to produce new viral DNA and protein molecules. Indeed, the host cell DNA is degraded to such a degree that only the circularised viral DNA is replicated and only new phage proteins are synthesised. Under the control of viral genes, these are then assembled into new phage particles. Specific viral genes then dictate the production of an enzyme which degrades the bacterial cell membrane, causing lysis and the release of new viral particles which are capable of infecting further bacteria.

LYSOGENIC LIFE CYCLE

Temperate or lysogenic phages do not destroy the host bacterium during replication (Fig. 39.3). They may however be induced to revert to a lytic cycle under certain external conditions. Lysogenic cycles involve integration of the phage DNA into the host bacterial chromosome. At this stage the viral DNA is termed a prophage since the genes encoding new phage structural proteins are repressed allowing the bacterium to function normally. The prophage is then replicated along with the host bacterial chromosome.

Lysogenic phages may however produce changes in the properties of a host cell (termed lysogenic conversion). This occurs by virtue of a process called transduction when a prophage is induced to enter a lytic cycle. It results from the fact that the new phage particles often contain portions of the original host bacterial DNA in place of their own. Consequently, bacterial DNA may be transferred into the genome of the new host bacterium during infection by the newly released phages. This genetic recombination by lysogenic phages is widely utilised in recombinant DNA methods.

FILAMENTOUS PHAGE REPLICATION

Once inside the bacterium, the single-stranded DNA of filamentous phages adopts a double-stranded replicative form (termed RF) which rapidly multiplies. When about 100 RF molecules have been formed, the production of a viral-encoded single-stranded DNA-binding protein

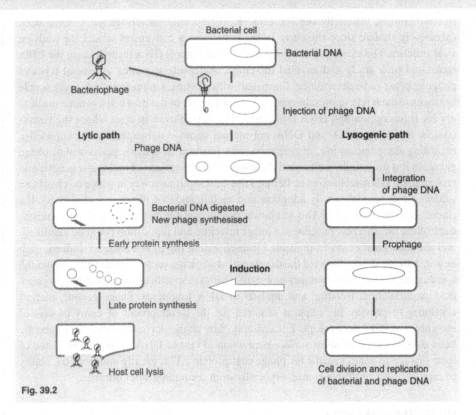

Fig. 39.2

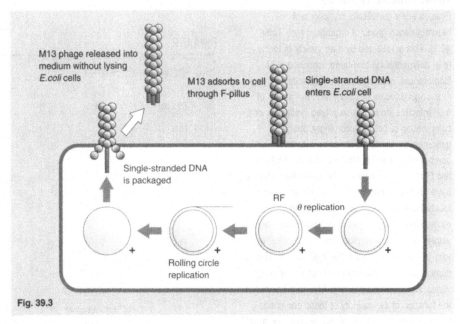

Fig. 39.3

ensures that only single-stranded DNA molecules are subsequently generated. Infection does not lyse the bacterium which still grows and divides. However, up to a 1000 new phage particles may be released per generation by extrusion into the external medium. Such phages have now been widely exploited for the development of cloning vectors.

107

40. BACTERIOPHAGE GENETICS

The variety of capsid forms, host cell specificities and life cycles of phages is reflected in their different genomes, which similarly vary in size, gene number, gene composition and gene organisation.

Bacteriophages, especially the lysogenic *E. coli*-specific bacteriophage λ, have been extensively studied since the early 1930s, initially as a convenient model for studying viral genetics. The expansive development of recombinant DNA techniques in the 1970s facilitated their study and enabled the entire nucleotide sequence of several types of phage genome to be determined. Consistent with all viruses, phages are relatively simple organisms which rely upon subversion of at least a part of the host cell's genetic machinery for transcription and DNA replication. This is true even in cases where the viruses contain their own RNA and DNA polymerase genes. The acquisition of knowledge regarding the composition, organisation, and function of specific genes within phage genomes has profoundly influenced the course of development of molecular genetic and recombinant DNA technologies. Perhaps the most significant way in which this has been achieved has been the early adoption and modification of phages as vectors for the cloning of DNA fragments. This was however only possible once the genetic mechanisms controlling the different processes of phage infection and replication could be identified, and the individual genes responsible localised within the phage genome. Indeed, each new insight has driven forward the development of phage vector systems. Consequently, a wide range of derivative vectors now exist which are specifically suited to the propagation, manipulation, isolation and analysis of DNA fragments. Phage genetic analysis continues to provide the impetus required for the development of many aspects of recombinant DNA technology. For example, new phage-derived vectors have recently been developed which allow surface expression of cloned DNA fragments by virtue of their linkage to genes specific for phage coat proteins. This greatly enhances the ability to rapidly select desirable clones, especially from recombinatorial libraries.

THE SMALLEST FORMS OF LIFE?

Viruses were, until recently, believed to be the smallest and simplest life forms on the planet, being composed solely of protein and nucleic acid. Indeed, their unusual composition makes it difficult to accurately describe them as cells. All viruses, be they specific for prokaryotic or animal cells, are obligate parasites. In other words, they are dependent upon at least part of the host cell machinery for replication and gene expression. Consequently, their genes must be matched to those of their host cell, restricting each virus in terms of the types of cell they can successfully infect, and replicate within. These features have led many biologists to question if viruses are true living organisms. This issue has been brought into sharp focus by the recent discovery of even smaller infectious and replicatable particles termed prions. These are composed solely of protein, and have been suggested to be linked with bovine spongiform encephalitus (BSE) in cattle, scrapie in sheep, and Creutzfeldt Jacob disease (CJD) in humans.

PHAGE GENOME VARIATION

Phages are a genetically complex and heterogeneous group of organisms (see Table 40.1). This is reflected by their variety of forms (e.g. polyhedral tail and head, icosohedral or filamentous), restricted host cell specificities (e.g. to a single species or strain), and the nature of their infective life cycles (e.g. lytic, lysogenic, or a combination of both). Accordingly, phage genomes are usually, but not exclusively, composed of a single RNA or DNA molecule. A few RNA phages have also been isolated which have segmented genomes (i.e. their genes are located amongst several separate RNA molecules). Phage nucleic acid molecules may be single- or double-stranded, linear or circular, and vary in size from about 1.6 to over 150 kb. The number of genes encoded by each phage also varies, from just three to over 200. In many cases the function of the majority of these genes has been determined. Nevertheless, even in the best characterised phages several genes have been identified but not as yet ascribed a function. For example, phage λ contains 63 genes based on sequence analysis, although the functions of only 48 have been determined.

Table 40.1

Phage genome variation

Phage	Host	Genome	Size(Kb)	No. genes
M13	*E. coli*	ssDNA	6.4	10
φX174	*E. coli*	ssDNA circle	5.4	11
λ	*E. coli*	dsDNA	49.5	48
T2/T4	*E. coli*	dsDNA	166	150
T7	*E. coli*	dsDNA	40	55
PM2	*P. aeruginosa*	dsDNA	13.5	–
SPO1	*B. subtilis*	dsDNA	150	–

Icosahedral	Filamentous	Head and tail
φX174, PM2	M13	λ, T2, T4, T7, SPO1

ORGANISATION OF PHAGE GENOMES

Complete nucleotide sequencing of differently sized phage genomes has revealed much about their organisation (see Fig. 40.1). Accordingly, small genomes contain only a few genes. For example, the genome of the filamentous phage M13 contains only 10 genes encoding phage coat proteins and the enzymes involved in phage DNA replication. Some small genomes do however exhibit a complex pattern of organisation. Thus, like many viruses, the phage φX174 makes use of overlapping genes to increase the amount of genetic information packaged into the genome. Consequently, the same nucleotide sequence is shared between two genes (e.g. E and D) to produce two different gene products by synthesising mRNA transcripts which are translated from different start points and hence different reading frames.

Larger phage genomes naturally contain more genes, often reflecting their more complex capsid structure and organisation. Thus, phage T4 has a genome of about 166 kb composed of some 150 genes of which a third are required solely for capsid construction. The isolated genome of phage λ consists of about 49.5 kb of linear double-stranded DNA and contains 48 genes clustered into functionally related groups. The phage λ DNA molecule has also been shown to have terminal cohesive ends. These are composed of 12 nucleotide overhangs which combine to form the *cos* site, thereby generating a circular DNA molecule following injection into the host cell. Genes encoding and controlling assembly of the capsid head and tail components are located on the left of the conventional linear map. Those to the right are sequentially involved with prophage integration and excision, early gene regulation, DNA synthesis, late gene regulation, and host cell lysis (Fig. 40.3). The central (b2) portion also appears to be non-essential for replication. This latter feature is crucial to the development of phage vectors since the central region (up to 22 kb) can be deleted or modified (e.g. to contain a polylinker sequence) and up to 25 kb of foreign DNA inserted without affecting phage packaging and lysis.

HOST CHROMOSOME INTEGRATION

Integration of the phage DNA into the host cell chromosome (a crucial feature of lysogeny) is achieved by a site-specific recombination event. In phage λ this results from the shared nucleotide homology between 15 base pairs of phage λ and *E. coli* DNA (Fig. 40.2). As a result the phage λ DNA is always inserted into the same place within the *E. coli* chromosome. Excision of the integrated prophage occurs by a reversal of the process, all of which is controlled by specific phage λ genes (e.g. *int*, *xis* and *exo*).

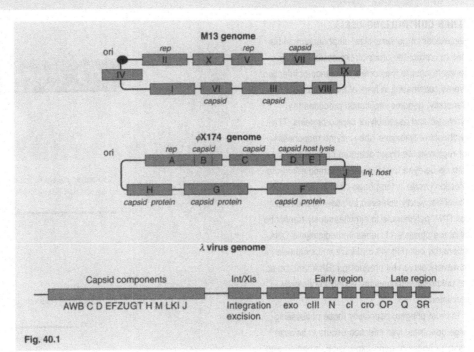

Fig. 40.1

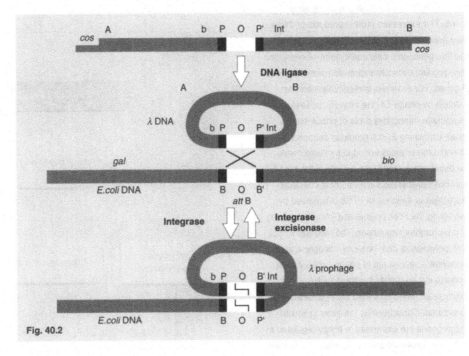

Fig. 40.2

OAs: 326–30, 587–9, 592–5.
RAs: 172–7, 214–15. SFR: 36, 78, 84, 101–2, 116–18, 120–2.

The processes of lytic and/or lysogenic infection cycles are dependent upon the actions of protein repressors which regulate the temporal and coordinated expression of sets of phage genes.

GENES CONTROLLING LYSIS

Regardless of genome size, all phages must be able to control the order of expression of their genes to ensure the correct sequence of infection events culminating in lysis of the host cell. Generally, genome replication precedes the synthesis and assembly of capsid proteins. The synthesis of lysozyme (the enzyme responsible for rupturing the host bacterium cell wall) must also be delayed until the very last moment in the infection cycle. In the phage φX174 this is straightforwardly achieved by allowing the host cell RNA polymerase to simultaneously transcribe all of the phage's 11 genes immediately its DNA enters the cell (Fig. 40.3). Lysozyme synthesis is however delayed by producing mRNA that, by as yet unknown mechanisms, is only very slowly translated.

In most phages, especially those possessing large genomes, lytic infection occurs in several distinct phases of gene expression. It begins with the transcription and translation of the early genes. The expressed protein products of these early genes then activate expression of middle and late genes in a cascade system whereby one gene product activates expression of another set of genes. For example, immediately following infection by phage T4, the host *E. coli* RNA polymerase transcribes a set of phage early genes containing *E. coli* promoter sequences. Amongst the proteins encoded by these genes are those which modify the activity of the host *E. coli* RNA polymerase such that host cell gene expression is switched off. This is achieved by preventing the RNA polymerase from recognising *E. coli* promoter sequences. The modified *E. coli* RNA polymerase can, however, recognise and transcribe a second set of phage genes whose products in turn further modify the RNA polymerase such that a third set of genes can be transcribed. Consequently, the order in which phage genes are expressed is tightly regulated *in vivo*. This allows molecular biologists to intervene in the process, e.g. to exploit phages for the cloning or expression of gene/cDNA fragments.

OAs: 326–30, 587–9, 592–5.
RAs: 89, 172–4, 214–15, 220–2, 224. **SFR:** 36–9, 78, 84, 101–2, 116–18, 120–2.

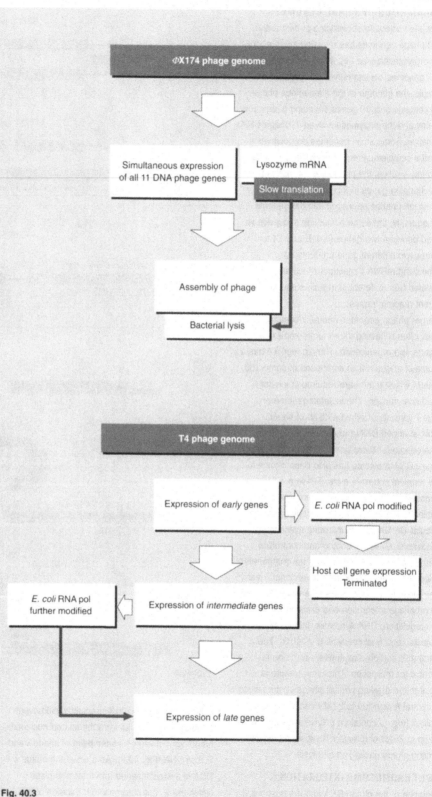

Fig. 40.3

GENES CONTROLLING LYSOGENY

The ability of phage λ to adopt lysogenic or lytic life cycles requires the interaction between several gene products and regulatory sequences (see Fig. 40.4). Lytic infection by phage λ occurs in three stages involving the sequential expression of the early, middle and late genes. The transcription of early genes depends upon the host RNA polymerase binding to two promoters, termed PL and PR, located on either side of a regulatory gene called *cl* (see Fig. 40.5). This gene encodes a repressor protein termed cl repressor which, even at very low levels, can block transcription from PL and PR by binding to the operators OL and OR. The inability to transcribe the early genes prevents the phage λ entering a lytic infection cycle and it consequently enters lysogeny. Lysogeny is maintained for many cell replications so long as cl repressor protein is still produced. In fact, cl repressor is continually produced since the binding of cl repressor proteins to OR not only prevents transcription from PR but stimulates transcription of the cl gene via its promoter, termed PM. Only when levels of cl repressor fall below a certain point will lysogeny be broken, leading to prophage excision, its replication by the rolling circle method, and ultimately lysis. This may occur spontaneously, or be induced by external chemical or physical stimuli which activate a specific protective mechanism, known in *E. coli* as the 'SOS' response. This response produces an *E. coli*-encoded protein called RecA which cleaves the cl repressor protein in half.

The race between repressors

When phage λ DNA enters into a host cell the decision to enter a lytic or lysogenic infection cycle is decided by a race between two repressor proteins: the cl repressor and the Cro repressor. If cl repressor protein accumulates more rapidly than Cro repressor protein, the binding of cl protein to OL and OR blocks early gene transcription preventing lytic infection and inducing lysogeny. If, however, the Cro repressor protein accumulates first, it switches off *cl* gene transcription by binding to its promoter (PM), allowing early gene transcription and lytic infection. Who wins appears to be due to the randomness of host cell RNA polymerase binding to the various phage λ promoters.

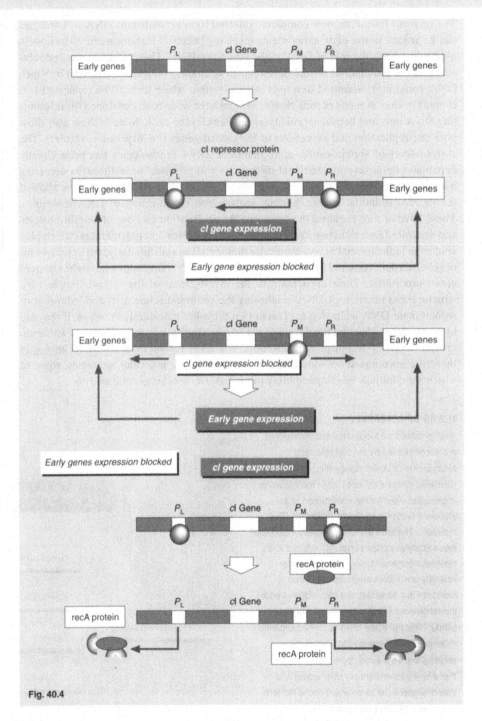

Fig. 40.4

ROLLING CIRCLE REPLICATION

During lytic infection several phages, including λ and M13, replicate their genomes via the rolling circle mechanism. In phage λ the excised prophage is initially replicated as a series of λ genomes joined head to tail. Termed 'concatamers', these are subsequently cleaved at the *cos* site by gene A endonuclease to yield single linear λ genomes which are packaged into new phage capsids and recircularised following injection into a host cell.

41. RECOMBINANT DNA TECHNOLOGY

Recombinant DNA technologies include any molecular genetic technique used for the *in vitro* construction, analysis and *in vivo* propagation (cloning) of recombinant DNA molecules.

The origins of what are now commonly referred to as recombinant DNA technologies can be traced to the early experiments involving bacterial transformation which were used to establish the role of DNA as the genetic material. These experiments undoubtedly laid the foundations for the development of cloning vehicles (or vectors) by which DNA could be transmitted to a host cell and within which it could be replicated (i.e. cloned) *in vivo*. A range of such cloning vectors have since been developed for transmitting DNA into, and between, prokaryotic and eukaryotic cells. Some of these also allow both the replication and expression of the cloned genes (i.e. expression vectors). The current level of sophistication of recombinant DNA technologies has been equally dependent upon several additional discoveries and technical breakthroughs occurring over the last 25 years. Amongst the most directly relevant are those which have allowed the *in vitro* isolation, cutting, joining, analysis and modification of DNA molecules. These have in turn required the isolation, characterisation and use of certain proteins and enzymes (e.g. restriction endonucleases, DNA ligases and polymerases) which play vital roles in the normal *in vivo* molecular biology of the cell. Similarly, the development of gel electrophoresis and hybridisation methods for DNA fragment analysis has been of great importance. Their application to the investigation of the *in vivo* and *in vitro* structure and function of DNA is allowing the continual refinement and extension of recombinant DNA technologies. This in turn provokes new questions which, if they are to be answered, demand the development of new approaches and techniques with even greater levels of technical sophistication. The term recombinant DNA technologies therefore encompasses a wide range of techniques and molecular strategies, some of which were initially developed purely to aid basic molecular genetic analysis.

TECHNOLOGICAL BREAKTHROUGHS

There have been, and continue to be, numerous technological breakthroughs which have vastly increased the level of sophistication of recombinant DNA technologies. Some of the critically important breakthroughs are:

1 Isolation and characterisation of enzymes such as restriction endonucleases, DNA ligases and DNA/RNA polymerases,

2 Isolation of self-replicatable bacterial plasmids and their initial development as cloning vectors,

3 Development of chemistries for the *in vitro* synthesis of oligonucleotides,

4 Development of gene probes and hybridisation-based detection assays,

5 Development of chemical or enzymatic nucleotide sequencing methods, and

6 Development of the polymerase chain reaction (PCR).

SFR: 119–23, 126–8, 161, 169, 213, 226.

TERMS OF REFERENCE

A large number of techniques and approaches are encompassed by the umbrella term recombinant DNA technology (Fig. 41.1). In the narrowest sense it refers to those techniques or approaches used for the construction of an artificially recombined (i.e. recombinant) DNA molecule. The term therefore naturally includes all those strategies often separately referred to as molecular or gene cloning. The processes whereby novel DNA constructs (and their products) are generated are also included under the recombinant DNA technology umbrella, although they are also often described separately in terms of gene manipulation or genetic engineering. To a large degree therefore, all these terms are essentially overlapping and interchangeable. In its broadest sense the term recombinant DNA technologies also encompasses a vast array of 'supporting' molecular techniques. These include methods for nucleic acid isolation and analysis (e.g. sequencing, hybridisation probing, polymerase chain reaction) which are also required for monitoring and assessing the success of each of the gene cloning and engineering processes.

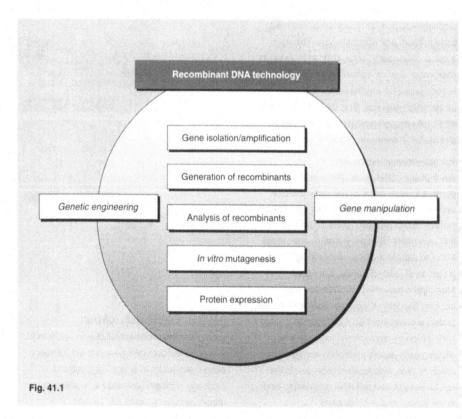

Fig. 41.1

ISOLATING NUCLEIC ACIDS

The ability to isolate nucleic acids from cells and tissues is essential not only to begin the process of gene cloning and manipulation (Fig. 41.2), but also throughout the various stages of such processes (e.g. excision of a vector insert for analysis). Still widely employed and the first to be developed are those methods which involve their physicochemical isolation. However, the recent development of the PCR has greatly enhanced the ability to isolate nucleic acids. Accordingly, the PCR can be used to rapidly and specifically provide large amounts of identical or heterogeneous nucleic acid molecules from just a single cell.

CUTTING AND JOINING DNA

Recombinant DNA strategies are also vitally dependent upon the ability to specifically cut and join DNA molecules. This is still generally achieved by the use of restriction enzymes to create cohesive or blunt ends which, following annealing, may be joined by DNA ligases.

CLONING DNA AND RNA

The ability to clone or replicate DNA and RNA molecules has only been possible through the development of independently replicatable cloning vehicles or vectors. Ironically, the most useful vectors have only been developed following their genetic manipulation. This has allowed the incorporation of features such as multiple cloning sites to facilitate foreign DNA insertion and selectable markers for detecting and retrieving desirable clones. A vast range of vectors are now available which allow the replication and expression of genes in almost any type of prokaryotic and eukaryotic host.

ANALYSING DNA

Recombinant DNA technologies were initially developed to facilitate gene analysis, however they themselves also rely heavily upon the use of analytical techniques. These are required to monitor the success and accuracy of the steps involved in gene cloning and engineering. Hybridisation-based techniques such as Southern blotting are therefore widely used to screen gene or cDNA libraries for the presence of vectors bearing the desired insert. Other techniques such

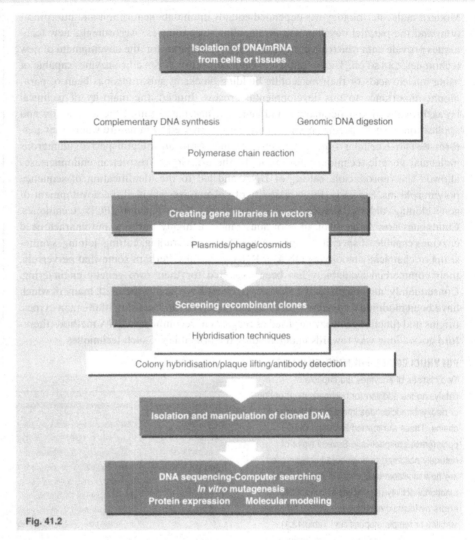

Fig. 41.2

as nucleotide sequencing are also required to provide primary data, to ensure that the inserted DNA retains its original sequence, or to monitor the success of any *in vitro* mutagenesis steps performed.

MANIPULATING DNA

In its simplest sense, gene or molecular cloning refers only to the replication of a specific gene or cDNA fragment. However, the processes involved often result in the replication of a truncated or incorrectly orientated fragment (e.g. reversed or out of reading frame), preventing its accurate expression. Artificial mutation of native sequences, or the correction of such defective inserts, by the processes of genetic engineering

can only be accomplished through the use of a range of approaches often referred to as *in vitro* mutagenesis strategies.

IN VITRO SYNTHESIS OF DNA/RNA

The isolation of polymerase enzymes for the *in vitro* synthesis of DNA and RNA molecules has also played a significant part in the development of several recombinant DNA techniques. A substantial proportion of these are also dependent upon the chemical synthesis of nucleic acids. For example, *in vitro* amplification of DNA by the PCR, dideoxynucleotide sequencing, and many forms of *in vitro* mutagenesis and gene targeting strategies are reliant upon the use of synthetic oligonucleotides.

42. ENZYMES COMMONLY USED IN MOLECULAR BIOLOGY METHODS

A wide range of enzymes which catalyse the formation, cleavage, linking or chemical modification of RNA and DNA polynucleotides are used in methods for gene analysis, labelling, isolation and manipulation.

Modern molecular biology has depended equally upon data acquisition and interpretation, and the parallel development of 'enabling' technologies. Accordingly, new techniques provide data which may form the basis of, or impetus for, the development of new techniques, and so on. The isolation and characterisation of specific enzymes capable of using nucleic acids or their nucleotide building blocks as substrates has been of paramount importance to this developmental process. Indeed, the majority of technical breakthroughs in genetic analysis and gene engineering required the availability and application of such specific enzymes. Originally borrowed from nature where they perform essential cellular functions, these enzymes form an integral part of numerous molecular genetic techniques. For example, the isolation of restriction endonucleases allowed the reproducible cutting of DNA and led to the identification of sequence polymorphisms. Similarly, the availability of polymerases enabled the development of gene cloning, dideoxy sequencing and polymerase chain reaction (PCR) techniques. Catalogues now bulge with an enormous range of highly purified and characterised enzymes capable of specifically (or non-specifically) degrading, cutting, joining, synthesising or chemically modifying DNA or RNA molecules. Perhaps somewhat perversely, their commercial availability has been exploited for their own genetic engineering. Consequently, new recombinant forms of enzymes have been generated, many of which have been modified to possess novel characteristics. An appreciation of the many types, origins and functions of enzymes used as reagents in recombinant DNA methods therefore goes a long way towards understanding the principles of such techniques.

REPORTERS AND RECOMBINANTS

In addition to manipulating or analysing nucleic acids, many enzymes have been used as 'reporter' molecules for detecting, analysing or monitoring the fate of nucleic acids in vivo and in vitro. For example, β-galactosidase is often used as a reporter molecule for the primary detection of recombinant clones in genomic or cDNA libraries. Many biotechnology and industrial processes also utilise enzymes. It is therefore not surprising that enzymes have themselves been subject to genetic engineering. Indeed, the genetic manipulation and production of new forms of recombinant enzymes is currently a growth area. Furthermore, the increasing demand for new enzyme activities, and the need to increase their utility in vitro and in vivo, has led to the genes of other proteins (e.g. antibody genes) being genetically altered to contain enzyme active sites (e.g. to generate catalytic antibodies).

POLYNUCLEOTIDE-FORMING ENZYMES

Two classes of enzymes are capable of catalysing the addition (or polymerisation) of ribo- or deoxyribonucleotides into polynucleotide chains. These are termed RNA and DNA polymerases respectively. Several types of naturally occurring or engineered polymerases are now available which vary in terms of their functional activity (e.g. ability to correct base errors by intrinsic exonuclease activity), thermal stability or template specificity (Table 42.1). These are widely used in molecular biology, especially for gene probe generation and labelling, dideoxynucleotide sequencing and in vitro DNA amplification.

Both DNA and RNA polymerases require two things besides the availability of a pool of nucleotide triphosphates and certain cofactors such as Mg^{2+} ions. These are: (i) a preformed (DNA or RNA) polynucleotide chain to serve as a template, and (ii) a hybridised 'priming' (RNA or DNA) polynucleotide with a free 3′ end to which nucleotides can be added by phosphodiester bond formation. Each type of polymerase may be distinguished/classified according to their template requirement and the type of polynucleotide formed. For example, DNA Pol I is a DNA-dependent DNA polymerase, whilst T3 RNA pol is a DNA-dependent RNA polymerase.

Another class of polymerases isolated from retroviruses are the reverse transcriptases. These RNA-dependent DNA polymerases are able to produce a DNA polynucleotide from an RNA template. This ability is exploited in a variety of molecular cloning techniques for the creation of complementary DNA (cDNA) strands from messenger RNA (mRNA) templates (e.g. as a prelude to PCR amplification or cDNA library generation).

Table 42.1

Example Enzyme	Functional characteristics or use
DAN Poll	DNA-dependent DNA polymerase containing 5′-3′ and 3′-5′ exonuclease activity
Klenow	Fragment of DAN Poll lacking 5′-3′ exonuclease activity
T4 DNA Pol	Bacteriophage T4 DNA-dependent DNA polymerase lacking 5′-3′ exonuclease activity
Taq DNA Pol	Heat stable DNA-dependent DNA polymerase isolated from *Thermus aquaticus* (perfectly used in vitro DNA amplification by the polymerase chain reaction: PCR)
T7 DNA Pol	Bacteriophage 17 DNA-dependent DNA poltmerase (particularly used in sequencing)
SP6 RNA Pol	DNA-dependent DNA polymerase with high affinity for Sp6 promoter sequences (used for in vitro translation and RNA probe generation)
T3 RNA Pol	DNA-depenent RNA polymerase with high affinity for T3 promoter sequences (used for in vitro translation and RNA probe generation)
T7 RNA Pol	DNA-dependent RNA polymerase with high affinity for T7 promoter sequences (used for in vitro translation and RNA probe generation)
Qβ replicase	RNA-dependent RNA polymerase (particularly used for in vitro amplification of specialised RNA probe)
AMV RT	RNA-dependent DNA polymerase (reverse trancriptase) isolated from avian myeloblastosis virus (used in cDNA synthesis)
M-MLV RT	RNA-dependent DNA polymerase (reverse transcriptase) isolated from moloney-monkey leukaemia virus (used in cDNA synthesis)

OAs: 301, 331–41, 343–5, 358–60, 441–6, 449, 490–508. **RAs:** 178–85, 195, 196, 205, 207. **SFR:** 50, 62, 73, 103, 104, 120–8, 136–9.

CUTTING OR DEGRADATIVE ENZYMES

Several classes of enzymes, termed nucleases, are capable of catalysing the breakdown of RNA (ribonucleases or RNases) or DNA (DNases). They are used in many molecular genetic methods (e.g. gene probe labelling by nick translation). However in some instances they have to be carefully avoided (e.g. in RNA extraction and isolation methods).

Nucleases may be divided into two groups (see Table 42.2) according to where they cleave: (i) those which cleave nucleotides from one end of a polynucleotide (termed exonucleases), and (ii) those which cleave internal phosphodiester bonds (termed endonucleases). Exonucleases may be specific for single- and/or double-stranded polynucleotides, the former being particularly useful in the mapping of homology regions within homo- or heteroduplexes. Most exo- and endonucleases are not sequence specific and, under optimum conditions, will continue to cleave/ degrade polynucleotides unless the phosphodiester bond is protected, e.g. by bound protein. This feature is exploited for the detection of protein–DNA interaction sites, e.g. by nuclease protection assays.

A specific group of bacterial endonucleases termed restriction endonucleases are however extremely specific, only recognising and cleaving at defined sequences. A wide range of these enzymes are available which vary in their recognition sequences, type of cleavage produced, and the effects of base methylation upon sequence recognition. They have been invaluable in gene analysis (e.g. restriction fragment length polymorphisms) and gene manipulation (e.g. gene cloning).

ENZYMES LINKING POLYNUCLEOTIDES

DNA ligases are widely employed in genetic manipulation methods for linking double-stranded DNA fragments (e.g. those generated by digestion with restriction endonucleases). Ligases catalyse the formation of phosphodiester bonds between the free 5′ phosphate groups on one end of a molecule and the free 3′ hydroxyl group of another or the same DNA molecule. This leads to the formation of a chimeric molecule, or results in circularisation, respectively. Ligases have also been employed in an analytical *in vitro* DNA amplification assay termed the ligase chain reaction.

Table 42.2

Example Enzyme	Functional characteristics or use
Exonnuclease III	Performs 3'-5' stepwise removal of nucleotides from double or single-stranded DNA
Mung bean nuclease	Degrades single-stranded DNA to mononucleotides and oligonucleotides
DNase I	Non-specific endonuclease that randomly cleaves double-stranded DNA used in many situations including probe labelling by ik translation
Bal 31	Exonuclease which removes nucleotides from DNA 3' and %' termini
RNaseA	Single-stranded RNA-specific exonuclease used in mapping studies
RNaseH	RNA-specific exonuclease used in second strand cDNA synthesis
S1 nuclease	Specifically cleaves single-stranded DNA, used in mapping studies
T4 DNA ligase	Bacteriophage T4 derived enzyme which is used to link 5' phosphate and 3' hydroxyl groups on the same or differnet double-stranded DNA molecules via formation of a phosphodiester bond
T4 polymerase-kinase	Bacteriophage T4 derived enzyme which transfers terminal phosphate groups from nucleotides to 5' hydroxyl groups of polynucleotides, often used in labelling the 5' ends of gene probes
Alkaline phosphatase	Removes 5' phosphates from DNA and RNA, e.g. in preperation for 5' end labelling and as a detection reagent in many gene probe assays
Terminal deoxynucleotide transferase	Adds homopolymer tails to the 3' ends of double-stranded DNA, particularly used in cloning and for the 3' end labelling of DNA probes
Eco RI methylase	Adds methyl group (N6 position) of adenines within the DNA recognition sequence of the restriction endonuclease *Eco* RI, preventing cleavage
Hae III methylase	Forms 5-methylcytosine within the DNA recognition sequence for *Hae* III restriction endonuclease, preventing cleavage

NUCLEOTIDE-MODIFYING ENZYMES

Several types of enzymes are also widely used during gene manipulation to chemically modify the nucleotides within, or at the ends of, polynucleotides. For example, methylases can be used to add methyl groups to certain bases to prevent the recognition and degradation of specific sequences by endonucleases. Transferases add single or multiple nucleotides (e.g. homopolymer tails) to the 3′ end of DNA fragments facilitating gene probe labelling and cloning respectively. Phosphatases and kinases are also used to facilitate cloning and probe labelling by modifying the 5′ groups of DNA or RNA.

OTHER USEFUL ENZYMES

Most recently the base-specific enzyme uracil DNA glycosylase or UDG which cleaves dUTP containing DNA polynucleotides at positions where it is incorporated has found application in emerging techniques. For example, UDG has particularly been employed for avoiding PCR carry-over contamination and to obviate the need for restriction endonucleases in PCR-based cloning. Enzymes termed agarases which are capable of degrading the polymerised agarose used in the gel electrophoretic separation of nucleic acids have also been used to facilitate the recovery of DNA fragments prior to analysis or further *in vitro* manipulation.

43. RESTRICTION ENDONUCLEASES

The power and central role played by various type II restriction endonucleases (or simply restriction enzymes) in molecular biology stems from where and how they cut double stranded DNA.

A significant problem which constrained investigations of genome organisation for many years was how to reproducibly cut DNA into fragments small enough to handle and analyse. By the 1970s it was a relatively easy matter to extract DNA and randomly cut it using chemical or mechanical methods (e.g. sonication). However, even when sufficient amounts of each fragment could be isolated for analysis, it was generally impossible to establish the original order of the fragments within the starting DNA. The breakthrough came with the discovery of bacterial enzymes which could be purified from homogenised bacteria and subsequently used *in vitro* to cut DNA in a highly specific fashion. In bacteria, these restriction endonucleases (so termed because they cut at restricted sites within double-stranded DNA) were part of its machinery for protecting against infection by viruses. Type II restriction enzymes in particular provided scientists with the necessary tools to rapidly extend the frontiers of DNA analysis and manipulation by enabling the generation of consistently sized fragments (e.g. for cloning, sequencing and mapping genes). Indeed, without restriction enzymes it is unlikely that many recombinant DNA technologies could have been developed. The fact that several types of restriction enzymes were soon described which varied in where and how the DNA backbone was cleaved, was an additional and significant bonus for molecular biologists. Restriction enzymes are consequently central to many recombinant DNA approaches ranging from Southern blotting to *in vitro* mutagenesis techniques used for the creation of genetically modified proteins (or even entire organisms). In their own right they may also be used as genetic tools for localising genes and the detection of mutations based upon differences in the lengths of the fragments generated following digestion (i.e. restriction fragment length polymorphisms or RFLPs).

METHYLATION PROTECTION SYSTEM.

During evolution bacteria have developed many similar methods for protecting their genomes from invading viral DNA sequences. The two most common systems employ sets of bacterial enzymes termed methylases. These add a methyl group (i.e. methylate) to either adenine or cytosine bases within the bacterium's DNA. Since restriction enzymes do not recognise sequences containing methylated bases, only the invading, non-methylated viral DNA is digested and thus eliminated. These two methylation modification systems, designated dam and dcm (for DNA adenine/cytosine methylase), are also effective in recognising and eliminating experimentally introduced DNA. Consequently, most recombinant cloning techniques employ bacterial strains which do not possess these methylation modification systems (designated dcm⁻ dam⁻).

OAs: 334, 341–50, 495, 576–8, 580–1, 684, 773. RAs: 178–80, 304, 306–8. SFR: 62, 67, 120–8, 160–3, 167, 174–81.

HOW THEY CUT DNA STRANDS

Restriction endonucleases cleave DNA by hydrolysis of the phosphodiester backbone. Depending upon the particular restriction enzyme, this may generate two types of fragments: (i) those with 'flush ends' where the two DNA backbones are cleaved at positions directly opposite one another (Fig. 43.1a), or (ii) those with so-called cohesive (also staggered or sticky) ends where the cuts are displaced relative to one another (Fig. 43.1b). Cohesive ends are therefore complementary and consequently retain the capacity to specifically hybridise or re-anneal together. They also contain a stretch of single-stranded sequence termed an overhang. Two types of overhang may be generated termed 5' and 3' where the single-stranded sequence ends with a 5' phosphate or 3' hydroxyl group.

Compatible cohesive ends may be specifically annealed and permanently joined (ligated) using another enzyme, DNA ligase. The capacity of ligases to join blunt or cohesive ended fragments of DNA is central to recombinant DNA methods. Consequently, cohesive ended fragments have been particularly useful for inserting fragments into cloning vectors or for joining two or more DNA fragments together to form novel proteins (e.g. to create chimeric genes or proteins).

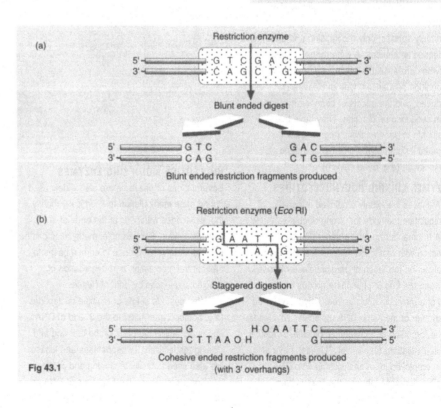

Fig 43.1

RECOGNITION SEQUENCES

DNA sequences recognised by restriction enzymes vary in length from 4 to 20 bases. This generates different types of cut ends of varying length and differently orientated overhangs (Table 43.1). Recognition sequences are, however, generally palindromic in nature (i.e. the sequence reads the same in one direction on one strand as it does in the opposite direction on the complementary strand). The longer the recognition sequence, the less likely it is to occur in any given piece of DNA, resulting in the generation of longer fragments.

The frequency of cutting affects the utility of restriction enzymes in both gene mapping and genetic engineering. In these cases, the ideal sized fragments are those resulting from cuts either side of, rather than within, the protein-encoding sequence of the gene of interest. Some type II enzymes, such as Hgal, do not cut within the recognition sequence itself but cut at a measured distance from one side of the palindromic sequence. Restriction enzymes isolated from different sources which recognise the same sequence and produce the same (e.g. Hpall and Mspl) or different (e.g. Xmal and Smal) cut ends are termed isoschizomers.

Where they cut DNA strands

Different enzymes (apart from isoschizomers) will generate a different pattern of digestion fragments from the same genomic DNA sample. Monitoring of the digestion patterns at different times, and comparison of the patterns obtained using combinations of different restriction enzymes, allows the generation of a restriction map. This is constructed by ordering the fragments according to their original position within the starting DNA. Having isolated a fragment containing a gene of interest, it can then be localised within the restriction map of the original genome.

The early application of restriction enzymes to generate restriction maps of human genomes quickly revealed differences in the restriction patterns obtained for individual human DNA fragments. These are termed restriction fragment length polymorphisms or RFLPs. RFLP analysis has been used for the initial localisation of genes (e.g. cystic fibrosis) and provided the early means of detecting and tracking genes and their mutations.

Nomenclature

The first restriction enzyme was discovered in 1970 in the laboratory of Hamilton Smith at Johns Hopkins University, USA. Isolated from the bacterium *Haemophilus influenzae*, strain d, subtype II this enzyme was consequently named *Hind*II. The naming of the wide variety of enzymes has generally followed this convention. However, a new convention may be required to discriminate between natural and recombinant/engineered forms.

Recognition requires access

The recognition and cleaveage of the sugar–phosphate backbone by a restriction endonuclease requires that the DNA double helical structure undergoes relaxation in order to make the recognition site accessible. Present evidence indicates that binding of the restriction enzyme dimer leads to the introduction of a 12° bend in the DNA double helix. This causes enlargement of the surface area, and a localised unwinding of the helix by 20°.

Table 43.1

Restriction endonuclease	Recognition sequence	Termini following digestion	
Hae III	5'-GGCC-3' 3'-CCGG-5'	5'-GG 3'-CC	CC-3' GG-5'
Hpa II (Msp I)	5'-CCGG-3' 3'-GGCC-5'	5'-C 3'-GGC	CGG-3' C-5'
Bam HI	5'-GGATCC-3' 3'-CCTAGG-5'	5'-G 3'CTTAA	GATCC-3' G-5'
Eco RI	5'-GAATTC-3' 3'-CTTAAG-5'	5'-G 3'CTTAA	AATTC-3' G-5'
Hind III	5'-AAGCTT-3' 3'-TTCGAA-5'	5'-A 3'-TTCGA	AGCTT-3' A-5'
Not I	5'-GCGGCCGC-3' 3'-CGCCGGCG-5'	5'-GC 3'-CGCCGG	GGCCGC-3' CG-5'
Xma I	5'-CCCGGG-3' 3'-GGGCCC-5'	5'-C 3'-GGGCC	CCGGG-3' C-5'
Sma I	5'-CCCGGG-3' 3'-GGGCCC-5'	5'-CCC 3'-GGG	GGG-3' CCC-5'
Hga I	5'-GACGC-3' 3'-CTGCG-5'	5'-GACGCNNNNN 3'-CTGCGNNNNNNNNNN	NNNNNN-3' N-5'
Xmn I	5'-GAANNNNTTC-3' 3'-CTTNNNNAAG-5'	5'-GAANN 3'-CTTNN	NNTTC-3' NNAAG-5'
Hinc II	5'-GTPyPuAC-3' 3'-CAPuPyTG-5'	5'-GTPy 3'-CAPu	PuAC-3' PyTG-5'

Pu = Purine, *Py* = Pyrimidine, *N* = Any nucleotide

44. RESTRICTION FRAGMENT LENGTH POLYMORPHISMS

RFLPs may be used to identify, localise and track gene markers according to how and where the polymorphisms occur within a genome, and are fundamental to the process of genetic fingerprinting.

The realisation that visible phenotypic differences were determined by differences within their inherited genomes was fundamental to the development of all genetics. Indeed, comparing gene differences remains central to molecular genetic approaches for defining genome organisation and the role of individual genes in regulating normal or abnormal processes. Accordingly, most of our current knowledge regarding gene functions arises from locating (or creating) differences in gene sequences and relating these to *in vivo* or *in vitro* alterations in gene and/or protein expression. The discovery of restriction endonucleases provided the first means to detect naturally occurring genetic differences (termed polymorphisms) by providing the means to generate restriction maps. Given the broad variation in physical characteristics between individuals, it is probably unsurprising that the first human restriction maps demonstrated many polymorphisms between individual genomes. These are derived from alterations in the nature or position of restriction endonuclease recognition sequences, and hence the profile of digestion fragments obtained. The term restriction fragment length polymorphism (RFLP) was consequently coined to describe this phenomenon. Combined with probe hybridisation technologies RFLPs provided the front-line approach to localising genes conferring a selectional advantage (e.g. antibiotic resistance) or disadvantage (e.g. early onset disease) upon the cell or organism. More recently these technologies have been used to track naturally occurring or engineered gene polymorphisms through populations, or to identify polymorphisms specific to each individual (e.g. by genetic fingerprinting).

FIRST CLINICAL APPLICATION

The first useful RFLP was described in 1978 by Kan and Dozy who showed a difference in the pattern of digestion with the restriction endonuclease *Hha*I between DNA samples from normal individuals and patients with sickle cell anaemia. This polymorphism was later shown to be the result of a single base substitution in the mRNA for β-globin which changed a GAG codon specific for glutamine to the codon GUG specific for valine. The presence of this mutation is now routinely detected by RFLP analysis employing digestion of DNA samples with the enzyme *Mst*II followed by Southern blot hybridisation using a β-globin cDNA probe. Accordingly, *Mst*II recognises and cleaves the normal haemoglobin (Hb) gene sequence CCT-GAG-G but not the sickle cell haemoglobin (HbS) gene sequence CCT-GTG-C.

DETECTING RFLPs

Restriction maps were initially developed by monitoring the digestion of DNA segments by single or multiple restriction enzymes (Fig. 44.1). Comparison of the lengths of fragments obtained then allowed their relative positions within the starting DNA fragment to be deduced. Any mutation which creates, destroys or moves the recognition sequence for a restriction enzyme leads to a RFLP. An RFLP can therefore be detected by examining the profile of restriction fragments generated during digestion.

Conventionally, this required the purification of the original starting DNA sample before digestion with single or multiple restriction enzymes. The resultant fragments were then size-separated by gel electrophoresis and visualised as discrete bands by staining with ethidium bromide. Routine RFLP analysis of genomic DNA samples generally also involves *in vitro* hybridisation with labelled gene probes (e.g. by Southern blotting) to detect a specific gene fragment(s). However, simple electrophoretic separation and visualisation is all that may be required using cloned or *in vitro* amplified DNA samples (e.g. produced by the polymerase chain reaction).

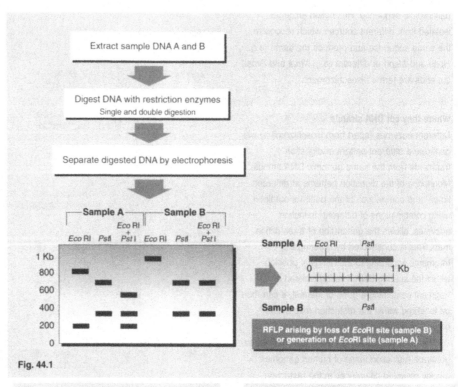

Fig. 44.1

OAs: 84–8, 132.,137, 349, 457, 475, 707–13, 739–41. RAs: 178, 179, 249, 260–3, 304–8. SFR: 62, 67, 120–8, 160–3, 166–8, 174–81.

HOW THEY ARISE

RFLPs may arise by any mutation which alters the relative position of restriction endonuclease recognition sequences (Fig. 44.2). Although most polymorphisms appear to be randomly distributed throughout a genome, there are certain regions where a particularly high concentration of polymophisms exist. Termed 'hypervariable regions', these have been found in regions flanking structural genes from several sources, including humans and microbes. In the latter case their analysis may be used to follow how polymorphic forms arise and are perpetuated within populations (e.g. resulting in different pathogenic strains).

In 1978, Alec Jeffreys at Leicester (UK) first described differences in the numbers of short repeated sequences, termed mini-satellites. These occurred within the genomes from different individuals as evidenced by RFLP analysis using specific gene probes. Several types of mini-satellite sequences such as variable numbers of tandem repeats (VNTRs) have now been described. Their detection has consequently allowed the development of genetic fingerprinting. This utilises the fact that the distance between two given restriction sites varies according to the number of repeat sequences present.

WHERE THEY OCCUR

Maximally only about 10% of a eukaryotic genome is known to encode expressed proteins. Many RFLPs therefore occur outside the protein-encoding regions and consequently have no effect upon the normal morphology or biochemistry of a cell or organism. RFLPs occurring outside a gene are termed extragenic whereas those that occur within a gene are termed intragenic. Generally, extragenic RFLP markers have been used to initially locate a gene to a specific chromosomal region. Once the gene has been fully characterised, intragenic RFLP markers (if they occur) are then generally used to track that gene within a given population. Both intragenic and extragenic RFLPs may however be used to locate and track disease-associated genes, or monitor the incorporation of genetically engineered genes following transfection into animal or plant cells. The accuracy of any such approach does however require that the extragenic marker is closely linked to the gene and consequently co-segregates following recombination during chromosomal pairing at meiosis. RFLPs co-segregating with a gene are said to be in linkage disequilibrium and have played important roles in identifying several disease-associated genes (e.g. the gene for cystic fibrosis). RFLPs are also still used for routine diagnostic analysis of genomic or polymerase chain reaction-amplified DNA samples.

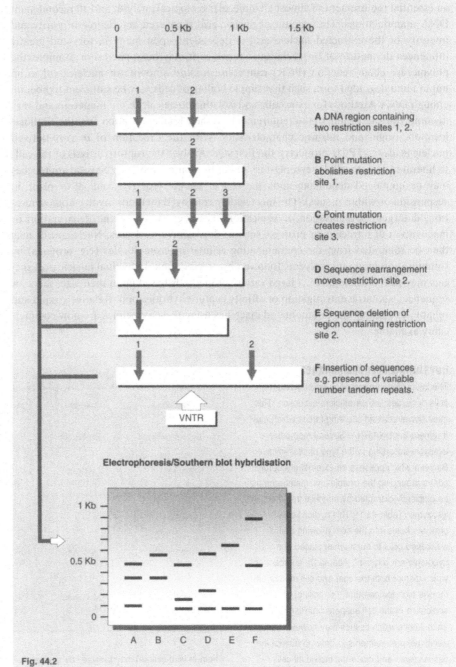

0   0.5 Kb   1 Kb   1.5 Kb

A DNA region containing two restriction sites 1, 2.

B Point mutation abolishes restriction site 1.

C Point mutation creates restriction site 3.

D Sequence rearrangement moves restriction site 2.

E Sequence deletion of region containing restriction site 2.

F Insertion of sequences e.g. presence of variable number tandem repeats.

VNTR

Electrophoresis/Southern blot hybridisation

1 Kb

0.5 Kb

0

A   B   C   D   E   F

Fig. 44.2

45. ISOLATION OF NUCLEIC ACIDS FROM CELLS AND TISSUES

All nucleic acid extraction methods require the physical or biochemical lysis of cell walls and membranes, and the inactivation, degradation and/or inhibition of intracellular proteins, especially nucleases.

The ability to isolate nucleic acids from cells in sufficient quantity, purity and integrity is an essential requirement of almost all molecular biological analyses and all recombinant DNA manipulations. The amount of nucleic acid recovered and degree of purity and integrity of the extracted nucleic acid is dependent upon many factors and greatly influences the nature of the techniques to which it may be applied. For example, the polymerase chain reaction (PCR) can utilise minute amounts of nucleic acid in an impure and degraded form such that simple boiling of cells may be sufficient to liberate enough DNA. Methods for extracting and isolating nucleic acids are numerous and vary in complexity according to the requirements of their final application, and their cellular location, origin and intrinsic characteristics. With the exception of *in vitro*-derived nucleic acids (e.g. PCR products), the first step involves the rupture or lysis of the cells to liberate the intracellular cytoplasmic and/or nuclear components. Several approaches may be employed depending upon the cell or sample type (e.g. animal or plant, in suspension or within tissues). The most widely employed methods involve homogenisation, detergent solubilisation or sonication. The second step is the denaturation or inactivation of intracellular proteins such as degradative nucleases. Nucleic acids may then be separated from the contaminating cellular macromolecules (e.g. proteins) by solubilisation in organic solvents followed by precipitation. Individual nucleic acid species may also be isolated by a large variety of methods based upon their size, shape or sequence, e.g. ultracentrifugation or affinity capture hybridisation. Relatively rapid and reliable extraction kits or automated machines using these principles are now commercially available.

UV EXTINCTION ESTIMATIONS

Absorption of UV light by nucleotide aromatic rings is often used to estimate the DNA/RNA concentration in soluble extracts. Thus, an absorbance value of 1.0 (1 cm pathlength; 260 nm) is obtained from a 50 µg/ml pure solution of double-stranded DNA, a 33 µg/ml solution of single-stranded DNA, and a 40 µg/ml solution of single-stranded RNA. Nucleic acid extracts may however, contain mixtures of DNA or RNA, each present to different degrees of integrity or double-strandedness. This greatly affects the accuracy of the method. The change in absorbance when a double-stranded nucleic acid becomes single stranded can however be used to monitor the process of denaturation or to determine melting temperatures. Since aromatic amino acids (especially tryptophan residues) maximally absorb light at 280 nm, an estimate of protein contamination may also be obtained based on the ratio of absorbances at 260/280 nm.

RUPTURING CELL MEMBRANES

The initial step in extracting nucleic acids from cells or tissues is their efficient disruption. This must however avoid shearing forces which may fragment the DNA/RNA. Several methods are available according to the type of cells involved. Bacteria which possess an external glycoprotein coat surrounding the cytoplasmic membrane may be efficiently disrupted by an initial treatment with lysozyme (Table 45.1). This hydrolytically punches holes into the coat allowing the weakened cells to burst when placed in a hypotonic environment. Alternative approaches which disrupt both the coat and cell membrane include high temperature (i.e. boiling) or sonication of the cell suspension. Plant and animal cells within tissues may however require additional pretreatments in order to disrupt the tissue matrix and make the individual cells available for lysis. These commonly involve enzymatic digestion and/or mechanical homogenisation under liquid nitrogen (which maximally preserves nucleic acid integrity).

Cells lacking an outer coat and/or in solution may be treated less harshly in order for efficient lysis to occur. Consequently mammalian cells in solution (i.e. separated from tissues, blood or

Table 45.1

Technique	Example	Membrane effect	Cell type
Enzyme based	Lysozyme	Produces membrane holes producing hypotonic state	Bacteria
Solution	Hypotonic salt solution	Produces hypotonic environment rupturing cell membranes	Bacteria/general
Detergent	SDS	Solubilises lipid bilayer	Cultured animal cells
Sonication	Sound waves	Disrupts membranes by high frequency sound	Bacterial cells
Homogenisation	Mechanical shearing	Disrupts membranes by physical force	Plant tissue/biopsies

from *in vitro* culture) may be lysed by several different detergents (e.g. sodium dodecyl sulphate, SDS). These are commonly used to solubilise cell and nuclear membranes. They also assist in the denaturation of nucleases and the dissociation of nucleoprotein complexes. Some fragmentation of the DNA is inevitable in all protocols. Yields of high molecular weight DNA can however be maximised by performing the

lysis and protein inactivation steps whilst the cells are embedded in a gel matrix such as agarose.

The glycoprotein coat from bacterial and plant cells can also be removed by a modified lysis approach to leave a membrane-bounded cytoplasm (termed a protoplast). Such protoplasts are frequently used in gene cloning since they enhance the efficiency of *in vitro* gene transfer (or transfection) processes.

DENATURING CELLULAR PROTEINS

Cell lysates naturally contain a complex mixture of macromolecules, including lipid membrane fragments and a wide spectrum of intracellular proteins. Amongst the most important of these are the degradative nuclease enzymes (i.e. DNases and RNases) and the DNA-binding proteins. These greatly affect the amount of nucleic acid damage and/or the amount of nucleic acid finally recovered.

All nucleic acid extraction methods are therefore performed at low temperatures which minimise protein activity. Most DNA and RNA extraction methods also employ additional reagents or incubation steps (Fig. 45.1). These involve either (i) the incorporation of enzyme inhibitors and/or 'decoy' substrates during or immediately following lysis, or (ii) incubation of lysates with protein denaturing reagents. A wide range of chemical compounds which retard DNase activity are also often included in DNA extraction buffers. Accordingly, the chelating agents citrate and EDTA (ethylenediaminetetraacetic acid) are commonly used since they effectively removes the divalent cations (e.g. magnesium ions) required by nucleases for activity. Compounds such as perchlorate which dissociate DNA-binding proteins from DNA–protein complexes may also be included in the buffer used for fractionating DNA from lysates. Phenol/chloroform/isoamyl alcohol is also often used to further denature nucleases and coagulate proteins during DNA and RNA extractions.

Successful RNA extraction also relies upon the effective inhibition of endogenous and exogenous ribonucleases (i.e. RNases). Exogenous RNases can be avoided by autoclaving all apparatus in diethylpyrocarbonate, the wearing of sterile gloves, the use of sterile reagents, and the performance of all manipulations below ambient temperature (e.g. on ice). Several, often combined, strategies may be used to inhibit or inactivate endogenous RNases which vary in abundance between different cell types and organisms (Fig. 45.2). For example, many methods include SDS in the lysis buffer to inhibit RNases, although RNases can often renature during the manipulation of the lysate. Ribonucleosides complexed to vanadyl (termed

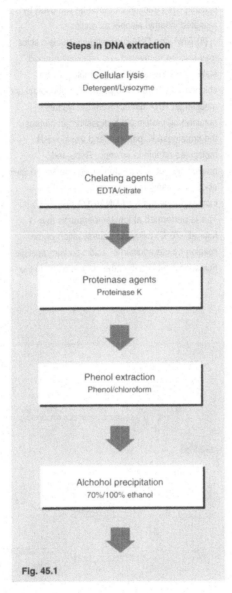

Steps in DNA extraction

Cellular lysis
Detergent/Lysozyme

Chelating agents
EDTA/citrate

Proteinase agents
Proteinase K

Phenol extraction
Phenol/chloroform

Alchohol precipitation
70%/100% ethanol

Fig. 45.1

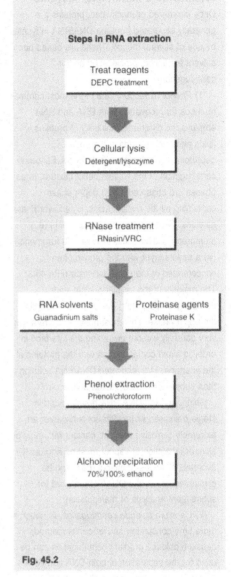

Steps in RNA extraction

Treat reagents
DEPC treatment

Cellular lysis
Detergent/lysozyme

RNase treatment
RNasin/VRC

RNA solvents
Guanadinium salts

Proteinase agents
Proteinase K

Phenol extraction
Phenol/chloroform

Alchohol precipitation
70%/100% ethanol

Fig. 45.2

vanadyl ribonucleoside complexes; VRCs) are therefore often included as 'decoys' for RNases since they may be degraded in preference to the cellular RNA species. A potent RNase inhibitor called RNasin (a protein derived from rat liver), or the protein denaturing compounds guanidinium isothiocyanate and β-mercaptoethanol may also be included in the extraction buffer to combat endogenous RNase activity.

An alternative approach commonly employed for deproteinisation and the reduction of RNase activity is incubation of the lysate with proteinase K. This is a very active non-specific proteolytic

enzyme extracted from the culture medium of *in vitro*-propagated *Tritirachium album limber*. It remains active over a wide pH range, at temperatures up to 65°C, and is unaffected by EDTA, cation availability, or the presence of SDS (all of which are common components of nucleic acid extraction buffers). Incubation of the cell lysate with proteinase K therefore reduces most proteins (including the nucleases) to short, inactive peptides. These are removed relatively easily from the lysate by subsequent solvent extraction, precipitation and/or ultracentrifugation.

Nucleic acids may be isolated from other components of cell lysates by several methods based upon organic solvent solubility, gradient centrifugation, and/or adsorption to solid phase particles.

SEPARATING DNA/RNA AND PROTEINS

Once denatured or inactivated, proteins are generally separated from the DNA/RNA in lysates by one of several means. These are based upon solvent extraction, precipitation and/or centrifugation.

(1) *Phenol extraction.* One of the most common methods for recovering both DNA and RNA employs the differential solubility of proteins, lipids and nucleic acids in the potent deproteinising organic solvent phenol. Following centrifugation of the lysate/phenol mixture, three phases are observed: (i) an upper phase containing all the nucleic acids, (ii) a lower phase containing hydrophobic cell components (e.g. membrane and intracellular organelle fragments), (iii) a middle layer with the protein often concentrated at the upper interface (Fig. 45.3). The majority of the cellular nucleic acids (hopefully of full length) may be recovered from the upper phase. This must however be removed very carefully without disturbing the interface in order to avoid contamination with the proteins at the interface. The recovered DNA/RNA solution is then subjected to precipitation and alcohol washing, and the DNA/RNA resuspended in sterile deionised water. Phenol is however an extremely corrosive reagent; contact with eyes or skin, or inhalation of vapour must be rigorously avoided. All traces of phenol must also be eliminated before the DNA/RNA is used for subsequent analysis or manipulation.

(2) *Caesium chloride centrifugation.* Although a more time-consuming and laborious method, caesium chloride gradient centrifugation can be used for the separation of both DNA and RNA from cell lysates. It is especially useful for the high purity isolation of plasmids from chromosomal DNA. This method is based upon the intercalation of dyes such as ethidium bromide (EtBr) which reduce the bouyant density of DNA by increasing the distance between base pairs within the DNA helix. This relies upon linear chromosomal DNA molecules binding more EtBr per unit length than the highly supercoiled closed circular conformation (ccc) plasmid DNA. Centrifugation in a caesium chloride gradient at 125*g* for 40 hours therefore generates two distinct bands with the ccc plasmid DNA in the lower band (Fig. 45.4). The intercalated EtBr in the ccc

plasmid DNA fraction can then be removed by repeated isoamyl alcohol extractions.

(3) *RNA and DNA precipitation.* Nucleic acids may also be recovered in a relatively purified state by precipitation. For example, lithium chloride may be used to precipitate ribosomal and messenger RNA directly from the lysates obtained by proteinase K digestion. In contrast, the proteinase K, peptides and transfer RNA molecules remain in solution. Repeated precipitation of *in vivo*- and *in vitro*-derived nucleic acids by ethanol is also a common method for increasing the purity of DNA/RNA preparations. This is performed at low temperatures (e.g. 1 hour at –70°C) following neutralisation of the solution by the addition of acidic sodium acetate. Following centrifugation the nucleic acids in the

OAs: 351–69, 601–4, 1139, 1140.
RAs: 186, 204, 216, 264, 304–8.
SFR: 126–8, 137, 148.

pelleted precipitate are air dried and resuspended in deionised water prior to use.

COMPOSITION OF PHENOL SOLUTION

Phenol used in the extraction of nucleic acids requires prior redistillation in m-cresol (adding an anti-freeze property), the addition of 8-hydroxyquinone (to inhibit oxidation) and saturation with water (to prevent reduction of the aqueous layer during subsequent phase partitioning).

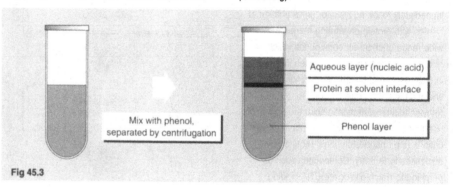

Mix with phenol, separated by centrifugation

Aqueous layer (nucleic acid)

Protein at solvent interface

Phenol layer

Fig 45.3

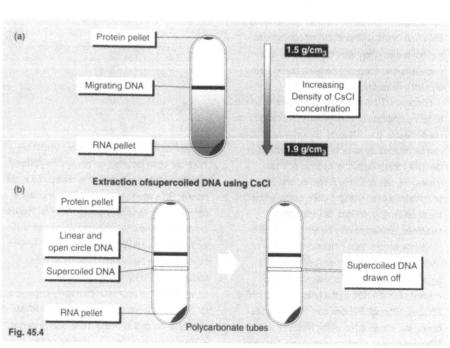

(a)

Protein pellet

Migrating DNA

RNA pellet

1.5 g/cm₃

Increasing Density of CsCl concentration

1.9 g/cm₃

Extraction ofsupercoiled DNA using CsCl

(b)

Protein pellet

Linear and open circle DNA

Supercoiled DNA

RNA pellet

Supercoiled DNA drawn off

Polycarbonate tubes

Fig. 45.4

KITS AND AUTOMATED EXTRACTIONS

Several commercial kits and automated machines have now been developed to facilitate the rapid isolation of nucleic acids (i.e. often within 1 hour). Most of the kits, and all the automated machines, utilise the same basic principles of lysis, protein inactivation/degradation and solvent extraction. These kits do however essentially differ in terms of the scale of extraction and/or the means by which the nucleic acids are finally isolated from cellular lysates. Accordingly, many kits utilise random adsorption of the nucleic acids to a solid phase particle (e.g. paramagnetic beads, silica resins or glass microspheres). These can be easily separated from the lysate, washed to remove any passively absorbed proteins, and the bound nucleic acids recovered by an elution step (e.g. incubation in low salt buffer).

Alternatively, particular oligonucleotides or DNA-binding proteins may be attached to the particles to allow specific nucleic acid species to be separated and recovered. For example, most mRNA extractions are based upon affinity capture using cellulose or paramagnetic beads coated with poly(dT) oligonucleotides (Fig. 45.5). These bind to the 3′ poly(A⁺) tails of mRNAs, thereby enhancing the recovery of full length or 5′ deleted mRNA transcripts. Similar principles of nucleic acid capture by hybridisation are also used in many gene probe assays, and for the isolation of *in vitro*-derived nucleic acids.

IN VITRO-DERIVED NUCLEIC ACIDS

In vitro-derived nucleic acids (e.g. generated by the polymerase chain reaction) may require separation from other potentially inhibitory molecules prior to their application in analytical or gene engineering steps. This can be achieved by solvent extraction and centrifugation/precipitation, or ultrafiltration in columns with differing molecular weight cut-off points (i.e. so-called spin columns) (see Fig. 45.6). In some cases the isolation of individual species of nucleic acids from the total population of nucleic acid reaction products may also be required. This is commonly achieved following size separation by gel electrophoresis and excision of the appropriate band. The DNA can then be recovered from the gel fragment by electro-elution, capillary transfer, or digestion of the gel (e.g. digestion of agarose by the enzyme agarase) followed by solvent extraction or ultrafiltration.

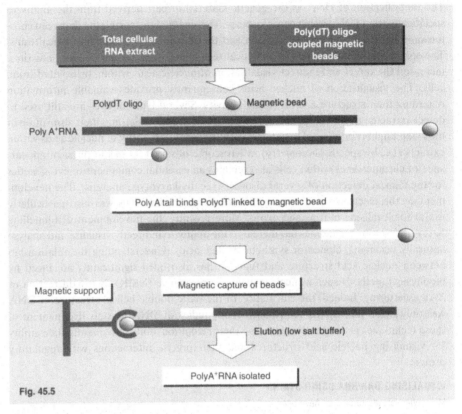
Fig. 45.5

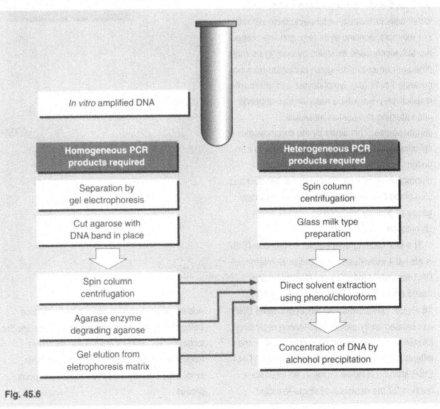
Fig. 45.6

46. VISUALISING NUCLEIC ACIDS

DNA/RNA may be directly or indirectly visualised within cells or following their isolation, by methods ranging from simple staining to high resolution electron microscopy or X-ray diffraction.

The identification of DNA as the genetic material in part derived from the ability to specifically stain DNA within chromosomes. This enabled the correlation between chromosome behaviour during cell division and the transmission of simple physical traits. The sophistication and power of analytical techniques in molecular genetics has since increased the direct and indirect visualisation of nucleic acids within, or isolated from, cells. The visualisation of nucleic acids consequently provides valuable information regarding their structure and functional activity. Several types of dyes are still used to detect extracted nucleic acids (e.g. following gel electrophoresis), and their absorption of ultraviolet light is often used to determine the quantity and purity of nucleic acids within extracts (i.e. by spectrophotometry). Microscopic observation of the physical appearance of chromosomes within cells also remains an essential component of cytogenetics for the clinical detection of several disorders (i.e. by karyotype analysis). The development of the electron microscopy by Kleinschmidt in the 1960s was also particularly useful for localising introns and exons. More recently, the development of tunnelling X-ray microscopy techniques has increased the ability to indirectly visualise and analyse naturally occurring, cloned or synthetic nucleic acids. Understanding the relationship between nucleic acid structure and function has also been significantly advanced by biophysical methods such as nuclear magnetic resonance (NMR), X-ray diffraction or X-ray scattering. Indeed, the elucidation of the basic double helical structure of DNA was only made possible by X-ray diffraction studies of DNA crystals. Refinement of these techniques is constantly improving their resolution, and consequently their utility for visualising nucleic acid structures and their precise interactions with regulatory proteins.

CHROMOSOME 'PAINTING'

The development of in situ hybridisation using differentially fluorochrome-labelled gene probes (often termed fluorescent in situ hybridisation or FISH) has allowed individual chromosomes or regions of chromosomes to be 'painted'. FISH has been successfully used to identify chromosomal regions containing specific genes or transcriptionally active gene clusters. The ability to 'paint' individual chromosomes within living cells has also been used to identify and isolate cells bearing karyotypic differences by flow cytometry. This utilises the same principles routinely used to separate phenotypically distinct cells based upon the labelling of specific cell surface proteins. The use of sex chromosome-specific 'paints' also allows the FISH technology to be used for selecting fertilised embryos of a given sex prior to implantation. FISH may also be used to isolate individual chromosomes for use in gene or genome mapping projects.

VISUALISING DNA/RNA USING STAINS

Many chemical stains or dyes have been developed and used to visualise nucleic acids within cells or tissues, or following their extraction. For example, acridine dyes (e.g. acridine orange) are still widely used as stains by cytologists (and others) in order to distinguish nucleated from non-nucleated cells (e.g. lymphocytes and erthrocytes, respectively). Karyotype analysis also depends upon staining to visualise individual chromosomes. This aided by the chromosome-specific banding patterns resulting from the non-uniform uptake of these dyes by individual chromosomal regions. Chromosome banding has also aided the primary localisation of individual genes or actively transcribing regions within chromosomes.

The intercalating agent ethidium bromide (EtBr) is also still widely used to visualise all manner of DNA fragments following gel electrophoresis based on its fluorescence under UV light (Fig. 46.1). EtBr visualisation of electrophoresed DNA can consequently provide information regarding its size, composition and conformation (e.g. the different degrees of supercoiling of plasmid DNA). EtBR staining is however rather an insensitive method for the detection of single-stranded

Fig. 46.1

nucleic acids. It is also potentially hazardous, being both a mutagen and a carcinogen. Specific target nucleic acid molecules are therefore frequently visualised using indirect methods employing their hybridisation to labelled gene probes.

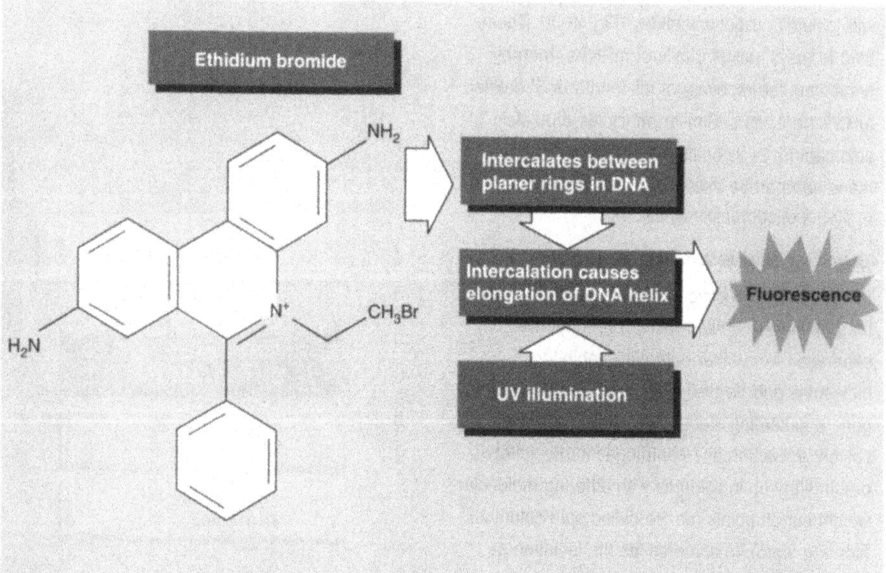

OAs: 16–17, 22–3, 29–37, 48, 63–5, 255–7, 734, 370–7. RAs: 3, 5–8, 10–15, 53, 187, 406–7. SFR: 5, 14–19, 23–30, 38–40, 43, 94.

ELECTRON MICROSCOPY

Nucleic acids can also be visualised by electron microscopy following their pre-treatment with reagents which ensure that they are spread in a monomolecular layer. This generally involves agents such as formamide which reduce clumping, and coating with a protein such as cytochrome c which further stabilises their structure (Fig. 46.2). Pre-treatments are also required to increase their dimensions and render them electron dense. Accordingly, the structurally stabilised nucleic acids are made electron dense by staining with uranyl acetate or phosphotungstic acid. They are then rotary shadowed with a metal such as platinum to enhance the contrast.

Electron microscopy has provided many initial insights into nucleic acid structures (e.g. the existence of nucleosomes). It may provide quantitative information regarding nucleic acid size, physical states, and nucleic acid concentrations within extracts, and can also provide information regarding the position and degree of homology between two nucleic acid strands. For example, the position and number of coding regions (exons) and introns in a cloned structural gene may be determined by the visualisation of loops within DNA–mRNA hybrids (termed R-loop mapping). Indeed, this provided some early clues to the existence of so-called 'split genes'. The location of large mutational differences can also be determined by the analysis of heteroduplexes (termed heteroduplex mapping).

INDIRECT BIOPHYSICAL METHODS

Several high resolution methods including X-ray diffraction, X-ray scattering, nuclear magnetic resonance, and X-ray tunnelling microscopy are increasingly providing detailed models of nucleic acid structures. A classic example of their utility was the elucidation of the double helical structure of DNA. This hinged upon the interpretation of the pattern and separation distances of spots on the X-ray diffraction photographs produced from crystalline DNA fibres by Franklin and Wilkins between 1950 and 1953 (see Fig. 46.3).

The ability of atoms to scatter X-rays has also been used to provide data for computer-based mathematical modelling of DNA and RNA molecules, either in isolation or complexed with specific DNA-binding proteins. The exact

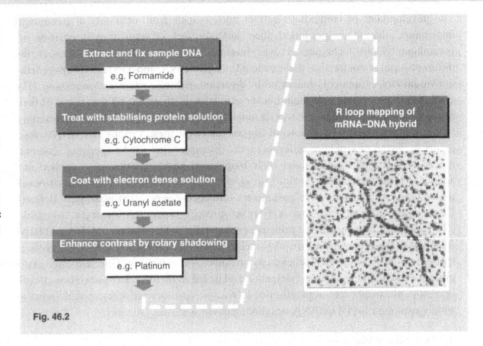

Fig. 46.2

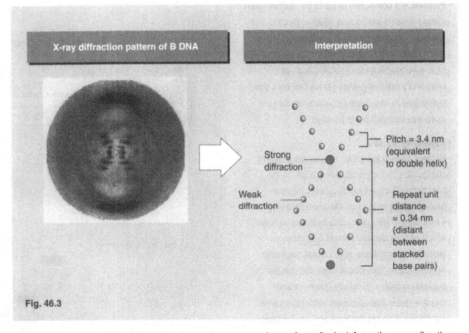

Fig. 46.3

orientation of histones in nucleosomes has only been possible using X-ray scattering techniques. X-ray diffraction analysis also provided the first details of the tertiary structure of tRNAs, thereby solving the mystery of how the two active arms remain at opposing ends of the molecule. Further insights into the structure of nucleic acids are being gained almost daily by the application of such increasingly sophisticated and accurate

analyses. Accordingly, information regarding the molecular basis of DNA–protein interactions is likely to provide answers to many fundamental questions concerning the functional activity of DNA. Such information is already providing new avenues for development, including the design of antisense or antigene reagents for the functional modification of specific genes *in vivo*.

125

47. ELECTROPHORESIS OF NUCLEIC ACIDS

The electrophoretic migration of nucleic acids is widely employed in molecular biological techniques to separate, isolate and manipulate DNA/RNA fragments based on their size, sequence and/or shape.

The development of methods to extract nucleic acids from cells was of paramount importance since they permitted their analysis and subsequent manipulation by recombinant DNA techniques. It was, however, difficult to distinguish between the different strands comprising the extracted cellular RNA and DNA, or between the individual DNA fragments generated by digestion with restriction endonucleases. The challenge of separating individual nucleic acid strands dissimilar only in terms of their nucleotide sequence and length was fortunately achieved at an early stage. Accordingly, before methods were available to distinguish amongst these various DNA/RNA fragments by virtue of sequence differences, the principles first used to separate complex mixtures of proteins based upon their biophysical characteristics were applied and adapted to nucleic acids. At the time, the electrophoretic separation of proteins was well understood, with many of the parameters influencing separation accurately defined. Consequently, of all the large variety of protein separation strategies available, electrophoresis became the first principle exploited for the separation of DNA and RNA fragments. The enduring nature of the electrophoretic separation of nucleic acids in modern molecular biology testifies to its simplicity and versatility. Thus, although many more specific and sensitive gene detection and isolation methods have since been developed, the electrophoretic separation of *in vivo-/in vitro*-derived nucleic acids remains widely exploited for DNA/RNA isolation, analysis and manipulation.

ELECTROPHORETIC MATRICES

Matrices used for the electrophoresis of nucleic acids are generally based upon polymerised agaroses or acrylamides. These vary in their handling and biophysical properties (e.g. acrylamide is neurotoxic). Their primary attribute is that they provide a porous medium which is tailored to survive the often harsh electrophoresis conditions employed (e.g. high voltage or temperature). The porosity of the matrix determines its resolving power and is generally a function of the degree of polymerisation (or cross-linking) between matrix monomers. Highly cross-linked acrylamide gels are therefore often used for high resolution analyses (e.g. nucleotide sequencing). Less highly cross-linked agarose gels are nevertheless widely used since they are easier to prepare, and greatly facilitate the subsequent recovery of the separated DNA/RNA.

NUCLEIC ACID 'HANDLES'

Proteins are composed of up to 20 biochemically distinguishable amino acid building blocks, and are often subjected to post-translational modifications (e.g. glycosylation). In the absence of *in vitro* modifications, nucleic acids are chemically relatively simple such that only size, conformation and duplex composition can be used as electrophoretic 'handles'.

Separation by size

Nucleic acids possess an overall net negative charge due to the phosphate groups on the backbone. They will consequently migrate towards the anode in an electrical field (Fig. 47.1). Under the appropriate buffer conditions which ensure denaturation of secondary or tertiary conformations, small linear fragments migrate faster than longer ones through the separating matrix. Similarly, single strands migrate faster than duplexes under the same conditions. Varying the type and concentration (i.e. degree of cross-linking) of the matrix provides different separation characteristics allowing the preferential resolution of larger or smaller fragments. Indeed, using the correct conditions of both buffer and matrix, it is possible to distinguish between DNA or RNA fragments varying in length by a single base (e.g. in nucleotide sequencing) or between identically sized fragments which vary in sequence by one base.

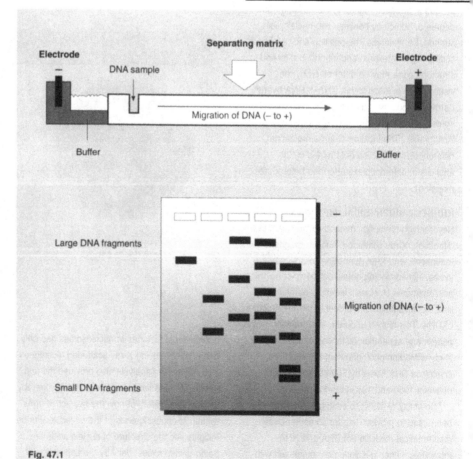

Fig. 47.1

SEPARATING DIFFERENT DUPLEXES

Several electrophoretic techniques have been developed employing gradients of chemical denaturants (i.e. denaturing gradient gel electrophoresis, DGGE), temperature (i.e temperature gradient gel electrophoresis, TGGE) and electrical gradients (e.g. pulse/inverse field gel electrophoresis, PFGE/IFGE) across the separation matrix. As duplexed nucleic acids migrate across the gel these gradients gradually denature the duplex according to the degree of complementarity between the two strands (Fig. 47.2). Heteroduplexes, composed of mismatched strands, will consequently separate at lower temperatures or denaturant concentrations than homoduplexes composed of perfectly complementary sequences. Accordingly, as the duplexes dissociate or denature, their electrophoretic mobility changes thereby dictating the final position of the DNA fragments within the gel. This general principle has been refined and manipulated to allow the development of a variety of electrophoretic techniques which are particularly suited to the detection/separation of small or large gene polymorphisms. PFGE and IFGE are therefore widely employed in genome mapping studies since they facilitate the separation of large DNA fragments.

Conformation detection

Under native conditions nucleic acids adopt a number of secondary and tertiary conformations. These are determined by their sequence, degree of complementarity between duplexed strands, and their interaction with other moieties (e.g. DNA-binding proteins). They may also adopt several conformations according to the degree of denaturation which is artificially induced. All these different conformational forms migrate at different rates during electrophoresis and can therefore be readily distinguished.

This principle is consequently widely used to distinguish between supercoiled, coiled and linearised plasmid DNA. The altered mobility of nucleic acids resulting from interaction with proteins is also widely used to investigate the mechanisms and roles of these interactions in regulating gene expression. The altered mobility of nucleic acids following interaction with labelled gene probes is also exploited in several additional analytical methods.

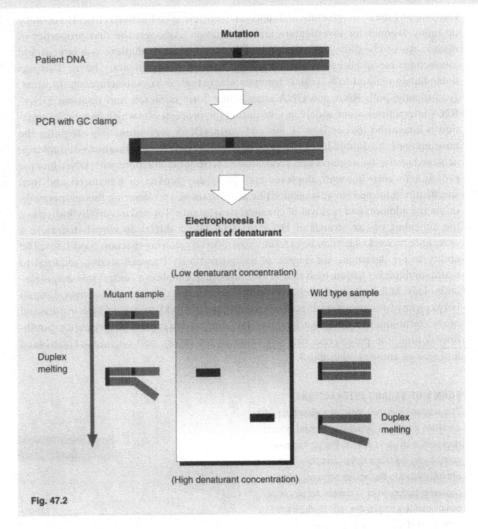

Fig. 47.2

Detecting electrophoresed DNA/RNA

Electrophoretically separated DNA or RNA fragments may be simply visualised as discrete bands within the separation matrix by staining with DNA intercalating dyes (e.g. ethidium bromide) which fluoresces under ultraviolet light. This is a rapid and still widely used method to visualise individual size-separated bands, detecting as little as 0.05 µg in a single band. The molecular size of the detected band can then be estimated by comparison to the relative mobility of commercially available 'marker' DNA or RNA fragments of pre-determined size. Other techniques utilise electrophoretic separation but rely upon the hybridisation of labelled gene probes to identify bands corresponding to gene sequences of interest (e.g. Southern and northern blotting).

Isolating electrophoresed DNA/RNA

Following detection, small-scale isolation of electrophoretically separated bands can be achieved by cutting out the relevant part of the gel. DNA/RNA within gel slices may then be recovered by the processes of electro-elution, enzymatic digestion of the separation matrix (e.g. digestion of agarose by agarase), or capillary transfer. These methods vary in efficiency but allow reasonably pure preparations to be obtained relatively quickly. These are then suitable for subsequent analysis, manipulation and *in vitro* amplification.

OAs: 378–92, 420–2, 446–8, 454, 464, 474–8, 720, 1132. RAs: 304, 306–8, 188–90. SFR: 119–24, 126–30, 162, 175–80, 225.

127

48. *IN VITRO* HYBRIDISATION

In vitro hybridisation involving the formation of duplexes between two nucleic acid strands can be readily manipulated, and is central to many analytical and manipulative recombinant DNA methods.

Determining the basic structure and function of nucleic acids has been critical in opening up many avenues for investigation and exploitation. Amongst the first properties of nucleic acids to be discovered was their ability to form stable duplexes via inter-strand interactions mediated by hydrogen bonding between complementary bases. This was immediately realised to be critical for maintaining their *in vivo* structure and function. Accordingly, both RNA and DNA strands may form duplexes, and transient DNA–RNA interactions occur widely *in vivo* during the processes by which genetic information is converted into cellular action and during DNA replication. This requires the unwinding of the double helix and subsequent breaking (or denaturation) of duplexes, achieved *in vivo* by a complex and coordinated series of interactions with DNA-binding proteins. *In vitro* however, duplexes may be readily broken or denatured and then specifically reformed (or re-annealed) by simply raising and lowering the temperature, or via the addition and removal of chemical denaturants. Termed *in vitro* hybridisation, the annealing of two strands of DNA or RNA into a duplex *in vitro* is therefore a reversible process which can be carefully controlled by altering reaction conditions. The ability to pre-determine the degree of complementarity between strands required to maintain duplexes has ensured that *in vitro* hybridisation plays a central role in molecular biology. Manipulating the relatively simple parameters determining the hybridisation between any two given nucleic acid strands is therefore fundamental to both *in vitro* and *in vivo* techniques for detecting, isolating, targeting, and manipulating genes (e.g. Southern blotting, the polymerase chain reaction amplification, and oligonucleotide-based antigene or antisense therapies).

MELTING TEMPERATURES

The energy required to break the hydrogen bonds holding the duplexes together is a function of the number of hydrogen bonds that are present. This is determined by both length and G+C content of the sequences, since G–C interactions involve formation of three hydrogen bonds rather than the two involved in A–T interactions. The temperature at which 50% of specific duplexes break to form single strands is referred to as the 'melting' temperature, or T_m. The T_m can be measured experimentally (by detecting a hyperchromic shift) or calculated from the sequence length and relative number of G and C bases. Duplexes formed between strands varying by as little as one base will consequently have a lower T_m than completely complementary strands. This is widely exploited in many analytical techniques.

OAs: 364, 370, 393–423, 426–33, 548–51. RAs: 191–4, 198, 209, 250, 304, 308. SFR: 119–24, 126–8, 131.

FORMS OF STRAND INTERACTIONS

The similarities in their structures allow both DNA and RNA strands to hybridise or anneal to form duplexes *in vivo* and *in vitro*. Perhaps somewhat confusingly, the terms homo- and heteroduplexes are often used to distinguish between both duplexes composed of (i) complementary or non-complementary sequences, and (ii) duplexes composed of identical or non-identical polynucleotides (i.e. DNA–DNA and RNA–RNA, or DNA–RNA) (Fig. 48.1).

In vivo, nucleic acids are most commonly found as homoduplexes due to their inherent stability. Indeed, sophisticated mechanisms are required to unwind these duplexes (e.g. to allow access to the proteins involved in transcription or DNA replication). Whilst generally short-lived *in vivo* due to their susceptibility to degradation by nucleases, stable heteroduplexes may be formed and maintained *in vitro*. This however depends upon the degree of complementarity between the two strands, and the stringency of the hybridisation conditions used. Indeed, many assays exploit such differences in duplex composition as the basis of detection, isolation, analysis, modification and amplification of DNA or RNA sequences. It has recently been

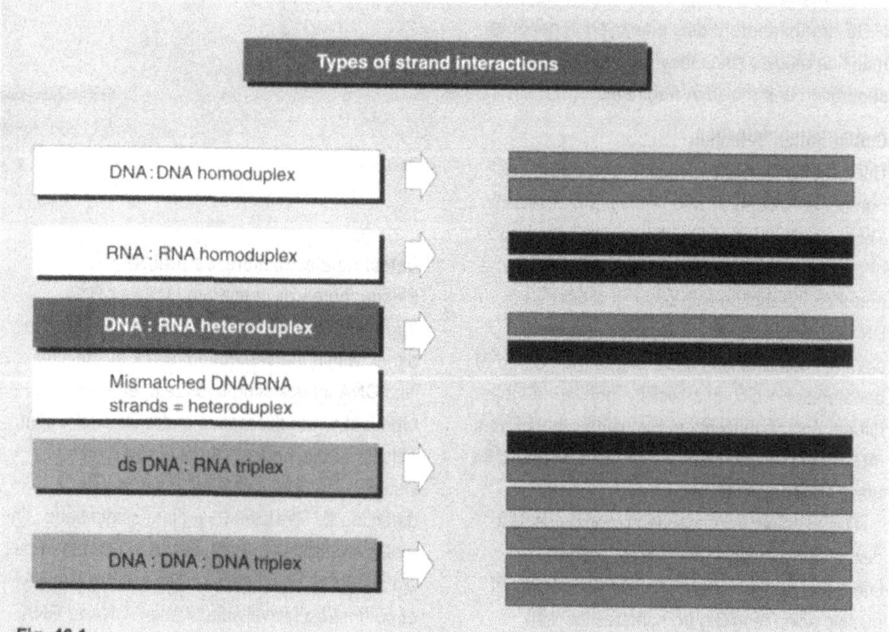

Types of strand interactions

DNA : DNA homoduplex

RNA : RNA homoduplex

DNA : RNA heteroduplex

Mismatched DNA/RNA strands = heteroduplex

ds DNA : RNA triplex

DNA : DNA : DNA triplex

Fig. 48.1

demonstrated that some *in vivo* processes also involve the transient formation of triplexes. Stable triplexes and even tetraplexes can also be formed *in vitro*, and have been used to investigate structural

mechanisms determining gene function. Triplex formation *in vivo* may be central to the successful development of antigene oligonucleotide-based gene therapies.

CONTROLLING HYBRIDISATION

The formation and disruption of stable homo- or heteroduplexes can be carefully controlled *in vitro* depending upon the 'stringency' of the conditions employed. Under stringent conditions of low salt, high temperature, and/or high concentration of chemical denaturants, only duplexes composed of perfectly complementary sequences may be formed and maintained (Fig. 48.2). Stable duplexes can however be readily formed between non-complementary sequences under less stringent conditions by simply altering the buffer composition or hybridisation temperature. Consequently, imperfectly matched or non-complementary sequences will anneal at lowered temperatures, at high salt concentration, and in the absence of a chemical denaturant (e.g. formamide, dimethylsulphoxide). Tailoring the hybridisation conditions in terms of ion or denaturant concentration and temperature therefore determines the degree of complementarity required to form a stable duplex.

UTILISING *IN VITRO* HYBRIDISATION

The capacity to determine the specificity of hybridisation between two nucleic acid strands by altering buffer or temperature conditions is central to most analytical and recombinant DNA techniques (Fig. 48.3). Hybridisation is used in all gene probe-based techniques, which essentially vary in terms of the type of assay format used and nature or location of the target sequence. Accordingly, gene probes may be hybridised to DNA or RNA within fixed cells or tissues (e.g. by *in situ* hybridisation), or following electrophoretic separation (e.g. in Southern or northern blotting). Many electrophoretic methods have also been developed which utilise temperature or denaturants to detect differences in duplex composition (e.g. homo- or heteroduplexes) according to their mobility within the separating matrix. Repeated denaturation and annealing is also at the heart of *in vitro* gene amplification by the polymerase chain reaction. Accordingly, altering annealing temperatures is used to determine the specificity of binding of the synthetic oligonucleotide primers. *In vitro* primer-mediated enzymatic copying of RNA or DNA (e.g. for the production of cDNA) is also dependent upon carefully tailored hybridisation conditions.

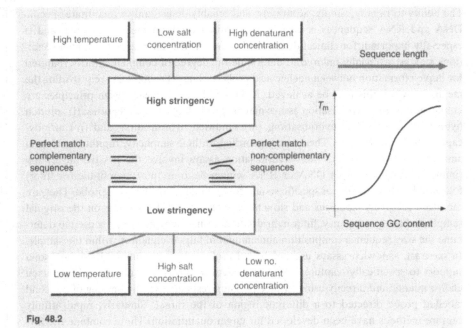

Fig. 48.2

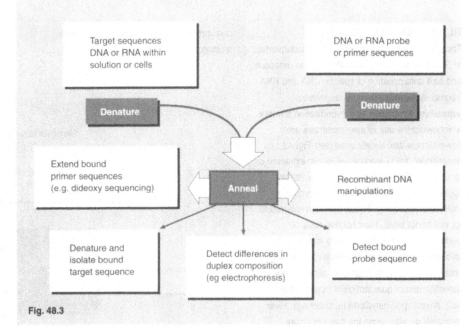

Fig. 48.3

Specific hybridisation is also used in dideoxynucleotide sequencing, and capture-based analytical techniques. Several nucleic acid isolation strategies also utilise the principles of *in vitro* hybridisation. For example, polythymine oligomers (termed oligo-dT) bound to cellulose or magnetic microspheres are used to specifically capture messenger RNA molecules (via their 3′ poly-dA tails) which are then recovered by reversing the process. The ability to hybridise complementary and non-complementary strands is also vital for gene manipulations, including the introduction of DNA into vectors via cohesive ends, and *in vitro* mutagenesis.

49. TYPES OF HYBRIDISATION ASSAY FORMATS

Four main forms of *in vitro* hybridisation assay have been developed based upon the use of immobilised targets or probes, or hybrid formation in solution, often followed by capture-immobilisation.

The ability to rapidly, simply, accurately, and reliably detect and/or quantitate specific DNA and RNA sequences is inceasingly required for fundamental research, and is especially important for clinical diagnosis. *In vitro* hybridisation assays in which conditions may be artificially tailored to determine the degree of complementarity required for duplex formation between nucleic acid probe and target sequences are providing the means by which this may be achieved. Although the same basic design principles are employed, *in vitro* hybridisation assays may be divided into four categories: (i) solution hybridisation, (ii) filter hybridisation, (iii) sandwich hybridisation, and (iv) affinity-capture hybrid collection. These vary according to their simplicity, rapidity, analytical and quantitative capacity. Filter hybridisation assays involve the passive adsorption (immobilisation) of target DNA or RNA sequences to an inert membrane (or filter) followed by the detection of specific sequences by binding to a labelled probe. They are therefore relatively laborious and slow to provide a result. Depending on the original sample and assay stringency, filter hybridisation formats may however be used to determine the size, sequence composition and amount of target sequences within the sample. In contrast, sandwich assays use unlabelled probe sequences immobilised to a solid support to specifically 'capture' target sequences. Hybrids are subsequently detected and/or quantitated directly using surface plasmon resonance or the binding of a second labelled probe directed to a different region of the target. Similarly, rapid affinity capture methods have been developed for target quantitation. These combine the favourable kinetics of solution hybridisation with the ability to separate target-bound probe and unbound probe offered by immobilisation-based strategies.

THE IMPORTANCE OF QUANTITATION

Semi-quantitative assays such as northern blotting and ribonuclease protection assays have proved useful in basic and applied research (e.g. for assessing transcriptional activity based on mRNA levels). Nevertheless, few rapid, simple, reliable and accurate quantitative methods are available. Such fully quantitative methods would find particular application in clinical diagnostics where quantitation of gene copy number or mRNA levels may be extremely informative. For example, the ability to quantify the copy number of amplified oncogenes may be useful in the diagnosis and prognostic evaluation of certain cancers. Quantitation of specific mRNA or RNA molecules may also be useful in the diagnosis and monitoring of a range of diseases (e.g. resulting from viral infection).

OAs: 403–21, 423, 430, 499, 692–6, 698, 722. **RAs:** 186, 191–4, 304, 306, 308. **SFR:** 126–7, 131, 181, 192–3.

FILTER HYBRIDISATION

The principle of filter hybridisation is fundamental to Southern and northern blotting for the detection and size determination of specific DNA and RNA fragments following their electrophoretic separation. Alternative filter hybridisation formats which avoid the use of electrophoresis are however now also widely used (see Fig. 49.1). Accordingly, small amounts of *in vivo*-extracted or *in vitro*-amplified nucleic acid may be directly applied and fixed to filters as discrete dots or slots prior to probing. These are often referred to as slot or dot blots. Their relative ease of performance and their potentially high sample throughput makes them attractive assays for both research and clinical diagnostic, allowing the detection and/or quantitation of target sequences (e.g. based upon densitometry scanning). Filter hybridisation also forms the basis of colony blotting, widely used for the detection of specific DNA/RNA sequences within *in vitro*-cultured microbes. This approach has found widespread application, especially for the screening of genomic and cDNA libraries for recombinant bacterial or bacteriophage clones. Thus, small amounts of cultured microbes may be transferred to a filter, and treated to both release and immobilise their nucleic acids prior to probing.

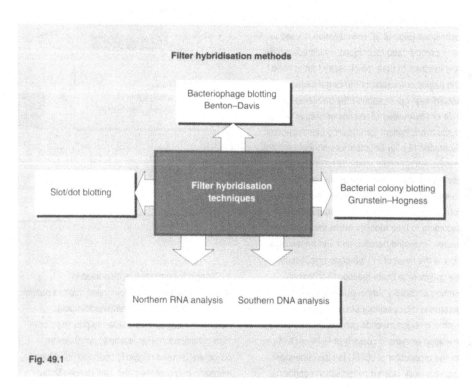

Filter hybridisation methods

Bacteriophage blotting
Benton–Davis

Slot/dot blotting

Filter hybridisation techniques

Bacterial colony blotting
Grunstein–Hogness

Northern RNA analysis   Southern DNA analysis

Fig. 49.1

SOLUTION HYBRIDISATION ASSAYS

The formation of duplexes in solution has been demonstrated to follow more favourable (i.e. rapid) kinetics than when using immobilised target (or probe) sequences. Solution hybridisation also obviates the problems of high background and inaccessibility of immobilised sequences encountered in filter hybridisation.

Despite these obvious advantages the development of solution hybridisation assays has not been straightforward. In particular, problems arise from the need to be able to distinguish between target-bound and unbound probe. Several types of sophisticated solution hybridisation assay formats have therefore been, or are currently being, developed which circumvent this problem. These are based upon the use of novel signal-generating systems. For example, two differentially labelled probes may be used, each directed to adjacent sites on the target molecule such that a detectable signal is only generated once they are brought into close juxtaposition (i.e. following hybridisation with the target).

SANDWICH AND CAPTURE ASSAYS

Although less sophisticated, sandwich hybridisation assays are more readily available. These also take advantage of the faster kinetics of solution hybridisation, and utilise two discretely binding oligonucleotide (oligo) probes (Fig. 49.2). Most commonly, an unlabelled oligo probe directed to a discrete region of the target sequence is immobilised to an inert solid support (either by passive adsorption or by chemical linkages which retain their accessibility for hybridisation). Specific hybrids formed in solution by hybridisation of targets with a labelled probe may then be captured (in the form of a 'sandwich') onto the surface of the solid support. The use of solid particles (e.g. paramagnetic microspheres) then allows their separation from unbound probe sequences for subsequent signal detection.

Immobilisation of unlabelled probe is also used in capture assays which obviate the need for a second labelled probe (e.g. for the separation and detection of intrinsically labelled targets produced by *in vitro* amplification). Alternatively, unlabelled targets may be captured by hybridisation with unlabelled surface-immobilised probes followed by detection using surface plasmon resonance.

AFFINITY-CAPTURE-BASED ASSAYS

A range of so-called affinity-capture, or hybrid collection, assays have also been developed which exploit the rapid kinetics of solution hybridisation (Fig. 49.3). These do not involve the direct use of immobilised probes for capture. They do however involve the formation of duplexes between target molecules and two discretely binding probes, one bearing a detectable label (e.g. enzyme or isotopic marker), and one carrying a ligand molecule (e.g. biotin) (Fig. 49.3). The latter probe is then used for immobilisation based upon interaction of the ligand with an immobilised receptor molecule (e.g. streptavidin). The labelled hybrids can then be separated for detection. Affinity-capture methods are increasingly being used in both clinical and research assays. For example, they are widely employed for the detection of microbial pathogens, and for the isolation of polymerase chain reaction (PCR) products. This facilitates both their quantitation and analysis (e.g. by dideoxy sequencing).

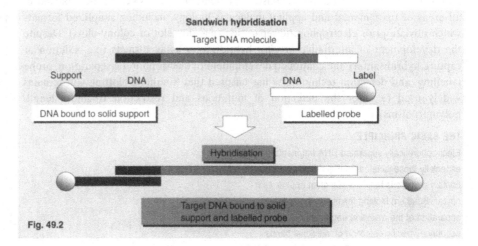

Sandwich hybridisation

Target DNA molecule

Support — DNA — DNA — Label

DNA bound to solid support | Labelled probe

Hybridisation

Target DNA bound to solid support and labelled probe

Fig. 49.2

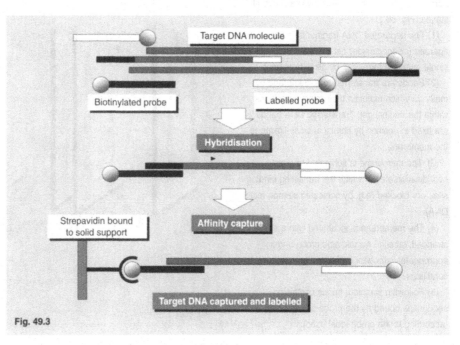

Target DNA molecule

Biotinylated probe | Labelled probe

Hybridisation

Affinity capture

Strepavidin bound to solid support

Target DNA captured and labelled

Fig. 49.3

50. SOUTHERN BLOTTING

Southern blotting allows specific DNA sequences to be detected and characterised by hybridisation of a DNA probe to electrophoretically size-separated fragments following transfer to a membrane.

The ability to reproducibly generate defined fragments of DNA by digestion with restriction enzymes, and separate them according to their sizes by electrophoresis, revolutionised DNA analysis. Indeed, both techniques remain widely exploited in analytical molecular biology. Until the 1970s, however, a significant problem facing molecular geneticists was how to specifically detect those fragments derived from a particular gene or containing sequences of interest amongst the complex mixture of fragments generated from genomic DNA samples. The solution was provided by Edwin Southern working in Edinburgh, UK, who developed a simple method, now known as Southern blotting. This method is based on transferring the separated bands from the gel to a membrane and then detecting specific bands by hybridisation with a labelled gene probe. This also served as the prototype for the subsequent development of similar methods which allow the analysis of RNA and proteins (i.e. northern and western blotting), and combination assays such as South-western blotting where labelled DNA probes bind to electrophoresed proteins (or vice versa). The principles of membrane immobilisation and detection by hybridisation established by Southern have since been employed in all areas of fundamental and applied molecular biology, including simplified formats which obviate prior electrophoretic separation (e.g. dot, slot or colony blots). Despite the development of alternative *in vitro* hybridisation assay formats (e.g. solution or capture hybridisation), the continued development of electrophoretic separation, probe labelling, and detection technologies has ensured that Southern blotting still remains widely used (e.g. for the detection of mutations and restriction fragment length polymorphisms).

THE BASIC PRINCIPLE

Electrophoretically separated DNA fragments, especially those isolated from genomic samples, contain numerous bands which often appear as a smear. Southern blotting maintains the spatial separation of the digestion fragments but allows sequence-specific detection of individual bands from amongst this smear. This involves several stages (Fig. 50.1).

(1) The separated DNA fragments in an agarose (or acrylamide) gel are denatured into a single-stranded form, e.g. by alkali treatment.

(2) Bands are transferred to a membrane by methods which maintain their relative positions within the original gel. Transferred DNA bands are fixed in position by chemical cross-linking to the membrane.

(3) The membrane is subjected to a pre-hybridisation step in which the remaining binding sites are blocked (e.g. by sonicated salmon testis DNA).

(4) The membrane is incubated with a single stranded, labelled nucleic acid probe under appropriately stringent buffer and temperature conditions.

(5) Following washing, bands containing sequences bound by the probe are detected (according to the probe label used).

TYPES OF MEMBRANES

Southern blotting was originally developed using nitrocellulose membranes, whereas diazobenzyloxmethyl (DBM) paper was used in northern blotting since this has the necessary surface groups (absent from nitrocellulose) for chemically binding RNA molecules. A range of synthetic (generally nylon-based) membranes are however now available for Southern and northern blotting. These have improved handling properties (i.e. being less brittle), and pose less problems of non-specific or background probe binding. These membranes are also reusable, since bound probe (but not the fixed sample bands) may be removed and the membrane rehybridised with a second probe. Indeed, this process can often be repeated as many as 10 times without significant loss of sensitivity or increased background binding problems. This allowed the development of multiplex sequencing approaches used by genome mapping projects.

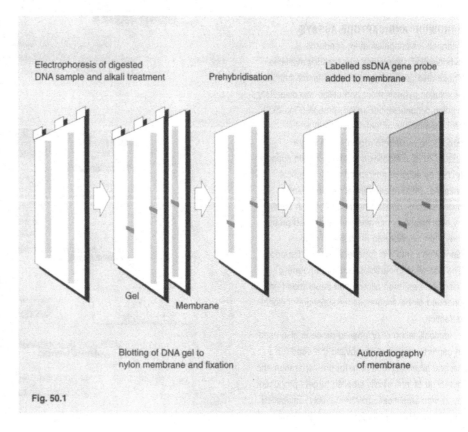

Electrophoresis of digested DNA sample and alkali treatment

Prehybridisation

Labelled ssDNA gene probe added to membrane

Gel

Membrane

Blotting of DNA gel to nylon membrane and fixation

Autoradiography of membrane

Fig. 50.1

Denaturation and fixation

All hybridisation assays require that the target DNA is present in single-stranded form in order to be accessible for probe binding. In Southern blotting this is achieved by alkali denaturation of the DNA in the gel before transfer. Following transfer, the positions of the bands are then maintained by permanently fixing the single-stranded DNA to the membrane in such a way as not to subsequently render the fixed DNA inaccessible to the probe. This is generally achieved by chemical cross-linking, baking at high temperatures (e.g. 80°C), or exposure to ultraviolet irradiation. The choice of a particular fixation method is largely dependent upon the type of membrane and whether the target is DNA or (in the case of northerns) RNA.

Transfer methods

Transfer of DNA from gels to nitrocellulose originally involved capillary action in which the gel is placed on top of buffer-saturated filter papers, with a piece of dry membrane placed on top of the resulting stack (Fig. 50.2). The stack is completed with dry filter paper, layers of absorbent tissue paper, and finally a weight. Buffer is consequently drawn up the stack by capillary action allowing the DNA bands to migrate onto the membrane in the same spatial arrangement as in the original gel. Since this may take many hours (typically overnight), several more rapid methods are now available. These commonly involve downward flow, vacuum-assisted or electrical (i.e. electroblotting) transfer protocols.

Pre-hybridisation

Regardless of the type of membrane used, pre-hybridisation steps are always required in order to reduce non-specific binding of labelled probe to remaining free sites. This commonly involves incubating the fixed membrane in Denhardt's solution, comprising buffer containing 0.2% Ficoll (a synthetic sucrose polymer), 0.2% polyvinyl-pyrrolidone, and 0.2% bovine serum albumin. An

OAs: 403–9, 420–5, 430.
RAs: 191, 304, 306, 308.
SFR: 120–4, 126, 127, 131, 174.

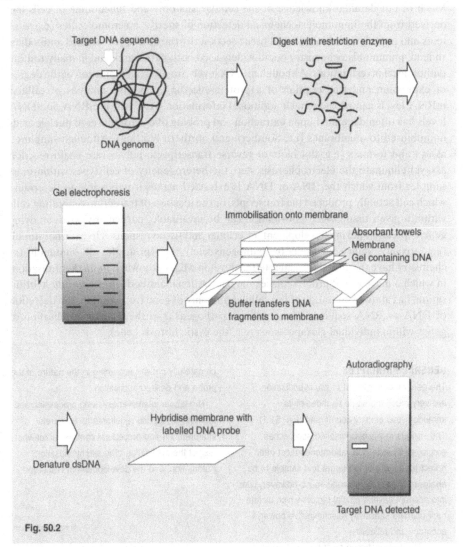

Fig. 50.2

irrelevant nucleic acid (e.g. tRNA or sonicated salmon testis DNA) is also often included to further reduce probe background binding.

HYBRIDISATION AND DETECTION

Polymerase chain reaction-amplified, cloned, or chemically synthesised DNAs and RNAs may all be used as probes. The essential requirement is that the probe is single stranded, or at least made single stranded (e.g. by heat denaturation), before use in the hybridisation step. This is required to allow their binding to complementary target sequences. Like all types of hybridisation assay, the temperature and buffer conditions used will determine the degree of specificity of probe binding (i.e. stringency). Southern blotting originally relied upon the binding of radioactive

probes detected by autoradiography (i.e. exposure to X-ray film). Because exposure times can be as much as several days, alternatives to autoradiography have been developed. For example, phosphor imagers can rapidly detect and quantitate the amount of radioactive phosphorous present in a specific portion of the membrane. These have the advantage of speed, and on-line computer analysis options.

The availability of non-radioactive gene probe-labelling molecules has allowed a wide variety of end point detection systems to be used, according to the degree of sensitivity required. For example, probes labelled to give a chemiluminescent end point can produce the equivalent of an autoradiograph, with the same detection sensitivity, but in only a few minutes.

51. *IN SITU* HYBRIDISATION

In situ hybridisation employs a variety of nucleic acid probes to semi-quantitatively detect and spatially locate DNA or RNA targets (e.g. mRNA transcripts) within cells or tissue sections.

Much of our detailed knowledge of the cellular anatomy and functioning of cells has derived from the immunohistochemical detection of specific macromolecules (e.g. proteins and peptides) within cells and tissue sections by the binding of labelled antibodies. Indeed, immunohistochemistry is still widely used both in research and in many routine pathological investigations. Although mRNA levels are often an indication of the degree of expression and transcription of a protein-encoding gene, the analysis of cellular mRNA levels can provide much additional information. Analysis of mRNA and DNA levels has often depended upon extraction and probing of electrophoresed nucleic acids immobilised to membranes (i.e. Southern and northern blotting). Although simplified assays and formats (e.g. dot blots or reverse transcriptase polymerase chain reaction assays) eliminate the electrophoresis step, the heterogeneity of cell types within most samples from which the RNA or DNA is extracted makes it impossible to determine which cell actually produced the transcripts, or the location of transcript-generating cells within a given tissue. Such information can be invaluable, particularly when studying gene expression during development, or cellular and tissue responses to external stimuli (e.g. toxicity, injury, invasion by pathogens etc.). The principles of immunohisto-chemistry have therefore been adapted to develop what is known as *in situ* hybridisation, in which a nucleic acid probe replaces the conventional antibody probe. *In situ* hybridisation, has allowed the study of the cellular anatomy of gene transcription, the detection of DNA or RNA sequences from cellular pathogens, and the primary localisation of genes within individual chromosomes (e.g. the cystic fibrosis gene).

SAMPLE AND TARGET TYPES

In situ hybridisation has been performed on a wide range of cells and tissues obtained by biopsy, post-mortem sampling or following *in vitro* cell culture. Tissue sections are generally processed whilst frozen or once paraffin-embedded. Improved permeabilisation methods also permit the analysis of cells in solution, allowing isolation of labelled cells and quantitation of bound probe per cell (e.g. by flow cytometry). Both DNA and RNA targets may be detected by *in situ* hybridisation. The detection of mRNA transcripts is perhaps the most frequent application, although it is often also used to detect RNA and DNA sequences from invading pathogens. *In situ* hybridisation using fluorescently labelled probes has also aided the primary chromosomal localisation of some genes.

OAs: 426–9, 431–3, 916, 1142, 1143.
RAs: 191–2, 151, 153, 158–9, 308–9.
SFR: 124, 126–7, 132–5, 175, 180–1.

GENERAL PRINCIPLES

The general principles of *in situ* hybridisation are very much the same as those of its immunohistochemistry counterpart (Fig. 51.1). The various *in situ* hybridisation approaches employed by individual laboratories most often reflect the nature of the target and sample to be analysed. All assay protocols must, however, take into account several crucial factors which dictate their ultimate sensitivity, discriminative power, accuracy and reliability.

(1) The cells or tissue must be pre-treated in such a way as to maintain cellular morphology, and the integrity and position of the target. The treatment must also make the cells permeable enough to allow the nucleic acid probe to enter and for any unbound probe to be removed.

(2) The appropriate type (i.e. DNA/RNA, single or double stranded), sequence, and length of nucleic acid probe must be used. This effectively determines the overall specificity and sensitivity of the method.

(3) The type of label and detection system must also be carefully selected since this also determines the degree of sensitivity, specificity, ease of detection, and accuracy of target quantitation that can be achieved.

(4) The hybridisation conditions must similarly be carefully chosen according to the nature of the probe and desired application.

The labour intensiveness, long processing and analysis times, and requirements for careful interpretation and procedural controls, somewhat restrict the application of *in situ* hybridsation. Automation, and the development of computer-assisted imaging/analysis apparatus, may however increase the accuracy of target quantitation, and its general applicability in research or clinical diagnosis.

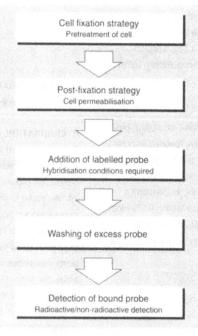

Fig. 51.1

- Cell fixation strategy
  Pretreatment of cell
- Post-fixation strategy
  Cell permeabilisation
- Addition of labelled probe
  Hybridisation conditions required
- Washing of excess probe
- Detection of bound probe
  Radioactive/non-radioactive detection

FIXATION AND PROBE PERMEABILITY

Successful *in situ* hybridisation is dependent upon the effective fixation and permeabilisation of cells or tissues. Fixation in paraformaldehyde is most commonly used in order to prevent the degradation of targets by endogenous nucleases, especially when studying mRNA. Excessive fixation will however often decrease probe binding (and hence sensitivity) by preventing entry of the probe. Consequently, post-fixation permeabilisation strategies often involve the use proteases or detergents (e.g. sodium dodecyl sulphate, Triton X-100). These aid probe penetration, improve signal intensity, and reduce non-specific probe binding. Acetic anhydride also reduces non-specific probe binding by acetylating cellular proteins, thereby reducing their potential for forming electrostatic interactions with the probe. Sample pre-treatment may not however be required when using cultured cells or short probes.

TYPES OF PROBES

DNA and RNA probes obtained by cloning, *in vitro* amplification or chemical synthesis have all been used for *in situ* hybridisation (Fig. 51.2). Several factors influence the type of probe used, including availability, desired sensitivity and specificity, and the nature of the target. Single-stranded probes ranging in length from 200 to 500 bases generally give optimum results since they can better penetrate the cells. They are also more sensitive than double stranded probes which are prone to self-annealing rather than target hybridisation. RNA probes are often favoured because of the greater stability of RNA–RNA hybrids over RNA–DNA hybrids. Both RNA and DNA oligonucleotide probes are relatively cheap to synthesise, can be produced from nuclease-resistant nucleotides (e.g. in the form of peptide nucleic acids), or intrinsically labelled during synthesis. Since they rarely exceed 50 bases in length they easily diffuse into both permeabilised and non-permeabliised cells. RNA probes recognised and amplified by Qβ-replicase enzyme may also be used. These eliminate the need for separate probe labelling.

PROBE LABELLING AND DETECTION

Both radioactive and non-radioactive probe labels have been used, primarily depending on the sensitivity, speed and resolution required (Fig. 51.2). Radioactive probes commonly utilise ^{32}P, ^{35}S and ^{3}H (tritium) labels. ^{32}P requires the shortest autoradiography time (3–6 days) but gives poor resolution, whereas tritium gives a high resolution but requires exposure times of 3–6 weeks. Hybrids within individual cells or areas of tissue may consequently be detected using X-ray film (macro-autoradiography), or photographic emulsions applied to the slide (micro-autoradiography). Biotin and digoxigenin are now commonly used non-isotopic labels. Bound probe is then detected using fluorescent or enzyme-labelled anti-biotin and anti-digoxigenin antibodies. Both give sensitivities equal to that of radioactive labels and may be combined to detect dual targets within the same tissue or cell.

HYBRIDISATION CONDITIONS

The conditions used for *in situ* probe binding are similar to those required for any hybridisation assay. Assay temperature is based upon probe length and G:C content, but in general a good starting point is 25°C below the probe T_m. Whilst most hybrids will form within a few hours, the hybridisation step is usually performed overnight. Non-specific probe binding requires the use of buffers containing Denhardt's, 'irrelevant' nucleic acid and high salt. The inclusion of dithiothreitol, dextran sulphate and formamide also enhances assay specificity and binding kinetics.

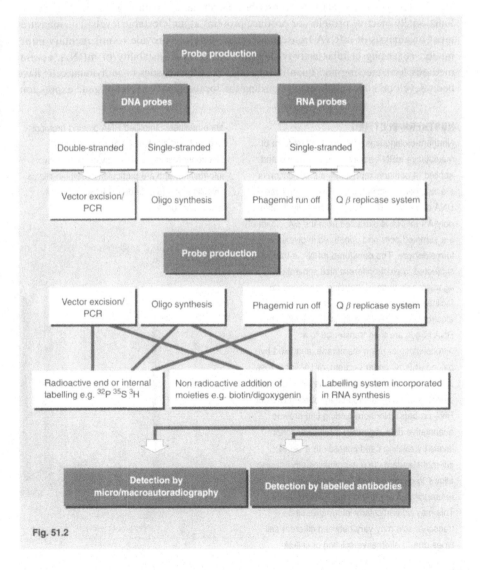

Fig. 51.2

52. MEASURING TRANSCRIPTIONAL ACTIVITY VIA MESSENGER RNA

Tanscriptional activity within cells or tissues may be determined by detection and quantitation of mRNA transcripts using a variety of hybridisation, mRNA labelling, or *in vitro* amplification-based assays.

Living cells are dynamic entities, often likened to mini-factories, continually producing new macromolecules for repair, division or for secretion/export. The rate of protein production therefore correlates with the functional activity of a given cell. The level of production of any particular protein varies at different stages within the lifetime of a cell (e.g. during development), or may fluctuate in response to external stimuli. Measuring the level of production of particular proteins has therefore been of enormous importance in understanding both normal and pathological cellular processes. This is reflected by the enormous effort expended upon developing quantitative protein assays. The quantitative measurement of some proteins within, or exported by, cells is however difficult or complicated because they are often produced at extremely low levels. Furthermore, a given protein type may be produced by several different cell types, and undergo post-translational processing or modifications which render them undetectable to conventional probes (e.g. monoclonal antibodies). Since protein production is a function of the transcriptional activity of a given gene, the measurement of the amount of short-lived messenger RNA transcripts generated by transcription can often be used as a superior alternative to protein quantitation. Indeed, this may identify particular disease-associated genes which are actively transcribed at an earlier pathological stage, generate functionally inactive proteins, or produce proteins at undetectable levels. The measurement or analysis of mRNA transcription may therefore provide complementary information regarding cellular activity. Despite the relative instability of mRNA, several methods for detecting and quantitating mRNA within tissues or individual cells have been developed to provide data regarding the location and/or level of gene expression.

AVOIDING RNA DEGRADATION

Total and messenger RNA (mRNA) can be extracted from cells using several methods. mRNA is often referred to as 'short-lived' RNA, being relatively unstable and rapidly degraded by cytoplasmic enzymes (RNases). Consequently, mRNA extraction protocols usually include RNase inhibitors in all steps where these enzymes may be present. Analysis or manipulation of mRNA must also be performed under conditions designed to avoid degradation. With the exception of commercial reagents which are supplied 'RNase-free', all apparatus, buffers and plastic disposables must be made RNase-free. This is commonly achieved by autoclave sterilisation employing, for example, diethylpyrocarbonate (DPEC). RNases are ubiquitously present in bodily secretions, therefore all *in vitro* manipulations should be kept to a minimum, where possible be performed below ambient temperature (i.e. on ice), and protective sterile latex gloves should be worn.

OAs: 403–4, 411–12, 418, 430–7, 479, 509–11, 916–19. RAs: 201–4, 281, 304, 306, 308, 321. SFR: 126, 131, 133–4, 137, 140, 180–1, 193.

NORTHERN BLOTTING

Northern blotting was amongst the first form of quantitative mRNA assay to be developed and applied. It primarily represents a modification of the Southern blotting method used to analyse DNA sequences (Fig. 52.1). Essentially, total or poly(A+) mRNA is extracted from the cytoplasm of the sampled cells and denatured in glyoxal or formaldehyde. The denatured mRNA is then subjected to electrophoretic size separation in agarose gels under denaturing conditions (to facilitate transfer and binding of the electrophoresed RNA to the solid matrix). The RNA bands are then transferred to a nitrocellulose or nylon membrane, and fixed by baking at 80°C under vacuum, or UV light cross-linking respectively. Specific transcripts are then detected using an appropriately labelled DNA or RNA probe(s). Northern blotting can provide quantitative data regarding mRNA transcript levels by relating band intensity to a set of internal standards (e.g. by densitometry). It also allows transcript size to be determined by reference to a set of molecular weight markers. This may be particularly informative since transcript size may vary between different cell types due to alternative splicing or cellular

abnormalities. Simplified RNA blotting protocols (e.g. dot and slot blots) which avoid the electrophoresis step also provide quantitative information, and are particularly useful in clinical investigations (e.g. to detect transcriptionally active oncogenes).

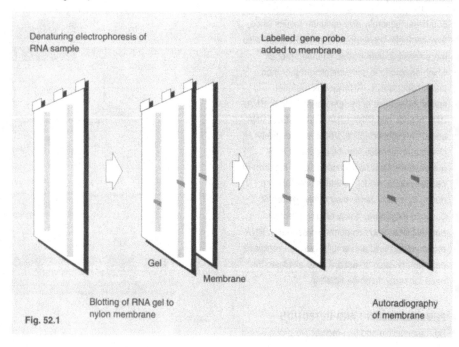

Denaturing electrophoresis of RNA sample

Labelled gene probe added to membrane

Gel

Membrane

Blotting of RNA gel to nylon membrane

Autoradiography of membrane

Fig. 52.1

23. Transcription.
24. Processing of mRNA transcripts.
26. Transcription control mechanisms.
30. *In vivo* translation.
31. Protein transport sequences.
45. Isolation of nucleic acids.
47. Electrophoresis of nucleic acids.
48. *In vitro* hybridisation.
50. Southern blotting.
51. *In situ* hybridisation.
53. Converting mRNA to cDNA.
55. Polymerase chain reaction.
58. Types of gene probe.
60. Nucleotide analogues.
61. Probe labelling methods.
76. Identifying protein-coding regions.
93. Oncogenes and cancer.
94. Cancer diagnosis and therapy.

SAMPLE HETEROGENEITY

With the exception of cell lines or purified cell subsets, most samples contain a heterogeneous population of cells. The extracted RNA consequently contains large amounts of 'irrelevant' RNA species which effectively 'dilute' the concentration of a specific transcript. This substantially reduces the sensitivity of blot hybridisation assays by increasing background (i.e. non-specific) binding of the probe, and preventing the original cellular source of the transcript from being identified.

In situ hybridisation

In situ hybridisation can be used on a wide range of samples to detect specific mRNA transcripts within individual cells. Thus, a labelled probe may be hybridised to specific mRNA transcripts fixed within cells or tissues (Fig. 52.2). The time-consuming and semi-quantitative nature of the assay, and the need for careful control and interpretation, has however largely limited this assay to research or specialised applications.

Nuclear run-off assay

Often used to discriminate between altered transcriptional rates or changes in mRNA stability, this assay, involves mixing isolated nuclei with radiolabelled dNTPs. This allows the elongation of initiated transcripts but does not allow the initiation of new transcripts. The amount of specific mRNA transcripts elongated can then be detected by hybridisation to an immobilised complementary probe.

RIBONUCLEASE PROTECTION ASSAY

In principle, the ribonuclease protection assay (RPA) involves incubating the extracted RNA with a labelled RNA probe in solution (see Fig. 52.3). This allows the formation of specific RNA–RNA (target–probe) hybrids. The mixture is then subjected to digestion with ribonucleases A and T1 which specifically degrade any single-stranded RNA molecules. This results in degradation of any non-specific sample RNAs and unbound probe, whilst leaving the double-stranded (target-probe) RNA–RNA hybrids intact. Following electrophoresis, the amount of hybridised target can then be determined from the intensity of the band representing undegraded probe.

Since the ribonuclease step eliminates background hybridisation problems, RPA is generally more accurate and sensitive. The improved kinetics of RNA–RNA hybrid formation in solution also makes it a more rapid assay than its blot hybridisation counterparts. The RPA can also be used to detect cell- or species-specific mRNA transcripts representing splicing or sequence differences. These are detected by the partial cleavage of probe sequences within hybrids at the non-base-paired regions.

Similar approaches utilise DNA probes and S1 nuclease digestion of single stranded probes (Fig. 52.4). RNA targets may thus be detected based upon degradation of single stranded probe sequences within or extending beyond the hybridised regions. In addition to mRNA quantitation both RPA and S1 nuclease may be used in transcript analysis to detect 5′ and 3′ ends, or for identifying intron–exon boundaries. Reverse transcriptase mapping, whereby a DNA probe hybridised to an mRNA coding region is enzymatically extended, can also be used for mRNA quantitation. It is, however, generally used analytically to map transcript mRNA 5′ ends.

POLYMERASE CHAIN REACTION (PCR)

Several PCR-based approaches have been developed for the quantitation of mRNA following the generation of a cDNA copy (i.e. using reverse transcriptase). These assays often employ a reporter gene approach. This involves the co-amplification of a known amount of an 'internal standard' sequence. The level of amplification of the 'internal standard' can then be used to back-calculate the initial target level. Internal standards include 'housekeeping' genes (e.g. rRNA encoding) which are transcribed at constant levels. Alternatively, sequences which have been *in vitro* manipulated to contain specific mutations, or which are shorter or longer than the target mRNA, can be used. Generally however, quantitative PCR assays require further development/refinement, but such assays promise increased rapidity, reliability and sensitivity (down to the single cell level).

Fig. 52.2

Fig. 52.3

Fig. 52.4

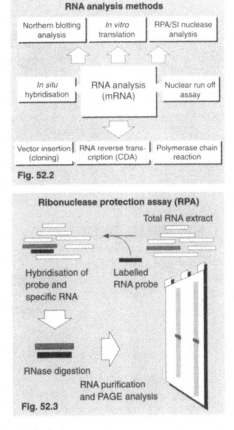

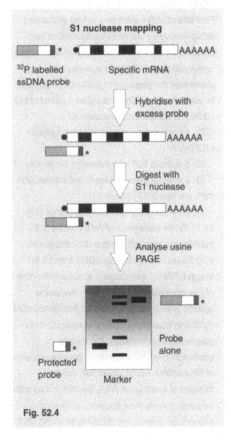

53. CONVERTING MESSENGER RNA INTO COMPLEMENTARY DNA

Reverse transcriptase can generate a first strand cDNA copy of an mRNA transcript, which can then be used in several ways to produce double stranded DNA suitable for inserting into cloning vectors.

The ability to extract messenger RNA (mRNA) from cells and tissues has greatly aided the investigation of gene structure. mRNAs are also ideal targets for cloning and *in vitro* manipulation by genetic engineering because they represent only the protein-encoding regions of a gene. The short half-life and inherent instability of mRNA however precludes its direct cloning, and complicates its use as a template for nucleotide sequencing. Cloning and sequencing of mRNA therefore requires the initial conversion of mRNA molecules into complementary DNA copies, termed cDNAs. The resultant single-stranded cDNA may then be directly sequenced or used as the substrate for the generation of double-stranded DNA suitable for introduction into cloning vectors. The first stage of the process requires RNA-dependent DNA polymerases, termed reverse transcriptases, derived from RNA-containing viruses (i.e. retroviruses). Termed reverse transcription, this process leads to the synthesis of a first strand cDNA copy of the mRNA by the extension of an annealed oligonucleotide primer. Primers may either be specific for a single mRNA species (e.g. annealing to a portion of the coding sequence), or capable of annealing to all mRNAs (e.g. to the poly(A$^+$) tail), thereby enabling cDNA library construction. Conventionally, double-stranded DNA is generated by a second strand synthesis reaction using two strategies, either: (i) the use of DNA polymerase I and S1 nuclease, or (ii) the use of RNAase H, DNA polymerase I and DNA ligase. More recently, the polymerase chain reaction (PCR) has been widely adopted as a means to generate multiple double-stranded DNA molecules from first strand cDNA molecules generated by reverse transcription. This increases the number of double-stranded DNA molecules that can be obtained from a single to facilitate cloning and analysis.

VIRAL REVERSE TRANSCRIPTION

First discovered in 1970, reverse transcriptases are RNA-dependent DNA polymerases found within certain RNA-containing animal cell viruses (termed retroviruses). These viral-encoded enzymes are vital to the process of viral replication. Thus, following infection, the few reverse transcriptase molecules present in the viral capsid convert the viral RNA genome into a DNA copy. This enables its incorporation into the host cell genomic DNA by the process of illegitimate recombination. Following integration viral genes are then activated to produce all the components needed for the assembly of new viral particles. This retroviral function is now being exploited for the development of retroviral vectors for use in gene therapy and the creation of transgenic mammalian cells and organisms.

OAs: 331, 332, 434, 438–45, 479, 691. **RAs:** 183, 195, 196, 202, 250, 251. **SFR:** 119–28, 136–9.

FIRST STRAND SYNTHESIS

First strand cDNA synthesis is best performed using relatively pure mRNA preparations (e.g. obtained by affinity chromatography) which have preferably been isolated in a manner that maximises the presence of full length transcripts. In addition to the pool of denatured mRNAs, four additional components are required:

(1) a pool of deoxynucleotide triphosphates (dNTPs) in;

(2) a suitable buffer to maximise the activity of;

(3) a reverse transcriptase to synthesise DNA from the free 3′ OH group of;

(4) an annealed oligonucleotide primer (Fig. 53.1). Since eukaryotic mRNAs contain a 3′ poly(A$^+$) tail, a poly(dT) oligonucleotide primer may be used to synthesise cDNA from all full length mRNAs. Alternatively, a mixture of primers of random sequence, or a single sequence-specific primer, may be used for first strand cDNA synthesis from prokaryotic or eukaryotic mRNA preparations. In all cases, the reverse transcription reaction ultimately generates a pool of RNA–cDNA hybrid molecules. Following removal of the original RNA, the cDNA may either be used in nucleotide sequencing, or for the subsequent generation of double-stranded DNA.

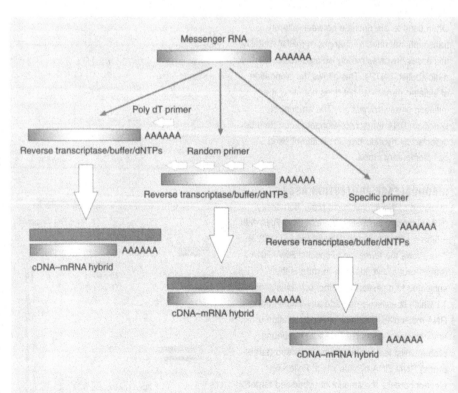

Fig. 53.1

SECOND STRAND REACTIONS

Two approaches for second strand synthesis are conventionally used. Both methods produce blunt-ended double-stranded DNA molecules which may either be introduced into a cloning vector directly, or following the addition of linkers, adaptors, or homopolymer tails. They vary primarily in the way in which the original RNA template is removed from the resulting RNA–cDNA hybrid, and how synthesis of the new DNA strand is primed.

(1) Under certain conditions the reverse transcriptase extends the synthesis of the first strand cDNA past the 5′ end of the original mRNA molecule to create a hairpin loop (Fig. 53.2). Following removal of the original mRNA molecule by mild alkali hydrolysis (e.g. using sodium hydroxide), this hairpin loop can serve as a primer for the synthesis of a second strand DNA molecule by DNA polymerase I. The hairpin loop can then be removed by digestion with S1 nuclease. This step is however hard to control and may result in considerable loss of additional nucleotides at the 5′ end.

(2) First strand cDNA molecules are generated in the presence of 4 mM sodium pyrophosphate which efficiently inhibits the formation of hairpin loops (Fig. 53.3). The resulting RNA–cDNA hybrids are then subjected to limited digestion with the enzyme RNase H which introduces random breaks (or 'nicks') within the RNA backbone. This produces a number of RNA fragments bearing free 3′ OH groups which can be used by DNA polymerase I to synthesise short DNA fragments. These can then be joined (ligated) by DNA ligase. Any protruding 3′ ends are finally removed from the double-stranded DNA molecules by a brief incubation with T4 DNA polymerase. Importantly, however, this method will also lead to limited loss of 5′ nucleotides. This results from the exposure of a few nucleotides at the most 5′ nick, which makes them susceptible to removal by the 3′–5′ exonuclease activity of DNA polymerase I.

Reverse transcriptase PCR

The PCR has now been widely adopted for the generation of double-stranded DNA molecules using first strand cDNAs produced in a separate reaction by reverse transcriptase. Often referred to as RT-PCR, this results in a dramatic increase

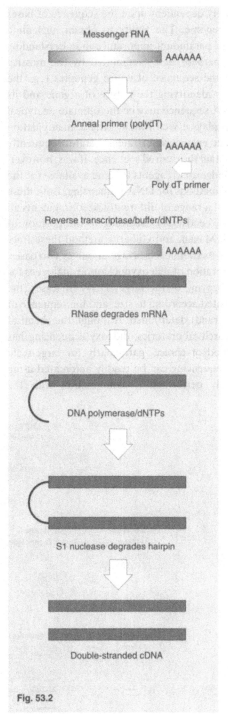

Fig. 53.2

in the number of DNA molecules produced. RT-PCR can also be tailored to allow the introduction of terminal restriction enzyme recognition sequences to facilitate cloning, or used to directly detect or analyse mRNA transcripts (e.g. produced by differential splicing). More recently,

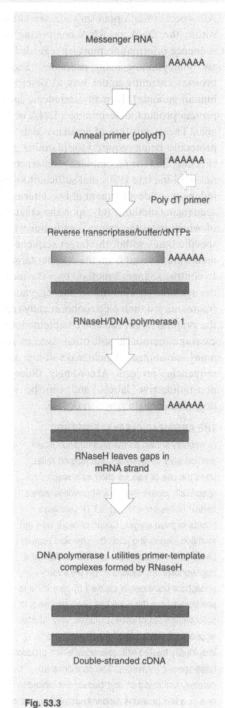

Fig. 53.3

several of the available heat-stable DNA polymerases used in PCR have been genetically engineered to contain intrinsic reverse transcriptase activity. Their application consequently eliminates the need for separate first strand cDNA synthesis reactions.

54. METHODS FOR DETERMINING DNA NUCLEOTIDE SEQUENCES

Two methods of nucleotide sequencing have been developed which utilise base-specific chemical cleavage, or base-specific termination of the enzymatic extension of DNA chains by dideoxy analogues.

All aspects of an organism's life are ultimately dependent upon the sequence of bases within the DNA or RNA comprising its genome. The ability to obtain nucleotide sequence information must be regarded as of paramount importance in understanding and manipulating living processes. This is perhaps best exemplified by the massive projects currently under way to determine the sequence of entire genomes (e.g. the human genome). In many instances, such as identifying the nature of a gene and its protein product, determining a DNA or RNA sequence may be the ultimate analytical goal. The acquisition of sequence data also plays a vital role in genetic manipulation protocols, being required for planning genetic engineering strategies and subsequently checking that the recombinant constructs contain the desired sequence. It was, however, not until the late 1970s that sufficient knowledge and reagents became available for the independent development of two alternative methods for DNA sequencing. Both these sequencing methods rely upon the creation of a range of differently sized fragments all of which have the same 5′ ends but whose 3′ ends vary according to the position of specific bases within the target sequence. In Maxam and Gilbert's method these fragments are derived from the original target by its chemical cleavage at one or two bases. In contrast, Sanger's method uses the incorporation of dideoxynucleotide analogues for the specific termination of the enzymatic copying of the target DNA. In cases, the fragments are then electrophoretically separated according to size, and the sequence of the original target (and its complementary strand) determined. Although the chemical cleavage method is still often used in research laboratories, dideoxy sequencing has many advantages which make it the method of choice, particularly for large-scale sequencing projects. Accordingly, dideoxy sequencing can be readily automated using non-radioactive labels, and can be rapidly performed in combination with the polymerase chain reaction (PCR).

GEL ELECTROPHORESIS

Both sequencing methods ultimately rely upon gel electrophoresis to size-separate the individual fragments in each reaction. This generates what is termed a 'sequencing ladder' in which each rung of the 'ladder' represents a fragment which varies in length from its neighbour by only one nucleotide. The sequence of the target DNA can therefore be read (starting from the shortest fragments at the bottom of the gel) according to the order of the rungs. Sequencing therefore depends upon the resolving power of the electrophoretic step. This is generally achieved using a highly cross-linked acrylamide gel under denaturing conditions (e.g. in 7 M urea and at about 70°C). This ensures that the fragments remain single stranded and do not adopt secondary structures which may influence electrophoretic mobility. A variety of detection methods, including sophisticated automated gel scanners and readers, are now available depending upon the strategy employed to label the fragments.

THE CHEMICAL CLEAVAGE METHOD

Commonly known as the Maxam and Gilbert method after its inventors, this method relies upon the use of various chemicals which specifically cleave the DNA at positions where certain bases appear (Fig. 54.1). Cleavage occurs in three stages: (i) one or two bases are modified, depending upon the chemical reagents used; (ii) the altered base is removed from the sugar-phosphate backbone; (iii) the sugar–phosphate backbone is cleaved by piperidine at positions where the base has been removed. In practice, the target DNA is radioactively labelled at one end, divided into four aliquots and individually treated with chemicals which produce base-specific cleavages. The conditions are carefully controlled so that cleavage is limited to one or a few positions randomly distributed within each target DNA molecule. A range of different sized fragments is therefore generated. The fragments from each reaction are then size-separated by gel electrophoresis and, following autoradiography using X-ray film, the sequence of the target DNA is determined from the relative positions of the fragments within the gel.

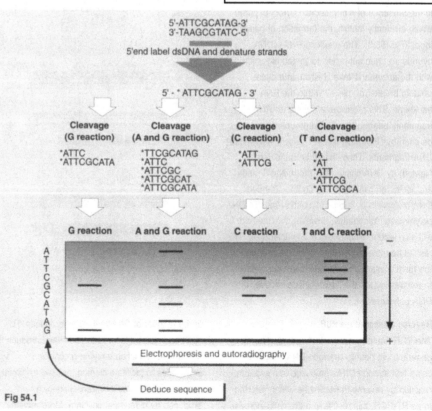

Fig 54.1

THE DIDEOXY METHOD

The dideoxy method was developed by Sanger and colleagues in 1977. It relies upon the production of a range of DNA fragments by enzymatic copying of the target sequence using DNA polymerase (Fig. 54.2). These fragments are generated by base-specific termination of the growing chain by the incorporation of dideoxynucleotide analogues (i.e. ddNTPs) which lack a 3′ hydroxyl group. This enzymatic copying requires the initial binding of a short oligonucleotide primer to a complementary sequence within the target. This is followed by the extension of the free 3′ ends of the primers such that each fragment has the same 5′ end. In its original/standard form, four individual reactions are performed in which limited amounts of one of the four dideoxys (i.e. ddATP, ddCTP, ddGTP, ddTTP) are included with an excess of dNTPs, target DNA, the primer, and DNA polymerase in an appropriate synthesis buffer. This generates different sized fragments which vary at the 3′ end according to the position at which a ddNTP is incorporated. Following a brief incubation, the fragments are electrophoretically size-separated to generate a sequence ladder.

One strand or two?

Chemical cleavage sequencing can only be performed on double-stranded DNA targets (e.g. restriction digest fragments or isolated plasmid DNA). In contrast, single-stranded templates were preferred for dideoxy sequencing. This prevented the preferential reannealing of the two target strands which can block primer annealing resulting in sequencing artefacts or short reads. This generally required prior cloning of the target DNA into appropriate single-stranded DNA generating vectors (e.g. M13). More recently, *in vitro* single stranded DNA generation and strand separation techniques have been used. The inclusion of low concentrations of duplex destabilising agents (e.g. dimethylsulphoxide, formamide or single-stranded binding proteins), and PCR 'cycle' sequencing protocols, now allows dideoxy sequencing of double-stranded

OAs: 336–9, 445–9, 479, 889.
RAs: 184, 197, 426.
SFR: 119–30, 136–9, 145, 181.

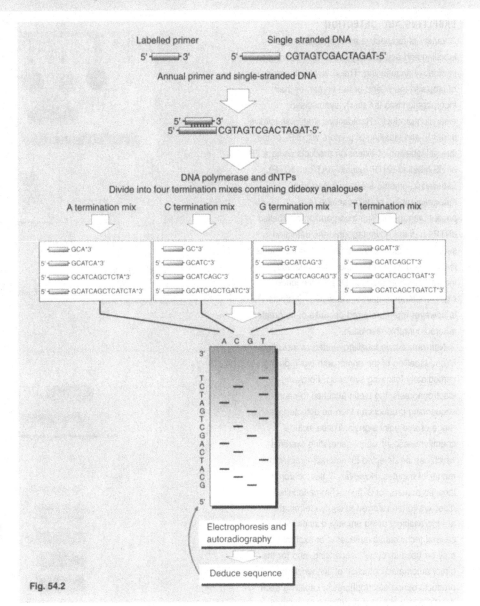

Fig. 54.2

DNA. It must however be remembered that only one strand is sequenced in any one reaction.

Types of DNA polymerase

Dideoxy sequencing originally employed the Klenow fragment of *E. coli* DNA polymerase I, since it lacks 5′→3′ exonuclease activity. This would otherwise recognise and remove the dideoxynucleotide. The T7 bacteriophage DNA polymerase (known as sequenase) is now frequently used in modified protocols, although the thermo-stable *Taq* DNA polymerase (used in

PCR) is increasingly finding favour, especially for double stranded templates. *Taq* DNA polymerase also enables sequencing reactions to be performed at high temperatures (65–70°C) which avoids many artefacts caused by DNA secondary structures. Indeed, protocols based on PCR 'cycle' sequencing have also been developed which allow total automation of the sequencing process. It is also possible to use the dideoxy principle to directly sequence RNA templates by substituting reverse transcriptase for the DNA polymerase.

141

Many modified dideoxy sequencing protocols and reagents (e.g. dye terminators) have been developed which enhance the reliability, speed, accuracy, analysis and automatization of DNA sequencing.

LABELLING AND DETECTION

A variety of radioactive and non-radioactive labelling and detection systems have been used in dideoxy sequencing. These involve the addition of labels to the 5′ end of the primer, or their incorporation into the newly synthesised extension products. Radioactive strategies involve either 5′ end labelling of primers with ^{32}P- or the internal labelling of extension products using a ^{32}P or ^{35}S-labelled dNTP (usually dATP or dCTP). Labelled fragments are then detected by autoradiography, or gel scanners (e.g. phosphorimagers). The incorporation of labelled dNTPs has the advantage that the detection sensitivity of the less abundant longer fragments is increased, since more than one label may be incorporated per product. Despite the longer exposure times required, the weaker β-emitter ^{35}S is however often preferred because of its greater autoradiographic resolution.

Non-radioactive labelling strategies include the 5′ end labelling of the primer with biotin during or immediately following synthesis. Following electrophoresis, the biotin attached to each sequencing product can then be detected using a range of end point signals. These include chemiluminescent signal-generating systems which can be detected by autoradiography in a matter of minutes. However, 5′ fluorochrome-labelled primers, or 3′ fluorochrome-labelled dideoxys (often referred to as dye terminators) are increasingly being employed since they offer several technical advantages. For example, both may be used in 'cycle' sequencing, and for the direct automated detection of sequencing products during electrophoresis. Labelling each ddNTP with a different fluorochrome emitting at wavelengths 20 nm apart also enables sequencing to be performed in a single reaction rather than the standard four reactions. Furthermore, they eliminate the need to synthesise and label each sequencing primer. Dye terminators also avoid the problems associated with false termination products (e.g. enzyme 'stuttering'). Available DNA polymerases do not however incorporate each ddNTP with

OAs: 445–56, 538, 544, 731–2, 889, 1155–62. **RAs:** 198, 211, 422, 424, 426. **SFR:** 119–30, 136–9, 145, 181.

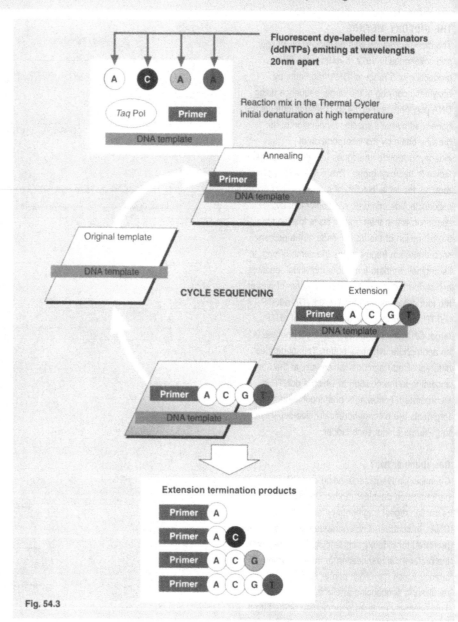

Fluorescent dye-labelled terminators (ddNTPs) emitting at wavelengths 20nm apart

Reaction mix in the Thermal Cycler initial denaturation at high temperature

Annealing

Original template

CYCLE SEQUENCING

Extension

Extension termination products

Fig. 54.3

equal efficiency, often resulting in variable signal intensities.

PCR 'CYCLE' SEQUENCING

The advent of the PCR has had a two-fold effect upon dideoxy sequencing. Firstly, it has provided a rapid means of producing large amounts of both double- and single-stranded DNA (or RNA) sequencing templates. Secondly, since both PCR and dideoxy sequencing rely upon enzymatic primer extension, the two techniques have

recently been combined to generate what has been termed 'cycle' sequencing (Fig. 54.3). Cycle sequencing involves the linear formation of sequencing termination products by simply adding all the sequencing reagents and Taq polymerase to a small amount of DNA template and subjecting it to rounds of PCR thermal cycling. Consequently, both double- and single-stranded templates can be used with equal efficiency. Sequencing reactions can also be performed in a single tube using dye terminators.

AUTOMATION OF DNA SEQUENCING

The development of new primer and ddNTP labelling chemistries has been paralleled by the development of machinery for automating part or all of the sequencing process. Several types of apparatus are therefore currently available. The first to be developed were the automated gel or autoradiograph scanners which detect bands by directly measuring their radioactivity or by densitometry respectively. These also incorporate on-line computerised sequence interpretation facilities.

A range of single unit apparatus has also been developed to combine gel electrophoresis, band detection, and sequence reading (Fig. 54.4). Consequently, sequencing reactions are performed manually employing fluorochrome-labelled primers or dye terminators and the products applied to the top of a pre-cast gel. As the gel is electrophoresed the labelled termination product bands are detected individually as they pass a scanning charged coupled device (CCD) camera detector or laser excitation-detector unit. Using dye terminators, the sequence of up to 48 samples can be analysed in one gel within a few hours and at a relatively low cost (important factors in large-scale sequencing projects). These apparatuses consistently provide up to a kilobase of sequence data per sample per day with an accuracy of >99%.

The final goal of technologists is full automation of both the sequencing and detection processes. In other words, the addition of sample DNA to the apparatus is all that will be required for the acquisition of sequence data. The development of dye terminators and cycle sequencing protocols has meant that single-unit automated work stations combining automatic robotic reagent additions, 'cycle sequencing', gel electrophoresis, band detection and computer analysis will shortly be available. Although initially expensive to purchase, their cost will soon be recouped by the time saved in producing and analysing templates.

All of these apparatuses, however, ultimately depend upon the sophistication and reliability of their associated computer software packages. These must incorporate features which allow on-line interpretation and editing to correct mis-reads/mis-assignments caused by sequencing

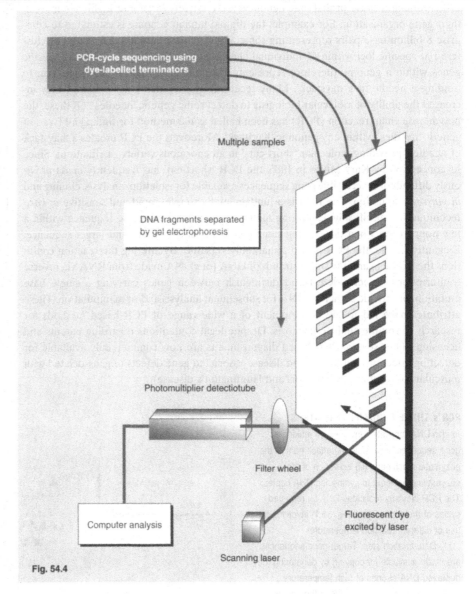

Fig. 54.4

artefacts (e.g. band compressions arising from DNA secondary structures).

Computer databases

Approaching 100 million base pairs of sequence are already contained within international computer databases. Many data still remain to be added and the figure increases daily, especially from projects aimed at determining the sequences of entire genomes. Making sense of such massive amounts of DNA sequence data requires the continued development of software analysis packages. Without these the tasks of sequence comparison, gene mapping, searching for restriction sites, determining open reading frames (i.e. protein-encoding regions), deriving consensus sequences for regulatory sequences (e.g. promoters and enhancers) and analysing secondary structures (e.g. hairpin loops) would be beyond the human brain. It is therefore likely that molecular biologists of the future must include computer literacy as an equally important part of their skill repertoire.

55. THE POLYMERASE CHAIN REACTION

The PCR is a simple and rapid *in vitro* method capable of synthesising millions of identical DNA copies of a defined sequence from as few as one DNA, cDNA or RNA strand.

Eukaryotic and some prokaryotic genomes are relatively large and highly complex in their gene organisation. For example, the diploid human genome is estimated to comprise 8 billion base pairs representing some 50000 genes randomly distributed or clustered at specific loci within 46 individual chromosomes. Isolating or detecting specific genes within a genome therefore represents a formidable technical challenge akin to 'finding a needle in a haystack'. Many technical developments have dramatically increased the ability of molecular biologists to detect some genetic 'needles'. Of these, the polymerase chain reaction (PCR) has been hailed as *the* method for finding all types of genetic 'needles' within any genomic 'haystack'. Moreover, the PCR creates a 'haystack of needles' providing molecular 'short-cuts' in an enormous variety of situations. Since its conception by Kary Mullis in 1985, the PCR 'short cut' has frequently made previously difficult or inaccessible gene sequences available for isolation, analysis, cloning and *in vitro* manipulation. The PCR is a fundamentally simple, rapid and sensitive *in vitro* technique for specifically generating millions of copies of a specific sequence within a few hours, even when the starting sample contains only one original target sequence. Elegantly simple, the PCR is also remarkably versatile. By altering the reaction conditions the PCR can amplify single strands of DNA (or cDNA made from RNA via reverse transcription), directly detect and distinguish between genes carrying a single base mutation, or provide sufficient DNA for subsequent analysis and/or manipulation. These attributes have led to the development of a wide range of PCR-based methods for research or routine clinical diagnosis. Despite legal contentions regarding patents and licensing rights, several PCR-based diagnostic kits are now commercially available for detecting infectious pathogens and disease-associated gene defects (e.g. associated with muscular dystrophy, cystic fibrosis and Huntington's disease).

TOO HOT FOR MOST POLYMERASES

The technical breakthrough required to develop a usable PCR method was the isolation of a heat-stable DNA polymerase found in a microbe, *Thermus aquaticus*, naturally inhabiting hot water springs. This so-called *Taq* polymerase is capable of remaining active throughout the high denaturation temperatures required at the beginning of each amplification cycle. Prior to this, fresh heat-labile DNA polymerase had to be added at the beginning of each amplification cycle. Ironically, the PCR has since been used to isolate and/or genetically engineer additional heat-stable DNA polymerases. These often possess altered properties (e.g. with proof-reading capacity, increased fidelity, or intrinsic reverse transcriptase activity) and are defined by their own 'commercially derived' names (e.g. *Vent* DNA polymerase).

OAs: 337–8, 457–65, 685–7, 706–18, 756–62. RAs: 199–204, 243, 251–4, 264–6, 304–8. SFR: 120–8, 136–40, 166–9, 173–81.

PCR's THREE FUNDAMENTAL STEPS

In vivo DNA replication involves: (i) making a gene sequence accessible; (ii) attachment of the enzymatic machinery; (iii) activation of the enzymatic machinery to synthesise DNA copies. The PCR similarly replicates DNA by repeated cycles of three steps (see Fig. 55.1) involving the use of different reaction temperatures.

(1) *Denaturation step*. Target gene sequences are made available for copying by denaturing the duplexed DNA strands at high temperature.

(2) *Annealing step*. The ends of the particular gene sequence to be copied are defined by the hybridisation (or annealing) of two short synthetic oligonucleotides, termed amplification primers or amplimers. This allows the subsequent binding and activation of the DNA polymerase. This is normally performed at a temperature close to the amplimer's melting temperature to ensure highly specific primer hybridisation.

(3) *Extension step*. The specific gene sequences are copied, normally as double-stranded DNA, by the addition of new bases to the free 3' ends of the primers by a heat-stable DNA polymerase. This is usually performed at

Fig. 55.1

around 70°C which represents the activity optimum of the DNA polymerase (i.e. the temperature at which it is most active).

Under certain circumstances the PCR can be performed using only two temperature cycles.

Accordingly, the annealing step may be omitted if the melting temperatures of the primers are sufficiently high (i.e. close to the activity optimum of DNA polymerase).

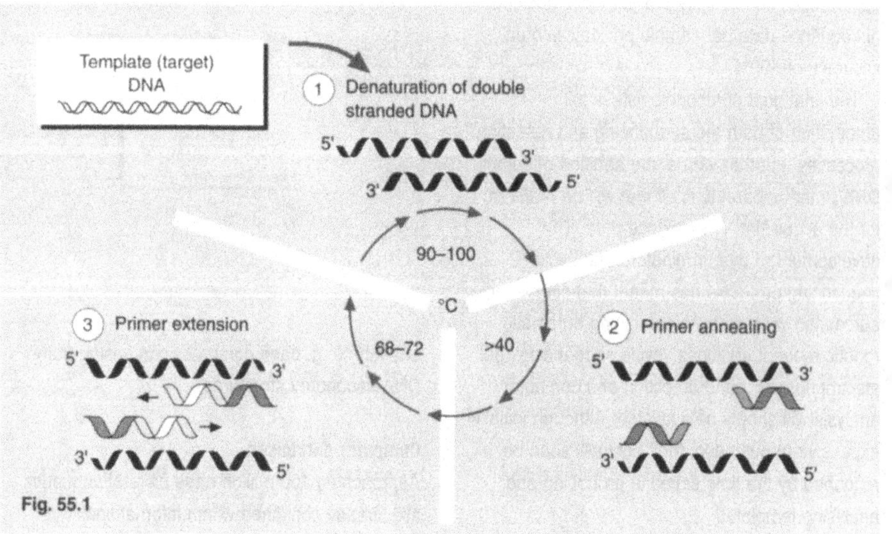

PCR's SIMPLE RECIPE

Essentially, five reaction components are required for PCR amplification: (i) a buffer tailored to suit the DNA polymerase; (ii) a pool of nucleotide triphosphates; (iii) a heat-stable DNA polymerase; (iv) at least one DNA or cDNA strand containing the target sequence to be copied; and (v) a pair (or more) of synthetic oligonucleotide primers (amplimers). Once combined, the reactants are subjected to 25 to 40 amplification temperature cycles in a simple programmable heating block which ensures accurate temperatures and timings (Fig. 55.2). After the first cycle, new strands of DNA have been generated, each of which have their 5′ ends defined by the amplimers. In subsequent cycles these serve as templates for the exponential accumulation of DNA fragments of equal size and sequence. Both the 3′ and 5′ ends of these fragments are now defined by the two amplimers. Each PCR cycle takes only a few minutes. Consequently, millions of identical and readily detectable copies of the target sequence may be obtained within a few hours.

PCR PRODUCT DETECTION/ISOLATION

A single PCR can generate sufficient DNA to be detected by ethidium bromide staining following gel electrophoretic separation (Fig. 55.2). This is often performed to assess the specificity of the amplification process (i.e. by checking product size against a series of markers). Alternatively, the amplified DNA can be detected and assessed by gene probe hybridisation (e.g. Southern blotting). Indeed, gene polymorphisms are often detected by Southern blotting employing enzymatically or radioactively labelled allele-specific oligonucleotide probes. Hybridisation formats employing probe-capture (sandwich) approaches and enzyme-labelled probes are also widely used. These are generally easier to perform, safer, and enable the quantitation of initial template numbers (often important in clinical diagnosis).

Alternatively, the amplified DNA is isolated to serve as the template for RFLP or nucleotide sequencing analysis. Having confirmed its fidelity, the purified PCR-amplified DNA is labelled and used as a gene probe and may also be used as the substrate for cloning and manipulation (e.g. *in vitro* mutagenesis).

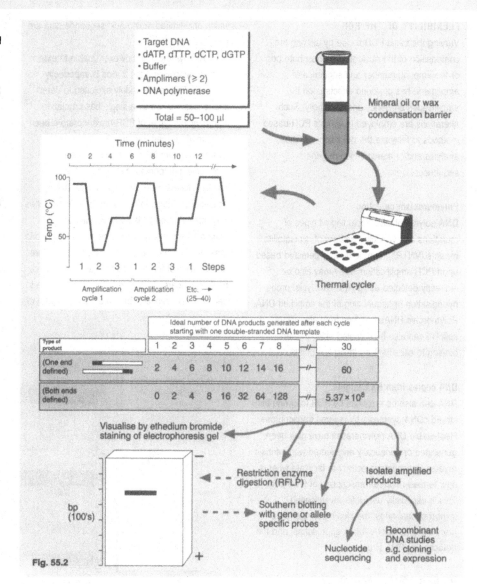

Fig. 55.2

Amplification sensitivity

The PCR can successfully amplify from a single strand of DNA (or cDNA). A single cell, or very small number of cells (e.g. obtained by chorionic villus sampling, mouth washes or swabs) can therefore provide sufficient DNA for PCR amplification. In fact, the PCR may often fail when there is too much DNA in the starting sample. The incredible sensitivity of the PCR does have drawbacks. Accordingly, any cross-contamination with another sample or previously amplified DNA can give false positive results unless appropriate avoidance strategies are used.

Size limits to amplification

In practice, the size of DNA amplified by the PCR is limited to a few thousand base pairs (i.e. Kilobases). This is partly due to the kinetics of the enzyme-mediated polymerisation process, and problems such as the premature release of bound DNA polymerase. In most instances, target sequences are a few hundred base pairs in length. These can consequently be easily amplified from even badly degraded or long-term preserved DNA samples. Combined with its sensitivity, this makes the PCR ideal for many clinical, forensic, and archaeological investigations.

The PCR is now widely used as a 'short-cut' technique for the isolation, analysis and manipulation of DNA and RNA sequences due to its rapidity, sensitivity, simplicity, and versatility.

FLEXIBILITY OF THE PCR

Varying the basic PCR recipe by altering the composition of the nucleotide triphosphate pool, or the amount, number and sequence of amplimers, has profound effects upon the amplification efficiency and specificity. Such alterations are employed in various PCR-based methods to facilitate the detection, isolation, analysis and/or manipulation of gene sequences.

Polymorphism detection

DNA polymorphisms, including all types of mutations and variable numbers of minisatellite repeats (VNTRs), can be directly detected based upon PCR amplification. They may also be indirectly detected using RFLP analysis, probe hybridisation or sequencing of the amplified DNA. Polymorphic RNAs resulting from alternative splicing can also be detected using primers binding to sites flanking the splice points.

DNA copies from RNA targets

RNA can also be amplified by PCR following first strand cDNA synthesis by reverse transcriptase. Heat-stable DNA polymerases have now been generated or genetically engineered with intrinsic reverse transcriptase activity to facilitate what is now termed reverse transcription or RT-PCR. This is especially useful for investigating transcriptional rates, the cloning of rare gene transcripts via cDNA library approaches, and the detection of RNA viruses.

PRIMER COMPLEMENTARITY

The specificity and efficiency of PCR amplification is dependent upon the degree of amplimer–target complementarity since this determines the strength of amplimer binding (Fig. 55.3). Perfect specificity is obtained if the primers are completely complementary to the target sequence and the PCR is performed using annealing steps at a temperature close to their melting temperatures. Incompletely complementary amplimers (often termed 'degenerate' primers) may however also bind and initiate amplification if annealing is performed at lower temperatures. Degenerate amplimers are particularly important since they allow the PCR to amplify related or homologous gene sequences, and sequences for

which only limited amino acid sequence data are available.

DNA polymerase is however unable to extend a bound amplimer if the 3' end is imperfectly matched. This fact is widely exploited to detect sequences varying by a single base mutation. Two principal types of PCR have therefore been developed, termed the amplification refractory mutation system (ARMS) or competitive oligonucleotide primed (COP) PCR. These employ allele-specific oligonucleotide (ASO) amplimers containing one or more 3' mismatches. An amplified product is consequently only generated if the fully complementary sequence is present. The reaction conditions, and the nature and location of the mismatches, are crucial and have to be carefully tailored. It is also possible to detect the presence of both (i.e. heterozygote) or

a single allele (i.e. homozygote) in a sample in a single PCR using differentially labelled ASOs.

Amplimer sequences may also be readily modified at the 5' end to enable attachment of moieties (e.g. biotin) to serve as markers for detection, or 'handles' for isolation of the amplified DNA. Furthermore, recognition sequences for restriction enzymes can be incorporated at the 5' end, or within the primers, facilitating the cloning of the amplified DNA. It has even been possible to introduce 5' terminal promoter sequences recognised by RNA polymerases in order to secondarily generate RNA copies from the amplified DNA (i.e. for ribonucleotide sequencing, or for the production of polypeptides by coupled *in vitro* transcription–translation).

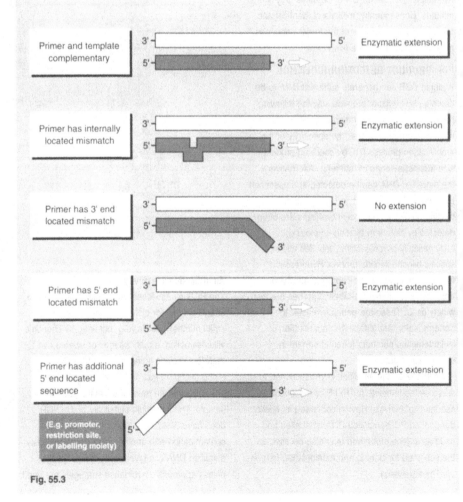

Fig. 55.3

ALTERING PRIMER ABUNDANCE

Decreasing the concentration of one primer relative to the other (by 50:1 to 100:1) can also be used to generate an abundance of single-stranded DNAs in a process called asymmetric PCR. Accordingly, after a limited number of amplification cycles the less abundant primer is exhausted (Fig. 55.4a). This allows the subsequent linear amplification of single-stranded DNA molecules from the remaining primer. This has been particularly useful for producing templates for dideoxynucleotide sequencing or generating single-stranded gene probes. Multiple primer pairs recognising different genes or polymorphisms between genes (e.g. alleles) may also be used simultaneously in PCR. Termed multiplex PCR (Fig. 55.4c), this method has been widely used in detecting gene polymorphisms and gene linkage analysis. Increased amplification specificity (and efficiency) can also be obtained if previously PCR-amplified DNA is used as the template for subsequent rounds of PCR amplification using primers designed to bind to 'nested' sequences (i.e. downstream of the original primer binding sites) (Fig. 55.4b).

ALTERING THE NUCLEOTIDE POOL

Many different base analogues (with or without labelling moieties attached) can be incorporated in addition to the normal four bases during PCR amplification. This can be used to introduce base mutations, to overcome difficulties such as tertiary structures formed within GC-rich sequences, or to label the amplified DNA for subsequent use as a gene probe. The inclusion of dideoxynucleotides in the cyclic amplification process has also been used to develop 'cycle sequencing' approaches to increase the rapidity of DNA sequence data acquisition.

Furthermore, the inclusion of dUTP is now extensively utilised as a means of eliminating problems of cross-contamination with previously amplified PCR products (Fig. 55.4d). Termed 'carry-over prevention', it relies upon the inclusion of dUTP into the PCR products and its

OAs: 465–90, 495–511, 529–30, 516, 545, 791–3, 849–60, 915–33, 1088, 1147. RAs: 424, 392–6. SFR: 136–40, 166–9.

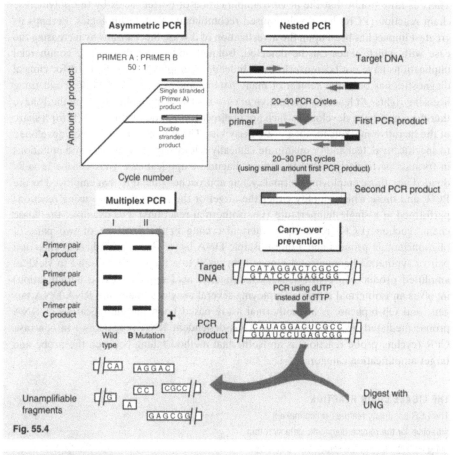

Fig. 55.4

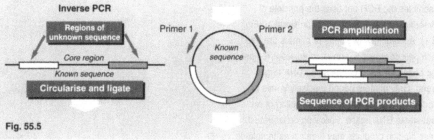

Fig. 55.5

subsequent elimination by digestion with an enzyme, uracil-N glycosylase (UNG). Performed prior to any PCR analysis, it ensures that any previously amplified DNA contaminant is eliminated.

Variations on a theme

Many variant PCR methods have now been developed. These enable the amplification of sequences for which there is only sufficient sequence data to design a single primer (e.g. anchored PCR, and rapid amplification from

cDNA ends (RACE)), or to amplify unknown DNA sequences flanking a region of known sequence by (e.g. inverse PCR (Fig. 55.5)).

Alternative amplification techniques have now been developed such as the ligase chain reaction (LCR), and replicatable RNA probes (Qβ-replicase system) which may also have great utility (e.g. in clinical diagnostics). Presently, however, they are less well developed or as versatile as the PCR and therefore unlikely to be as widely applied in recombinant DNA studies.

56. ALTERNATIVES TO PCR-BASED *IN VITRO* DNA/RNA AMPLIFICATION

A number of isothermal and thermal cycling-based assay strategies have now been developed for the detection of nucleic acid sequences based upon *in vitro* amplification of RNA and/or DNA targets.

There is little doubt that the *in vitro* amplification of nucleic acids by the polymerase chain reaction (PCR) has revolutionised recombinant DNA technologies. Perhaps its greatest impact has been upon the investigation of disease mechanisms by increasing the ease with which genes can be detected, isolated and manipulated. The commercial implications have not been overlooked. Indeed, the application of the PCR for clinical diagnostics has been the subject of many patents and several legal battles concerning licensing rights. At least ten alternative *in vitro* nucleic acid amplification methods have therefore now been developed which are in direct competition with the PCR for a share of the lucrative markets for diagnostic assay kits. These have primarily been developed to specifically detect and/or quantitate clinically relevant gene sequences and mutations in formats particularly suited to clinical diagnostic applications. They can be broadly divided into those employing thermal cycling approaches similar to that employed by the PCR, and those which amplify either the target or the probe sequence using reactions performed at a single temperature (i.e. isothermal reactions). For example, the ligase chain reaction (LCR) relies upon thermal cycling in the presence of two pairs of oligonucleotide primers and a thermostable DNA ligase. This joins adjacent ends of a pair of synthetic oligonucleotides once hybridised to a target DNA strand to yield an amplified product. In contrast, NASBA (nucleic acid sequence-based amplification) involves an isothermal reaction employing several enzymes to amplify RNA/DNA targets, and Qβ-replicase is an isothermal assay based upon the amplification of RNA probes mediated by a phage-derived RNA-dependent RNA polymerase. In contrast, CPR (cycling probe reaction) is an isothermal method falling between the probe and target amplification categories.

MARKET FORCES AND DEVELOPMENT

The size of the anticipated market is a primary commercial consideration in developing any new diagnostic assay technology. New technologies based upon target or probe amplification must therefore be judged worthwhile in terms of anticipated sales. In other words, a new technique must be likely to capture a reasonable percentage of the market for a particular diagnostic test kit. This requires careful market research/analysis since long periods may elapse between primary assay development and production of an adequately evaluated assay suitable for clinical adoption. Commercially available assay kits are therefore predominantly targeted at diseases for which large diagnostic markets already exist. These include the detection of infectious microbial pathogens, several types of cancers, and common genetic disorders (e.g. cystic fibrosis, Duchenne's and Becker's muscular dystrophies, and Huntington's disease).

THE LIGASE CHAIN REACTION

The LCR has many features which make it attractive for the routine diagnostic detection (but not quantitation) of target sequences. It closely resembles the PCR but uses the principle of primer ligation as the amplification strategy (Fig. 56.1). Accordingly, two pairs of primers are used, each pair designed to anneal to adjacent regions one base pair apart on the same DNA target strand. Only primers bound precisely in this way are then joined (or ligated) by the action of a heat-stable DNA ligase. Following denaturation, these ligation products may then act as templates in subsequent cycles to give rise to the exponential amplification of the target sequence. In clinical diagnostic formats, LCR utilises capture and detection ligands attached to opposite ends of the pairs of primers such that only the ligated products are subsequently captured and detected. This allows the LCR to be performed using currently available automated instruments (i.e. based upon conventional immunoassays).

The LCR is particularly effective for detecting multiple point mutations based upon the fact that mismatches either side of the ligation site effectively prevent ligation. Similarly, insertional or deletional mutations which alter the primer annealing positions may also be detected. Following a reverse transcription reaction, the LCR can also be used to detect alternatively spliced transcripts (e.g. for viral strain typing).

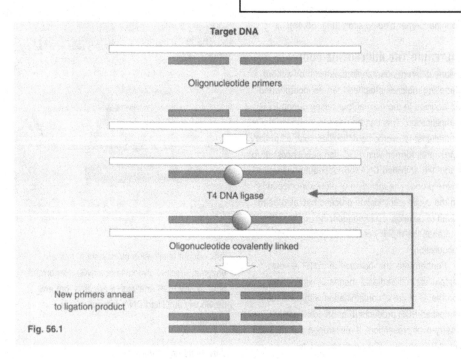

Fig. 56.1

Target DNA

Oligonucleotide primers

T4 DNA ligase

Oligonucleotide covalently linked

New primers anneal to ligation product

OAs: 457, 491–5.
RAs: 178–80, 182, 205, 206.
SFR: 126, 136–40.

THE PCR/LCR COMBINED

The specificity of the LCR is relatively high due to the proportionately longer primers used (which effectively reduces non-specific binding). The blunt-end ligation process employed can however result in significant background amplification problems. Consequently, two companies (Imclone and Abott Laboratories) have developed the polymerase/ligase chain reaction (P/LCR) or repair chain reaction (RCR) (Fig. 56.2). This overcomes the background problems associated with mispriming and the blunt-end ligation associated with the PCR and LCR respectively.

Similar to conventional LCR, the P/LCR also utilises four primer probes. These are however designed to anneal at sites which leave a gap of several nucleotides between the ends of each pair of primers. This gap is carefully chosen such that only three of the usual four dNTPs are required, and only these three dNTPs are added to the reaction mixture. The DNA polymerase is then able to extend one of the two primers of a pair (i.e. in a 5′ to 3′ direction) to fill the gap. This permits ligation by the DNA ligase, whereas non-specific extension of primers is prevented by the absence of the fourth dNTP. Denaturation then allows the ligation products to act as further templates in the next cycle of P/LCR. Competition between ligation and extension reactions can also be eliminated by performing the reactions at different temperatures.

The increased sensitivity of the P/LCR makes it suited to clinical diagnostic application. It is, however, a yes/no assay capable of detecting the presence of target sequences, but poorly suited to target quantitation.

Other amplification methods

Several additional methods employing a thermal cycling strategy for target amplification have also been developed for use in clinical diagnostics. For example, the boomerang DNA amplification (bDNA) technique combines thermal cycling amplification with restriction endonuclease digestion of target sequences. Digestion products produced at each amplification round are then ligated to adaptors. A single primer complementary to the adaptor is then annealed. This is then extended by a DNA polymerase.

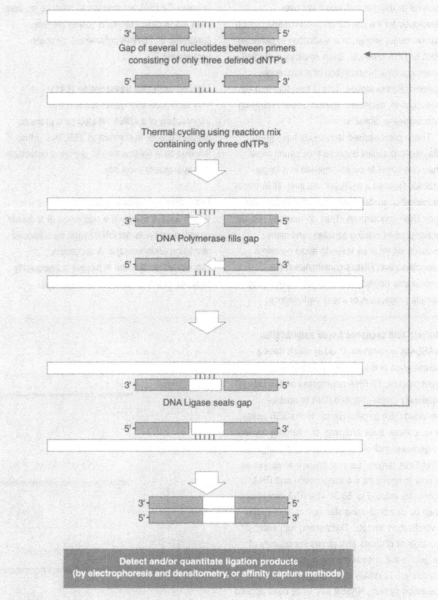

Gap of several nucleotides between primers consisting of only three defined dNTP's

Thermal cycling using reaction mix containing only three dNTPs

DNA Polymerase fills gap

DNA Ligase seals gap

Detect and/or quantitate ligation products (by electrophoresis and densitometry, or affinity capture methods)

Fig. 56.2

Non-thermal cycling methods

Several non-thermal cycling (i.e. isothermal) systems are now commercially available or under development for clinical diagnostic application. These may be divided into two categories: (i) those that result in amplification of the target sequence; and (ii) those that result in amplification of the probe sequence.

There is, however, one exception to this classification termed the cycling probe reaction (or CPR) which falls between signal and target amplification. This uses a DNA–RNA–DNA probe which anneals to the target sequence and is then cleaved by RNase H to form two DNA fragments. These fragments do not act as templates in subsequent rounds of CPR reactions. Their generation and detection can however be used to indicate the presence of targets in the original sample.

Several isothermal nucleic acid target/signal amplification systems have been developed for clinical diagnostic use, including the 3SR, NASBA, LAT, RAMP, SDA, Qβ-replicase and bDNA systems.

TARGET AMPLIFICATION SYSTEMS

Several isothermal methods have been developed for the clinical diagnostic market which involve target sequence amplification to generate RNA or DNA products. Each employs strategies which combine hybridisation of a sequence-specific oligonucleotide. This is then followed by transcription, extension, ligation and/or restriction endonuclease digestion.

These primer-based techniques have great diagnostic potential because they permit more than one target to be co-amplified in a single reaction (termed a multiplex reaction). RNA target amplification systems are also often faster than their DNA counterparts. They do however require highly purified starting samples, and many precautions must be taken to avoid problems associated with RNase contamination (i.e. undesirable degradation of target RNAs during sample preparation or assay performance).

Nucleic acid sequence based amplification

NASBA is an isothermal assay which uses a combination of three enzymes (reverse transcriptase, T7 RNA polymerase and RNase H) to amplify single-stranded RNA or double-stranded DNA targets. Similar to the 3SR assay, it uses these three enzymes and flanking primers to generate multiple RNA copies of the original RNA/DNA targets. Each of these then serves as a new template for the transcription and DNA synthesis steps (Fig. 56.3). The RNA copies can then be detected using standard capture-probe hybridisation formats. Comparison with internal controls or dilutions also allows the quantity of targets in the original starting sample to be calculated. Originally developed as a virus detection system, NASBA has since been applied to the diagnostic detection of several types of clinical markers. Accordingly, it can be used to detect RNA viruses and retroviruses, to specifically detect live microbial pathogens, and can also be used to monitor levels of cancer cell proliferation markers (e.g. cytokine/oncogene mRNAs). Indeed, NASBA has been shown to be as sensitive and practical as PCR for determining the level of HIV-1 viruses in clinical samples.

Self-sustained sequence replication (3SR)

3SR also involves reverse transcription of RNA targets to DNA using reverse transcriptase. It also employs T7 RNA polymerase to produce multiple RNA amplification products (using primers containing 5′ T7 RNA polymerase promoter sequences).

Ligation-activated transcription (LAT)

This technique also depends upon the transcription of a cDNA initiated by a promoter oligonucleotide. In contrast to 3SR, this is then followed by a ligation step to generate detectable and quantifiable products.

Restriction amplification (RAMP)

RAMP assays involve the cleaveage of a double stranded probe–target DNA hybrid by a specific restriction endonuclease. A secondary oligonucleotide present in excess subsequently attaches to the cut end of the target and

OAs: 495–511. RAs: 205, 206. SFR: 126, 136–40.

reconstructs the originally cleaved target sequence. Annealed probe is therefore repeatedly cleaved, and the breakdown products accumulate exponentially.

Strand displacement activation (SDA)

SDA involves the hybridisation of an oligonucleotide containing a 5′ *Hin*cII site to the target DNA sequence. The primer is then extended in the presence of sulphur-containing adenine analogues. *Hin*cII cuts only the unmodified strand which is extended downstream from the nick, thereby displacing the original primer. This allows the repeated annealing and extension of primer.

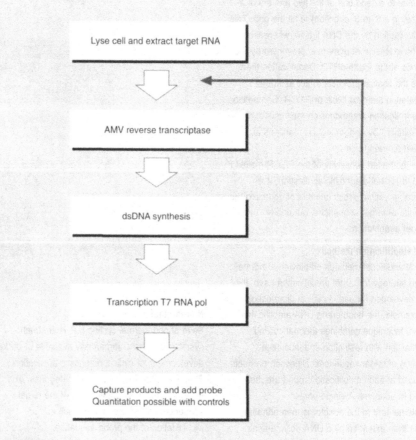

Lyse cell and extract target RNA

AMV reverse transcriptase

dsDNA synthesis

Transcription T7 RNA pol

Capture products and add probe
Quantitation possible with controls

Fig. 56.3

SIGNAL AMPLIFICATION SYSTEMS

Two principal probe/signal amplification strategies are commercially available: (i) Qβ-replicase or QβR, and (ii) branched DNA amplification or bDNA. Both methods are particularly suitable for target quantitation and form the basis of kits marketed for microbial infection detection, cytokine mRNA quantitation and detection of cancer markers.

Q-β replicase system

The Qβ R system is based upon the use of a bacteriophage-derived RNA-dependent RNA polymerase (termed Qβ-midivariant RNA replicase). This enzyme is capable of replicating a particular type of RNA molecule termed a midivariant-1 RNA (or MDV-1 RNA) (Fig. 56.4). Probes are generated by inserting target-specific (i.e. target complementary) sequences into the MDV-1 RNA sequences at a specific point in the MDV-1 RNA. This allows its presentation at the end of one of the stem-loop structures, but does not affect its ability to be replicated *in vitro* by the Qβ-replicase. Therefore, following probe binding and separation of the target–probe hybrids (e.g. using affinity capture methods), the MDV-1 RNA probe is replicated to easily detectable levels. Similar to PCR and LCR, the Qβ reaction is exponential with each newly synthesised MDV-1 RNA molecule serving as a template. This allows quantitation of targets based on the rate of MDV-1 RNA accumulation. Although quantitative and extremely rapid (capable of billion-fold probe amplification within 30 minutes), the QβR system has low specificity due to the co-amplification of non-specifically hybridised probe.

Branched DNA system

bDNA amplification involves a two-step hybridisation approach to capture and amplify DNA (Fig. 56.5). The first step involves capture-hybridisation of the target sequences by unlabelled probes immobilised to a solid phase (e.g. the walls of a tube or microwell). A second probe linked to a highly branched synthetic 'amplifier' DNA molecule is then hybridised to the target sequences. Each amplifier is about 1 kb in length and each branch can be attached to thousands of labels (e.g. chemiluminescent tags). The many labels present in the final complex

gives the assay great detection sensitivity (e.g. sub-attomole levels of viral DNA in clinical specimens). The bDNA system can also be used with relatively impure DNA samples. Indeed, second generation bDNA probes are claimed to be able to detect as few as 62 target molecules present in a millilitre of clinical sample.

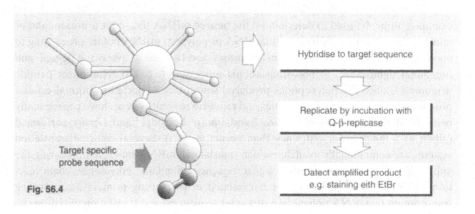

Target specific probe sequence

Hybridise to target sequence

Replicate by incubation with Q-β-replicase

Datect amplified product e.g. staining eith EtBr

Fig. 56.4

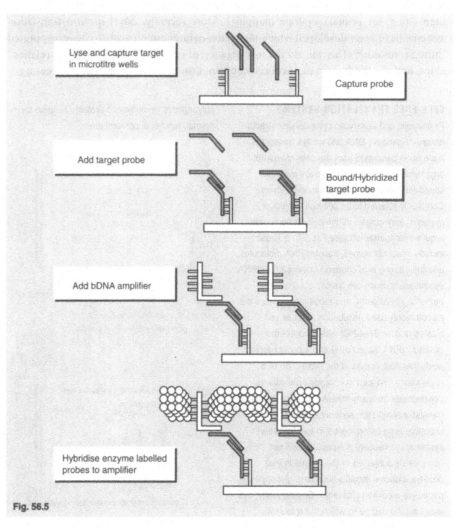

Lyse and capture target in microtitre wells

Capture probe

Add target probe

Bound/Hybridized target probe

Add bDNA amplifier

Hybridise enzyme labelled probes to amplifier

Fig. 56.5

57. *IN VITRO* TRANSLATION METHODS

Polypeptides may be generated by cell-free *in vitro* translation of isolated mRNA transcripts, or using defined RNA molecules generated by *in vitro* amplification using the polymerase chain reaction.

The *in vivo* production of proteins encoded by genomic DNA is a complex and highly regulated process generally involving coordinated transcription and translation. The complexity of translation has favoured transfection of host cells with DNA inserted into specific vectors (termed expression vectors). Recombinant proteins can then be produced following the induction of host cell-mediated translation. Under certain circumstances, however, *in vitro* translation of isolated DNA or RNA may be useful. For example, it may be used to determine if the desired mRNA transcript is present and/or intact within an extract of total cellular RNA or poly(A$^+$) mRNA before proceeding to make a cDNA library. *In vitro* translation may also be used to assess the degree, and functional significance, of post-translational processing (e.g. glycosylation or peptide 'trimming'), since the polypeptides produced will not be subjected to normal cellular processing events. Several approaches and requisite reagent systems have consequently been developed which allow *in vitro* translation to be simply and rapidly performed (albeit at a much lower efficiency than occurs *in vivo*). Several cell-free translation systems are commercially available which translate an mRNA molecule(s) bearing the appropriate translational initiation signal sequences. Similarly, polymerase chain reaction (PCR)-amplified DNA may be translated *in vitro* using primers incorporating transcription (e.g. RNA polymerase promoter sequences) and translation initiation signals. This allows the rapid production of small amounts of defined polypeptide fragments (e.g. for primary epitope mapping). More recently, novel *in vitro* translation systems have been developed which allow non-natural amino acids to be incorporated during translation. This will aid the investigation of protein structure–function relationships, and the production of novel recombinant proteins (i.e. by protein engineering).

INCORPORATING NOVEL AMINO ACIDS

An intriguing research area currently benefiting from *in vitro* translation system development, is the evolution of systems capable of introducing non-natural amino acids into proteins during *in vitro* (and *in vivo*) translation. Altering the nature of codons using nucleotide analogues with unusual bases is suggested to have great potential in protein engineering. Accordingly, they may increase the range of structural changes that can be introduced beyond those possible using standard *in vitro* mutagenesis techniques. Known as site-directed non-native amino acid replacement (SNAAR), this requires the generation of new bases and new tRNAs. It also requires the development of new enzymes (i.e. aminoacyl-tRNA synthetases) capable of recognising the new codons and charging the tRNAs with the non-natural amino acids.

OAs: 197, 339–40, 512–26.
RAs: 83, 200, 207, 274, 275.
SFR: 119, 120, 124, 125, 136–9.

CELL-FREE TRANSLATION SYSTEMS

Prokaryotic and eukaryotic cytoplasmic extracts devoid of genomic DNA and mRNA transcripts have been generated from the cells of several organisms. Many are commercially available specifically for *in vitro* translation experiments. Commonly referred to as cell-free translation systems, they contain all the cellular components required for translation (see Fig. 57.1). These include intact ribosomes, transfer RNA molecules, and their amino acid charging enzymes (i.e. tRNA synthetases). Such cell lysates obtained from bacteria, wheat germ, and rabbit reticulocytes are all commonly used. Incubation of these cell lysates (e.g. at 37–42°C) with a pool of the desired mRNA transcript(s) and a pool of amino acids therefore results in the production of a population of nascent (i.e. newly synthesised) polypeptides from any mRNA transcript bearing a translational initiation sequence. Similar to the situation using heterologous *in vivo* translation systems, the nascent polypeptides will not however be subjected to the normal *in vivo* post-translational modifications (e.g. glycosylation, peptidase-mediated cleavage). Consequently, this may limit the degree to which the resultant polypeptides or multimeric proteins assume their normal, functional conformations.

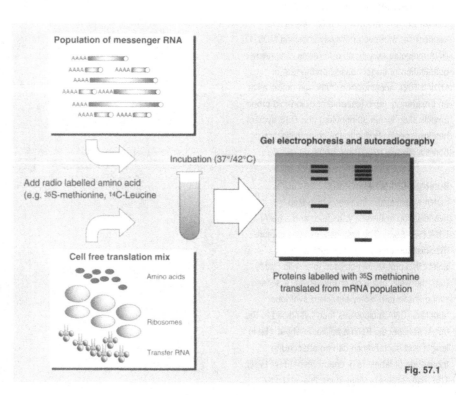

Population of messenger RNA

Add radio labelled amino acid
(e.g. ^{35}S-methionine, ^{14}C-Leucine)

Incubation (37°/42°C)

Cell free translation mix

Amino acids

Ribosomes

Transfer RNA

Gel electrophoresis and autoradiography

Proteins labelled with ^{35}S methionine translated from mRNA population

Fig. 57.1

DETECTING NASCENT POLYPEPTIDES

Nascent polypeptides formed by *in vitro* translation can be readily detected by including a radiolabelled amino acid(s) (e.g. ^{14}C-leucine) into the amino acid pool (Fig. 57.2). This results in the intrinsic (i.e. internal) labelling of the polypeptide. The degree of labelling is therefore proportional to the frequency of the labelling amino acid within the translated sequence. The presence, amount and molecular weight(s) of the polypeptides can then be determined by electrophoretic separation of the reaction components followed by autoradiographic detection of the labelled band(s). Unfortunately, the inherent instability of mRNA transcripts often gives rise to a smear of labelled polypeptides. Nevertheless, this in itself may still be useful in indicating the degree of integrity of the initial mRNA transcripts. Western blotting employing specific antibody probes, for example, may be used to more definitively establish if a specific polypeptide is produced. This may indicate if the original starting material contains intact mRNA transcripts encoding the desired protein(s) (e.g. prior to cDNA cloning).

TRANSLATION OF PCR-AMPLIFIED DNA

An obvious limitation of *in vitro* translation is the availability of sufficient quantities of mRNAs bearing intact 5′ sequences. Thus, a polypeptide can only be synthesised from transcripts bearing the requisite 5′ translational initiation and ribosome binding sequences. The advent of the polymerase chain reaction (PCR) has increased both the amount and range of polypeptides that can be obtained by *in vitro* translation. The technique known as RNA amplification with *in vitro* translation (RAWIT) allows any single DNA or RNA fragment bearing a coding sequence to be translated. Accordingly, a forward PCR oligonucleotide primer is synthesised which contains an additional RNA polymerase recognition sequence (e.g. for bacteriophage T4 and T7 RNA polymerases), a ribosome binding sequence, and possibly a methionine initiation codon (located respectively from 5′ to 3′). Standard PCR amplification using an unmodified reverse primer then results in the production of a pool of identical double-stranded DNA products which may be used as templates for *in vitro* RNA synthesis. The pool of identical RNA molecules, if

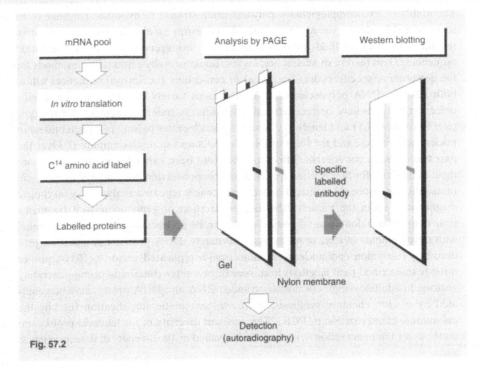

Fig. 57.2

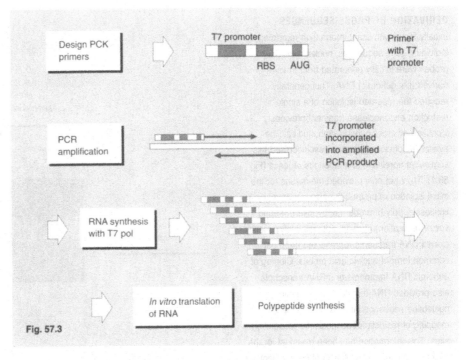

Fig. 57.3

generated in an open reading frame, can then be used for *in vitro* translation using any cell-free lysate and the standard intrinsic labelling strategy.

RAWITs has several possible applications,

particularly for rapidly generating polypeptide fragments for primary epitope mapping by western blotting, or for the analysis of cell- or tissue-specific transcription patterns.

58. TYPES AND METHODS OF GENE PROBE GENERATION

Single or double stranded RNA or DNA probes may be generated by transcription or insert excision from vectors, chemical synthesis or *in vitro* amplification by the polymerase chain reaction.

The ability of two complementary polynucleotide strands to hybridise (or anneal) to form a stable duplex is at the heart of numerous analytical and manipulative recombinant DNA methods ranging from *in vitro* mutagenesis to dideoxynucleotide sequencing. This feature of nucleic acids is also fundamentally exploited by methods for the detection of specific coding (i.e. exon) or non-coding (i.e. intron) sequences within both RNA or DNA polynucleotides. An enormous variety of such sequence-specific detection methods have been developed. Each differ in their requirements for the type (e.g. RNA or DNA) and length (i.e. short oligonucleotides to long polynucleotides) of nucleic acid probes, and the form of labelling/detection strategies employed. Over the past two decades, considerable effort has therefore been expended upon the development of methods for the rapid, simple, and reliable production of a range of nucleic acid (and antibody) probes in sufficient quantities to enable repeated analyses (e.g. in clinical diagnostics). From the initially laborious preparation of genomic-derived fragments from restriction endonuclease digests, the art of probe production has evolved in parallel with developments in cloning methods. Accordingly, DNA probes can either be produced by restriction endonuclease excision from a replicated vector, or RNA probes directly transcribed (and labelled) from specific promoter-containing probe-generating vectors. In addition, single- or double-stranded RNA and DNA probes may be generated by *in vitro* chemical synthesis, or *in vitro* enzymatic amplification (e.g. by the polymerase chain reaction or PCR). The resultant diversity of nucleic acid probes and methods for their generation is, perhaps, only rivalled by the diversity of detection labels used.

ANTIBODY-BASED PROBES

Antibody probes may also be used to directly or indirectly detect nucleic acids in either a non-specific or sequence-specific manner. Radioactively (e.g. ^{125}I) or non-radioactively (e.g. peroxidase) labelled, these may be used to specifically bind to different forms of RNA or DNA targets. For example, polyclonal or monoclonal antibodies have been generated which will specifically detect RNA–DNA hybrids, single- or double-stranded polynucleotides, and chemically modified or protein-bound nucleic acids (i.e. for detecting protein binding motifs by South-western blotting). Labelled antibody probes are also often used directly to detect a specific gene fragment within a cDNA library. This is based upon recognition of specific translation products generated from cDNA inserts within expression vectors. Antibodies may also be used as indirect probes by virtue of their ability to recognise labels (e.g. biotin or digoxigenin) previously attached to nucleic acid probes.

DERIVATION OF PROBE SEQUENCES

Initially faced with scant information regarding individual gene sequences, nucleic acid gene probes were initially generated from *in vivo*-derived (i.e. genomic) DNA. This generally required the repeated isolation of a single restriction endonuclease fragment previously identified by cross hybridisation studies. The availability of cloning vectors revolutionised this somewhat unreliable and laborious process (Fig. 58.1). They not only provided the means for the initial isolation of probes, but also a potentially limitless supply of probe (i.e. by insert-bearing vector replication). Indeed, until relatively recently, cloned DNA fragments represented the most common form of nucleic acid probes. Cloning of genomic DNA fragments or mRNA transcripts also provided DNA templates for obtaining nucleotide sequence information and the precise mapping of restriction endonuclease recognition sites. This information has been essential for the design and chemical synthesis of oligonucleotides for direct or indirect applications as probes. Two principal vector-based methods have been used to generate nucleic acid probes. These are referred to as insert excision, and promoter 'run-off' methods.

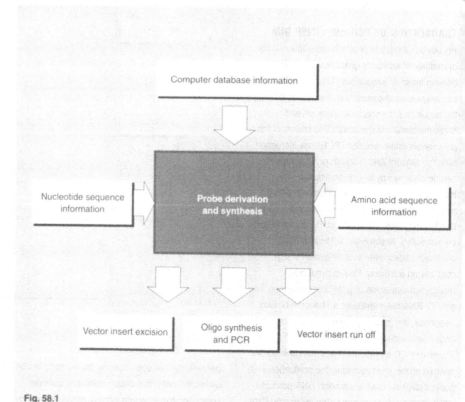

Fig. 58.1

Vector insert excision

DNA fragments are generally inserted into the multiple cloning sites of vectors (i.e. areas of sequence containing multiple, vector-unique, restriction endonuclease recognition sites). This involves the ligation of their respective blunt or cohesive ends generated by prior restriction endonuclease digestion. Replication of the vector in a suitable host cell allows the subsequent isolation of large amounts of the vector DNA (e.g. by ultracentrifugation). The double-stranded insert sequences (or their subfragments) may then be excised by digestion with the appropriate restriction endonuclease(s), and separated from the vector DNA prior to labelling (Fig. 58.2a).

Promoter 'run-off' vectors

Several f1 or M13 phage-based vectors are now commercially available specifically designed for the generation of single-stranded DNA or RNA probes. These also permit internal labelling during probe synthesis by the addition of labelled (radioactive or non-radioactive) nucleotide analogues to the synthesis reaction mixture. Two vector strategies are generally used for producing RNA probes.

(1) The probe sequence is inserted into a plasmid vector at a position flanked on either side by specific polymerase promoter sequences (e.g. one for T7 RNA polymerase and one for T3 RNA polymerase) (Fig. 58.2b). Following linearisation of the vector, either promoter may be used to direct the synthesis of the inserted probe in either orientation (i.e. complementary to the coding or non-coding strand of the insert).

(2) The probe sequence is inserted into a phage vector capable of adopting a single-stranded form during its life cycle. The sequence is inserted between a downstream (3′) primer binding site and an upstream (5′) polymerase promoter sequence. Annealing of a primer oligonucleotide to the single-stranded vector DNA subsequently allows copying of the DNA sequence by Klenow fragment. The partial duplex formed is then recognised by RNA polymerase which leads to transcript synthesis (or 'run-off'), terminated by the position of the annealed primer.

SYNTHETIC OLIGONUCLEOTIDES

Both single-stranded RNA and DNA oligonucleotide probes may be synthesised (and 5′ end labelled) *in vitro*. Synthetic oligonucleotides can be used directly as probes, or as primers for probe production by enzymatic extension (e.g. in the PCR). Whilst endless supplies of defined oligonucleotides may be chemically synthesised, their direct use as probes is limited by the relatively short lengths (generally up to 100 bases) that can be reliably achieved. This is due to the inevitable accumulation of errors arising from inefficient base incorporation during each synthesis cycle.

PCR-BASED PROBE PRODUCTION

A series of chemically synthesised oligonucleotides may however be generated and subsequently joined together to create a long artificial nucleic acid probe. This may either be achieved by ligation, or more reliably and conveniently, using the PCR. The long probe can

then also either be replicated by PCR, or by insertion into an appropriate vector. Alternatively, a potentially endless supply of double stranded DNA probes ranging from a few hundred bases to several kb may be generated directly from genomic or cloned DNA by the PCR. Furthermore, these may also be labelled at the 5′ end or internally labelled during the amplification process obviating the need for separate labelling reactions. The incorporation of RNA polymerase promoter sequences at the 5′ end of either oligonucleotide primer also allows the *in vitro* generation of single-stranded RNA probes from the amplified DNA products.

OAs: 527–31, 536–40, 546, 564–5, 636–7. RAs: 204, 11, 215, 218, 321. SFR: 120–7, 131, 136–45.

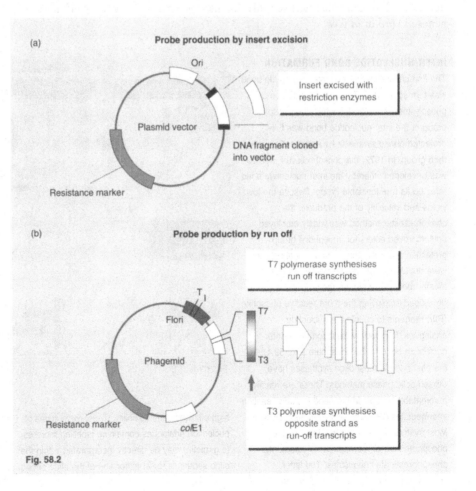

Fig. 58.2

(a) **Probe production by insert excision**

Ori

Plasmid vector

Insert excised with restriction enzymes

DNA fragment cloned into vector

Resistance marker

(b) **Probe production by run off**

T1

Flori

Phagemid

Resistance marker

colE1

T7

T3

T7 polymerase synthesises run off transcripts

T3 polymerase synthesises opposite strand as run off transcripts

59. CHEMICAL SYNTHESIS OF OLIGONUCLEOTIDES

RNA and DNA oligonucleotides may be chemically synthesised in a 3'–5' direction by the coupling of phosphoramidate nucleotide analogues, for use as probes or primers for gene analysis and manipulation.

Short stretches of defined DNA or RNA sequence (termed oligonucleotides or oligos) are vital to many recombinant DNA methods. Accordingly, they are used directly as probes for gene analysis (e.g. Southern blotting), or as primers for the manipulation (e.g. the polymerase chain reaction and gene cloning) and modification (e.g. *in vitro* mutagenesis) of specific gene sequences. The relatively simple chemical composition of nucleic acids has enabled organic chemists to provide a range of chemically modified building blocks (i.e. nucleotide analogues). It has also enabled the development of chemistries for the *in vitro* synthesis of both RNA and DNA oligos. This involves a radically different approach to that used in normal *in vivo* processes of DNA or RNA synthesis which rely upon the enzymatic addition of bases in a 5'–3' direction to the ends of a priming polynucleotide annealed to an existing polynucleotide template. Chemical synthesis does not therefore require a template, and proceeds in a 3'–5' direction starting from an initial nucleotide analogue chemically bonded to an immobilising solid support. Originally produced using phosphodiester and phosphotriester methods, most oligo syntheses now rely upon the use of phosphite triester (phosphoramidate) methods. Using automated solid phase methods employing repeated cycles of solvent-mediated elongation and deprotection, it is now possible for any laboratory to cheaply and simultaneously synthesise micromole quantities of multiple oligonucleotides, ranging in size from 10–100 bases, within a few days. Furthermore, a wide range of modified nucleotide analogues may be used to 5' label the oligo during synthesis, and generate novel forms of nucleic acids (e.g. peptide nucleic acids) for altering both the sequence and function of genes *in vitro* or *in vivo*.

PEPTIDE NUCLEIC ACIDS

Researchers at the University of Copenhagen, Denmark, have recently invented a new class of synthetic molecules which look and act like DNA, termed peptide nucleic acids (or PNAs). PNAs have a peptide backbone incorporating normal nucleic acid bases in the desired sequence. This molecule has a neutral rather than negative net charge, and does not contain labile phosphate–oxygen–sugar bonds. Consequently, PNAs survive much longer within cells than their conventional oligonucleotide counterparts. They have also been demonstrated to bind 50–100 times more tightly to complementary sequences than 'natural' nucleic acids. PNAs therefore look set to revolutionise some DNA-based diagnostic test formats. Moreover, despite concerns about the potential *in vivo* side-effects of their longevity, PNAs could be immensely useful for both antisense and antigene based therapies.

INTER-NUCLEOTIDE BOND FORMATION

The first chemically induced inter-nucleotide bond was formed in 1955. This was based upon a phosphotriester method in which the phosphoryl group of the inter-nucleotide bond was fully protected during synthesis by esterification with a third group. In 1972, the phosphodiester method was developed whereby the inter-nucleotide bond retained its free ionisable group. Despite the low yields and solubility of the products, the phosphodiester method was widely employed until improved esters for phosphoryl group protection and solid phase synthesis approaches were developed. Shortly thereafter, phosphite triester (phosphoramidate) methods were developed employing the more reactive phosphite (PIII) reagents to couple the nucleotide analogues, followed by oxidation of the inter-nucleotide bond to the phosphate (Fig. 59.1). In the past decade most oligo syntheses have utilised solid phase methods. These are readily automatable and avoid the need to purify reaction intermediates between each elongation cycle. Most syntheses are now based upon either phosphotriester or, perhaps more commonly, phosphoramidate chemistries. The latter chemistries predominate due to their relatively

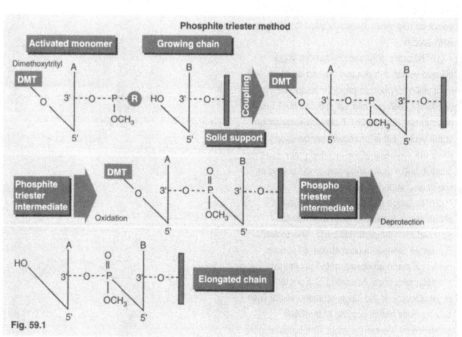

Fig. 59.1

high efficiency and flexibility. Thus, many types of nucleotide analogues containing labelling moieties (e.g. biotin) may be directly incorporated within the oligo sequence, or at either end of the oligo during synthesis.

OAs: 456, 531–42, 580–1, 1120–8, 1059–64. **RAs:** 210–12, 406–8. **SFR:** 142–5, 181, 207–10.

SOLID PHASE METHODS

Oligonucleotide syntheses by both phosphoramidate and phosphotriester methods are now routinely performed using solid phase methods (Fig. 59.2). These involve the initial coupling of the 5′ end of the primary nucleotide to an inert solid support. Porous glass microspheres (controlled pore glass; CPG) are commonly used because of their excellent handling properties (e.g. they do not swell or break) and because they provide a large surface area for loading of nucleotides (of the order of 30 μmol/g). The immobilised nucleotides are also freely exposed to the flow of solvents and reagents. Following linkage of the primary nucleotide, the chain is elongated in a 3′–5′ direction by interaction of the 5′ phosphoryl group on the incoming nucleotide with the deprotected 3′ hydroxyl (OH) group of the immobilised nucleotide, and so on.

Protection and deprotection

All successful oligodeoxynucleotide syntheses require the protection of the exocyclic amino groups on adenine, cytosine and guanine (Fig. 59.3). They also require the protection of the sugar OH groups and the phosphoryl group. The 2′ OH group of ribonucleotides must also be protected, a fact which significantly delayed the development of methods for oligoribonucleotide synthesis. A range of analogues are now available in which the vulnerable exocyclic amino groups and other potentially reactive groups (e.g. the potential phenol at 0–6 of guanine) are protected. Accordingly, most methods utilise phosphoramidates with β-cyanoethyl-protected phosphoryl groups (Fig. 59.4). New analogues containing additional reactive groups or labels are also available for the synthesis of a range of RNA or DNA oligos with novel properties (e.g. resistance to degradation or formation of branched oligonucleotide multimers). These are likely to be valuable for fundamental research into the structure–function relationship of nucleic acids, or for direct diagnostic and therapeutic application.

Just as importantly for oligo synthesis however, protecting groups must be able to be removed following the completion of synthesis without damaging the oligonucleotide product. The 5′ hydroxyl group on the incoming nucleotide must also be protected and then deprotected at the

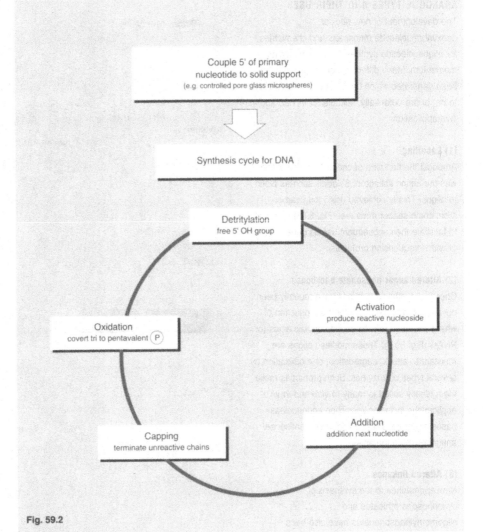

Fig. 59.2

Fig. 59.3

Fig. 59.4

end of the cycle to expose it for the formation of the next inter-nucleotide bond. This is generally achieved by treatment with acid in an organic solvent. Once synthesis is complete, the bases and phosphate groups are totally deprotected and the oligo is cleaved from the CPG solid support using concentrated aqueous ammonia.

The availablity of a range of chemistries and analogues now allows the synthesis of RNA and DNA oligonucleotides with novel activities/ properties with potential for application *in vivo* and *in vitro*.

ANALOGUE TYPES AND THEIR USES

The development of new ribo- or deoxyribonucleotide monomers and chemistries for oligonucleotide synthesis is rapidly gaining momentum. Many different monomers have now been generated which have been demonstrated to be, or are potentially, suitable for *in vitro* and *in vivo* applications.

(1) Labelling

Amongst the first type of chemistry modification was the amino linkage of 5′ labels such as biotin to oligos. This is achieved using long carbon atom chain spacer arms (see Fig. 59.5) in order to facilitate their subsequent use as gene probes, or within sequencing protocols.

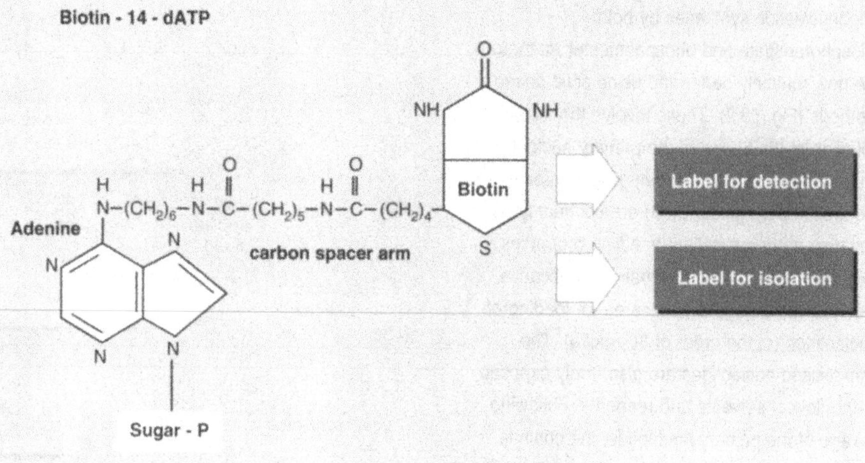

(2) Altered sugar-phosphate backbones

Chemical synthesis methods have recently been developed (and patented) for the production of what have been termed peptide nucleic acids (or PNAs) (Fig. 59.6). These modified oligos are resistant to attack, degradation, or modification by several types of enzymes. Such properties make them ideally suited to many *in vitro* and *in vivo* applications including restriction endonuclease-based analytical assays, and *in vivo* antisense/ antigene-based therapies.

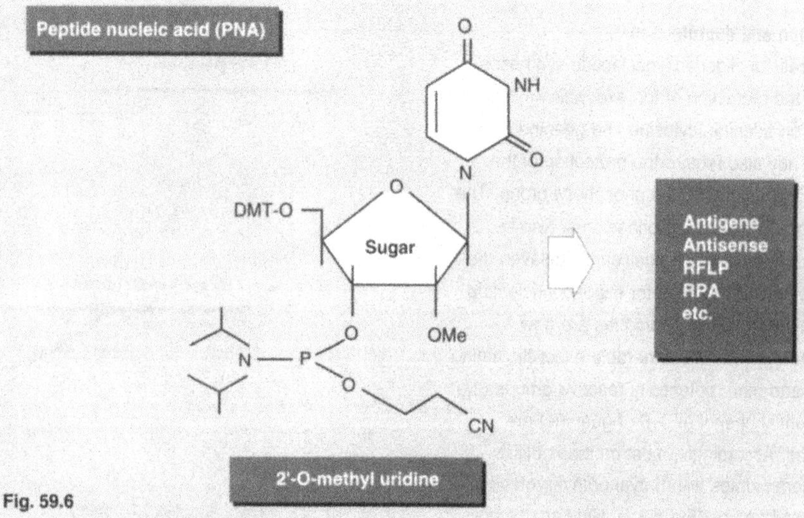

(3) Altered linkages

New approaches to the synthesis of oligophosphorothioates and oligomethylphosphonates have also been developed. New analogues such as phosphorodithioates have also been developed. Both have been used to create nuclease digestion-resistant oligos. Similar properties have also been attributed to oligos constructed with sulphonamide and sulphonate linkers replacing the normal phosphodiester bonds, or oligoribonucleotides containing 2′-5′ rather than 3′-5′ linkages. Whilst the biological activity of many of these new constructs has not been fully tested, they could also be valuable as *in vivo* antisense reagents.

(4) Terminal modifications

Similarly, modifications of oligo termini by alkylating, intercalating or cleaving agents (e.g. bleomycin) may also provide reagents protected

against exonuclease attack. They may also improve cellular uptake, or be used to induce irreversible changes in the target sequence to which they hybridise. The introduction of cholesterol into oligos has particularly been suggested to promote cellular uptake during antisense or gene-targeted approaches by allowing the oligo to bind and be endocytosed by cell surface lipoprotein receptors.

(5) Ribozyme modification.

Although at an early stage of development, research has also been invested in the chemical

synthesis or modification of ribozymes. For example, some replacements of ribonucleotides with 2′-NH2 or 2′-F analogues in hammerhead ribozymes is reported to protect them from nucleases without altering their biological activity. Such a purpose-built ribozyme has already been shown to be able to repair a defective piece of *lacZ* mRNA in *E. coli* by catalysing *trans*-splicing and thereby generating a translatable mRNA. This type of ribozyme-mediated *trans*-splicing is suggested to provide an alternative to gene replacement-based approaches for the therapeutic correction of defective genes.

Fig. 59.5

Fig. 59.6

BRANCHED OLIGONUCLEOTIDES

An area of significant development has been the generation of protecting groups which can be removed under the milder conditions preferred for RNA oligomer synthesis. In addition, attention has been focused upon ways to produce branched oligonucleotides. For example, branched oligoribonucleotides with the potential to inhibit *in vivo* splicing machinery have been developed by linkage of one ribonucleotide to three others by 5′, 3′ and 2′ linkages. Branched or 'dendritic' oligomers have also been developed for *in vitro* diagnostic use (Fig. 59.7). Accordingly, the attachment of a single label to each of the branches produces a probe with many labels. The resultant high level labelling of each target–probe hybrid provides great detection sensitivity (e.g. down to sub-attomolar levels of virus in a clinical specimen).

DEGENERATE OLIGONUCLEOTIDES

Most oligos are synthesised based on experimentally determined or defined sequences. In some situations however (e.g. library screening), only the amino acid, and not the nucleotide sequence, of the target may be known. So-called degenerate oligos may therefore be synthesised comprising a heterogeneous population of oligo sequences reflecting the degeneracy of codons used by the genetic code (see Fig. 59.8). This is simply achieved by adding two or more nucleotide 'monomers' in the synthesis reaction at the appropriate cycles. It must however be realised that this heterogeneous population is less effective or sensitive in terms of target sequence recognition since the concentration of the one authentic sequence is effectively diluted. Furthermore, some nucleotides may be more efficiently incorporated than others.

Efficiency constraints

All oligo syntheses, regardless of the method used and the sophistication of the monitoring process, are less than 100% efficient. Consequently, the more nucleotides added to the growing chain, the greater the percentage of oligos which are produced containing base errors (i.e. incorrect bases). Although this affects only an exceedingly small percentage of oligos per cycle, the accumulation of base errors effectively

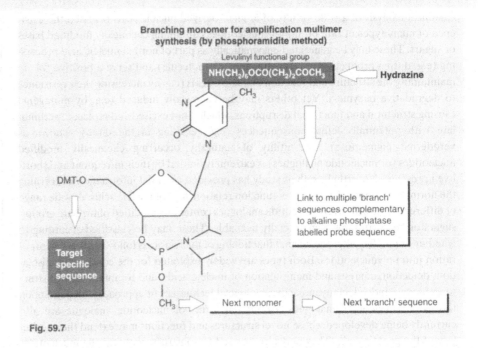

Fig. 59.7

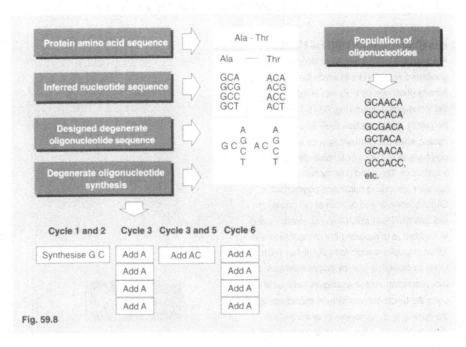

Fig. 59.8

limits the usable length that can be achieved to the order of a 100 bases. Furthermore, differences in base incorporation may, in some rare cases, lead to total failure in synthesising the desired sequence.

OAs: 178, 531–42, 546–7, 547–52, 1059–64, 1120–8. RAs: 210–12, 406–8.
SFR: 142–5, 181, 207–10.

60. TYPES AND APPLICATIONS OF NUCLEOTIDE ANALOGUES

Nucleotide analogues with modified phosphate groups, sugars or bases are widely used for nucleic acid labelling, manipulation, and synthesis, or to expand the nature of the genetic code.

Chemical analysis of a wide variety of *in vivo* -derived nucleic acids revealed the existence of many types of rare or 'minor' nucleotides containing chemically modified bases or sugars. These may be generated enzymatically as part of normal nucleic acid processing (e.g. during production of a functional RNA molecule) and serve a positive role in maintaining the structure and function of nucleic acids (e.g. by increasing their resistance to degradative enzymes). Yet others may be chemically induced (e.g. by mutagens) causing structural and functional disruptions, which if not excised and replaced, accumulate with potentially lethal consequences (e.g. resulting in the disease known as xeroderma pigmentosa). The utility of naturally occurring chemically modified nucleotides (or nucleotide analogues) is extremely limited by their infrequent and short-lived presence. Nevertheless, their study has provided essential information concerning the normal and abnormal structure–function relationships of nucleic acids. A wide range of different *in vitro*-derived nucleotide analogues containing modified phosphate groups, sugars or bases are now commercially available. These may be classified according to whether or not they alter the normal functioning of nucleic acids following their incorporation into polynucleotides. Both types are widely exploited for the construction, isolation, detection, analysis and manipulation of nucleic acids, and to mimic natural structures *in vivo*. Indeed, many molecular biological techniques or approaches depend upon the use of nucleotide analogues. Additional forms of nucleotide analogue are also currently being developed whose novel structures and functions may extend the vocabulary of the genetic code. They may also be used to construct synthetic oligonucleotides which are protected from intracellular degradation during gene targeting.

PHOSPHATE GROUP MODIFICATION

Amongst the first nucleotide analogues to be generated were dNTPs in which the radio-isotopic form of phosphorous (^{32}P) was substituted into the γ phosphate group (Fig. 60.1). These are still frequently used to radioactively label the 5′ end of nucleic acids using enzymes such as T4 polynucleotide kinase or terminal deoxynucleotidyl transferase. The γ and β phosphate groups are however lost during nucleotide polymerisation. Consequently, 3′ end or internal radiolabelling requires dNTPs in which the α phosphate group is modified (e.g. replacing the phosphorous with ^{32}P or the radio-isotopic form of sulphur (^{35}S)). Once incorporated (e.g. by primer extension or nick translation), these analogues have no effect upon the functional or structural characteristics of the nucleic acid. Analogues in which the α phosphate group oxygen is replaced by sulphur (i.e. α-thionucleotides) may also be incorporated into the 3′ end of polynucleotides. This protects them from exonuclease digestion (e.g. for the creation of 5′ deletion mutants). Another important class of analogues are the phosphoramidates used in the *in vitro* chemical synthesis of oligonucleotides in a 3′ to 5′ direction.

OAs: 527, 536–40, 544–8, 552–65, 771, 775, 944–7, 1120–8. RAs: 210–12, 274–5, 378, 406–7. SFR: 10–13, 136–9, 142–5, 207–10.

NATURALLY ALTERED NUCLEOTIDES

A wide range of post-transcriptional modification processes are used *in vivo* to alter specific bases particularly those within RNA molecules. These are enzymatically mediated to ensure the correct structure and appropriate stability of the molecules (e.g. the cloverleaf structure of long-lived transfer RNAs or tRNAs). DNA may also be subject to post-replication enzymatic modification, especially in bacteria where methylase enzymes methylate specific bases, thereby protecting it from digestion by certain endogenous restriction endonucleases. Several nucleotide alterations may also occur due to the action of external mutagens such as heat, various chemical compounds, and high or low energy forms of ionising radiation.

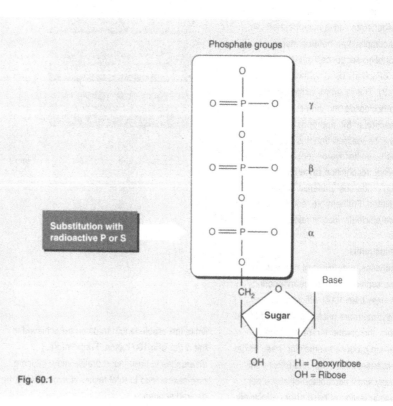

Fig. 60.1

MODIFICATION OF SUGAR RINGS

Non-isotopic labels such as biotin may be covalently linked to one of the sugar ring carbon atoms (via a long chain carbon spacer) without affecting the polynucleotide into which it is incorporated (Fig. 60.2a). This type of analogue is used for both gene probe labelling and the isolation of specific gene sequences. In contrast, dideoxynucleotides in which the 3′ OH group is replaced with a hydrogen atom (Fig. 60.2b) will prevent the further addition of nucleotides once incorporated at the 3′ end of a growing polynucleotide. This is crucially exploited in the process of nucleotide sequencing by the Sanger chain termination method. The capacity of dideoxynucleotides to block DNA replication is also exploited in anti-cancer and anti-HIV therapies. Nucleotides containing no sugar or drastically modified sugar moieties may also be vital for the development of *in vivo* antisense oligo or gene targeting approaches since the oligos are extremely resistant to DNA/RNA degrading enzymes.

BASE MODIFICATIONS

Nucleotides containing modified bases are rarely utilised *in vivo* since alterating the base can profoundly affect the base pairing specificity of the resulting polynucleotide. Nucleotide analogues bearing modified bases, such as deoxyinosine (dITP) and deazaguanine, have however found widespread application in several *in vitro* molecular biology techniques (e.g. they can be incorporated into polynucleotides by the polymerase chain reaction). Due to their unusual base pairing properties, such analogues may be used to introduce mutations or overcome problems associated with the tertiary structure of polynucleotides respectively. Similarly, the nucleotide analogue 5-bromouracil (5bU) may be used to introduce mutations *in vivo* or *in vitro* since it is capable of rapidly undergoing a tautomeric shift from the keto to the enol form (Fig. 60.3). This results in a change in base pair specificity from A to G respectively. Nucleosides containing novel bases (e.g. iso-cytosine and iso-guanine) have also been synthesised which are capable of being incorporated and replicated. These have great potential for protein engineering by allowing non-natural amino acids to be incorporated during *in vitro* and *in vivo* translation.

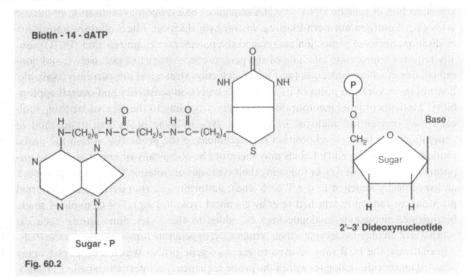

Fig. 60.2

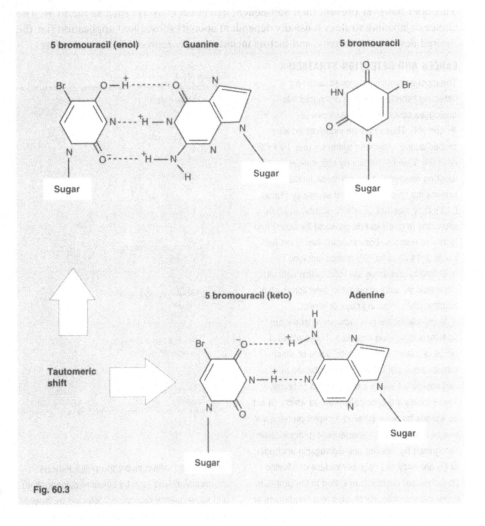

Fig. 60.3

61. METHODS FOR LABELLING GENE PROBES

A range of radioactive and non-radioactive labels may be uniformly incorporated within the sequence of double or single-stranded nucleic acid probes by nick translation or primer extension methods.

The detection of specific DNA or RNA sequences based upon hybridisation of nucleic acids (e.g. Southern/northern blotting, *in situ* hybridisation, filter/solution/capture hybridisation, nuclease protection assays, and the polymerase chain reaction (PCR)) usually requires appropriate labelling of the gene probe. A range of radioactive and non-radioactive labelling molecules (and hence detection strategies) are currently available. Each of these varies in terms of its achievable detection sensitivity and overall applicability. Methods of label incorporation also vary according to the type of labelling molecule (e.g. nucleotide analogue or enzyme), type of gene probe and its method of generation, and the desired position of attachment to the probe (i.e. within the probe sequence or at either end). Labels may therefore be chemically or enzymatically incorporated into probes during or following their synthesis or isolation (i.e. genomic probes), and specifically attached to the 3′ or 5′ ends, uniformly incorporated at many internal positions, or randomly attached (e.g. by chemical cross-linking). For example, a single biotinylated nucleotide analogue may be added to the 5′ terminus during chemical synthesis of an oligonucleotide which is then used to generate longer probes by the PCR. Alternatively, the PCR may be used to generate gene probes which incorporate many labelled nucleotide analogues within the probe sequence (i.e. internally labelled). Radioactive and non-radioactive labels may also be incorporated at the 3′ end of gene probes. This does however prevent their subsequent extension in assays such as the PCR. The choice of labelling strategy is usually dependent upon its subsequent application (i.e. the desired detection sensitivity, and factors including assay reproducibility and safety).

DETECTION END POINTS

The type of labelling moiety used obviously determines the way in which bound probe may be detected (i.e. the detection end point). Radioactive probes are generally detected by exposure to X-ray film or photographic emulsions (i.e. by autoradiography). Radioactivity imaging systems have however been developed which both detect the position, and quantitate the amount of bound probe. Labels capable of generating a chemiluminescent (i.e. chemically generated light) end point may also be detected by autoradiography, or, like fluorochrome labels, directly detected by photometry. Many analytical methods employ enzyme labels. These can generate different end points (e.g. colour, light or fluorescence) depending on the substrate used, or utilise signal amplification based upon the build up of secondary and tertiary linking molecules.

LABELS AND DETECTION STRATEGIES

The most commonly employed labels for detecting hybridised probes are nucleotide analogues containing radioisotopes (e.g. ^{35}S, ^{32}P, ^{3}H and ^{125}I). These may be incorporated into probes during enzymatic synthesis (e.g. by PCR) or at the 3′ and 5′ termini by enzyme-catalysed labelling reactions. Although these labels currently provide the greatest detection sensitivity (Table 61.1), they may not be wholly suitable in some situations (e.g. clinical diagnostics) for safety and technical reasons. For example, their short half-life (e.g. 14 days for ^{32}P) makes storage, quantitation and assay standardisation difficult. Conventional autoradiographic detection is also lengthy, often requiring days or weeks.

Many alternative non-radioactive labels are now therefore being exploited. These afford a range of detection end points, some of which provide equivalent detection sensitivities to radiolabels but within a few minutes or hours. These labels are principally proteins which: (i) act as ligands for other labelled receptor proteins via one or more layers of reactants (e.g. digoxigenin recognised by labelled anti-digoxigenin antibody), or (ii) are enzymes (e.g. peroxidase or alkaline phosphatase) directly conjugated to the probe. In some instances fluorochromes (e.g. rhodamine or fluoroscein)

Table 61.1

Label type (substrate)	Label	Labelled probe	Probe in assay
	Detection limit (amol)		Hybridised
Fluorescein	20 000	100 000	500 000
Rhodamine	5000	20 000	100 000
Phosphorous 32	50	50	50
Alkaline phosphatase (PNPP)	5000	5000	5000
Alkaline phosphatase (BCIP)	200	200	200
Horsradish peroxidase (Luminol)	1000	1000	1000
Horsradish peroxidase (DAB)	50	50	50

may be used. Alternatively the probe may be chemically altered (e.g. by cytosine sulphonation) and subsequently recognised/detected by labelled antibodies specific for the altered nucleotide.

OAs: 410, 414, 424–9, 529, 538–40, 546–59. **RAs:** 191, 194, 204–6, 208–10, 213. **SFR:** 119, 122–8, 131–3, 136–43, 145–7, 179–81.

UNIFORM (INTERNAL) LABELLING

Essentially two approaches are routinely used to enzymatically incorporate labels uniformly within the sequence of nucleic acid probes. These are termed nick translation and primer extension.

Nick translation

Nick translation (Fig. 61.1) was amongst the first methods to be developed, and is still commonly used for labelling double-stranded DNA. This method is well suited to large-scale labelling of even relatively long probes to a high specific activity (i.e. to detect single copy genes in Southern blotting). It is also able to introduce biotinylated nucleotides (e.g. biotin-dUTP) into double stranded DNA probes.

The method is based upon limited treatment of the DNA with DNase I which creates widely separated breaks or nicks within the duplex exposing free 3′ hydroxyl (OH) groups. These nicks can then be repaired by incubation with DNA polymerases which contain 5′–3′ exonuclease activity (e.g. DNA pol I from *E. coli*). In the presence of one or all four radioactively labelled nucleotide triphosphates (e.g. ^{32}PdCTP), the polymerase will therefore progressively incorporate the labelled dNTPs. The 5′ terminus of the existing nucleotide will also be simultaneously hydrolysed by the enzyme's intrinsic 5′–3′ exonucleolytic activity leading to the release of 5′ mononucleotides. This results in movement or translation of the nick along the length of the duplex producing uniform labelling of the duplex. (This is unrelated to the process of protein synthesis also termed translation.)

Primer extension

Primer extension methods utilise the ability of DNA polymerases lacking 5′–3′ exonuclease activity (e.g. the Klenow fragment of DNA pol I or reverse transcriptase) to synthesise a complementary copy from a single-stranded or denatured double-stranded template, beginning from the free 3′ hydroxyl group of a hybridised oligonucleotide primer. Two approaches are used: (i) random priming, using a mixture of primers of random hexanucleotides, and (ii) the use of a sequence-specific primer (Fig. 61.2). Both approaches can produce multiply-labelled probes of high specific activity but only in relatively low

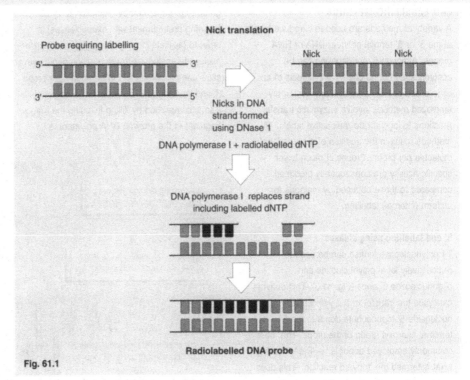

Fig. 61.1

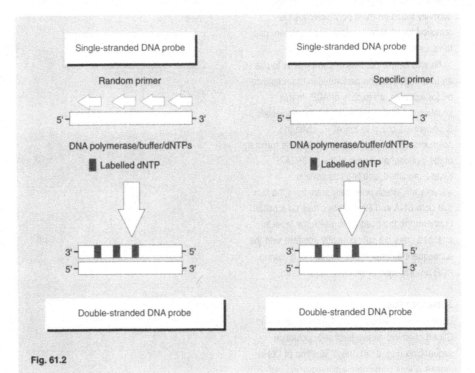

Fig. 61.2

yields compared to nick translation methods. The principle of primer extension is also used to produce, and simultaneously incorporate, one or more labels within nucleic acid probes by the PCR.

61. METHODS FOR LABELLING GENE PROBES/Contd

END LABELLING METHODS

A variety of methods are used to introduce labels at the 3' or 5' termini of linear DNA or RNA probes. Although 5' end labelling can be accomplished during chemical synthesis of an oligo probe or PCR primer, many commonly employed methods involve enzymatic transfer reactions to incorporate radioactive labels. All methods result in the addition of a single label molecule per probe. Probes of much lower specific activity are consequently produced compared to those obtained by methods for uniform (internal) labelling.

5' end labelling using kinases

T4 polynucleotide kinase can be used to radioactively label polynucleotide and oligonucleotide 5' ends (Fig. 61.3). This enzyme catalyses the transfer of the γ-phosphate from a nucleoside 5'-triphosphate donor to the 5' terminal hydroxyl group of the probe. The most commonly employed donor is [γ-^{32}P]dATP using what is termed the 'forward reaction'. This does however require that the 5' phosphate group normally found on most polynucleotides is removed by alkaline phosphatase treatment prior to its use.

An alternative, but less efficient way is to use an exchange reaction performed in the presence of, for example, an excess of ADP. In this situation the kinase transfers the polynucleotide's 5' phosphate group to the ADP, whilst the polynucleotide is rephosphorylated by the transfer of the γ-phosphate group of the [γ-^{32}P]dATP. Kinase-mediated labelling has several advantages, which principally stem from the fact that both DNA and RNA probes may be labelled. Furthermore, the position to which the label is attached does not subsequently interfere with the subsequent extension of the probe (e.g. during PCR amplification or dideoxy sequencing).

3' end labelling

Used for the 3' labelling of DNA for Maxam and Gilbert chemical cleavage-based nucleotide sequencing (Fig. 61.4), the 3' labelling of DNA probes is less commonly achieved using calf thymus terminal deoxynucleotidyl transferase (TdT). This enzyme adds homo-polymer extensions to DNA containing a free 3' hydroxyl

group (e.g. generated by restriction enzymes or following pretreatment with exonucleases) in order to facilitate cloning. Supplying radioactive nucleotide analogues to the reaction can however allow one or more labels to be added per probe. 3' end labelling of 3' recessed DNA fragments is also accomplished by 'filling in' using the Klenow fragment of the enzyme DNA polymerase I.

OAs: 529, 538–40, 546–8, 559–71.
RAs: 191, 194, 204–6, 208–10, 213.
SFR: 119, 122–8, 131–3, 136–42, 145–7, 179–81.

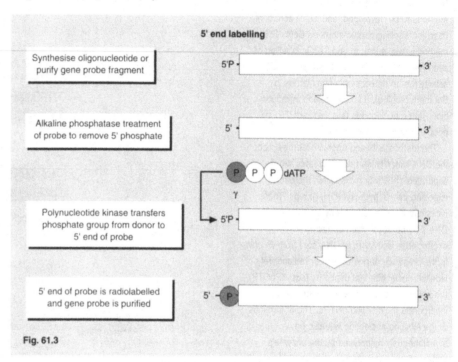

Fig. 61.3

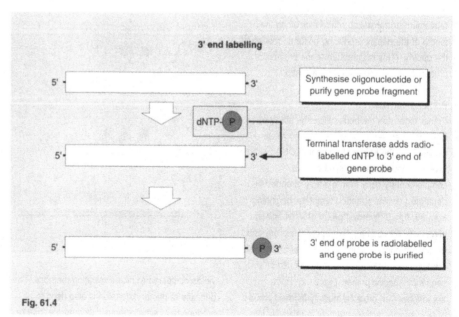

Fig. 61.4

PCR-BASED LABELLING

The PCR is frequently used to provide 5′ end labelled DNA or RNA probes (Fig. 61.5). This is accomplished by extending sequence-specific oligonucleotide primers which have previously been 5′ end labelled with non-radioactive labels such as biotin, digoxigenin or fluorochromes during chemical synthesis. Such labelled probes may be used for both the isolation and/or detection of target sequences (e.g. in capture hybridisation assays). Both radioactive and non-radioactive nucleotide analogues may however also be uniformly incorporated along the length of the synthesised probe. For example, biotin-dUTP analogues (composed of dUTP linked by a spacer arm of a defined number of carbon atoms to a biotin molecule) may be used to replace some or all of the dTTP in the PCR reaction, and hence the PCR product. Although high levels of label may be incorporated, this can in some instances reduce the subsequent hybridisation efficiency of the probe.

VECTOR-BASED LABELLING

Vectors used to generate single stranded RNA (or DNA) probes may also uniformly incorporate labelled nucleotides during probe synthesis by simply adding the nucleotide analogue into the synthesis reaction mixture (Fig. 61.6).

CROSS-LINKING METHODS

In addition to the enzymatic incorporation of biotinylated nucleotide analogues, biotin may be stably cross-linked to single- and double-stranded polynucleotide probes. This is normally accomplished by brief irradiation in visible light using a photoactivatable biotin analogue. Biotin, digoxigenin, and several enzymes (e.g. peroxidases, alkaline phosphatases) have also been successfully covalently linked directly to the sugar–phosphate backbone of polynucleotide probes using chemical cross-linking agents. For example, horseradish peroxidase and alkaline phosphatase are initially cross-linked to polyethyleneimine by *p*-benzoquinone (Fig. 61.7). The resulting complex is then covalently linked to the probe by incubation with glutaraldehyde. This method is a much simpler and rapid way of uniformly labelling probes. Furthermore, enzyme labels also allow the use of several different types of detection end points.

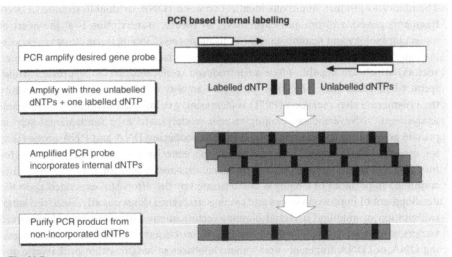

PCR based internal labelling

- PCR amplify desired gene probe
- Amplify with three unlabelled dNTPs + one labelled dNTP
- Amplified PCR probe incorporates internal dNTPs
- Purify PCR product from non-incorporated dNTPs

Labelled dNTP   Unlabelled dNTPs

Fig. 61.5

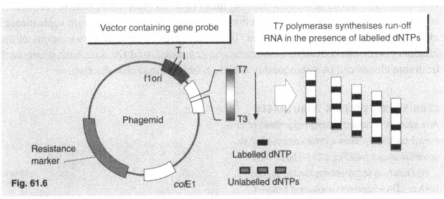

Vector containing gene probe

T7 polymerase synthesises run-off RNA in the presence of labelled dNTPs

Phagemid
f1ori
Resistance marker
T7
T3
*col*E1

Labelled dNTP
Unlabelled dNTPs

Fig. 61.6

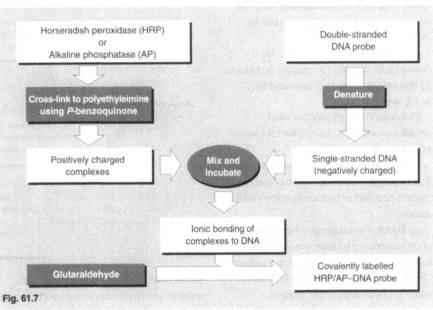

Horseradish peroxidase (HRP) or Alkaline phosphatase (AP)

Double-stranded DNA probe

Cross-link to polyethyleimine using *P*-benzoquinone

Denature

Positively charged complexes

Mix and incubate

Single-stranded DNA (negatively charged)

Ionic bonding of complexes to DNA

Glutaraldehyde

Covalently labelled HRP/AP–DNA probe

Fig. 61.7

165

62. FUNDAMENTAL PRINCIPLES OF CLONING

Cloning of DNA or cDNA fragments can be achieved in several ways, each based upon their replication and/or expression following initial insertion into an appropriate vector and introduction into suitable host cells.

The ability to produce numerous identical copies of DNA or double-stranded cDNA fragments generated from messenger RNA by reverse transcription is at the heart of molecular biology and recombinant DNA technologies. Cloning is a multi-stage process involving the insertion of DNA fragments into specialised DNA molecules (termed vectors) which are capable of being introduced, replicated and/or expressed within specific living host cells. The ability to *in vitro* amplify DNA (and RNA) fragments by the polymerase chain reaction (PCR) is increasingly being used to identify, isolate and analyse genes. Nevertheless, cloning remains widely used as a fundamental step in protein and genetic engineering approaches for producing DNA and RNA probes (e.g. for use in *in vitro* assays), and for introducing genes into living organisms (e.g. for producing genetically modified and transgenic organisms, and for gene therapy). The continued importance of cloning is demonstrated by the effort still expended upon the development of improved vectors and cloning strategies. Consequently, since the initial exploitation of modified bacterial plasmid vectors, many different types of improved vectors have now been developed for specifically introducing, replicating and/or express-ing DNA or cDNA fragments up to many kilobases in length within prokaryotic and eukaryotic cells. Although cloning strategies based upon the construction and screening of genomic and cDNA libraries containing many millions of different gene sequences are still widely employed (especially in genome mapping studies), many more sophisticated cloning approaches are now also used. These strategies essentially vary in terms of the efficiency with which desirable clones can be generated, and the ease with which such desirable clones can be subsequently detected, isolated and characterised.

DEVELOPING VECTOR VARIATION

A wide range of different types of prokaryotic and eukaryotic cloning vectors have already been developed. Each is capable of replicating and/or expressing DNA inserts in different types of host cells. These include vectors developed from bacterial plasmids, bacteriophage genomes, yeast chromosomes, engineered hybrids (e.g. cosmids and phagemids), and specialised vectors suitable for use in animal and plant cells. Although basically similar in composition and function, they vary in the size of DNA which can be inserted and replicated, how desirable recombinants may be detected, and their range of potential applications. Vectors capable of harbouring ever larger inserts are still being developed, especially for use in animal cells (e.g. mammalian artificial chromosomes or MACs).

RAs: 214–24, 241–5, 249–52, 376–9, 392–7, 418–24. **SFR:** 36–9, 115–28, 136–9, 149, 154–5, 158–62, 169–70, 197–200, 219–22, 224–6.

CLONING INVOLVES SIX VITAL STAGES

Any successful cloning strategy, regardless of its overall complexity, uses a cycle composed of six essential stages (see Fig. 62.1). These involve:

(1) Obtaining and preparing double-stranded DNA or cDNA fragments containing 5′ and 3′ termini compatible with sites created within the cloning vector (e.g. overhangs generated by restriction enzyme digestion).

(2) Insertion of the DNA fragments into a cloning vector (i.e. within the multiple cloning site) by annealing and ligation of vector and DNA 3′ and 5′ termini.

(3) Introduction of the newly formed recombinant vector (i.e. bearing the DNA insert) into an appropriate host cell.

(4) Replication of the recombinant vector within transformed (prokaryotic and plant) or transfected (animal) host cells by their propagation in *in vitro* culture.

(5) Selection and expansion of individual clones of cells harbouring the desired recombinants (i.e. harbouring vectors containing inserts of interest).

(6) Isolation and analysis of the DNA inserts, and/or their expressed protein products.

In many cases these newly isolated DNA clones may be used in further cloning cycles, a process often described as subcloning. For example, subcloning may be required following *in vitro* mutagenesis to transfer the mutated DNA into a different vector for expression in a different type of host cell, or for the generation of single-stranded DNA/RNA (i.e. for use as sequencing templates or as gene probes).

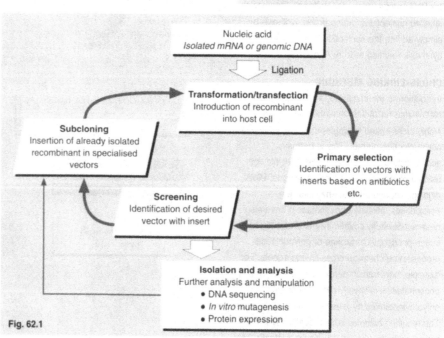

Fig. 62.1

DIVERSITY OF CLONING STRATEGIES

The availability of many vectors with different functional characteristics and limitations has resulted in the generation of many cloning strategies. Initial cloning strategies were based upon the creation of genomic or cDNA libraries (Fig. 62.2). These are composed of thousands (or millions) of different clones created by the insertion of large numbers of heterogeneous DNA fragments generated by restriction digestion of genomic DNA samples, or cDNA fragments produced from total cellular or poly(A+) RNA. The difficulty in detecting desirable sequences in such libraries (e.g. due to low copy abundance) has however led to alternative cloning strategies based upon producing 'restricted' or 'gene-enriched' libraries (e.g. based upon subtractive hybridisation or the PCR). The PCR has also been used to generate combinatorial libraries by repertoire cloning, whereby an array of clones are produced bearing new randomised gene combinations within expression vectors. The PCR has also been particularly used as a means of specifically cloning or subcloning genes or gene fragments (e.g. in antibody gene engineering).

Detecting recombinant clones

Irrespective of the type of cloning strategy employed, the source of inserts (i.e. cell-derived or PCR amplified), or the type of host cell used, the ability to detect and isolate specific clones ultimately dictates the overall success of any cloning approach. The first step in this process involves distinguishing between infected and uninfected cells, and those harbouring recombinant and non-recombinant vectors. This is generally achieved using selectable markers introduced into vectors during their original construction. Accordingly, antibiotic resistance markers are often used to prevent the growth of uninfected cells (i.e. containing no vector). Furthermore, recombinants may be distinguished from non-recombinants based upon insertion-mediated enzyme inactivation. For example, insertional inactivation of β-galactosidase caused by the introduction of DNA into the *lacZ* gene of a vector yields white recombinant colonies in substrate-containing medium. In contrast, non-recombinant vectors continue to express β-galactosidase thereby generating a blue colony.

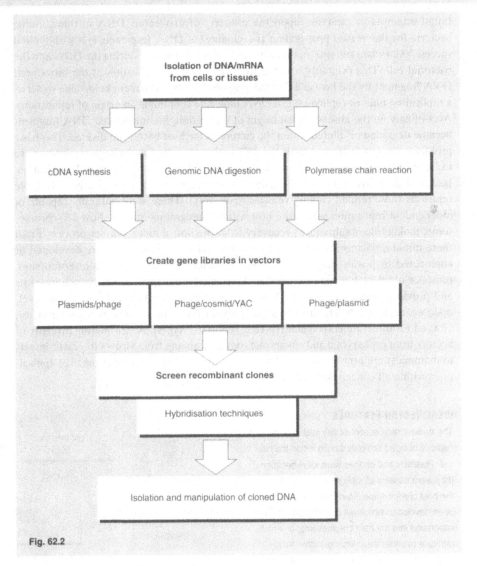

Fig. 62.2

Isolation of DNA/mRNA from cells or tissues

cDNA synthesis — Genomic DNA digestion — Polymerase chain reaction

Create gene libraries in vectors

Plasmids/phage — Phage/cosmid/YAC — Phage/plasmid

Screen recombinant clones

Hybridisation techniques

Isolation and manipulation of cloned DNA

Detecting desirable clones

Recombinant clones containing desired inserts can be detected in several ways based upon the analysis of DNA inserts or their expressed products. Consequently, hybridisation of labelled gene probes generated by previous cloning cycles or by *in vitro* chemical synthesis based upon partial sequence data is widely used to detect specific DNA insert sequences within genomic or cDNA libraries. Indeed, the former types of probes are often used in chromosome 'walking' and 'jumping' strategies for isolating hitherto unknown genes starting from linked polymorphic markers (e.g. restriction fragment length polymorphisms). More recently, the PCR has been successfully employed to rapidly screen clones based upon amplification and direct analysis of DNA inserts. Desired clones may then be removed, expanded by limited culture, rescreened, and finally propagated in bulk culture. Alternatively, desirable cDNA clones may be detected using antibody-based immuno-probes directed to an epitope on the expressed protein product, and similarly isolated for bulk propagation. The availability of surface display vectors has however greatly facilitated the detection and isolation of desirable cDNA clones based upon immuno-affinity chromatography.

63. THE NATURE OF CLONING VECTORS

Several types of vectors have been designed to facilitate the *in vivo* propagation (cloning), expression, functional analysis and selection of DNA fragments within prokaryotic and eukaryotic host cells.

Initial attempts to capitalise upon the property of exogenous DNA in transforming bacteria for the *in vivo* propagation (i.e. cloning) of DNA fragments met with limited success, even when suitable methods became available for introducing the DNA into the bacterial cell. This primarily resulted from the fact that replication of the introduced DNA fragment by the host cell depended upon the DNA fragment being able to act as a replication unit, or replicon (i.e. a DNA molecule containing an origin of replication). Accordingly, in the absence of an origin of replication, the introduced DNA fragment became degraded or diluted within the culture as the host bacterium divided. This basic problem was not overcome until the 1970s when bacterial and phage replicons were isolated and into which DNA could be inserted for host cell-mediated replication. Bacterial plasmids and phages specific for *E. coli* provided the first types of usable replicons (now termed cloning vehicles or vectors). These were naturally capable of independent replication within the host without integrating into the host cell chromosome, making their subsequent recovery from the host a much simpler process. From these initial replicons a whole range of plasmid and phage vectors were developed or engineered to possess improved cloning characteristics (e.g. multiple insertion sites, presence of selectable markers) making them suitable for specific cloning requirements and providing them with broader host cell specificities. Vectors have also been developed which allow DNA replication and expression (i.e. expression vectors), or which can be used for direct functional analysis (e.g. reporter vectors). A bewildering diversity of vectors from prokaryotic and eukaryotic sources (ranging from viruses to yeasts, insects to mammals) are now already available, or being developed, for all manner of applications within all conceivable cell types.

INTRODUCING DNA INTO CELLS

The ability to introduce a recombinant DNA fragment into a host cell for propagation is a vital requirement for DNA cloning. Retroviral and bacteriophage vectors achieve this by the normal processes of viral infection. Molecular biologists have however been required to develop several additional 'artificial' mechanisms by which to introduce other vector-borne (e.g. plasmid) or 'naked' (e.g. antisense oligonucleotides) DNA fragments into cells. Additional techniques or approaches have also been specifically devised to overcome problems encountered with animal and plant cells, where the DNA may have to cross both the external cell membrane and internal nuclear membrane. Methods range from those involving fusion or permeabilisation of cell membranes to those involving micro-injection, the firing of DNA-containing 'micro-projectiles' into the cell.

IDEAL VECTOR FEATURES

The primary requirement of any vector is that it is capable of being replicated once inside the host cell. Plasmids and phages were ideal because they were capable of replication independent of the host chromosome. Many vectors have now been developed which will only replicate following integration into the host chromosome, or which replicate in either way. Amongst other things, such vectors are extremely useful for studying normal chromosomal DNA function and for the development of several gene therapies.

Of great practical importance is that vectors must contain at least one unique restriction endonuclease recognition site into which the DNA fragment to be cloned may be introduced (Fig. 63.1). Indeed, the successful development of many vectors has required their engineering to contain DNA sequences termed polylinkers or multiple cloning sites (MCSs), which contain a grouping of multiple vector-unique restriction enzyme recognition sites. These MCSs allow greater cloning flexibility in terms of the types of DNA restriction fragments which can be inserted by ligation. Although not essential, it is nevertheless desirable that vectors exhibit two

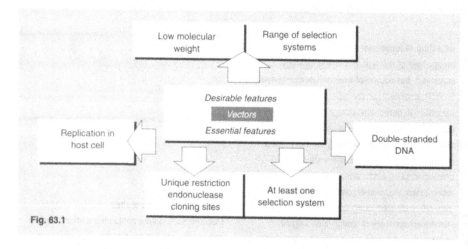

Fig. 63.1

additional attributes. Accordingly, in most circumstances they should be of relatively low molecular weight to allow them to both accommodate larger DNA fragment inserts, and to be maintained at high copy number within the host cell (thereby facilitating maximum recovery and yields of replicated DNA fragments). Ideal vectors should also contain at least one gene capable of conferring a selectable phenotype upon the host cell (e.g. antibiotic resistance).

Given the relatively low efficiency of transformation achievable with even the most efficient vectors, the presence of selectable markers greatly increases the ability to subsequently detect cells transformed with insert-bearing vector molecules amongst the host cell population. A range of selectable markers and alternative strategies, often in combination, are now commonly used to further enhance clone selection, especially from gene libraries.

DEVELOPING DESIRABLE VECTORS

Originally vectors were purely designed for the cloning (i.e. bulk propagation by *in vivo* replication) of DNA fragments to facilitate subsequent analysis (e.g. by nucleotide sequencing). Vector development has since taken several complementary and intertwined paths leading to the generation of a wealth of vectors which satisfy the various requirements of research scientists, clinicians and biotechnologists alike (Fig. 63.2).

Size of inserted DNA

New vectors are continually being developed which allow the successful propagation of larger and larger DNA fragments. Initially, only low molecular weight prokaryotic vectors were used, however, even these did not permit sufficiently large fragments to be propagated. Consequently, several forms of plasmid vector (e.g. yeast artificial chromosomes, YA) have been successfully developed which potentially increase the size of insert which can be accommodated to those in excess of 40 kb (often representing an entire gene(s)). Such vectors have found particular application in large-scale genome mapping exercises, and for studying specific gene functions.

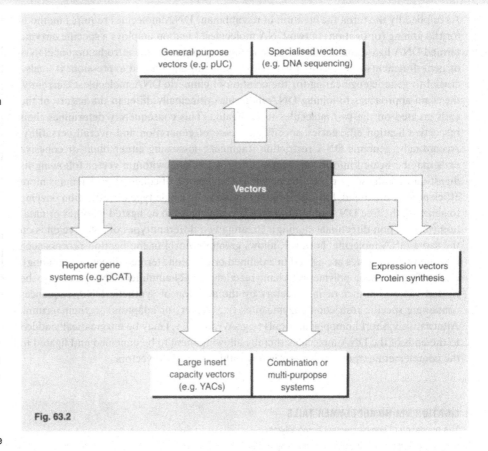

Fig. 63.2

Expression of DNA fragments

The availability of vectors which combine the cloning of DNA fragments with the ability to express the insert-encoded protein has been of immense importance to both basic and applied research. Expression vectors, as they are termed, now vary enormously from systems for expressing small amounts of protein to facilitate detection of appropriate recombinant clones, to those specifically designed for the bulk production of recombinant proteins. The development of expression vectors remains an actively ongoing and fruitful area of research directed at meeting the biotechnological needs for the bulk expression of functionally active recombinant proteins; in other words, to allow expression in

the correct cellular environment to ensure correct folding, assembly and secretion of the recombinant polypeptide chain(s) for subsequent analysis or application (e.g. as therapeutic agents).

'Reporter' gene systems

Many vectors containing reporter genes have been developed in order to provide an easy method of recombinant selection, or to provide additional information concerning the functions of specific DNA sequences. For example, recombinants may be recognised if reporter genes encoding readily detectable molecules (e.g. luciferase or β-galactosidase) are activated or inactivated upon insertion of the foriegn DNA fragment. Similarly, the rate of expression of the reporter gene may be altered by the insertion of a DNA fragment which contains a transcription regulatory sequence(s) (e.g. promoter or enhancers). Vectors employing reporter gene systems are also increasingly providing models

for studying gene function and *in vitro* alternatives to animal-based toxicity testing.

Combination vectors

Perhaps an obvious line of development has involved the engineering of vectors which combine elements from several types of vectors. Of significance are the so-called shuttle vectors which allow DNA fragments to be cloned and manipulated in prokaryote hosts and subsequently transferred into a eukaryote host for expression. A wide range of vector molecules are now available which combine the positive attributes of several types of vector into one molecule. For example, sequences from bacterial plasmids may be combined with specific sequences from phage vectors to generate cosmid and phagemid vectors. Such prokaryotic plasmid or phage vector sequences have also been combined within eukaryotic vectors (e.g. yeast-derived plasmid vectors).

64. INSERTING FOREIGN DNA INTO VECTORS

Foreign DNA molecules may be inserted into vectors by DNA ligases acting upon blunt or cohesive ends generated by restriction enzymes, or via complementary homopolymer tails (e.g. polydA, polydT).

As implied by the term, the creation of recombinant DNA molecules requires methods for the joining (or ligation) of two DNA molecules. Ligation employs a specific enzyme termed DNA ligase and is vital to all areas of gene cloning for the introduction of cDNA or gene fragments into vectors for replication, manipulation and expression. It is also crucial to genetic engineering for the creation of chimeric DNA molecules. Currently, the main approaches to joining DNA molecules principally differ in the nature of the ends created on the two molecules to be ligated. This consequently determines their respective ligation efficiencies, specificities, ease of generation and overall versatility. Accordingly, genomic DNA restriction fragments possessing either blunt or cohesive ends may be ligated into the appropriate restriction sites within a vector following its digestion with the same restriction enzyme. The ligation of cohesive fragments is more efficient than blunt end, ligation, however the use of a single type of restriction enzyme to generate the free DNA ends allows the two molecules to be ligated in either orientation (termed non-directional cloning). Creating two different types of cohesive ends on the same DNA molecule, however, allows greater control of the ligation process such that the two molecules are ligated in a defined orientation (termed directional cloning). cDNA fragments or polymerase chain reaction (PCR)-amplified DNA may also be ligated to one another or into vectors by the addition of synthetic DNA sequences containing specific restriction enzyme sites (e.g. linkers or adaptors) to their termini. Alternatively, short homopolymer tails (e.g. AAAAA . . .) may be enzymatically added to the ends of the DNA molecules thereby allowing them to be annealed and ligated to the complementary homopolymer regions within specialised vectors.

CHIMERIC MOLECULES

The term chimera is derived from a mythical creature with a lion's head, goat's body and serpent's tail. It is most frequently used to describe recombinant molecules created by the joining of two or more different DNA segments. Accordingly, a vector containing a foreign DNA insert may be referred to as a chimeric (or hybrid) vector. More commonly the term is used to describe the recombinant protein product generated by the fusion of two different gene fragments. For example, a mouse antibody which has been 'humanised' by genetic engineering such that it contains human antibody constant regions and mouse variable regions is often termed a chimeric antibody.

OAs: 333, 347, 443–4, 563, 591, 572–84. RAs: 178–82, 210, 214–15, 218–21, 241–5. SFR: 36, 67, 119–28.

LIGATION VIA HOMOPOLYMER TAILS

The annealing of complementary homopolymer tails is a general method which is commonly used to join DNA molecules during cloning. For example, a homopolymer polydA (or polydC) tail may be added to a cDNA or genomic DNA fragment. These fragments are then capable of annealing to the polydT (or polydG) tails added to the ends of a linearised plasmid vector. The gaps left following annealing may then be filled in by the action of gap-filling enzymes such as DNA polymerase and the free ends joined by DNA ligase (see Fig. 64.1).

Homopolymer tails can be added to any DNA fragment with an exposed 3′ OH group, such as generated by exonuclease digestion or certain restriction enzymes (e.g. PstI). DNA fragments bearing a 5′ overhang in which the 3′ OH group is obscured may also be used under certain conditions. In all cases however, the addition of the homopolymer tail is mediated by enzymes known as terminal deoxynucleotidyl transferases (TdT). If presented with a single deoxynucleotide triphosphate source, these enzymes are capable of repeatedly adding the nucleotides to the free terminal 3′ OH groups on the DNA molecule.

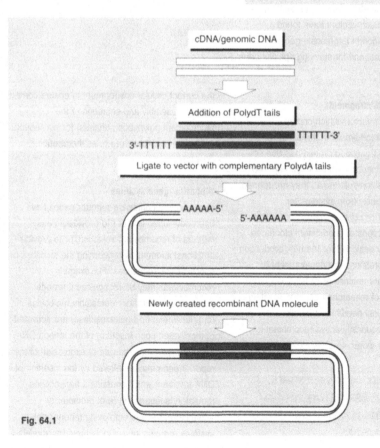

cDNA/genomic DNA

Addition of PolydT tails

3′-TTTTTT   TTTTTT-3′

Ligate to vector with complementary PolydA tails

AAAAA-5′   5′-AAAAAA

Newly created recombinant DNA molecule

Fig. 64.1

LIGATION VIA COHESIVE ENDS

Ligation via cohesive (or sticky) ends generated by certain restriction enzymes is a widely used and highly efficient process which has revolutionised gene cloning techniques. Accordingly, molecules with compatible cohesive ends, bearing either 3′ or 5′ overhangs, will rapidly anneal through complementary base pairing (Fig. 64.2). The remaining nicks in the sugar–phosphate backbone may then be simply closed by the action of DNA ligase. The use of two different types of cohesive ends at either end of the DNA molecule (whether naturally occurring, added during *in vitro* amplification or by the use of linkers and adaptors), also allows the orientation of DNA insertion into a vector to be controlled. Termed directional cloning, this helps to ensure that the foreign DNA molecule can then be correctly expressed as protein (i.e. due to insertion in an open reading frame).

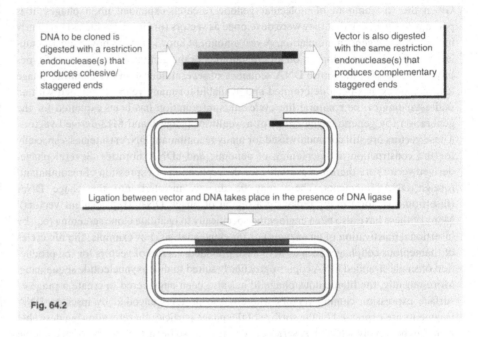

DNA to be cloned is digested with a restriction endonuclease(s) that produces cohesive/staggered ends

Vector is also digested with the same restriction endonuclease(s) that produces complementary staggered ends

Ligation between vector and DNA takes place in the presence of DNA ligase

Fig. 64.2

LIGATION VIA BLUNT ENDS

DNA fragments with blunt ends may be generated by physically breaking DNA or by digestion with certain restriction enzymes such as *Hind*II (Fig. 64.3). Blunt ended fragments may then be joined using the specialised DNA ligase of bacteriophage T4 (T4 DNA ligase). This process is however a very inefficient process which often results in the non-specific joining of fragments, thereby creating many unwanted products (e.g. self-ligation to form dimers or concatemers, or insertion of the fragment in the wrong orientation). Ligation of blunt ended fragments can however be successfully used to add linkers or adaptor molecules to DNA to facilitate the creation of cohesive ended fragments for more efficient ligation and directional cloning.

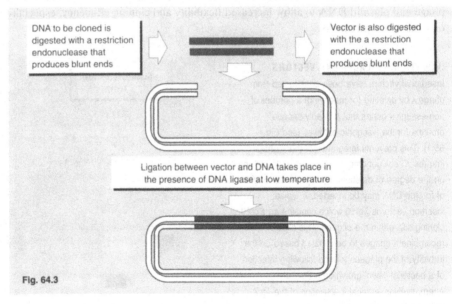

DNA to be cloned is digested with a restriction endonuclease that produces blunt ends

Vector is also digested with the a restriction endonuclease that produces blunt ends

Ligation between vector and DNA takes place in the presence of DNA ligase at low temperature

Fig. 64.3

Linkers and adaptors

Linkers are synthetic pieces of blunt ended double-stranded DNA which contain recognition sequences for defined restriction enzymes. These may be ligated to blunt ended DNA fragments by T4 DNA ligase and subsequently digested with the appropriate restriction enzymes to create cohesive ends. Most cloning protocols employ two sets of linkers to facilitate directional cloning. Adaptors are also chemically synthesised DNA fragments which contain one blunt end used for ligation but which also possess one preformed cohesive end.

Reversing the process

An important feature of ligation via blunt and cohesive ends created by restriction enzyme digestion is that, in almost all situations (except where random base substitution mutations have been inadvertently introduced), the ligation process regenerates the original restriction enzyme recognition sequence. It is therefore possible to subsequently redigest the recombinant DNA molecule such that the original two molecules may be recovered (e.g. to allow excision of the inserted fragment from a vector following host cell-mediated replication).

65. THE DEVELOPMENT OF BACTERIOPHAGE VECTORS

A range of insertional and replacement bacteriophage-derived vectors have been developed by genetic engineering to allow foreign DNA to be efficiently inserted, cloned, expressed, and selected.

Given the vast amount of molecular genetic research expended upon phages, it is perhaps unsurprising that they were developed as vectors (or cloning vehicles) very early in the history of gene manipulation. A vast amount of knowledge has now been accumulated concerning the composition and functioning of the genomes of a number of types of phages. Indeed, the entire DNA sequence of several helical and filamentous phage 'chromosomes' has been determined and available for many years. Combined with an understanding of their natural life cycles, this information has been exploited for the generation (by genetic engineering) of a wealth of phage λ- and M13-derived vectors. These vectors are still commonly used for many recombinant DNA strategies, especially for the construction and screening of genomic and cDNA libraries. Several phage-derived vectors are therefore available for the replication or expression of recombinant/foreign DNA fragments. These may be directly inserted into the phage DNA (insertional vectors) or by replacing a region of phage DNA (replacement vectors). Many of these have also been engineered specifically to facilitate clone screening (e.g. by insertional inactivation of an enzyme) or for clone analysis. For example, the life cycles of filamentous coliphages such as M13 have provided a series of vectors for the production of single stranded DNA copies particularly suited to dideoxynucleotide sequencing. More recently, the filamentous phage fd has also been engineered to create a range of surface expression cloning vectors whereby the proteins encoded by inserted DNA fragments are expressed on the surface of the phage particles thereby allowing desirable clones to be rapidly selected. A series of vectors has also been developed which combine phage and plasmid DNA to allow increased flexibility and cloning efficiency, especially of large DNA fragments.

PLASMID CONNECTIONS

The identification of sequences which determine phage and plasmid functions (e.g. those recognised by the phage packaging system) has led to the development of two types of hybrid vectors. Termed cosmids and phasmids (or phagemids), they combine phage DNA sequences with plasmids allowing the virtues of both vectors to be exploited. Cosmids represent plasmids containing a fragment of phage λ DNA including the *cos* site. This allows them to be used for cloning large pieces of DNA whilst also maintaining the ability to undergo *in vitro* packaging into new phage particles. Phasmids combine plasmid vector DNA with phage λ attachment sites allowing a plasmid to be excised from the phage genome. This allows phasmids such as λZAP to be propagated in *E. coli* host strains as either plasmids or as phages, followed by release of the replicated plasmid vectors.

INSERTION/REPLACEMENT VECTORS

Insertional vectors have been constructed from phage λ by deleting (or modifying) a number of non-essential genes that primarily encode proteins for the lysogenic pathway (see Fig. 65.1). This prevents integration of the λ phage into the *E. coli* genome. Consequently, depending on the degree of deletion of the gene, up to 10 kb of foreign DNA may be inserted. A typical insertion vector is λgt10 which contains an *Eco*RI cloning site within the *cI* gene. This allows recombinant phages to be isolated based on the turbidity of the plaques formed following infection of a bacterial 'lawn' grown on a Petri dish. Alternatively insertional inactivation of the *lacZ* gene has been utilized in λ insertional vectors (e.g. λ charon 16A) allowing a blue/white selection process to be applied.

Replacement vectors are particularly useful for genomic library construction because they are designed to clone larger fragments of DNA (up to 25 kb) by containing a removable region termed the 'stuffer' fragment. A number of these, such as EMBL3 and EMBL4, also have a number of unique restriction sites for inserting DNA fragments. Accordingly, the restriction enzymes

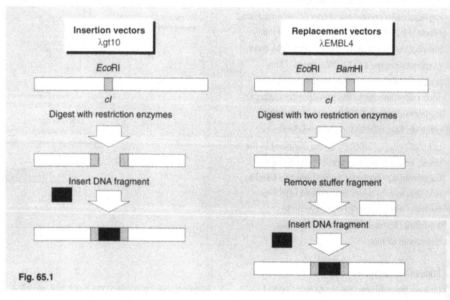

Fig. 65.1

Insertion vectors
λgt10

*Eco*RI

cI

Digest with restriction enzymes

Insert DNA fragment

Replacement vectors
λEMBL4

*Eco*RI   *Bam*HI

cI

Digest with two restriction enzymes

Remove stuffer fragment

Insert DNA fragment

*Eco*RI, *Bam*HI or *Sal*I may be used alone, or in combination, to both remove the stuffer fragment and provide appropriate ends for ligation. A potential problem associated with replacement vectors is the subsequent religation of the stuffer fragment. This may however be prevented by preparative gel electrophoresis of the vector arms, or the use of stuffer fragments with multiple restriction sites which allow its extensive cleavage followed by their removal using isopropanol precipitation.

OAs: 585–90, 633–42, 656–7.
RAs: 172–5, 214–15.
SFR: 36, 101–2, 116–24, 126–7, 213.

INSERTIONAL INACTIVATION VECTORS

Initial phage λ vectors contained two selectable markers (e.g. antibiotic resistance genes) to facilitate detection of recombinant clones. This selection approach however requires a two-step process: (i) the selection of clones containing phage DNA, followed by (ii) the selection of those clones harbouring a foreign DNA insert (i.e. recombinants).

In order to simplify and speed up the process of selecting recombinant DNA-containing clones, derivative phage λ vectors were developed which employ insertional inactivation strategies.

Although several other enzymes have since been used, the first and still most widely used system involves the modification of the *lacZ* gene encoding the enzyme β-galactosidase. Vectors utilising insertional inactivation of *lacZ* were first developed by engineering several restriction sites into the 3′ terminal end of the *lacZ* gene (subsequently referred to as *lacZ′*) to allow the subsequent insertion of foreign DNA fragments (see Fig. 65.2). These engineered restriction sites do not alter the expression and functional activity of the encoded β-galactosidase which can hydrolyse the colourless substrate Xgal (5-bromo-4-chloro-3-indolyl-β-D-galactopyranoside) to yield a blue insoluble precipitate in appropriate culture medium. Consequently, infection of bacteria by *non-recombinant* phages in media containing the substrate Xgal and the *lacZ′* expression inducer IPTG (isopropyl-thiogalactoside) produces plaques with a blue colour. However, insertion of a DNA fragment into the *lacZ′* gene results in the expression of a non-functional β-galactosidase molecule (in the form of a fusion protein with the insert-encoded protein) which is unable to hydrolyse Xgal and therefore forms a white plaque. Accordingly, blue plaques on a bacterial lawn represent non-recombinant clones capable of the expression of functionally active β-galactosidase, and white plaques represent recombinant clones in which DNA insertion has disrupted the activity of the β-galactosidase.

This *lacZ′* blue/white selection strategy has been incorporated into a number of plasmid (e.g. the pUC series), phage and phagemid vectors, including M13mp and pBluescript.

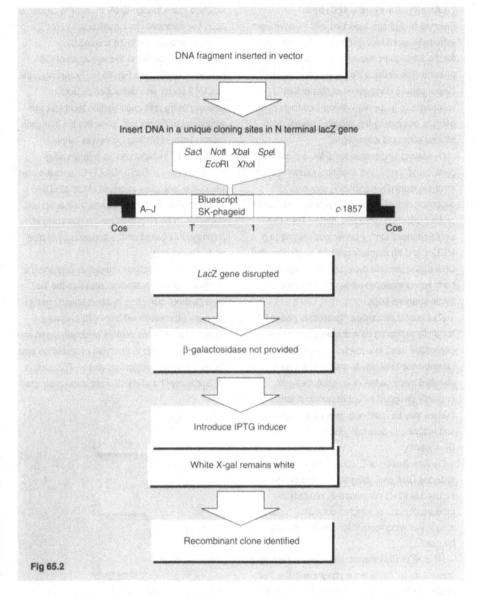

Fig 65.2

Fusion protein expression

IPTG induction of *lacZ* expression in, for example, λgt11 recombinant phages results in the synthesis of a fusion protein composed of the majority of the β-galactosidase molecule linked to the protein encoded by the DNA insert. If the foreign DNA is inserted in the correct open reading frame the fusion protein provides a means of selecting desirable clones in libraries (e.g. by recognition with specific antibodies). Fusion protein expressed from DNA inserted in the wrong reading frame, however, will not be recognised by specifc antibodies. Such clones are therefore best selected based upon gene probe hybridisation strategies. The nature of the DNA insert will also affect how much β-galactosidase protein is produced. For example, inserts which maintain the *lacZ′* open reading frame result in fusion protein 'sandwiches' with β-galactosidase protein at both amino and carboxy termini. An additional advantage is that the fusion proteins are expressed into the bacterial cytoplasm and can be readily isolated from recombinant bacterial lysates using antibodies specific for the amino terminal portion of β-galactosidase.

Filamentous phage-derived vectors have been developed to facilitate the generation of single-stranded DNA inserts, or for clone selection based upon surface expression of the insert-encoded proteins.

FILAMENTOUS PHAGE VECTORS

Plasmid- and phage λ-derived vectors have been extremely useful for creating gene libraries and for the subsequent expression of recombinant proteins. The unusual life cycles and relatively simple genetic composition of filamentous coliphages (e.g. the M13 genome consists of only 6407 bp comprising ten essential genes) offer several additional advantages.

(1) *Making single-stranded DNA*: a number of gene manipulation and analytical methods, such as chain termination (dideoxy) sequencing, oligonucleotide-directed mutagenesis and certain probe production strategies, require DNA in a single-stranded form. Filamentous phages (e.g. M13, f1 and fd) normally produce single-stranded DNA during their life cycle and have therefore been widely exploited as vectors for producing single-stranded DNA.

(2) *Plasmid similarities*: filamentous phage DNA is initially replicated as a double-stranded RF (replicative form) intermediate molecule. Its resemblance to plasmids means that a number of standard manipulative or isolation methods originally designed for application to plasmid vectors may be used (e.g. restriction digestion and mapping to determine the orientation of a DNA insert).

(3) *Transfection of* E. coli: both RF and single-stranded DNA containing filamentous phage may be used to infect competent *E. coli* cells to produce plaques or infected colonies. This allows a variety of recombinant screening techniques to be used.

(4) *Size of DNA inserts*: there are no packaging constraints for filamentous phages with the size of particle dependent upon the size of the viral DNA molecule. Although occasionally unstable, large amounts of DNA may consequently be inserted into the viral genome (e.g. viral DNA six times larger than the M13 genome has been successfully packaged).

BUILDING THE M13 VECTOR SERIES

Unlike phage λ for example, M13 has no non-essential genes for use as potential cloning sites. It does, however, contain 507 base pairs of DNA termed the intergenic region which does not encode protein and contains origins of replication. It also has a few unique restriction sites into which to insert foreign DNA. In order to capitalise upon the attractive cloning attributes of M13, a series of vectors has been created by engineering changes in the sequence of this intergenic region (see Fig. 65.3). Accordingly, the first M13 vector was developed by Jochim Messing using restriction ligation techniques to introduce the *E. coli lacZ* gene into the intergenic region. Termed M13mp1, this was further enhanced by mutagenesis to create a single restriction site for *Eco*RI (GAATTC) near the start site of the *lacZ* gene to create M13mp2. The inclusion of *lacZ* enabled insertional inactivation strategies to be employed for the screening of recombinants based on the appearance of blue/white phenotypes.

Insertion of a synthetic polylinker containing a number of unique restriction sites into the *lacZ* gene (without disrupting its open reading frame) generated the vector M13mp7. This was then used to construct the vectors M13mp8, mp10 and mp18 by increasing the number of restriction sites within a non-symmetrical polylinker. The vectors M13mp9, mp11 and mp19 were also constructed which contained the same polylinkers in a reverse orientation to M13mp8, mp10 and mp18 respectively, thereby allowing DNA insertion either way round.

A further useful modification incorporated into M13mp vectors is the siting of a universal priming site −20 or −40 bases from the start of the *lacZ* gene. This allows any insert to be sequenced using the same vector-specific primer annealing outside the polylinker and flanking the insert (i.e. rather than using insert-specific primers). The advent of the polymerase chain reaction also further reduces the time required to screen M13 recombinants since PCR primers can be designed to anneal to the polylinker or universal priming site. These may then be used to distinguish a recombinant from a non-recombinant clone based purely on the size of the resultant PCR product. A problem frequently encountered with M13mp vectors is however the relative instability of large inserts. Accordingly, insertion of 2–3 kb of foreign DNA often results in the loss of the M13 clone.

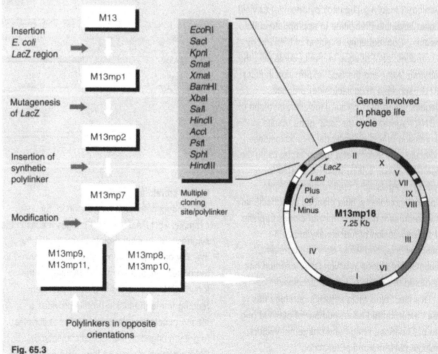

Fig. 65.3

Interchangeable polylinkers

The construction of M13mp vectors containing a polylinker was a significant step in recombinant DNA technology. Not only did it allow a variety of DNA fragments to be inserted in a defined orientation (i.e. directional cloning) but it also allowed the insert to be excised and inserted in the opposite orientation in another M13mp vector. This feature is not confined to M13 since Messing and colleagues also devised a means of introducing an identical polylinker in plasmids, thereby generating the pUC vector series. This facilitates the direct subcloning of a fragment from a general plasmid cloning vector to a single-stranded phage vector for more complex manipulations. The use of interchangeable polylinkers has also been extended to other plasmid, phage and phagemid systems (e.g. the pEMBL, λZap and pBluescript vector series).

SURFACE EXPRESSION VECTORS

Filamentous phages have also been used as the starting point for the development of a range of 'surface display' vectors which offer several advantages. Based primarily upon the coliphages M13 and fd, these vectors have been engineered to contain cloning sites associated with phage coat protein-encoding genes (e.g. encoding phage coat proteins 3 or 8). Linkage to a pel B leader sequence also ensures that the expressed insert is then targeted (i.e. transported) to the phage periplasmic space. Consequently, recombinant phage particles express the insert-encoded protein on the surface of the phage particle linked to coat protein 8 or 3 (Fig. 65.4).

Such surface display vectors have dramatically increased the ability to detect and isolate desirable recombinants. Accordingly, rather than having to screen many thousands or millions of clones within a conventional library (e.g. by hybridisation with a gene or antibody probe), desirable recombinants may be rapidly 'selected' from a surface display library. For example, affinity chromatography employing an antibody or other ligand may be used to recognise the recombinant protein on the phage surface (Fig. 65.5). The bound phages therefore represent a highly enriched population of recombinant phages containing desirable inserts. These can be eluted from the affinity chromatography column and then

further expanded and reselected following infection of new host bacteria. The higher copy number of coat protein 8 compared to coat protein 3 on the phage surface also means that vectors linking inserts to coat protein 8 are particularly suited to the selection of recombinant proteins with low affinity for a receptor or ligand.

The availability of surface display vectors has also driven the development of new cloning strategies. For example, random recombinatorial or hierarchical library strategies have been

developed. In particular, these strategies have dramatically increased the ability to generate and engineer monoclonal antibodies from a wide range of organisms, including humans.

OAs: 456, 591–600, 656–7, 1039, 1041, 1081–7. RAs: 172–5, 214–15, 392–7, 376–7, 402. SFR: 36, 101–2, 116–24, 126–7, 219.

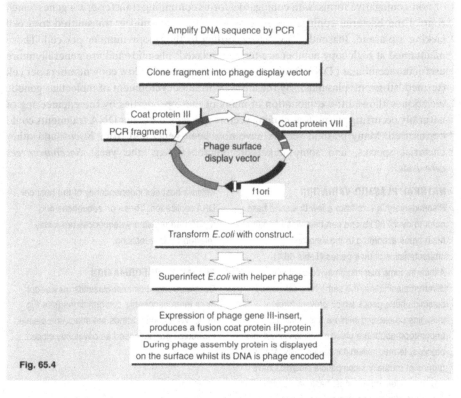

Fig. 65.4

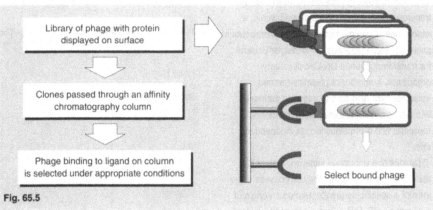

Fig. 65.5

66. PLASMIDS: DEVELOPMENT AS CLONING VECTORS

The ability to isolate naturally occurring bacterial plasmids possessing independent replication, cell transfer, and selectable marker genes made them suitable for development as cloning vectors.

Amongst the most useful discoveries arising from the analysis of bacteria such as *E. coli* was that many contain small mini-circular or (more rarely) linear DNA molecules called plasmids. Furthermore, these plasmids were demonstrated to be capable of replicating independently of the much larger single circular 'chromosomal' DNA molecule. Plasmids of different sizes, forms and characteristics are now known to be widely distributed amongst prokaryotes, and, although generally not essential to cellular survival, have several characteristics that make them ideally suited for their development as cloning vectors. One of the most important features is that plasmids contain genes encoding phenotypic traits which confer a selective advantage upon the cell (plasmids to which phenotypes have not been ascribed are termed 'cryptic'). Such phenotypes include antibiotic production or resistance, heavy metal resistance, enterotoxin and bacteriocin production, and degradation of aromatic compounds. Plasmids also exist in conjugative or non-conjugative forms, with conjugative forms containing transfer or *tra* genes which control the bacterial conjugation process whereby a plasmid is transmitted to a cell lacking a plasmid. Plasmids also vary according to their copy number per cell. Those maintained at high copy number are termed 'relaxed' plasmids and are generally more useful in recombinant DNA studies than those maintained at low copy numbers per cell (termed 'stringent' plasmids). By the early 1970s the development of molecular genetic techniques allowed the exploitation of many of these properties by the engineering of naturally occurring plasmids as cloning vectors into which foreign DNA fragments could be inserted. Many plasmid vectors have now been generated from *E. coli* and other bacterial species, and some eukaryote microbes (e.g. the yeast *Saccharomyces cerevisiae*).

NATURAL PLASMID VARIATION

Plasmids vary in size from a few thousand base pairs to over 200 kb and can be divided into five main types according to the functional characteristics of their genes (Table 66.1). Although some plasmids may co-exist with different plasmids in the same host cell, many plasmids have genes which prevent foreign plasmids co-existing within a host. Little is understood about the precise mechanism of this process, termed plasmid incompatibility. Many groups of mutually incompatible plasmids have now been defined amongst bacterial species. The members of some groups are however able to transfer between, and be maintained within, a wide range of diverse hosts. Termed promiscuous plasmids, these plasmids are particularly useful for transferring cloned DNA across many organisms. A number of plasmids (termed episomes) are also able to integrate themselves into the host chromosome, whilst yet others replicate, and are maintained, as independent units.

Despite this enormous variation, a common feature of plasmids is that they all possess some form of a specific sequence, termed an origin of replication (*ori*). This allows a plasmid to replicate within a host cell independently of the host cell DNA replication. These *ori* sequences and associated regulatory sequences are jointly referred to as a replicon.

PLASMID CONFORMATION

Double-stranded circular plasmids may adopt three interchangeable conformations (see Fig. 66.1). When both strands are intact, the plasmid molecules are described as covalently closed circles or CCC DNA (often referred to as relaxed CCC DNA). However, if one strand is broken (e.g by digestion with a nuclease) the molecule is described as an open circle or OC DNA. When isolated from cells, plasmids adopt a third

OAs: 36–8, 296, 601–4. **RAs:** 17, 130, 132–5, 216–18. **SFR:** 36, 67, 97–102, 106, 119–23, 126–8, 148–9.

NAMING PLASMIDS AND VECTORS

Naturally occurring plasmids are usually denoted by the phenotypic characteristic they confer on a host bacterium. For example, the ColE1 plasmid contains genes which encode for the production of the bacteriocin colicin E1 in the transformed host. As plasmids began to be developed as vectors, a new nomenclature system evolved based upon the original laboratory or investigator's name. Accordingly, the original plasmid vector designed by Stanley Cohen is termed 'p' for plasmid followed by his initials and the original laboratory code (i.e. denoted pSC101). Other plasmid-derived vectors have been named after the institution in which they were devised (e.g. pUC vectors originate from University of California). More recently however, plasmid vectors developed by biotechnology companies have often been given 'trade' or 'generic' names such as pEX (denoting a plasmid expression vector).

Table 66.1

Plasmid	Conjugative	Size (MDa)	No. Plasmid copies	Function
Col E1	No	4.2	10–15	Colicin E1 production
R6K	Yes	25	13–38	Ampicillin resistance
F	Yes	62	1–2	-
R1	Yes	62.5	3–6	Multi-drug resistance
RSF1030	No	5.6	20–40	Ampicillin resistance
EntP307	Yes	65	1–3	Enterotoxin production

'supercoiled' conformation created from CCC DNA by the loss of turns in the double helix. The interchange between these three forms is mediated/controlled by the action of enzymes such as endonucleases, ligases, DNA gyrases and various topoisomerases. The intercalation of supercoiled DNA with increasing concentrations of ethidium bromide (EtBr) also causes it to unwind into the OC form followed by rewinding in the opposite direction. This conformational behaviour of plasmids, and the altered mobility of OC and supercoiled forms during gel electrophoresis and centrifugation, is utilised by several plasmid purification methods.

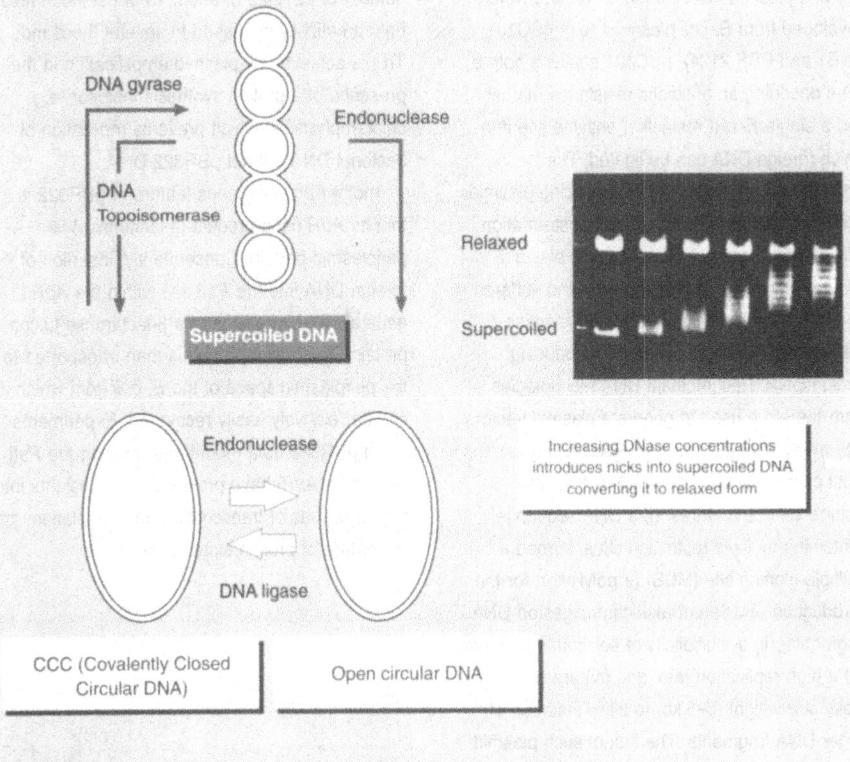

ISOLATING PLASMIDS FROM CELLS

The development and subsequent application of plasmid cloning vectors ultimately requires the availability of methods for their isolation and purification from host cells. Plasmid purification may be performed in a number of ways, but all approaches depend upon the initial lysis of plasmid-harbouring bacterial cells harvested from liquid media cultures. Although cell lysis is not difficult to achieve, it is a crucial step. If not performed successfully, lysis drastically reduces the efficiency with which plasmids are recovered by subsequent purification methods. Accordingly, incomplete lysis prevents the release of all the available plasmid, whereas total disruption of bacterial cells releases not only the plasmids, but also large amounts of contaminants including the chromosomal DNA, proteins, and RNA molecules. The favoured method therefore employs a gentle procedure whereby the cell wall of the bacterium is weakened by lysozyme followed by lysis with sodium hydroxide and non-ionic detergents (e.g. sodium dodecyl sulphate). Centrifugation then produces a 'cleared lysate' containing the plasmid, whilst the cell debris and high molecular weight chromosomal DNA forms a pellet at the bottom of the centrifuge tube.

Plasmid purification

Isolation of plasmids from cleared lysates is frequently achieved by alkaline denaturation and caesium chloride (CsCl) density gradient centrifugation. Both techniques rely upon the supercoiling of CCC plasmid DNA and the differential denaturation of linear chromosomal

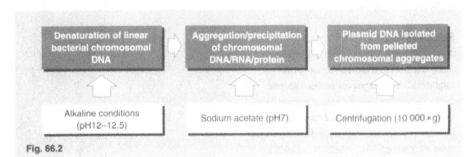

Fig. 66.1

Increasing DNase concentrations introduces nicks into supercoiled DNA converting it to relaxed form

DNA gyrase
Endonuclease
DNA Topoisomerase
Supercoiled DNA
Relaxed
Supercoiled

Endonuclease
DNA ligase
CCC (Covalently Closed Circular DNA)
Open circular DNA

Fig. 66.2

Denaturation of linear bacterial chromosomal DNA → Aggregation/precipitation of chromosomal DNA/RNA/protein → Plasmid DNA isolated from pelleted chromosomal aggregates

Alkaline conditions (pH12–12.5)
Sodium acetate (pH7)
Centrifugation (10 000 × g)

DNA compared to supercoiled plasmid DNA. Consequently, under carefully controlled alkaline conditions (pH 12.0–12.5), only the linear bacterial chromosomal DNA will denature (Fig. 66.2). Decreasing the pH back to 7.0 using sodium acetate then causes aggregation and precipitation of the chromosomal DNA, protein and RNA complexes. Centrifugation then allows the plasmid DNA (which remains in the supernatant) to be separated from the pelleted contaminants. Supercoiled plasmid may also be separated from a cleared lysate by isopycnic centrifugation through CsCl in the presence of EtBr. A little EtBr can bind supercoiled DNA, but EtBr freely binds the linear chromosomal DNA thereby increasing its density. Centrifugation on a CsCl density gradient saturated with EtBr therefore results in the banding of plasmid and chromosomal DNA at different positions, allowing their separate removal (e.g by fine needle aspiration).

A wide variety of 'improved' plasmid vectors have been genetically engineered from naturally occurring plasmids, thereby increasing their cloning flexibility, protein product expression and copy number.

NATURAL PLASMID VECTORS

The first plasmids to be used as vectors were developed from *E. coli* plasmids (e.g. pSC101, ColE1 and RSF 2124). pSC101 contains both a gene encoding an antibiotic resistance marker and a single *Eco*RI restriction enzyme site into which foreign DNA can be ligated. The development of methods for introducing plasmids into host cells greatly increased transformation efficiencies. However, these original plasmid vectors replicated very inefficiently and suffered from a lack of flexibility in terms of selection markers and restriction sites for introducing foreign DNA. Recombinant DNA technologies were therefore used to generate plasmid vectors with more desirable and versatile properties. The most commonly used plasmid vectors now include several features: (i) a DNA sequence containing multiple restriction sites, termed a multiple cloning site (MCS) or polylinker, for the introduction of different restriction digested DNA fragments; (ii) a multiplicity of selectable markers; (iii) a high replication rate; and (iv) are of low molecular weight (3–5 kb) to allow insertion of larger DNA fragments. The first of such plasmid-derived vectors was pBR322, from which many improved or specialised vectors were subsequently developed.

Plasmid pBR322

The construction of plasmid pBR322 involved many steps of recombination between DNA fragments. These fragments encoded specific genes derived from several natural plasmids and their variants. The final pBR322 construct was composed of 4363 bp, essentially derived from just three plasmids (pSC101, pSF2124 and pMB1), although several others (e.g. R1 and ColE1) served as carriers of an ampicillin resistance (*ApR*) transposon (see Fig. 66.3). The entire sequence of pB322 has been determined, allowing accurate mapping of all the restriction sites and the calculation of their product lengths. In all there are over 20 restriction enzymes with a single recognition site in pBR322, of which 12 occur in the *ApR* and tetracycline resistance (*TcR*) genes and their promoters. Cloning into these 12 sites makes selection of recombinants much simpler since it results in insertional inactivation of the antibiotic resistance genes.

Being a relaxed plasmid, the normal copy number of pBR322 of about 15 per cell can also be dramatically increased to several thousand. This is achieved by plasmid amplification in the presence of a protein synthesis inhibitor (e.g. chloramphenicol) which prevents replication of bacterial DNA but not pBR322 DNA.

Another advantageous feature of pBR322 is that its *ApR* gene product (β-lactamase) is a periplasmic protein. Consequently, insertion of foreign DNA into the *Pst*I site within the *ApR* gene allows expression of a β-lactamase fusion protein. The fusion protein is then transported to the periplasmic space of the *E. coli* from which it can be relatively easily recovered. Experiments using pBR322 as a model (e.g. moving the *Pst*I site within *ApR*) have provided many insights into the processes of transcription and translation, and the nature of protein signal sequences.

Plasmid map reading

For convenience, plasmids are most often schematically represented as a circle with the plasmid name in the centre. The positions of the various restriction enzyme sites are then given around the circle along with their position in the map according to the 5′ base of their recognition sequence. For the sake of simplicity and clarity however, many representations only show the positions of unique restriction sites. The numbering system moves clockwise starting from the top of the circle (12 o'clock). In the case of pBR322 the numbering begins with the first T in the single *Eco*RI recognition sequence, GAATTC. Features such as *ori* (the replication origin) and individual plasmid genes are also shown, often in the form of arrows in order to indicate the direction of replication or transcription respectively.

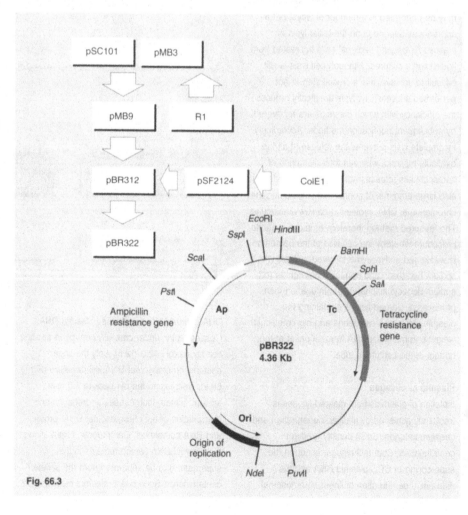

Fig. 66.3

Improved pBR322-derived vectors

Many different vectors have now been derived from pBR322 which are particularly suited to specific cloning requirements. Accordingly, vectors pBR324 and pBR325 have been constructed to allow insertional activation of selectable markers (derived from other plasmids or phages) using restriction sites for commonly employed enzymes such as *Eco*RI and *Sma*I. A commonly used pBR322 derivative, pAT153, has been generated by removal of a *Hae*II restriction fragment from pBR322. This increases the copy number of plasmids per host cell by up to 3 times compared to pBR322. It also increases the amount of plasmid-specific protein subsequently produced.

A series of pUC vectors have also been generated by inserting the polylinker sequences of phage M13mp vectors into the *lacZ* gene encoding for β-galactosidase (without affecting the *lacZ* gene's function). Consequently, the *lacZ* gene continues to produce β-galactosidase unless foreign DNA is inserted (Fig. 66.4). This insertional inactivation results in colonies of cells harbouring pUC recombinants bearing a white phenotype on Xgal-containing media, whilst non-recombinant colonies remain blue. This blue/white system is used as a basic selection feature of many plasmid and phage vectors.

DIRECT SELECTION VECTORS

Most plasmid vectors used today have been developed to contain two selectable markers, one used to select transformants and one used to select recombinants (via insertional inactivation). Although this enhances the ability to select recombinants, a few plasmid vectors have also been constructed which allow direct selection (i.e only transformed host cells harbouring vector containing a foreign DNA are selected based upon survival in certain media). For example, the vector pLX100 has been constructed containing the *E. coli* gene for xylose isomerase. Accordingly, *E. coli* cells transformed by pLX100

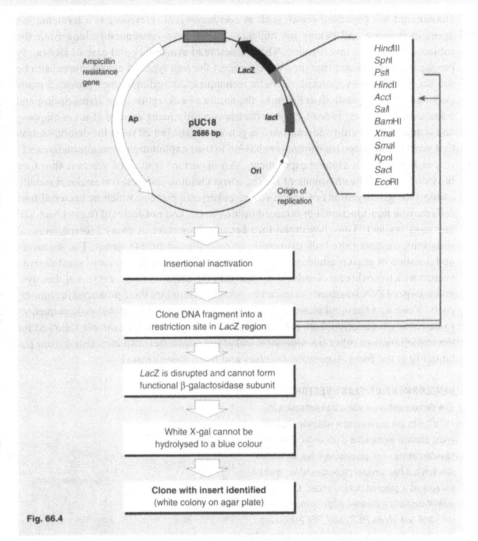

Ampicillin resistance gene

LacZ

Ap

pUC18
2686 bp

lacI

Ori

Origin of replication

HindIII
SphI
PstI
HincII
AccI
SalI
BamHI
XmaI
SmaI
KpnI
SacI
EcoRI

Insertional inactivation

Clone DNA fragment into a restriction site in *LacZ* region

LacZ is disrupted and cannot form functional β-galactosidase subunit

White X-gal cannot be hydrolysed to a blue colour

Clone with insert identified
(white colony on agar plate)

Fig. 66.4

can only survive in minimal media containing xylose if DNA has been inserted into the sites adjacent to the xylose isomerase gene.

Low copy and runaway vectors

Most vectors have been specifically engineered to increase plasmid copy number and the amount of cloned protein that can be produced per host cell. There are however a few instances where low copy number vectors are desirable. This is especially the case when high level expression of the encoded protein has a deleterious or lethal effect upon the host cell. A series of vectors have therefore been developed which retain the two selection markers and multiple cloning sites of

pBR322 derivatives, and the low copy number characteristic of pSC101. Although such low copy number vectors allow 'lethal' proteins to be cloned, they limit the amount of protein that can be expressed. In order to overcome this, so-called 'runaway' vectors have been developed. At the maintenance temperature (which depends upon the particular vector) the 'runaway' plasmid is present at low copy number in the transformed cell. Above this temperature however, the plasmid undergoes uncontrolled replication which allows the protein to be over-produced. After a few hours the cell stops growing and becomes non-viable, however by this time it contains significant amounts of both plasmid and expressed protein.

OAs: 251–2, 281–3, 296, 528, 605–19, 681–7, 692, 695. RAs: 17, 130, 132–5, 216–18. SFR: 36, 67, 97–102, 119–23, 126–8, 148–9.

67. YEAST-DERIVED PLASMID VECTORS

A variety of yeast-derived plasmid vectors have been constructed to allow the insertion, replication, host cell transfer, expression, and study of large DNA fragments within a eukaryotic cell environment.

Humankind has exploited yeasts, such as *Saccharomyces cerevisiae*, as a fermentation agent in brewing and baking for millenia. Yeasts have consequently long been the subjects of scientific investigation. Their widespread availability and ease of laboratory propagation also meant that they were amongst the first types of eukaryotic cells to be subject to analysis by molecular genetic techniques. Accordingly, they provided many initial answers to questions concerning the nature of eukaryotic gene transcription and translation processes. Indeed, major effort is currently being expended upon mapping and sequencing the entire *Saccharomyces* genome. In the last 20 years the demonstration that yeast can undergo transformation has led to their exploitation as an alternative to *E. coli* as host cells for cloning experiments. An important feature of yeasts is that they provide a eukaryotic environment for the correct folding and post-translational modification (e.g. glycosylation) of an expressed eukaryotic protein, which in bacterial host cells may be non-functional or retained inside the cell and not secreted (e.g. as bacterial inclusion bodies). They have therefore become important in many biotechnological situations, including the bulk expression of recombinant human genes. The discovery and isolation of yeast plasmids also allowed the development of a series of yeast-derived vectors which provided molecular biologists with the ability to clone, express and analyse much larger DNA fragments than can be accommodated by their prokaryotic counterparts. Yeast and bacterial plasmids have also been combined to enable gene sequences to be readily transferred from prokaryotic to eukaryotic cells; an example followed for the exploitation of other transformable eukaryotic microbes with biotechnological potential (e.g. the fungi *Aspergillus nidulans* and *Neurospora crassa*).

DEVELOPMENT OF YEAST VECTORS

The demonstration of yeast transformation in 1978 led to the isolation of a naturally occurring yeast plasmid termed the 2 μm circle. Present in several strains of *S. cerevisiae*, it has no known function but has several properties which make it suitable as a plasmid cloning vector. Only 6 kb in size, it contains a plasmid origin of replication (*ori*) and two genes *REP1* and *REP2* encoding the protein machinery required for replication. The *LEU2* gene encoding an enzyme (β-isopropylmalate dehydrogenase) involved in converting pyruvic acid to leucine has subsequently been cloned into the 2μm circle to serve as a selectable marker (Fig. 67.1). Accordingly, when used to transform a host yeast cell unable to synthesise its own leucine, only those cells containing the plasmid will grow on minimal media lacking amino acids.

Yeast vectors have been developed from the 2μm circle which contain similar features, such as high copy number (70–200) replication in *E. coli*, and a range of selectable markers used for isolation by complementation in *E. coli*. Most also contain antibiotic resistance markers for use in *E. coli* cloning and unique sites for a number of restriction enzymes.

OAs: 71, 383–4, 620–32, 688–9, 874, 972, 1118–19, 1147, 1154. **RAs:** 219–24, 419–24. **SFR:** 119–24, 126–8, 150–2, 162, 223–5.

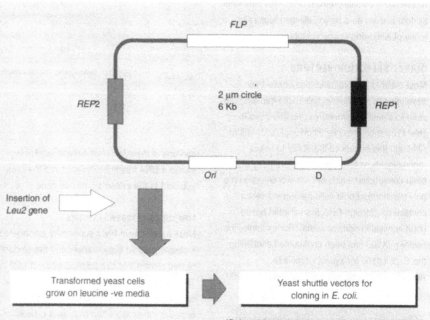

Fig. 67.1

Yeast episomal plasmids

Abbreviated to YEps, these were the first vectors to be derived from the 2 μm circle (in 1978) by the integration of a copy of the *E. coli*-derived plasmid pMB9 or pBR322 (Fig. 67.2). This allows YEps to be used to transfer DNA between *E. coli* and yeast cells. These so-called 'shuttle' vectors allow basic gene manipulations to be undertaken in *E. coli* cells, followed by the study of the cloned genes and their products in the eukaryotic environment of the yeast cell. Being episomal, YEps replicate independently of the host chromosomal DNA. However the homology between the YEp *LEU2* gene and its *E. coli* chromosomal equivalent may result in its integration into the host chromosome. YEps have a high transformation efficiency and may be present in high copy number. They may however be unstable, in many cases undergoing spontaneous gene rearrangement or loss of DNA.

Yeast artificial chromosomes

YACs are linear yeast vectors first developed in 1982 from a yeast plasmid containing a centromere combined with fragments of a rare linear plasmid isolated from the protozoan *Tetrahymena*. One of the common YAC vectors is pYAC2 constructed from pBR322 into which the selectable marker genes *URA3* and *TRP1* have been inserted. In YAC vectors, the *TRP1* gene also includes two other important elements, an origin of replication and the DNA from the centromere region of chromosome 4 known as the *CEN4* sequence (Fig. 67.3). The artificial yeast chromosome is completed by *TEL* sequences which act as mini-telomeres in the yeast host. During cloning, the YAC is usually restricted at *Sma*I and *Bam*HI sites and the foreign DNA fragments to be inserted are blunt end ligated to *Sma*I sites on the two arms to create an artificial chromosome. Only correctly constructed YACs can replicate in appropriate yeast strains on minimal media by virtue of their selectable markers. An advantage of YACs is that DNA fragments in the Mb range can be cloned, allowing complete genes and their control regions to be isolated and studied. Although difficult to map by standard techniques, YACs have become widely used in the human genome mapping project where DNA fragments for sequencing are initially located/isolated from YAC libraries.

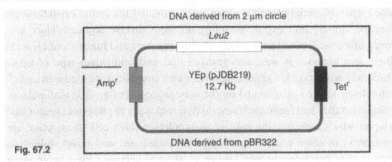

DNA derived from 2 μm circle

Fig. 67.2

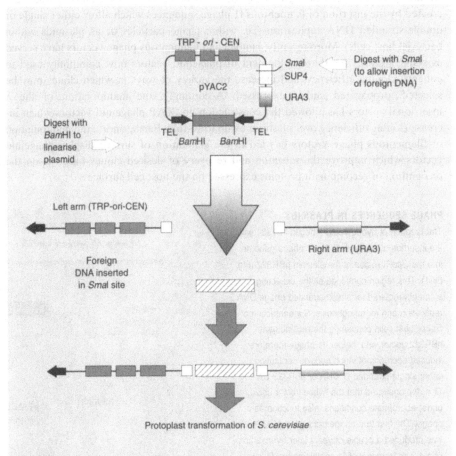

Fig. 67.3

YEAST INTEGRATIVE PLASMIDS

Abbreviated to YIps, these are bacterial plasmids based on the pBR family of vectors. They contain a yeast *URA3* gene which encodes orotidine-5-phosphate decarboxylase which is used in much the same way as the *LEU2* gene for selection. Unlike YEps, YIps replicate by chromosomal integration and produce extremely stable transformants. They do, however, lack the high transformation efficiencies characteristic of YEps.

YEAST REPLICATIVE PLASMIDS

Originally derived from a 1.4 kb yeast DNA fragment, YRps replicate independently. They do however also carry a chromosomal DNA sequence which includes a replication origin. A frequently used plasmid is YRp7 which is pBR-derived but carries a selectable marker *TRP1* gene (involved in tryptophan biosynthesis) located near an origin of replication. A disadvantage of YRps is the relative instability of the resultant transformants.

68. PHAGEMIDS, HYBRID PHAGE AND PLASMID CLONING VECTORS

Phagemids represent a generation of hybrid vectors genetically engineered to combine some of the advantageous features of both plasmid and phage-based vector cloning systems.

Vectors based upon bacterial plasmids and phages provided the initial breakthroughs required for the cloning and expression of genes from diverse sources. They have therefore been immensely useful in the study of gene structure and function, not least by allowing their own analysis. It was soon realised that each individual type of vector possessed uniquely advantageous genetic features which could be combined into a single type of vector (now termed phagemids) by the very processes of genetic manipulation whose development they had facilitated. Several different types of phagemid vector have been developed which combine the ease of manipulation, host cell transfection and single-strandedness of some phage vectors with the small size and insert stability of plasmid vectors. For example, the pUC series of plasmid vectors served as an initial starting point for the development of the pEMBL series of phagemids. These were created by the insertion of filamentous f1 phage sequences which allow either single or double-stranded DNA replication (i.e. within phage particles or as plasmids within bacterial host cells). More recently, lambda and filamentous phage vectors have served as starting points for the development of phagemid vectors now commonly used to enhance cloning efficiency, and increase the variety of ways in which clones may be selected, propagated and/or expressed. Accordingly, the manipulation of the λ insertional vectors has allowed the construction of λZAP phagemid vectors which increase cloning efficiency over plasmids by up to 16-fold. Furthermore, the manipulation of filamentous phage vectors has led to the generation of surface display phagemid vectors which improve the selection and recovery of desired clones based upon the recognition of recombinant proteins expressed on the host cell surface.

'HELPER' PHAGE SUPERINFECTION

Cis-acting phage elements within phagemids used to transfect host cells will only be activated to allow replication of the single-stranded plasmid DNA, assembly of new phage virion-like particles, and secretion into the culture medium following subsequent infection of the same host cell by a 'helper' phage. This process, termed superinfection, relies upon the 'helper' phage providing the necessary proteins to facilitate assembly of the new phage particles from the phagemid. Thus, in the case of F⁺ *E. coli* host cells harbouring phagemids containing f1 control sequences, superinfection utilises f1 phages as 'helper' phages to activate the phagemid-contained f1 origin of replication. Superinfection of host cells by helper phage is a common feature of most phagemid systems including λZAP and surface display vectors.

PHAGE SEQUENCES IN PLASMIDS

The creation of hybrid vectors began in 1981 with the insertion of a region of the f1 phage genome into the *Eco*RI region of the plasmid pBR322 (Fig. 68.1). This region contained all the *cis*-acting elements required for single-stranded phage DNA replication and morphogenesis. Superinfection of *E. coli* host cells containing the recombinant pBR322 vector with 'helper' f1 phage therefore induced secretion of virion capsids containing either single-stranded f1 DNA or pBR322 DNA. This demonstrated that the hybrid vector could, under appropriate conditions, also function as a phage. The first true phagemid vector, pEMBL, was produced a couple of years later by insertion of a 1.3 kb fragment of f1 containing the f1 *ori*, and morphogenesis elements, into a unique *Nar*I site of plasmid pUC8. The resulting phagemid retained all the features of pUC8 (i.e. plasmid *ori*, ampicillin resistance and Xgal blue/white colony selection) with the ability to generate single stranded plasmid DNA-containing virion particles following superinfection. Furthermore, by inserting the f1 fragment in both orientations to generate pEMBL8(+) and pEMBL8(–) vectors, virion particles could be secreted which contained either the non-coding or coding strands respectively of

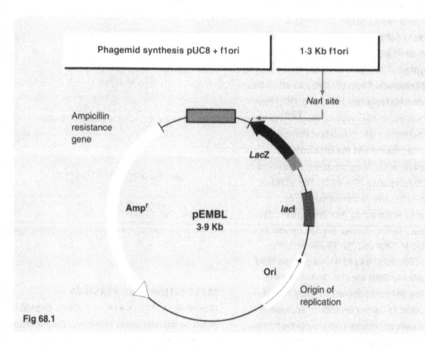

Fig 68.1

the polylinker containing β-galactosidase gene. Consequently, in the absence of helper phage, pEMBL replicates as a pUC plasmid allowing the replicative form (RF) double-stranded DNA to be isolated at higher yields than can be obtained from the M13mp series of vectors.

OAs: 514, 528, 633–42, 723–8, 737–8, 782–4, 1141–2, 1152. **RAs:** 255, 423–4. **SFR:** 36, 115, 119–24, 126–8, 225.

Phage lambda hybrid vectors

A sophisticated virus vector, termed λZAP, has recently been developed based upon a λ insertional vector with the addition of M13 biology. As may be expected, λZAP also has many of the advantageous features of λ phage vectors, such as: (i) high cloning efficiency, which thereby increases the chance of recovering the desired fragment from a library; (ii) a multiple cloning site capable of retaining up to 10 kb inserts; (iii) insertional inactivation of β-galactosidase for Xgal blue/white clone selection; and (iv) expression of β-galactosidase fusion proteins analogous to λgt11.

λZAP also has several additional advantageous features. Firstly, it allows the production of RNA transcripts from the inserted DNA (e.g. for use as gene probes). Directed from either of two phage RNA polymerase (T3 and T7) promoter sequences flanking the insertion site, RNA transcripts can be produced which represent either the coding or non-coding strand. The expression of eukaryotic DNA may also be accomplished in λZAP vectors containing a cytomegalovirus immediate early promoter sequence. Inserts may also be expressed as functional synthetic mRNA transcripts bearing 5′ caps for use in *in vitro* translation. They may also be used to study RNA processing events/mechanisms (e.g. following injection into *Xenopus* oocytes) by the incorporation of appropriate termination and polyadenylation signals within vectors.

λZAP also allows automatic *in vivo* excision of a 2.9 kb colony-producing phagemid termed pBluescript (Fig. 68.2). This technique is known as single-stranded DNA rescue and requires superinfection with helper phage (e.g. M13VCS or R408). Accordingly, the Bluescript SK phagemid (analogous to pEMBL phagemids) has been sited between the f1 initiator and terminator domains within the vector. This allows the recognition of the signal by the phage gene II protein followed by synthesis of a new strand of Bluescript DNA containing the insert. This displaces the existing strand, and the displaced strand is automatically circularised, packaged and secreted as a filamentous phage. Recombinant Bluescript plasmid colonies may therefore be obtained by infecting an F′ strain of host cells on ampicillin

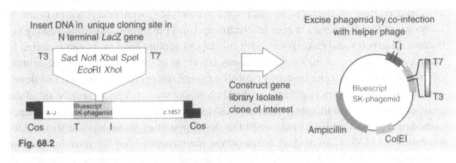

Fig. 68.2

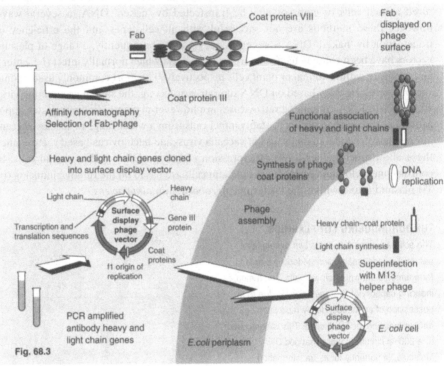

Fig. 68.3

plates where the ColE1 *ori* directs replication and ampicillin selects recombinants.

Surface display vectors

As a consequence of phagemid vector developments, and in order to reduce the amount of colony screening required to obtain the desired clone from libraries, vectors have been developed which link inserts to the expression of phage coat proteins (Fig. 68.3). Termed phage surface display vectors (e.g. Stratagene's SurfZAP phagemids) several have been developed based upon f1 or M13 phages. These vectors link the expression of inserts to the expression of phage coat proteins (e.g. gIII or gVIII). Some also allow switching between soluble or surface-bound

recombinant protein production by the use of an amber codon activated in the presence of an appropriate medium additive.

A distinct advantage of surface expression of inserts is the rapid selection of desirable clones from libraries. For example, phage expressing the desired recombinant antibodies or receptors can be recovered by passage of the library over affinity chromatography columns coated with the appropriate antigen or ligand. Only those viruses bearing the desired specific inserts will be retained and can subsequently be recovered by elution. Indeed, this technology has already proved to be of enormous value in antibody engineering, and the cloning and mutagenesis of a wide range of polypeptides.

69. VECTORS FOR USE IN PLANT AND ANIMAL CELLS

Plant cells may be stably transformed using modified bacterial Ti plasmids, whilst plasmid vectors developed from DNA- and RNA-containing viruses are commonly used to transfect animal cells.

The construction of appropriate eukaryotic vectors has been necessary in order to compensate for differences in post-transcriptional and post-translational processing between prokaryotes and eukaryotes. Such specialised vectors have also been required to facilitate the study and manipulation of eukaryotic genes within their normal cellular environment (e.g. for gene therapy or construction of transgenic organisms). The development of yeast vectors (e.g. yeast artificial chromosomes or YACs) greatly aided the analysis, expression and manipulation of eukaryotic genes. Although the experience gained in constructing YACs may assist the development of mammalian artificial chromosomes (MACs), yeast vectors are however inappropriate for the introduction and expression of foreign DNA within plant and animal cells. Furthermore, although cultured animal cells or embryos may be transfected by 'naked' DNA in several ways, however these methods are not successful with all cell types and the efficiency of transfection by 'naked' DNA is generally very poor. Consequently, a range of plasmid vectors have been constructed from viruses or bacteria which normally infect (i.e. enter), and replicate within, animal or plant cells respectively. The most commonly used animal vectors were originally based on DNA tumour viruses (e.g. the simian vacuolating virus 40 or SV40). RNA-containing retroviruses are however currently finding greater application because of their ability to stably infect cells from a wider range of species. Recent interest has also focused on the use of vaccinia virus- and baculovirus-based vectors since these allow much higher levels of expression of foreign DNA within animal cells. In contrast, most cloning vectors for use in plant cells are based on the tumour-inducing (or Ti) plasmid found within the bacterium *Agrobacterium tumefaciens*.

MOLECULAR BIOLOGY'S 'BIG MACS'

Mammalian artificial chromosomes (or MACs) have many potential applications in basic research and gene therapy. MAC development is however currently restricted by the incomplete characterisation of the functional elements of mammalian chromosomes required for their construction (e.g. telomeres, centromeres, origins of replication and nuclear protein scaffold attachment sites). Nevertheless, once successfully constructed MACs could allow very large DNA fragments to be replicated, segregated and expressed in mammalian cells whilst still being maintained and inherited as a single copy alongside the endogenous chromosomes. This would allow tissue-specific expression of DNA fragments containing an entire gene (or genes) for use in basic research and several gene-based therapies.

OAs: 643–55, 600–1, 803, 808–22, 826–30. **RAs:** 225–34, 240, 253, 281–90, 294–303, 320, 404. **SFR:** 156–9, 169, 120–4, 220–1.

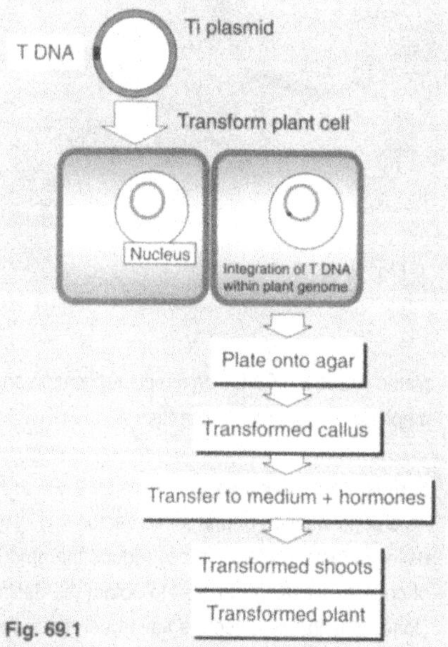

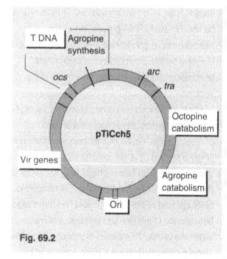

TUMOUR-INDUCING (Ti) PLASMIDS

The soil bacterium *Agrobacterium tumefaciens* naturally invades many dicotyledonous plants to form tumours known as crown galls. The tumour-inducing capacity of *A. tumefaciens* is due to its possession of an independently replicating tumour-inducing, or Ti, plasmid. This is composed of a 200 kb circular double stranded DNA molecule. Importantly for its transformation property and the development of plant cell vectors, Ti plasmids contain a region called transferred DNA, or T-DNA. During infection, this T-DNA is excised from the plasmid and integrates into the plant cell DNA by a process of recombination analogous to that observed in bacterial conjugation.

Several Ti plasmids have been developed as vectors for introducing foreign DNA into plant cells by deleting T-DNA-contained genes that lead to uncontrolled cell proliferation. These deleted regions are then used as sites for the insertion of the foreign DNA. The recombinant Ti plasmid bearing the foreign DNA insert may then be introduced into the plant by the nomal process of *A. tumefaciens* infection (Fig. 69.1). Integration of the T-DNA bearing the insert then leads to transformation. This process generally involves transformation of a single cell or protoplast (i.e. lacking the outer cell wall) which can be initially propagated as a callus on agar medium, and finally an intact plant.

Fig. 69.1

Fig. 69.2

An important feature of the range of Ti plasmid vectors currently available has been the inclusion of selectable marker genes (e.g. neomycin phosphotransferase II) into the T-DNA-flanking or border sequences (Fig. 69.2). This ensures that they are also transferred to the plant cell during infection. Such Ti plasmids are now frequently used in plant genetic analysis, and for the formation of a wide range of transgenic plants.

SV40-BASED VECTORS

Animal cell plasmid vectors based on the monkey DNA tumour virus SV40 were first constructed by substituting viral late antigen-encoding genes with the foreign DNA (Fig. 69.3). These recombinant molecules were then circularised and used to transfect *in vitro*-cultured monkey cells. Co-transfection with a second 'helper' viral plasmid lacking early antigen-encoding genes but containing the late antigen genes then allows co-operative replication of the two plasmids and their packaging into new virus particles. This mixed population of viral particles can then be harvested from the culture medium and used to efficiently infect a second set of monkey cells. This allows high level synthesis of the foreign DNA-encoded protein.

A different strategy is however based upon the use of a cell line (termed COS). These cells each contain a region of a defective SV40 genome which has been stably integrated into their chromosomes allowing the expression of the viral T antigen required for viral replication (Fig. 69.4). Foreign DNA is inserted into another viral plasmid containing the origin of replication which, following transfection of COS cells, leads to massive viral replication and high level protein production. The use of SV40 vectors is however limited to the infection of monkey cells with only small foreign DNA inserts, and viral replication also ultimately leads to cell lysis. Consequently, a range of different viral vectors with broader infection specificities are now routinely used.

VACCINIA AND BACULOVIRUS VECTOR

Both the cytoplasmic-replicating, DNA-containing vaccinia virus and baculovirus have been used to develop vectors for high level expression of foreign genes in mammalian and insect cells respectively. The most widely used vaccinia vector expression systems are based upon the use of recombinant plasmids carrying a bacteriophage (T7) RNA polymerase. Thus, following infection, expression of the bacteriophage T7 RNA polymerase essentially shuts down host cell protein synthesis so that insert-bearing viral mRNA transcripts are preferentially translated. This system has been particularly useful in proving that a normal wild-type cystic fibrosis gene can correct the defective

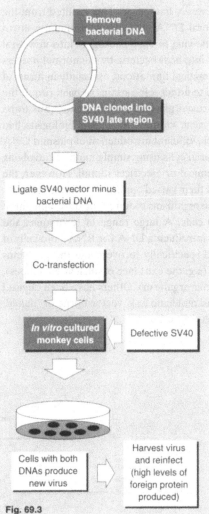

Fig. 69.3

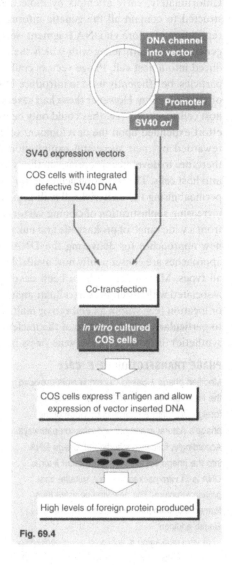

Fig. 69.4

ion transport within cells from individuals with cystic fibrosis.

RETROVIRAL VECTORS

Significant effort has been expended on the development of retroviral vectors (e.g. based on adenoviruses). Importantly, these RNA viruses (so called because they contain RNA-encoded genomes) do not cause host cell lysis. However, following infection, the viral RNA containing a foreign gene is converted into DNA and stably incorporated into the host genome. RNA viruses will infect virtually all types of mammalian cells and consequently are proving to be extremely

useful for the development of animal cell vectors. Accordingly, retroviral vectors have been constructed which contain a range of selectable markers, sites for foreign DNA insertion and which allow integration into host genomes. Crucially, all lack the ability to subsequently produce viral particles. Their ability to infect cells and induce full expression of a foreign gene has led to their widespread use in studying the cellular effects of genes (e.g. oncogenes and anti-oncogenes). They also promise to be very useful in the development of transgenic mammals and as vehicles for use in gene therapy.

70. DELIVERING DNA INTO CELLS

Many delivery systems have been developed to introduce DNA into cells based upon viral infection, protoplast or liposome fusion, electroporation, micro-injection and firing of nucleic acid coated microprojectiles into cells.

Unfortunately, early attempts by molecular biologists to utilise cloning vectors constructed to contain all the genetic information required to ensure host cell-mediated replication of a foreign DNA fragment were severely hampered. This resulted from the generally poor efficiency with which the vectoral DNA could be subsequently introduced into a host cell. Phage vectors could, following *in vitro* packaging into new viral particles, be efficiently used to introduce DNA into host bacteria by the normal process of viral infection. However these had several practical limitations, especially in terms of host cell specificity (i.e. they could only be used to infect some strains of bacteria). If the effort expended upon the development of improved plasmid cloning vectors was to be rewarded by their successful exploitation for gene cloning, molecular biologists had therefore to develop appropriate methods for the efficient introduction of plasmid DNA into host cells. This was soon achieved for bacterial cells using simple methods involving permeabilising the bacterial cell wall using chemical or electrical stimuli. However, the increasing sophistication of cloning vectors and their varied application within host cells from a wide range of prokaryotic and eukaryotic organisms required the development of new approaches for delivering the DNA into cells. A large range of techniques and approaches are consequently now available for introducing DNA (or RNA) into cells of all types. Many of these have been developed specifically to overcome the problems associated with differences in cellular anatomy (e.g. the existence of nuclear envelopes), or location (e.g. within an embryo or multicellular organism). Others have been devised to particularly suit the nature of the nucleic acid molecule (e.g. vector-borne or 'naked' synthetic) or its target (i.e. a gene or its mRNA transcript).

PHAGE TRANSFECTION OF *E. COLI*

Modified phage λ-based cloning vectors provided the first high efficiency method of introducing foreign DNA into *E. coli* cells by virtue of the phage's natural assembly and infection pathways. Accordingly, following insertion of foreign DNA into the phage vector, the recombinant λ virus DNA is *in vitro* packaged into a suitable coat protein structure. The new viral particles thus formed can then infect cultured *E. coli* cells by simple addition.

In vitro packaging is achieved using packaging extracts produced in one of two ways (Fig. 70.1).

(1) A single strain packaging mix may be synthesised using a λ virus possessing defective *cos* sites. This allows synthesis of the λ capsid proteins in the *E. coli* strain SMR10 but prevents packaging from occurring. The capsid proteins accumulated within the *E. coli* cells may then be isolated and purified.

(2) A two strain system can be used which utilises two *E. coli* strains which each produce incomplete sets of λ capsid proteins: strain BH2688 defective in the production of coat protein E, and strain BHB2690 defective for coat protein D. Neither strain can therefore assemble complete phage particles. However, the phage components from each strain may be purified and

combined for use as a packaging mix.

The strategy of using naturally infective bacterial pathogens (e.g. *Agrobacterium tumefaciens*) or viruses (e.g. SV40) as vectors for packaging and delivering nucleic acids into

THE VALUE OF 'INSIDE' INFORMATION

Information regarding the external composition of prokaryotic and eukaryotic cell membranes was vital to the development of strategies for introducing foreign DNA (or RNA) molecules into cells (e.g. those based upon membrane fusion or permeabilisation). The acquisition of knowledge regarding the internal infrastructure and biochemical functions within cells has also been important for many molecular biological applications. For example, the success of antigene and antisense oligonucleotide-based therapies hinges on being able to deliver nucleic acid molecules to specific targets within the cell. These approaches also depend upon the development of strategies which allow the oligonucleotides to evade recognition or degradation whilst they perform their desired function (e.g. down- or up-regulate gene expression). This has recently been aided by the *in vitro* synthesis of oligonucleotides incorporating novel 'protected/protective' nucleotide analogues or peptides.

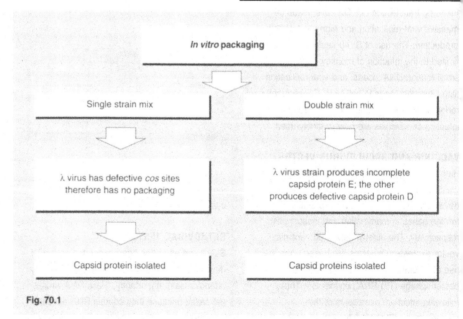

Fig. 70.1

In vitro packaging

Single strain mix | Double strain mix

λ virus has defective *cos* sites therefore has no packaging | λ virus strain produces incomplete capsid protein E; the other produces defective capsid protein D

Capsid protein isolated | Capsid proteins isolated

plant and animal cells has also been successful. These are currently the subject of ongoing development and are proving to be an area of considerable importance (e.g. for application in gene-based therapies).

BACTERIAL TRANSFORMATION

In the early 1970s, several research groups provided the solution to the low efficiency of *E. coli* transformation by 'naked' plasmid DNA. Studying the uptake of linear and circular plasmid DNA, they established that efficient transformation required: (i) an environment of calcium ions and a low temperature (0–5°C) to make the cells competent (i.e. capable of being transformed); and (ii) a heat shock step (37–45°C) to stimulate DNA uptake (Fig. 70.2). Although the precise mechanisms by which this is achieved have not yet been determined, the approach still forms the basis of a moderately efficient method for bacterial transformation routinely used today.

Essentially, log phase cultured *E. coli* cells are made competent by resuspending in ice-cold 50 mM calcium chloride followed by 15–30 minutes incubation on ice. Plasmid DNA is then added, the cells incubated for a further 15–30 minutes on ice and then the cells subjected to a 'heat shock' of a few minutes at 42°C. Although generally only applicable to bacteria, a similar protocol employing lithium chloride or lithium acetate has also been used to enhance the uptake of DNA by yeast cells.

Removing the cell wall barrier

The cell wall forms the primary barrier to DNA uptake in many organisms. This cell wall may however be enzymatically removed to generate what is termed a 'protoplast'. Protoplasts derived from yeast, bacteria or plant cells readily take up DNA from solution by adsorption. Following transfection the cell walls may then be regenerated around the protoplasts during *in vitro* culture.

DNA may also be efficiently introduced into protoplasts or animal cells by means of electrical gradients which momentarily induce holes in the plasma membranes. Known as electroporation, this method has been successful with several animal cell types which have been resistant to transfection by other means. Arresting eukaryotic cells in metaphase using drugs such as colchicine also increases the efficiency of DNA uptake, presumably because the nuclear membrane is absent or maximally permeable at this stage in the cell cycle.

Liposome fusion delivery

Although DNA precipitated onto animal cells is relatively readily taken up, liposome fusion-mediated delivery of DNA (termed lipofection) has also been adapted for use with animal cells and protoplasts. Lipofection was originally developed for cell fusion techniques (e.g. to create monoclonal antibody-secreting hybridoma cell lines), whereby non-ionic detergents allowed cell plasma membranes to fuse. It was subsequently discovered that exogenous DNA in solution is also spontaneously complexed or entrapped in liposomes (small single layer lipid vesicles) during their formation from non-ionic or cationic lipids. These DNA-laden liposomes can therefore deliver the DNA into the cell by binding to, and fusing with, the lipid bilayer of the cell plasma membrane.

Injection and microprojectiles

DNA may also be delivered into animal or plant cells by purely physical methods. Accordingly, microinjection allows DNA or RNA to be introduced directly into cell nuclei or cytoplasms by means of a very narrow bore glass capillary. This approach has also been used for removing cells or genomic DNA for analysis (e.g. from *in vitro* fertilised embryos for pre-implantation screening for genetic diseases such as cystic fibrosis) or for studying gene expression during development. Macro-injection of DNA solutions into the plant vasculature has also been successfully used for plant transformation. Microprojectile delivery has also been widely used for plant and animal cell transformation. This involves the use of microprojectiles (e.g. a gold or tungsten particle a few microns in size coated with DNA or RNA) which are fired at high velocity directly into untreated cells.

OAs: 281–4, 656–80, 794–5, 803, 825–7, 840. **RAs:** 225–32, 235–40, 281–90, 294, 404–7. **SFR:** 36, 120–4, 126–8, 154, 156–9, 220–1.

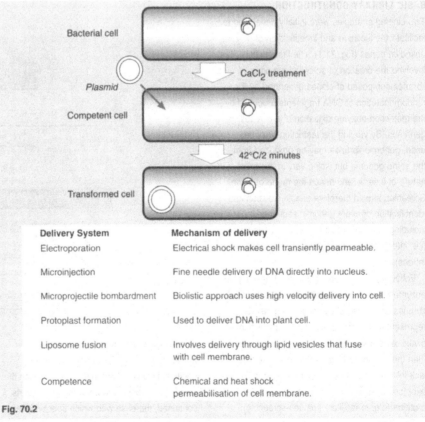

Delivery System	Mechanism of delivery
Electroporation	Electrical shock makes cell transiently pearmeable.
Microinjection	Fine needle delivery of DNA directly into nucleus.
Microprojectile bombardment	Biolistic approach uses high velocity delivery into cell.
Protoplast formation	Used to deliver DNA into plant cell.
Liposome fusion	Involves delivery through lipid vesicles that fuse with cell membrane.
Competence	Chemical and heat shock permeabilisation of cell membrane.

Fig. 70.2

71. PRINCIPAL GENOMIC AND cDNA CLONING STRATEGIES

Several approaches have been developed for the construction of libraries of clones bearing genomic and cDNA fragment inserts to enable the isolation, identification and mapping of specific genes.

The cloning (or subcloning) of genomic DNA and cDNA fragments by insertion and replication within vectors is an essential step in many recombinant DNA strategies. The need to increase the efficiency, reliability and rapidity with which genomic or cDNA fragments may be cloned and identified has resulted in the development of a variety of cloning strategies, each geared to specific applications. For example, gene mapping and identification initially relied upon the screening of millions of different clones present within conventional genomic and/or cDNA libraries. The development of improved screening and selection methods (often based upon the incorporation of new characteristics within vectors) has reduced the time and effort required to isolate desired clones from genomic and cDNA libraries. Alternative strategies have however also been developed to increase the ability to isolate and identify clones. Accordingly, subtractive hybridisation has been developed to greatly increase the representation (and hence the likelihood of identifying) specific gene and/or cDNA fragments. Similarly, differential display cloning using the polymerase chain reaction (PCR) is greatly assisting the analysis of differential gene expression and the isolation of differentially expressed mRNAs (i.e. as cDNA clones). Genomic cloning strategies have also been developed to specifically facilitate gene mapping (e.g. by chromosome walking or jumping). These methods utilise the insert from one clone as a probe to re-screen the library for clones harbouring inserts originally located in an adjacent position within a chromosome (or some distance away in one direction). The PCR has also been exploited for a variety of cloning and subcloning strategies, including the construction of recombinatorial libraries particularly useful in antibody engineering for creating arrays of novel chimeric genes/proteins.

BASIC LIBRARY CONSTRUCTION

Two cloning strategies were initially developed to facilitate the isolation and identification of unknown genes (Fig. 71.1). The first simply involved the creation of so-called genomic libraries composed of clones generated by the random insertion of DNA fragments produced by the restriction enzyme digestion of entire cell genomes. By varying the restriction enzymes used, genomic libraries may be produced from the same genome but which vary in the exact nature of inserts, and hence the range of clones. Screening should therefore enable isolation and identification of every genomic sequence, including both coding and non-coding sequences (i.e. intergenic and extragenic DNA, exons and introns).

Following the isolation of reverse transcriptase enzymes, it was also possible to generate cDNA libraries composed of clones bearing inserts representative of the mRNA transcripts within a given cell or tissue. These inserts are smaller than their genomic counterparts because they lack intronic sequences and, in an appropriate expression vector, can be used for producing proteins (e.g. to facilitate immuno-screening).

The technical difficulties encountered when screening genomic and cDNA libraries have however led to the development of many additional cloning differ in the way in which the libraries are constructed, the range of inserts contained, the ease with which specific insert-bearing clones can be isolated, and hence their ultimate utility.strategies. These essentially differ in the way in which the libraries are constructed, the range of inserts contained, the ease with which specific insert-bearing clones can be isolated, and hence their ultimate utility.

SUBCLONING: DEFINITION AND USE

The term subcloning is used to define any operation which involves the removal and transfer of a cloned DNA fragment from one vector to another. Subcloning procedures are widely used to facilitate gene analysis and manipulation (e.g. *in vitro* mutagenesis), or to overcome limitations imposed by a particular vector. For example, the expression of a cloned gene or cDNA fragment may require its excision from the orginal cloning vector used for its replication but which lacks the capacity to express an insert (e.g. lacking the appropriate transcriptional regulatory sequences), into another vector capable of expression. Subcloning may involve restriction enzyme-mediated excision, or PCR amplification of the insert (or a defined part of the insert) followed by its religation into another vector.

OAs: 438, 442–5, 630–2, 681–8, 713, 1081–5. RAs: 219–24, 241–2, 247–9, 392–7, 418–24. SFR: 113–15, 119–24, 126–8, 136–9, 158–63.

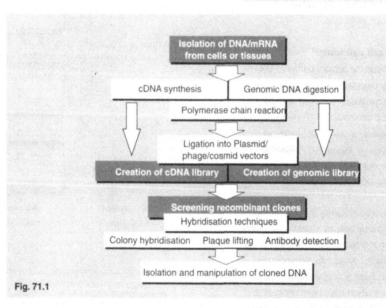

Fig. 71.1

GENE MAPPING STRATEGIES

Gene mapping has increasingly relied upon the development of cloning strategies which enable the determination of an insert's original chromosomal location. Often referred to as chromosome 'walking' or 'jumping', they rely upon the use of one clone insert (or part of its sequence) as a probe to identify other inserts originally adjacent to, or located some distance from, its original position within the chromosome. The new clone can then serve as a probe in subsequent rounds of library screening. Libraries for chromosome walking have been prepared in both phage and cosmid vectors. Their overall utility depends primarily upon their quality (i.e. the degree to which they contain large fragments representative of the whole chromosome(s)).

Jumping and linking libraries

Two types of 'jumping' libraries have been constructed: general jumping and specific jumping libraries. General jumping libraries are constructed in a manner which allows the identification of clones by travelling in a specific direction along the chromosome. In contrast, specific jumping libraries consist of clones which jump from a relatively rare restriction site (e.g. *Not*I) to the next adjacent site. The basic strategy (Fig. 71.2) for producing a jumping library initially involves partial digestion of high molecular weight genomic DNA with a restriction enzyme (e.g. *Mbo*I). This is followed by size-selection of particular fragments by pulsed field gel electrophoresis, and circularisation of the fragments by ligation. This brings together fragments of DNA which were originally located at large distances apart in the genome. Digestion of the DNA circles with a different restriction enzyme (e.g. *Eco*RI) and selective cloning of these junction fragments in standard vectors forms a library of jumping clones. Specific jumping libraries are prepared in a similar manner, however the initial digestion is carried out to completion and there is no size-selection of the resulting fragments.

Linking libraries are often combined with specific jumping libraries to enhance the detection of 'linking' clones. They are constructed in a similar fashion but use much smaller fragments such that clones are likely to cross or overlap

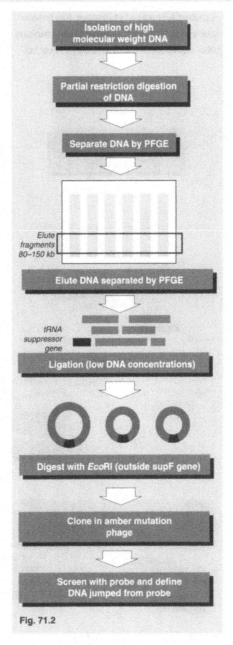

Fig. 71.2

with the rare restriction sites used in specific jumping library construction.

Inverse PCR as an aid to mapping

Inverse PCR has also been used to clone hitherto unknown sequences flanking a region of known sequence. This involves the use of primers which are designed to a known genomic sequence but are orientated on the target DNA with their 3' ends directed away from one another (Fig. 71.3).

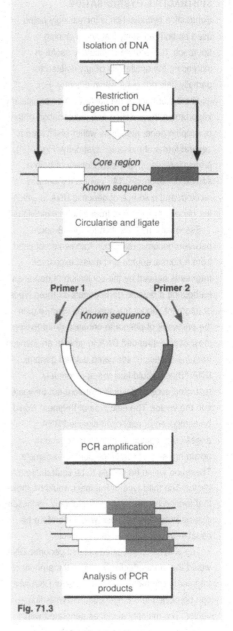

Fig. 71.3

An amplification product can therefore only be generated when the DNA has been cut upstream and downstream of the primer binding sites and circularised by ligation. Such amplification products may then be cloned producing a library composed of clones which represent the two flanking sequences defined by the restriction site used for circularisation. This process can be repeated to gradually move along a chromosome in both directions.

189

Subtractive hybridisation and differential display techniques have enhanced the ability to obtain specific gene and cDNA clones, and allowed the detection and analysis of changes in gene expression.

SUBTRACTIVE HYBRIDISATION

Subtractive hybridisation is increasingly being used as both an analytical and a cloning approach. It has been particularly useful in increasing the availability of any deleted (or partially deleted) gene within a library. For example, subtractive hybridisation was particularly important in the isolation and identification of the dystrophin gene, mutations within which lead to several forms of muscular dystrophy (Fig. 71.4). It primarily relies upon the availability of both a source of genomic DNA in which the gene is deleted, and a source of genomic DNA in which the gene is present (i.e. from a normal individual).

Essentially, hybridisation for 24–48 hours between genomic restriction fragments obtained from a normal source and a vast excess of fragments derived by the sonication of genomes harbouring a deleted gene generates three types of fragment. Under the conditions used (e.g. in the presence of phenol to enhance re-annealing) most double-stranded DNA fragments are derived from the excess of sonicated deleted genomic DNA. These would lack the appropriately restricted ends and would therefore not be ligated into the vector. The next type of fragment would be composed of restriction-digested DNA annealed to complementary (i.e. commonly occurring) sequences in the sonicated sample. These too would be unable to be ligated into the vector. The third type of fragment would be those that formed by the re-annealing of only restriction-digested fragments. These would therefore be capable of being ligated into the vector.

The excess of sonicated deleted genomic DNA would therefore effectively reduce the amount of fragments common to both sources of DNA which can be cloned into a restriction enzyme-digested vector. The majority of clones generated within the library would therefore represent DNA fragments from the normal source which are deleted in the other genomic sample, i.e. producing a library enriched for the deleted gene.

It is also possible to use this approach to generate and identify cDNA clones representing mRNA transcripts produced from differentially expressed genes. This relies upon the subtractive hybridisation between the cDNAs derived from two different cell populations. Indeed, the analysis of differentially expressed genes by many assays,

including subtractive hybridisation, has provided many insights into the fundamental mechanisms of cell development, differentiation and control.

OAs: 458, 472, 595–600, 689–91, 1037–40. **RAs:** 243–5, 250, 377, 392–7, 418–24. **SFR:** 113–15, 119–24, 126–8, 136–9, 158–63, 225.

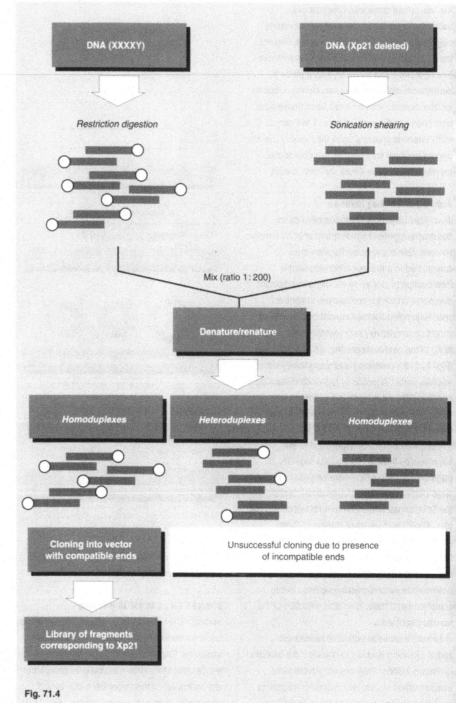

Fig. 71.4

CLONING BY DIFFERENTIAL DISPLAY

Whilst subtractive hybridisation techniques can be used to detect changes in gene expression between different cell types or cells at different stages of development or differentiation, they cannot quantify changes in gene expression. Techniques such as nuclear run-on assays, northern blotting, and nuclease protection assays which measure quantitative changes in expression can also only be used for the analysis of known genes.

The technique of differential display offers a potential solution since it allows both the isolation of mRNA species (i.e. as cDNA clones) produced from differentially expressed genes, and their subsequent identification (e.g. by sequencing or hybridisation with gene probes). Firstly, total cellular mRNA is extracted and then converted into cDNA by reverse transcription (Fig. 71.5). The cDNA is then systematically amplified by the PCR using an 'anchored' 3′ primer which anneals to the poly(A⁺) tail present in most eukaryotic mRNAs. This is used in combination with an 'arbitrary' 6–7 bp 5′ primer whose degenerate sequence allows it to anneal near the 5′ end of the cDNA. This latter primer anneals at many different positions relative to the 'anchored' primer depending upon the length and sequence of the cDNA.

This consequently results in the generation of multiple PCR products which can be separated by gel electrophoresis, and autoradiographically detected (based upon the incorporation of radiolabelled nucleotides during PCR amplification). Multiple primer sets can be used to generate reproducible patterns of PCR products which, when combined, represent most (if not all) of the mRNA species found in a particular cell type. Comparative analysis of the PCR products generated from different cell types therefore allows the identification and isolation of differentially expressed genes.

Differential display offers technical advantages. For example, a clone may be obtained within 5 days using this approach, compared to two months by other conventional library cloning methods. The use of the PCR also increases gene representation (and hence the chance of isolating it from the library) and enables the use of very small original samples (e.g. single cells).

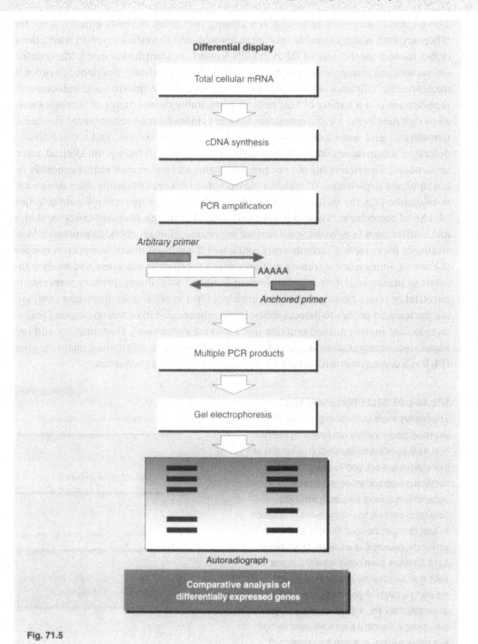

Fig. 71.5

PCR-based cloning strategies

The versatility and sensitivity of the PCR has inevitably led to its widespread use in both cloning and subcloning strategies. For example, primers may be designed from known sequences to allow specific fragments to be cloned. Alternatively, degenerate primers may be designed (i.e. based on limited amino acid sequence data) to enable the cloning of related genes or genes for which only limited sequence data are available. Anchored PCR can also be used to enable the cloning of mRNA transcripts or genes of unknown sequence.

The widespread use of PCR in gene and protein engineering has also facilitated the development of additional cloning strategies such as recombinatorial/repertoire cloning which have been widely used for antibody engineering, vaccine and drug development.

72. STRATEGIES FOR IDENTIFYING DESIRABLE RECOMBINANT CLONES

Selectable markers (e.g. antibiotic resistance) may be used to detect recombinant clones, whilst screening and/or selection-based approaches may be used to detect specific insert sequences.

The ultimate success or failure of any cloning procedure depends equally upon the efficiency with which recombinants are generated, and the efficiency with which those clones harbouring the desired DNA or cDNA insert are identified/isolated. Many different vectors and cloning strategies have therefore been specifically developed in order to maximise the efficiency with which foreign DNA may be inserted and subsequently replicated within a variety of host cells. A comparably diverse range of strategies have also been developed for discriminating between clones bearing recombinant (i.e. insert harbouring) and non-recombinant (i.e. devoid of insert) vectors, and for specifically detecting those recombinant clones harbouring vectors including the desired insert sequence(s). Discriminating between recombinants and non-recombinants generally relies upon the exploitation of features incorporated into vectors during their design and construction (e.g. the inclusion of selectable markers). The rapid and efficient selection of desired recombinants is also being made possible through advances in vector design and construction (e.g. based upon surface expression of insert-encoded proteins). Most strategies for detecting recombinants containing the desired insert, however, rely upon the use of appropriate screening methods which specifically recognise and analyse the insert sequence and/or its encoded protein product. Accordingly, primary screening of bacterial or phage libraries commonly involves filter hybridisation approaches employing nucleic acid probes to detect specific insert sequences, and/or antibody-based probes to recognise insert-encoded proteins (i.e. following expression). Both primary and secondary screening/analysis is also often achieved using the polymerase chain reaction (PCR) in conjunction with Southern blotting and nucleotide sequencing.

VARIATION IN CLONING EFFICIENCY

Cloning efficiency is determined by several factors. Perhaps the most important factor is the strategy employed to generate and/or insert the foreign DNA into a vector. Accordingly, the insertion of blunt-ended DNA fragments is much less efficient than the insertion of 'sticky' or cohesive ended fragments carrying 3′ or 5′ overhangs. Unidirectional cloning based upon the insertion of fragments with two different cohesive ends (i.e. produced by digestion with two different restriction enzymes) also increases the ability to obtain clones in which the foreign DNA is inserted in the correct orientation for subsequent expression. The use of rare-cutter restriction enzymes also increases the likelihood of obtaining full length inserts (e.g. containing an entire gene).

THE USE OF SELECTABLE MARKERS

The primary steps to identifying and isolating desirable clones involve discriminating between host cells transformed/transfected by vectors and those which are not, and between those containing recombinant or non-recombinant vectors. This is most frequently achieved using selectable markers whose presence or absence in host cells can be used to indicate infection and/or the presence of an insert. Most vectors have therefore been designed and constructed such that they contain one or more genes encoding a range of phenotypes to serve as selectable markers. Antibiotic resistance genes are amongst the most commonly used form of selectable marker, enabling the selection of vector-containing bacterial, plant or animal cells. Accordingly, vector-containing cells are able to survive in culture medium containing the antibiotic whilst non-infected cells (lacking the antibiotic resistance gene) are non-viable. Enzymes are also often used as selectable markers based upon their ability or inability to generate a detectable signal (e.g. colour change) in the presence of an appropriate substrate. For example, animal cells transfected by *lux* gene-bearing vectors may be detected based upon

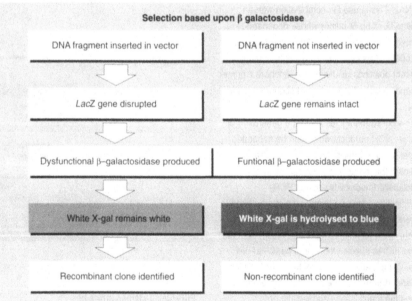

Selection based upon β galactosidase

DNA fragment inserted in vector	DNA fragment not inserted in vector
LacZ gene disrupted	*LacZ* gene remains intact
Dysfunctional β–galactosidase produced	Funtional β–galactosidase produced
White X-gal remains white	White X-gal is hydrolysed to blue
Recombinant clone identified	Non-recombinant clone identified

Fig. 72.1

their ability to generate a bioluminescent signal in the presence of luminol. Non-recombinant and recombinant bacterial colonies are also often distinguished by their colour (i.e. blue or white) in medium containing the substrate X-gal (Fig. 72.1). This system relies upon the inactivation of β-galactosidase caused by the insertion of the foreign DNA into the vector.

OAs: 595–9, 617, 692–706, 776, 1147, 1036–40. **RAs:** 210, 214–22, 241–2, 246, 376–7, 393–6. **SFR:** 119–24, 126–8, 136–9, 158–9, 181.

SCREENING FOR SPECIFIC INSERTS

The most widespread approach employed for the detection of microbial host cell clones harbouring vectors containing the desired insert relies upon filter hybridisation (Fig. 72.2). This involves placing a filter (e.g. nylon or nitrocellulose) onto the microbial colonies growing on solid agar medium in Petri dishes. After a few minutes, a small amount of each colony becomes attached to the filter. The filter is then treated in one of several ways such that the DNA from the host cells representing each colony is irreversibly bound to the filter, but remains available for hybridisation with a labelled nucleic acid probe. Specific insert-bearing clones are therefore indicated by the position of dots on the filter where the probe has bound to a specific insert sequence(s). These can then be related to the original position of the colony on the Petri dish. Depending upon the stringency of hybridisation and the probe, any clones harbouring full length and partial length sequences which have been inserted in either orientation may be detected.

Antibody probes can be similarly used to detect clones expressing protein encoded by the desired insert. This approach can however only detect clones in which the foreign DNA is inserted in the correct orientation and reading frame. It is also dependent upon the nature of the antibody probe (i.e. poly- or mono-specific), which may not recognise partial length inserts.

The PCR is also increasingly used for screening, both to detect and assess the nature of specific inserts in microbial, plant and animal host cells bearing recombinant vectors. Accordingly, a small portion of a bacterial colony may be removed and subject to PCR amplification using either (i) insert-specific primers (in cases where the sequence is known), or (ii) primers specific for vector sequences flanking the insert site. Although the latter approach can indicate the size of inserts, the definitive identification of inserts relies upon further analysis of the PCR products by Southern blotting, RFLP, and/or nucleotide sequencing. Nevertheless, the sensitivity and specificity of the PCR allows both approaches to be used, singly or in combination, for the detection/analysis of mammalian and plant cell clones (e.g. during the generation of transgenic lines). Thus, a single cell from a

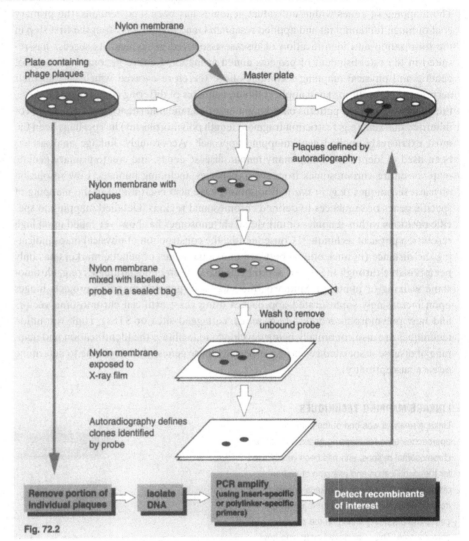

Fig. 72.2

fertilised embryo or a callus culture may provide sufficient DNA for PCR-based screening.

Selection-based approaches
The labour-intensive and time-consuming nature of screening methods (particularly for identifying specific clones within a genomic or cDNA library) has led to the development of specialised vectors which may be used for clone selection. Phagemid vectors have consequently been developed which enable the expression of insert-encoded proteins directly onto the surface of the phage particle. Phages within the library which harbour the desired insert in the correct orientation may therefore be selected (i.e. isolated) based upon

their recognition by, and binding to, immobilised ligand or receptor proteins. For example, phages bearing surface expressed functional recombinant antibody or receptor proteins may be selected using columns to which a specific ligand has been immobilised. Similarly, columns containing immobilised antibody or receptor proteins may be used to select phages bearing surface expressed recombinant clones encoding specific peptide/protein ligands. It should however be remembered that techniques such as the PCR and nucleotide sequencing are also subsequently required for the definitive analysis of any selected clones.

73. GENE MAPPING TECHNIQUES

Both genetic (e.g linkage analysis of polymorphic markers) and physical techniques (e.g. *in situ* hybridisation, chromosome walking/jumping) have been used for the mapping of single genes or entire genomes.

The mapping of genes within individual genomes has been, and remains, the primary goal of much fundamental and applied research. Gene mapping is often the first step in the localisation and identification of disease-associated genes, and its success has resulted in the establishment of projects aimed at mapping entire genomes. A range of genetic and physical mapping techniques have therefore evolved which vary in their accuracy (i.e. resolution) and applicability to genomes of differing complexity. Of these, the analysis of linkage patterns based upon observing the inheritance of two or more co-inherited markers (e.g. restriction fragment length polymorphisms) has perhaps been the most extensively applied gene mapping approach. Accordingly, linkage analysis has been used to identify and map many human disease genes, and create primary genetic maps of entire chromosomes from several species, including humans. Low resolution physical techniques (e.g. *in situ* hybridisation) have also been useful in the mapping of specific genes or sequences to defined chromosomal regions. Detailed mapping to specific positions within genomes or individual chromosomes has however relied upon high resolution physical techniques. Consequently, the construction of physical maps indicating the distance (in nucleotides) between individual genes or genetic markers has only been possible through the use of several cloning and screening strategies (e.g. chromosome walking or jumping). Similarly, the success of genome mapping projects hinges upon increasingly sophisticated approaches using yeast artificial chromosome vectors and new polymorphic markers (e.g. sequence tagged sites or STSs). High resolution techniques are also continually being developed to facilitate the identification and mapping of disease-associated genes (e.g. especially those genes which combine to determine disease susceptibility).

PATENTING RIGHTS (OR WRONGS?)

In recent years, some researchers have attempted to obtain patent rights over each newly mapped and sequenced human genomic DNA fragment. This has been motivated both by previous successes in patenting genes from other species, and the anticipation that many such cloned DNA fragments may contain genes (or their regulatory sequences) of ultimate commercial value. This has elicited great controversy within both the scientific and legal communities regarding the legal and ethical rights or wrongs of patenting naturally occurring sequences. Although far from settled, it is clear that any patent rights ascribed to such sequences are unlikely to be supported in light of the widely held view that no individual or set of individuals should have sole control over any specific human genes.

OAs: 428–9, 682–91, 866–72, 882–8, 890–3, 1141–4. **RAs:** 200, 224, 219–22, 241–5, 247–9, 419–25. **SFR:** 119–20, 161–3, 181, 223–5.

LINKAGE MAPPING TECHNIQUES

Linkage analysis was one of the primary approaches used for mapping genes to specific chromosomal regions, and has been widely used for the identification and mapping of human disease genes. In essence, linkage analysis involves determining the degree to which two genes are physically linked on the same chromosome. This is achieved by tracing how often different forms of two variable phenotypes or genotypes are co-inherited amongst the offspring of many generations. It can therefore allow the distance between two or more alleles or genetic loci to be calculated based upon the probability of cross-over events occurring between them during meiosis. This requires the analysis of many families with large numbers of siblings (or the products of repeated experimental crosses) in order to establish a recombination fraction and thereby calculate the likelihood (probability) of obtaining this result if the loci are linked or unlinked (reported in terms of a lod score).

Linkage analysis is frequently performed using polymorphic DNA markers such as restriction fragment length polymorphisms (RFLPs) (Fig.

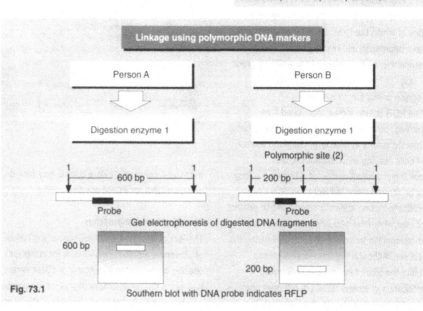

Fig. 73.1

73.1). Highly informative minisatellite repeats (e.g. STS and variable number of tandem repeats (VNTRs)) are now increasingly being used. These allow more direct calculation of inter-gene distances in terms of the physical length of intervening DNA. Moreover, their highly variable

nature allows the construction of more detailed (i.e. high density) chromosome maps. Linkage analysis using polymorphic markers is therefore finding particular application in the generation of physical maps of the human genome.

PHYSICAL MAPPING TECHNIQUES

A wealth of low and high resolution physical mapping techniques are now available to facilitate the identification of specific genes (e.g. human disease genes) and to enable the mapping of entire genomes (Fig. 73.2). Accordingly, *in situ* hybridisation involving the binding of cloned and fluorochrome-labelled DNA probes to separated chromosomes is a standard low resolution mapping technique which has been widely used for the primary localisation of human disease genes to specific chromosomal locations (e.g. those associated with cystic fibrosis and Huntington's disease).

High resolution mapping of such genes, however, relies upon the use of several genomic (and latterly cDNA) cloning strategies such as chromosome walking and jumping. These involve the initial generation of large libraries of overlapping genomic DNA fragments from isolated chromosomes by their partial digestion with restriction enzymes and insertion into phage or cosmid vectors. More recently, microdissection and microcloning of defined chromosomal regions has been achieved using focused UV laser microbeams to generate large numbers of genomic fragments. Starting from specific linked polymorphic markers, chromosome walking involves linear screening steps which allow the identification and ordering of overlapping clones moving in one direction along the chromosome. Each step can span large distances depending upon the type of library (i.e. up to 50 kb using cosmid libraries and up to 20 kb using phage libraries). Chromosome jumping also facilitates mapping in one direction from a single marker but involves the identification of fragments originally located at great distances apart within the genome. This necessitates the construction of specific jumping or linking libraries containing large numbers of cloned junction fragments.

The generation of physical maps covering individual chromosomes or entire genomes by such methods has required the use of vectors capable of bearing extremely large, overlapping inserts. Yeast artificial chromosomes (YACs) have been of particular importance since they can accommodate inserts of an average 300 kb and can be used to create libraries of continuously overlapping cloned fragments (termed contigs).

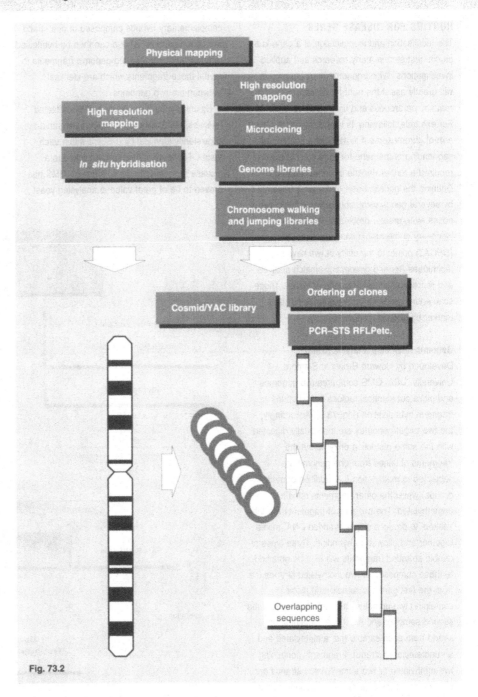

Fig. 73.2

Individual contigs may be identified and ordered following screening cycles based upon the use of several mapping techniques. Accordingly, a clone 'fingerprint' may be derived by restriction enzyme mapping or analysis of the presence of STSs. Contigs may also be ordered using *in situ* hybridisation, Southern blotting and polymerase chain reaction (PCR) amplification-based library screening methods. Ultimately, their position within genetic and physical maps may be determined by nucleotide sequencing and subsequent computerised alignment of the sequences based upon STS markers.

The identification, mapping and tracking of genes has been greatly enhanced by the development of genome scanning techniques for detecting similarities or differences between two (or more) genomes.

HUNTING FOR DISEASE GENES

The localisation and identification of a gene is a crucial first step in many research and applied investigations. Whilst genome mapping projects will greatly assist the hunt for disease genes, this remains an arduous and time-consuming task. For example, following its localisation to the short arm of chromosome 4 in 1983, the isolation and sequencing of the gene for Huntington's disease required a further decade of concerted effort. Defining the genetic basis of diseases influenced by several genes using conventional approaches poses even greater problems. The recent discovery of the breast cancer susceptibility gene (*BRCA2*) points to the utility of two new techniques (termed genomic mismatch analysis and representational difference analysis). These scan entire genomes in one step, rather than marker by genetic marker.

Genomic mismatch analysis (GMS)

Developed by Howard Brown at Stanford University, USA, GMS compares two genomes and picks out identical regions based upon fragment hybridisation (Fig. 73.3). Accordingly, the two target genomes are individually digested with the same restriction enzyme but the fragments obtained from one genome are subjected to methylation (i.e. addition of methyl groups) whilst the other fragments remain unmethylated. The two sets of fragments are then 'melted' to produce single stranded DNA, mixed together and allowed to hybridise. Three types of double stranded fragments will then be obtained: (i) those composed of two methylated strands (i.e. from the first genomic sample); (ii) those containing two unmethylated strands (i.e. from the second sample); and (iii) those containing one strand from each sample (i.e. a methylated and an unmethylated strand). Fragments containing two methylated or two unmethylated strands are then removed (by digestion). Consequently, the resultant population of fragments will all contain one methylated and one unmethylated strand. These may however differ in their degree of complementarity (i.e. some hybrids will contain mismatches). GMS then utilises a mixture of three *E. coli* proteins which recognise and nick mismatches in unmethylated strands allowing their digestion. This leaves only truly

complementary hybrids composed of one strand from each genome. These can then be hybridised to an array of immobilised genomic fragments to reveal those fragments which are identical between the two genomes.

By comparing the genomes of two affected relatives, GMS will reveal those regions that are consistently identical by descent within each family. These must therefore be linked to a disease susceptibility gene. Although GMS has proved to be of great value in analysing yeast

OAs: 707–19, 1145–53. RAs: 200, 219–22, 224, 450–4, 419–25. SFR: 119–20, 136–9, 161–3, 181, 223–5.

genomes, several technical problems need to be resolved before GMS can be widely adopted and used for the identification of disease susceptibility genes, or to identify genomic differences, within mammalian genomes.

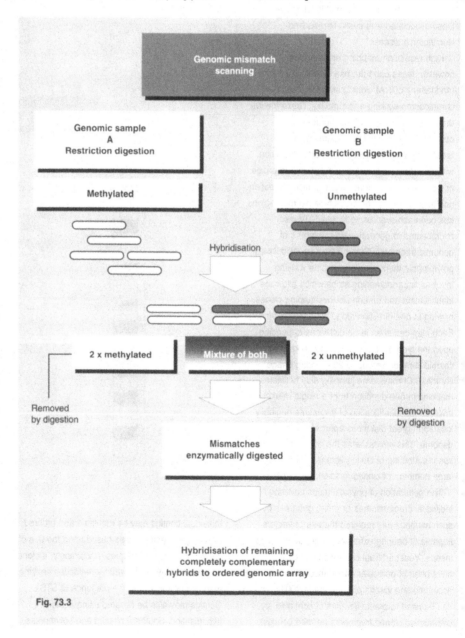

Fig. 73.3

Representational difference analysis (RDA)

Developed by Michael Wigler and Nikolai Lisitsyn at the Cold Spring Harbor Laboratories, USA, RDA is the simpler of the two techniques. RDA is also based upon hybridisation but enables the identification of regions which differ between two genomes. In contrast to subtractive hybridisation techniques which are difficult to perform on large complicated genomes, RDA is similar to differential display analysis of mRNA, being based upon the analysis of only a representative sample of the two target genomes (Fig. 73.4). These representative samples are generated by exploiting the PCR's preference for amplifying small DNA fragments over large fragments. Accordingly, the DNA isolated from two sample genomes (or two pools of genomes from affected and unaffected individuals respectively) is first digested with a restriction enzyme. These are then used as target templates for PCR amplification. This results in a collection of small PCR products which can readily hybridise to allow the identification and isolation of sequences unique to one genome. Although RDA has often been found difficult to perform it has already been successfully applied to the analysis of mammalian DNA. RDA is also already being used to search for mutations involved in certain cancers by comparing the DNA from tumour cells with the DNA derived from the patient's healthy, non-tumour cells.

MAPPING BY RANDOM PRIMER PCR

The PCR has been widely exploited as a mapping tool based upon amplification using primers designed to known sequences. Many PCR-based mapping techniques (collectively referred to as multiple arbitrary amplicon profiling (MAAP) techniques) have also been developed which use short primers of random or arbitrary sequence. Accordingly, abitrarily primed PCR (APPCR) or rapid amplification of polymorphic DNA (RAPD) uses primers 10 nucleotides in length, and DNA amplification fingerprinting (DAF) uses arbitrary primers as short as 5 nucleotides which contain 5′ mini-hairpin structures. Amplification of target nucleic acids with either type of primer generates characteristic profiles of PCR products (amplicons) which vary between different individuals and organisms. These can

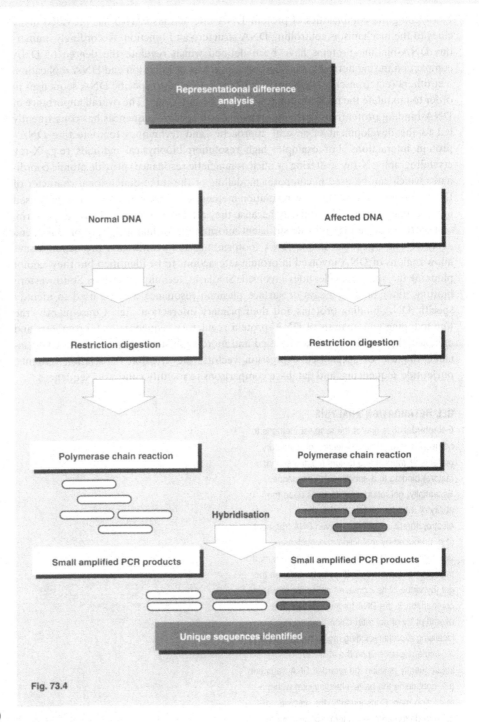

Fig. 73.4

then serve as potential genetic markers in mapping studies and general DNA fingerprinting. MAAP techniques have consequently been extensively applied to the DNA fingerprinting of genomes of varying complexity (i.e. bacterial, fungal, plant and animal genomes) both for gene

mapping and tracking. It is also likely that refinements in the design of MAAP techniques will result in their increased application for the analysis of sub-genomic fragments (e.g. derived from mitochondrial DNA and YACs).

197

74. DETECTING DNA–PROTEIN INTERACTION SITES

DNA-binding protein interaction sites may be detected and localised within DNA fragments by gel retardation, DNA footprinting, and exonuclease protection analysis of defined DNA–protein complexes.

Unravelling the complexities of protein–DNA interactions *in vivo* has provided many clues to the mechanisms controlling DNA structure and function. Accordingly, numerous DNA-binding proteins have been defined which regulate the degree of DNA compaction during normal cellular processes such as cell division and DNA replication. Yet others (e.g. transcription factors) are known to bind to specific DNA sequences in order to modulate the transcriptional activity of certain genes. The overall importance of DNA-binding proteins and their interactions with specific sequences has consequently led to the development of several approaches and techniques for detecting DNA–protein interactions. For example, high resolution biophysical methods (e.g. X-ray crystallography, X-ray scattering or nuclear magnetic resonance) provide atomic coordinates which can be used in computer modelling of the three-dimensional character of DNA–protein interactions. Low resolution methods are, however, more commonly used in molecular biology. This is partly because they are less complex and easy to perform, and partly because they provide sufficient information for many types of primary gene analysis. Thus, gel retardation, DNA footprinting, and exonuclease protection assays allow regions of DNA involved in protein interactions to be identified but they cannot pinpoint the specific nucleotides involved. Similarly, techniques such as Southwestern blotting, filter binding assays or surface plasmon resonance may be used to identify specific DNA-binding proteins and their primary interaction sites. Consequently, the fine mapping and analysis of DNA–protein regulatory sequences (e.g. promoters and enhancers) generally requires modified and more sophisticated approaches. These include the use of specialised expression vectors, the creation of deletion mutants, nucleotide sequencing, and database comparisons to identify consensus sequences.

CONSENSUS SEQUENCES

The analysis of DNA sequences from a wide variety of sources has enabled the identification of certain conserved sequence motifs within several different types of regulatory sequences. These are often referred to as consensus sequences because they are identical or very similar across a range of organisms or species. Several classes of promoter and enhancer sequences have therefore been identified which are conserved between eukaryote species, and in some cases, between eukaryotes and prokaryotes. Consequently, the ease with which nucleotide sequence data may be obtained and compared with computer database entries, allows many putative regulatory sequences to be identified, even in the absence of proven DNA–DNA, RNA–DNA or DNA–protein interactions.

OAs: 44–54, 230, 372, 376, 380, 720–2, 731, 780. **RAs:** 12–15, 194, 256–7. **SFR:** 27–8, 119–24, 126–8, 165, 169, 213.

GEL RETARDATION ANALYSIS

Gel retardation is one of the simplest methods to be used for the primary detection of regulatory proteins or *trans*-acting factors (i.e. transcription factors) binding to a specific DNA sequence. Essentially, gel retardation depends upon the ability of a bound protein to alter the electrophoretic mobility of a given DNA fragment (e.g. produced by restriction endonuclease digestion) (Fig. 74.1). This results in the retarded migration of a protein–DNA complex through the gel (by virtue of its increased molecular mass) in comparison to the DNA fragment alone. The overall utility of gel retardation analysis in localising a protein-binding region within a DNA molecule depends upon the ability to subsequently position the retarded DNA fragment (i.e. containing the protein-binding site) within a restriction map. Consequently, the precision of gel retardation analysis in terms of defining the exact location and nucleotides involved relies upon the accuracy of the restriction map, and the availability of appropriate restriction sites. Despite this potential drawback, gel retardation remains a primary method for detecting protein–DNA binding sites.

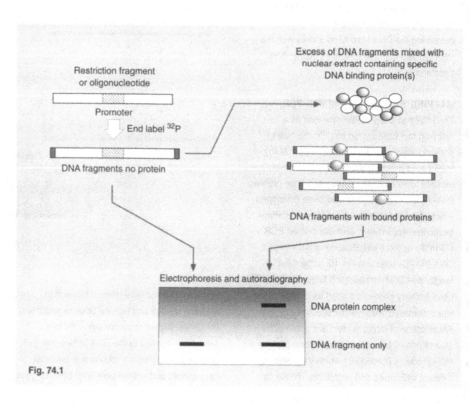

Fig. 74.1

Nuclease protection assays

The ability of bound proteins to protect DNA from digestion with nuclease enzymes has also been widely used for more precisely localising protein-binding regions with DNA. A classical example is a technique known as DNase I footprinting analysis, or 'DNA footprinting' (Fig. 74.2).

This method involves the partial digestion of DNA–protein complexes with the enzyme DNAase I. In practice, the DNA is radioactively end-labelled prior to incubation with the desired DNA-binding protein (generally as a crude protein extract but preferably in a relatively highly purified form). The resultant complexes are then incubated with DNase I under conditions which favour partial digestion. This ensures that many different labelled fragments are generated by cleavage at unprotected sites (i.e. at sites where no protein is bound). In contrast, regions to which proteins have bound remain undigested. The labelled fragments are then separated by high resolution polyacrylamide gel electrophoresis alongside control fragments produced from digests performed in the absence of any DNA-binding protein. If the desired partial digestion has occurred, autoradiography will reveal a ladder of bands representing each of the individual partially digested fragments. Holes, termed 'DNA footprints', will only appear within the ladder where bound protein has protected a specific DNA region from digestion. Comparison of the two ladders can therefore relatively accurately determine the position of the protein-binding region based upon the size of the digestion fragments either side of the 'footprint'. This allows the location (and subsequently the sequence) of the bound fragment to be more precisely determined than using gel retardation. DNA footprinting may not, however, be sufficient to precisely determine the nature of interaction sites, or the contribution of individual nucleotides.

An alternative, and potentially more precise method involves the use of the *E. coli* enzyme, exonuclease III. This enzyme progressively degrades double-stranded DNA from the 3′ end to the 5′ end until digestion is blocked by bound protein. This assay is also capable of providing binding information from crude nuclear extracts. Usually, however, not all the specific protein-binding sites are protected from exonuclease

digestion. Furthermore, the exonuclease may often pause or stop at regions of DNA secondary structure rather than due to the presence of bound protein alone. Such so-called 'exo-stops'

may also be induced by secondary structures imposed by the interaction of DNA-binding proteins acting in *cis*.

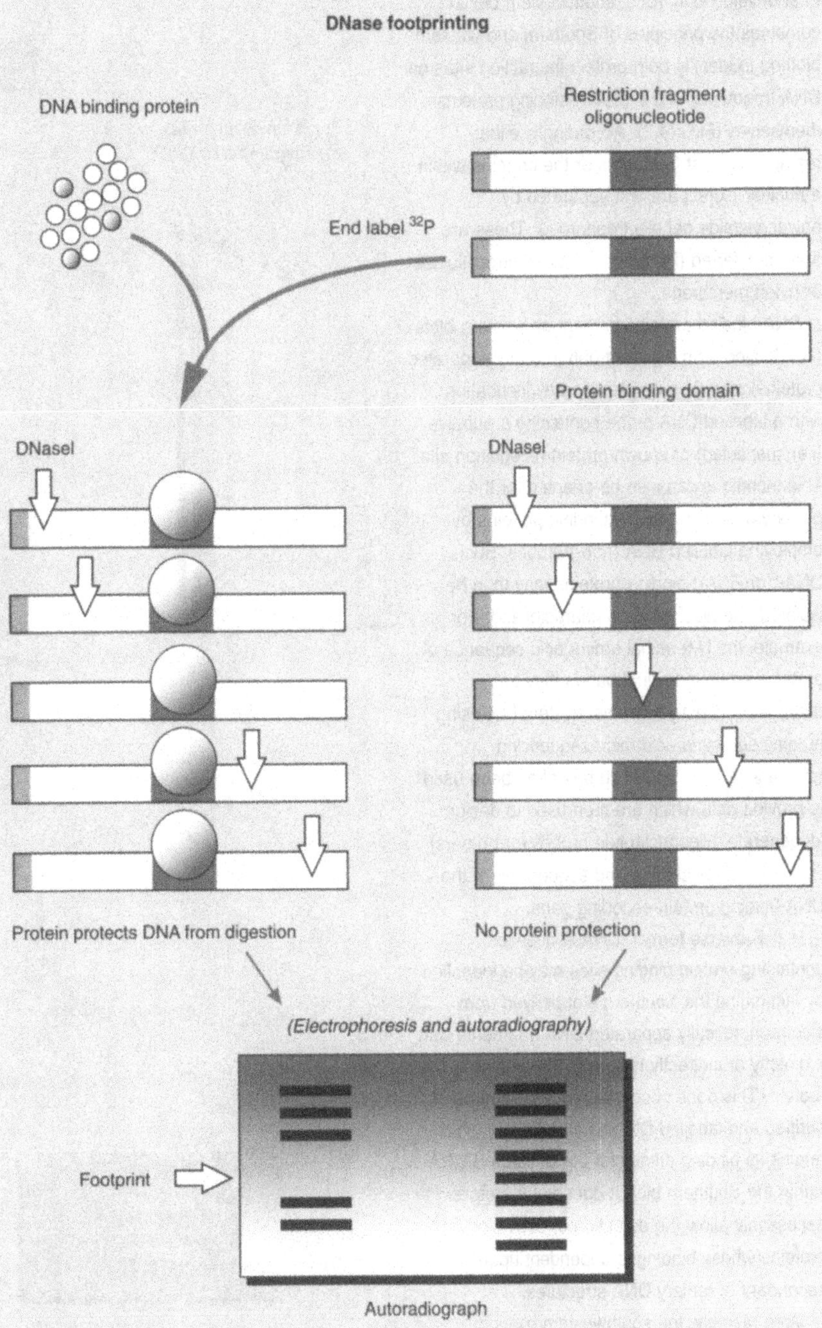

DNase footprinting

Fig. 74.2

Several methods, including south-western blotting, filter binding assays, and surface plasmon resonance techniques, can be used to detect, isolate and/or characterise different DNA-binding proteins.

Southwestern blotting methods

First developed in 1980, southwestern blotting combines the principles of Southern and western blotting to identify both protein interaction sites on DNA fragments, and the DNA-binding proteins themselves (Fig. 74.3). Accordingly, either restriction digest fragments or the proteins within a nuclear extract are first separated by polyacrylamide gel electrophoresis. These are then transferred (i.e. blotted) onto a nitrocellulose or nylon membrane.

DNA-binding proteins present on western blots (i.e. derived from the electrophoretically separated proteins) may then be identified by incubation with a labelled DNA probe containing a putative (i.e. suspected) or known protein recognition site. This technique can also be adapted for the primary detection of RNA-binding proteins by employing labelled RNA (ribo-) probes. Such DNA- (or RNA-) binding proteins may then be isolated and biochemically characterised. For example, the N-terminal amino acid sequence of proteins within individual bands may be determined directly from the western blot using automated amino acid microsequencing approaches. This approach has often been used to provide data which are then used to design 'degenerate' oligonucleotide probes (or primers) for the eventual cloning and sequencing of the DNA-binding protein-encoding gene.

In the reverse format, DNA sequences containing protein-binding sites may be identified by incubating the Southern blot derived from electrophoretically separated DNA fragments with a directly or indirectly labelled DNA-binding protein. This does of course require a source of purified and labelled DNA-binding protein which retains its binding affinity for the denatured DNA within the Southern blot. It does not therefore necessarily allow the detection of DNA-binding proteins whose binding is dependent upon secondary or tertiary DNA structures.

More recently, the southwestern blotting approach has proved useful for detecting exposed DNA regions susceptible to in vivo breakage (e.g. so-called fragile sites associated with fragile X syndrome). Accordingly, DNA–protein complexes may be chemically treated (e.g. alkylated) prior to digestion, separation and blotting. Areas of DNA unprotected by bound

protein (e.g. those involved in maintaining tertiary structures) will therefore be chemically modified. These can then be recognised by labelled antibody probes which can discriminate between natural and chemically modified DNA bases.

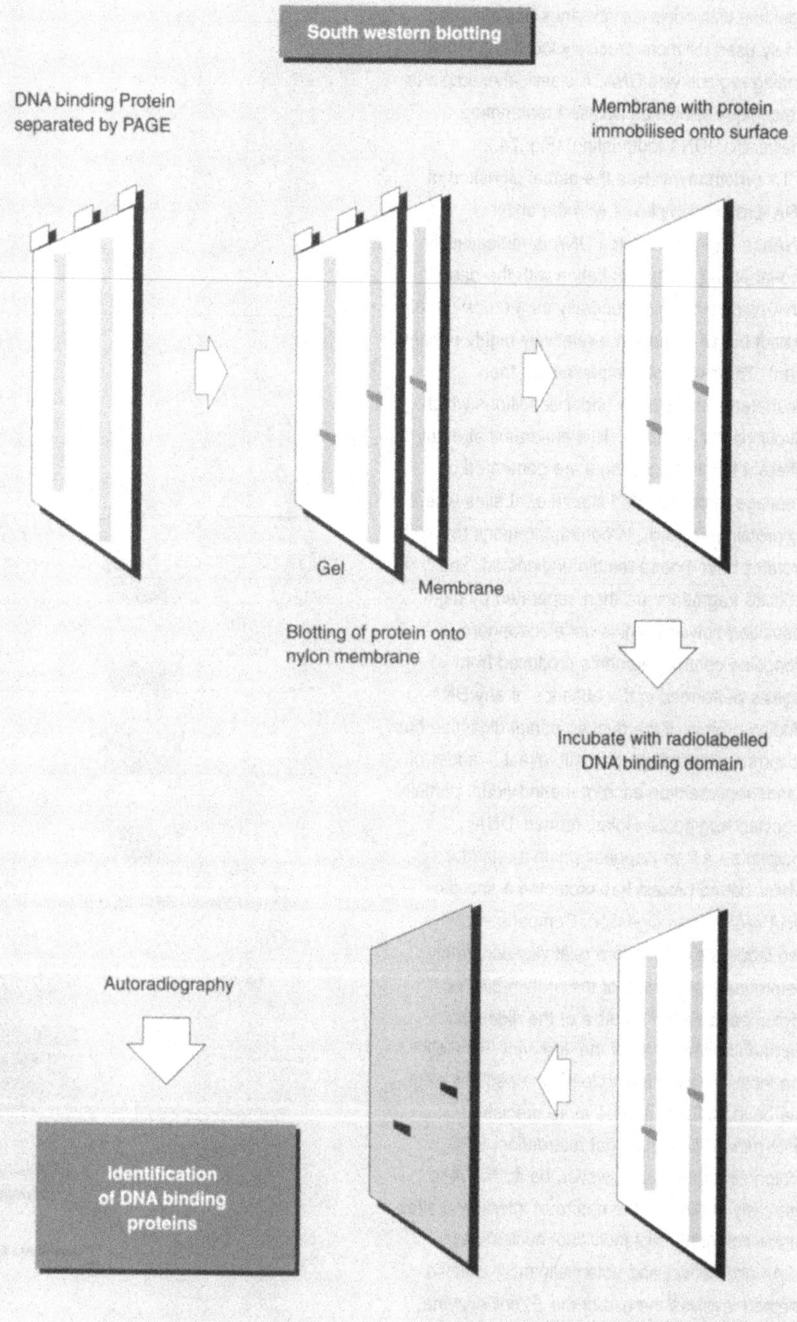

South western blotting

Fig. 74.3

OAs: 44–54, 372, 376, 380, 720–2, 731. RAs: 12–15, 194, 256–7. SFR: 27–8, 119–24, 126–8, 169, 213.

Filter binding assays

The ability of high concentrations of magnesium ions to prevent double-stranded DNA molecules from binding to nitrocellulose filters has also been exploited for the detection and isolation of DNA-binding proteins. Assays employing this strategy are often referred to as filter binding assays (Fig. 74.4). Accordingly, under high magnesium ion conditions, only protein or DNA–protein complexes will bind to nitrocellulose following filtration. This allows the kinetic characteristics (e.g. binding constants) and specificity of the interactions between DNA and proteins present in crude or purified preparations of protein or DNA fragments to be monitored or determined. Bound complexes can also be dissociated from the nitrocellulose filter, the complexes disrupted, and the individual proteins and DNA fragments separately analysed (e.g. by nucleotide or amino acid sequencing). Filter binding assays are therefore often used for the primary characterisation of DNA-binding proteins.

A significant drawback is however that only high affinity interactions can be detected by this method, since low affinity DNA-binding proteins present in crude nuclear extracts are unable to form sufficiently strong interactions to be maintained during the assay. The use of crude extracts also makes it difficult to directly determine the number of different proteins bound to each DNA fragment.

Surface plasmon resonance

A technique termed surface plasmon resonance (SPR) now also offers a rapid and sensitive approach to the detection of protein–DNA interactions (amongst other applications). SPR is based on the principle that changes in the thickness or shape of a monolayer of molecules coated onto a thin gold or silver surface (caused by either an interaction with, or an alteration in, the macromolecule conformation) causes the generation of an electromagnetic evanescent field (Fig. 74.5). Several manual and automated optical systems based on evanescent field technology are now available for detecting and monitoring the kinetics of DNA–protein interactions (amongst other macromolecular interactions). These may play a significant future role in applied molecular biology (e.g. clinical

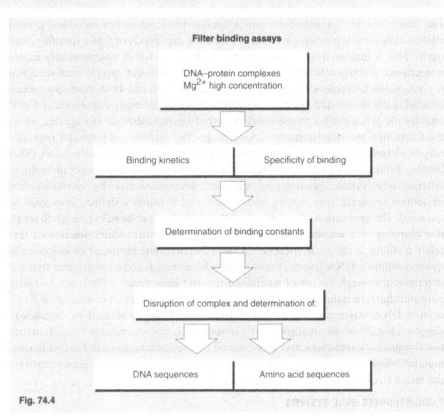

Fig. 74.4

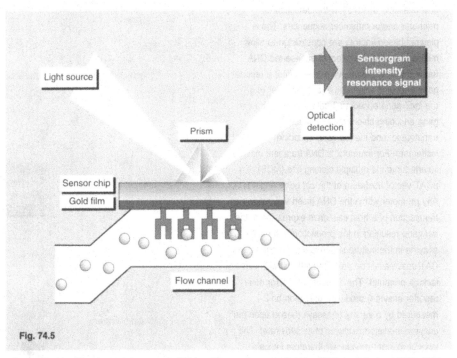

Fig. 74.5

diagnostics) and fundamental research. This rapid and sensitive technology is particularly finding application in basic research to analyse the key elements dictating DNA–protein (and DNA–DNA or DNA–RNA) interactions (e.g. between mutated proteins, and/or DNA targets).

201

75. DETECTING PROMOTER AND ENHANCER SEQUENCES

The existence, location, functional activity, and sequence of promoters or enhancers within a DNA fragment may be determined by deletion analysis and or their insertion into reporter gene vectors.

The ability of cells to regulate the expression (i.e. transcription) of individual genes is vital to their normal physiological function, and the expression of tissue-specific pheno-types. This is mediated by several genetic mechanisms which co-operatively control transcriptional activity through the presence and recognition of specific gene sequences (i.e. promoters and enhancers). The importance of promoter and enhancer sequences in regulating the *in vivo* and *in vitro* transcriptional activity of many eukaryotic and some prokaryotic genes has led to the development of several methods and approaches for their detection and characterisation. Accordingly, the existence of promoter sequences may be determined by many of the same assays commonly employed to detect DNA-binding proteins (e.g. DNA footprinting, gel retardation or Southwestern blotting). Alternatively, certain promoter or enhancer sequences may be identified from nucleotide sequence data (i.e. by comparison with previously defined consensus se-quences). The application of recombinant DNA techniques is, however, required for the fine mapping of transcriptional regulatory sequences and the determination of their relative ability to dictate transcriptional rates. Accordingly, promoter or enhancer se-quences within a DNA fragment may initially be detected, and their relative strengths determined, through the use of specialised reporter gene vectors. However, fine map-ping and characterisation generally involves the *in vitro* generation of a series of 5' or 3' deleted DNA fragments (often termed deletion mutants) generated by exonuclease digestion (i.e. by *in vitro* mutagenesis). Consequently, key nucleotides within transcrip-tional regulatory sequences may be identified by the sequencing of individual deletion mutants which do, or do not, modulate the expression of the reporter gene (relative to the intact DNA fragment) following reinsertion into the vector.

HARNESSING PROMOTER POWER

The characterisation of transcriptional promoter sequences in both eukaryotes and prokaryotes has provided data which are widely exploited in many areas of molecular biology. For example, RNA polymerase promoter sequences have been engineered into many different prokaryotic vector to facilitate transcription and gene expression. They have also been added to polymerase chain reaction primers to facilitate the *in vitro* translatic or RNA-based sequencing of PCR products. The ability to define and manipulate eukaryotic promoter (and enhancer) sequences is also vital to the development of expression vectors for transfecting eukaryotic cells, many of which may determine the ultimate success and efficacy of *in vivo* gene therapies.

OAs: 203–7, 295, 723–9, 777–84.
RAs: 82–88, 255, 257, 272, 278.
SFR: 27–8, 37–9, 72–5, 85, 119–24, 126–8, 165.

BASIC REPORTER GENE SYSTEMS

A number of so-called reporter gene vectors are now available which allow the detection of promoter and/or enhancer sequences. These plasmid-based vectors are constructed to allow measurement of the effect of an inserted DNA fragment upon the *in vivo* expression of a reporter gene encoding a specific enzyme marker (e.g. the *lacZ* gene encoding β-galactosidase, the *cat* gene encoding chloramphenicol acetyl transferase, and the *lux* gene encoding luciferase). For example, a DNA fragment may be inserted into the multiple cloning site (MCS) of a pCAT vector upstream of the *cat* gene (Fig. 75.1). Any promoter within the DNA insert will, following transfection of a host cell, drive expression of the *cat* gene resulting in the production of the CAT enzyme in the culture supernatant. In contrast, no CAT enzyme will be produced if the DNA insert lacks a promoter. The amount of CAT (or other reporter enzyme) produced may then be measured by a variety of assays (based upon the enzyme-mediated catalysis of its substrate). The level of reporter protein will therefore indicate both the presence of any promoter (or enhancer) sequence within the insert, and its relative efficiency in determining transcriptional activity.

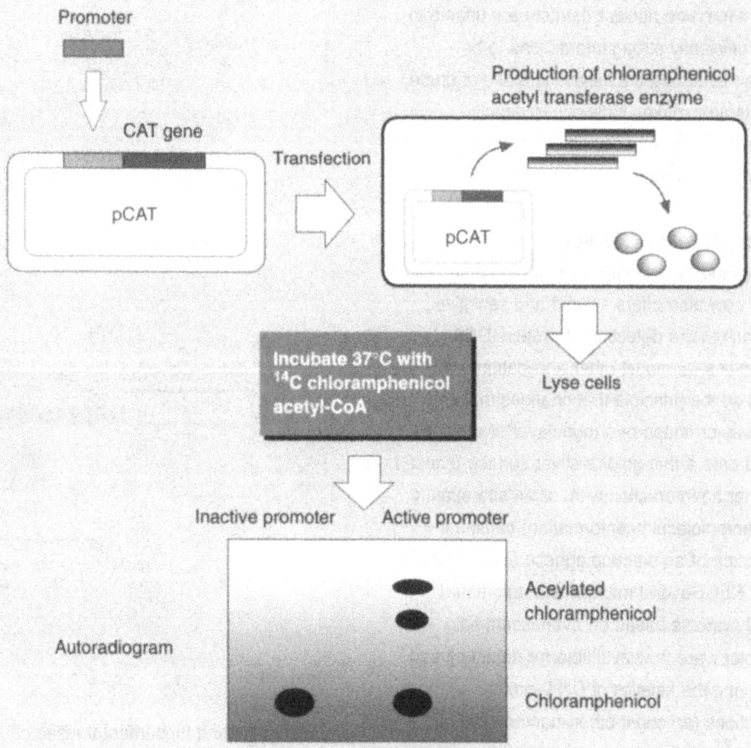

Fig. 75.1

Promoter and enhancer vectors

Additional vectors have also been made available which can detect promoter and enhancer sequences within DNA inserts, and simultaneously determine their functional activities. Promoter sequences may therefore be specifically detected using enhancer vectors which contain, for example, an SV40 promoter located downstream of the reporter gene (see Fig. 75.2). Insertion of a putative promoter-containing DNA insert upstream of the reporter gene therefore drives reporter gene transcription, whilst the vector-borne SV40 enhancer increases the rate of transcription to an easily detectable level. It should however be remembered that unlike enhancers (which function in either orientation, upstream and downstream), promoter sequences are only functional if inserted upstream of the structural gene, and in the correct 5′>3′ orientation. Promoters may therefore only be detected when the DNA is inserted in the correct orientation.

Promoter probe vectors have also been developed which contain, for example, an SV40 promoter upstream of the reporter gene. This allows its expression at a defined rate following transfection of a host eukaryotic cell. DNA inserted in any orientation, both upstream or downstream of the reporter gene, will then allow the detection of enhancer sequences based upon their effect upon normal transcription rates.

Detailed mapping of sequences

The precise location and nucleotide composition of promoter and enhancer sequences within DNA fragments can be determined in two principal ways. This most commonly involves limited exonuclease digestion of the DNA fragments to generate a series of DNA subfragments containing various 5′ or 3′ deletions (Fig. 75.3). These so-called deletion mutants may then be individually inserted into the reporter gene vectors, and their ability to subsequently initiate or modulate transcription can be evaluated. Similarly, a series of random or site-directed mutants may be generated. The sequences of mutants lacking promoter or enhancer activity can then be compared to those retaining such activity. Most commercially available reporter gene vectors consequently incorporate several

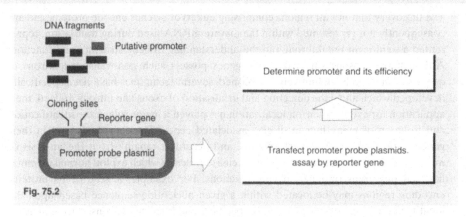

Fig. 75.2

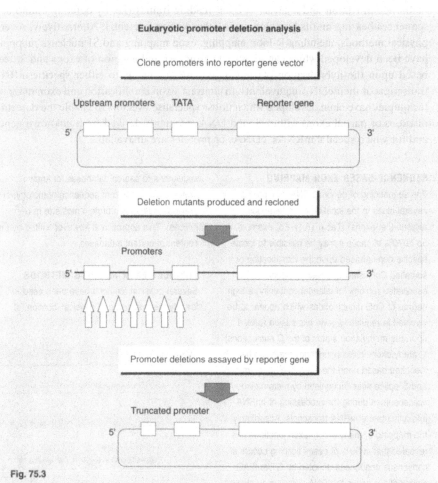

Fig. 75.3

additional features to faciliate the mapping and analysis of putative promoter/enhancer sequences. For example, most have the ability to produce single stranded DNA copies of inserts in order to facilitate dideoxynucleotide sequencing.

Most vectors also allow deletions or site-directed mutations to be made directly to the inserted DNA without the need for separate subcloning routines.

76. METHODS FOR IDENTIFYING PROTEIN-ENCODING REGIONS

The number and position of exons within genomic DNA may be determined by sequence analysis or several physical techniques based upon the formation of genomic DNA–mRNA heteroduplexes.

The discovery that not all regions comprising eukaryotic genes encode protein, and are consequently not represented within the mature mRNA used during translation, represented a significant breakthrough in the understanding of gene structure and function. Although the reasons why eukaryotic genes possess such non-translated intron sequences remain to be definitively established, several techniques have been specifically developed which allow the detection and localisation of exons and introns. Indeed, their application for exon and intron localisation has played a major role in the identification and isolation of many human disease-associated genes. These methods vary in their general approach, ease of performance, and accuracy, ranging from the analysis of nucleotide sequence data to the use of specialised vectors which exploit normal mammalian cell post-transcriptional splicing reactions. For example, a gene or its protein-encoding regions may be located within a given nucleotide sequence based upon the identification of upstream CpG islands, specific 5′ and 3′ splice site consensus sequences, or the identification and analysis of open reading frames (i.e. by defining amino acid sequences bearing motifs homologous to other known proteins). Alternatively, several physical methods, including R-loop mapping, exon mapping and S1 nuclease mapping, have been developed which can identify the number and location of exons and introns based upon the hybridisation of cloned genomic fragments to either specific mRNA transcripts or their cDNA equivalents. In contrast, exon amplification and exon mapping techniques have been developed which utilise specialised vectors to enable the detection of exons or partial exons within cloned DNA fragments derived from unknown genes, and for which specific mRNAs, cDNAs or proteins are unavailable.

THE SMALLEST GENE?

Short open reading frames (ORFs) are common to many genomic sequences but have hitherto largely been regarded as being of no biological importance. Analysis of microbial genomes has however shown that many short ORFs (especially those preceded by consensus ribosome-binding sites, RBS) may indicate the existence of very small genes encoding bio-active peptides. For example, the 22 codon *iad* gene of the bacterium *Enterococcus faecalis* is now known to encode the precursor of a competitive inhibitor of the sex pheromone cAD1. Similarly, the cytolytic toxin δ-lysin of *Staphylococcus aureus* is now known to be encoded by a 26 codon gene. Of the few small genes so far identified, the *E. coli* microcin C7 (*MccC7*) gene appears to be the smallest. This is composed of a 21 nucleotide ORF encoding a linear heptapeptide inhibitor of protein synthesis, preceded by an RBS located six nucleotides 5′ of the first codon.

OAs: 370–1, 380, 730–6, 788, 975–7. RAs: 29–32, 256–9, 272–3. SFR: 119–24, 126–8, 164–5, 223–5.

SEQUENCE-BASED EXON MAPPING

The sequencing of genomic DNA may provide several clues to the location of protein-coding regions (i.e. exons) (Table 76.1). For example, in up to 70% of cases it may be possible to locate specific genes based upon the identification of so-called CpG islands. These are gene-associated regions (or islands) containing a high degree of CpG dinucleotides which appear to be involved in regulating gene expression (based upon the methylation status of the C nucleotides).

Many exon–intron boundaries may also be identified based upon the existence of common 5′ and 3′ splice sites recognised by mammalian spliceosomes during the processing of hnRNA into fully mature mRNA transcripts. Accordingly, the mapping of many intron–exon junctions has revealed that in 90% of cases splicing occurs at consensus sequences denoted by GT and AG dinucleotides within the DNA. Exons may also be identified based upon the determination of open reading frames (ORFs). It is not however always easy to correlate ORFs with protein-encoding regions in the absence of additional corroborating information (e.g. the presence of upstream ribosome-binding sites). Nevertheless, it may be possible to use the deduced amino acid sequences to search databases for known proteins with significant sequence homology, or previously described protein motifs (e.g. α-helices). This approach is however limited by the content of current databases.

PHYSICAL EXON MAPPING METHODS

Several physical methods have been used to identify the number, and physical location, of exons and introns within genomic DNA. Many of these are based upon the formation of heteroduplexes by the hybridisation of genomic DNA fragments with mRNA or cDNA fragments

Table 76.1

Method/approach	Examples
Location of promoter consensus regions:	TATA, GC, CAT, enhancer
Identification of splice site elements:	GT–AG dinucleotides
Determination of CpG islands:	Undermethylation
Location of open reading frames (ORF):	Absence of stop codons
Homology to family related sequences:	Database searching

(i.e. in which the introns have been removed during *in vivo* splicing reactions). For example, R-loop mapping can be used to locate exons based upon the formation of single-stranded DNA loops within genomic DNA–mRNA heteroduplexes. Each loop therefore represents a region within the DNA which corresponds to an intron.

Alternatively, a restriction enzyme may be selected which cuts relatively frequently within a cloned genomic fragment, but which does not cut the corresponding cDNA (i.e. a restriction enzyme which only cuts within intronic sequences) (Fig. 76.1). Following digestion of the genomic DNA, the resulting fragments are size-separated by gel electrophoresis. Specific exons are then identified by Southern blotting using the labelled cDNA as a probe. The order of the exons within the gene can then be determined by sequence analysis. The accuracy of this method does however depend upon all introns being cleaved (i.e. some bands may contain two or more exons if the enzyme fails to cut within the separating introns).

Exons within cloned genomic fragments may also be precisely located by a technique termed S1 nuclease mapping. It also involves the formation of DNA–mRNA heteroduplexes but exploits the ability of mature mRNA transcripts to protect exonic sequences, but not intronic sequences, from degradation by S1 nuclease (Fig. 76.2). Indeed, it is similar to S1 nuclease protection assays used to locate the DNA transcriptional regulatory sequences (e.g. promoters and enhancers) recognised by DNA-binding proteins. S1 nuclease mapping, however, involves the hybridisation of the denatured cloned genomic DNA with the corresponding mRNA transcripts. This results in the formation of a single-stranded DNA loop corresponding to the unbound intronic sequences which is susceptible to degradation by S1 nuclease. Consequently, only the DNA sequences which correspond to the exons present in the mRNA remain intact following S1 nuclease digestion of the heteroduplexes. These exonic fragments can then be detected by gel electrophoresis and subsequently isolated for separate sequence analysis. The polymerase chain reaction (PCR) can also be used to detect exon–intron boundaries based upon comparing the PCR products generated from genomic and cDNA

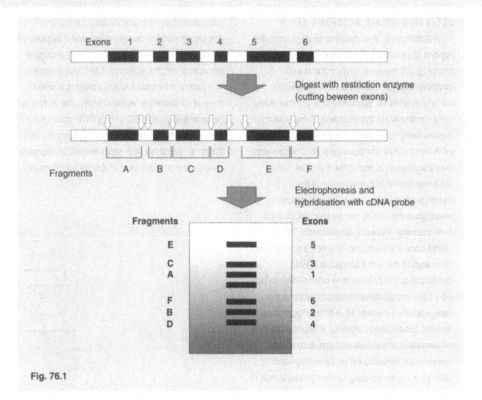

Fig. 76.1

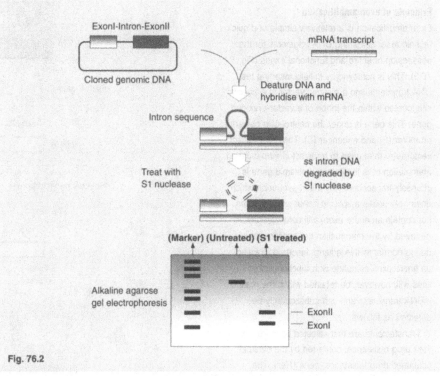

Fig. 76.2

templates (i.e. derived by reverse transcription of mRNA).

The detection of exons by the techniques of exon trapping and exon amplification is increasingly being used for the identification, isolation and analysis of previously unknown human disease genes.

DETECTING SPLICE ACCEPTOR SITES

The identification of transcribed protein encoding regions of genes by techniques such as R-loop mapping, S1 nuclease mapping and exon mapping assays requires access to specific mRNAs or cDNAs. Such methods are therefore not appropriate for the detection of transcribed regions within unknown genes for which specific mRNAs or cDNA are not available. Consequently, the identification of transcribed regions within unknown genes (e.g. the exons of the Huntington's disease gene) has increasingly relied upon the use of two techniques; termed exon trapping and exon amplification. These are based upon the detection of splice acceptor sites. They exploit the fact that mature mRNA transcripts are only formed in mammalian cells by the post-transcriptional removal of introns from heterogeneous nuclear RNA (hnRNA) precursors, and the subsequent religation of exons mediated by the action of the spliceosome. Exons may therefore be detected within DNA fragments based upon their splicing following transfection of mammalian cells.

Principle of exon amplification

Exon amplification is a relatively simple and quick technique for screening DNA fragments for the possession of entire and functional exons (Fig. 76.3). This is achieved by initially inserting test DNA fragments into a plasmid vector at a specific site located within the intron of a vector-encoded gene. This gene is under the control of a powerful promoter (P) and enhancer (E). The recombinant vectors are then used to transfect in vitro-cultured mammalian cells in which the plasmid gene is intensely transcribed. Any insert sequence which does not contain a splicing donor site (i.e. does not contain an entire exon) will consequently be removed by the mammalian cell spliceosomes during normal mRNA editing. Inserts containing an entire exon complete with functional donor sites will, however, be retained within the mature mRNA transcripts and can subsequently be detected as follows.

Transfectants are first selected based upon their drug resistance, conferred by the plasmid-contained drug resistance, gene (Res). The transcribed mRNAs are then isolated and reverse transcribed. The resultant cDNAs are then subjected to in vitro amplification by the PCR employing primers designed to anneal to plasmid-borne exon sequences (i.e. flanking the original intron into which the foreign DNA was inserted). Any inserted fragment which contains a whole exon (and therefore retained within the cDNA) will be amplified by the PCR. This PCR product may then be analysed by nucleotide sequencing and the data compared with other known DNA/protein sequences, or used as a probe for chromosome

walking/jumping. Alternatively, the PCR product may be subjected to in vivo or in vitro translation in order to generate a protein/peptide product for epitope mapping or as an immunogen for producing antibody-based probes.

OAs: 609, 737–8, 788, 891, 893–4. **RAs:** 29–32, 256–9, 272–3. **SFR:** 119–24, 126–8, 164–5, 223–5.

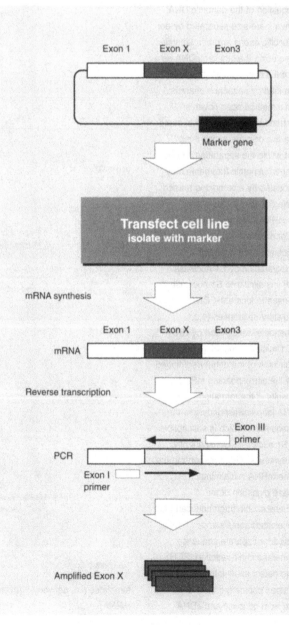

Fig. 76.3

The principle of exon trapping

Exon trapping is a general method for screening DNA sequences for splice acceptor sites based upon the use of a specialised exon trap vector. Although much more complicated and time consuming than exon amplification, exon trapping is the more reliable of the two methods. Accordingly, it can detect any DNA fragment containing an acceptor site and does not require insertion of entire exons.

The exon trap vector involved is a modified retrovirus-derived shuttle vector which is capable of replicating in mammalian cells in its RNA form, and in *E. coli* cells in its DNA form. This is possible because it possesses both an SV40 origin of replication and bacterial *ori* site. It also contains two drug resistance genes (*Res* I and *Res* II) which act as selectable markers, and long terminal repeats (LTRs) which allow it to function as a retrovirus (Fig. 76.4). Of central importance is the fact that the vector contains a downstream splice donor site, SD (but not a splice acceptor site, SA). It also contains an upstream multiple cloning site (MCS) polylinker sequence flanking a bacterial β-galactosidase gene. This gene allows transfected cells to be stained blue in the presence of the substrate Xgal (i.e. represents another selectable marker system).

Transcribed regions within DNA fragments can therefore be detected in the following way. The vectors are first grown in their DNA form by replication in bacterial cells and the foreign DNA test sequences are randomly inserted into the MCS. The recombinants produced are then converted into their retroviral form and the retroviruses used to transfect *in vitro*-cultured mammalian cells. Transfectants are then selected based upon their resistance to an appropriate drug (conferred by the *Res* II gene) and the vectors recovered. In instances where the inserted DNA contains a splice acceptor site (i.e. at the end of an exon), the mammalian cell spliceosomes will remove any of the retroviral RNA lying between the vector's SD site and the inserted fragment's SA site (i.e. the β-galactosidase sequence will be removed). The recovered *Res* II vectors are then returned to β-galactosidase-deficient *E. coli* cells grown on *Res* I-selectable medium. Any colony derived from a vector harbouring an insert with a SA site will

consequently be unable to express any β-galactosidase and will remain white in the presence of Xgal. The vectoral DNA of such β-galactosidase-lacking clones may then be

isolated and the inserts analysed (i.e. by nucleotide sequencing), and/or used as probes in subsequent cloning and screening procedures.

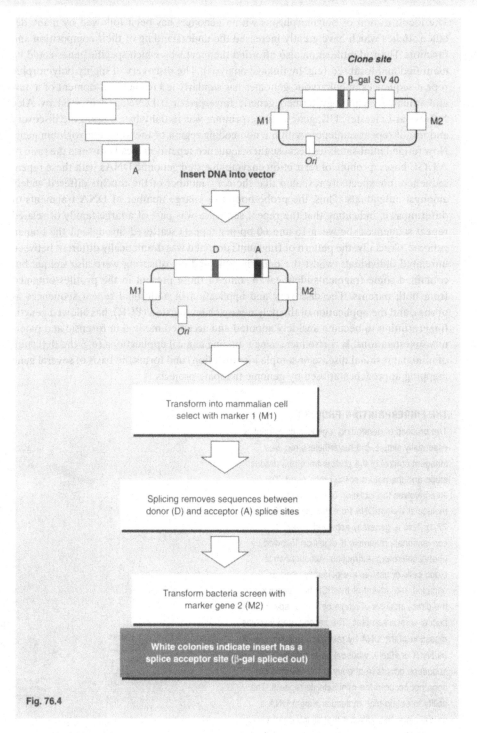

Clone site

Insert DNA into vector

Transform into mammalian cell select with marker 1 (M1)

Splicing removes sequences between donor (D) and acceptor (A) splice sites

Transform bacteria screen with marker gene 2 (M2)

White colonies indicate insert has a splice acceptor site (β-gal spliced out)

Fig. 76.4

77. GENETIC FINGERPRINTING

Genetic fingerprinting is based upon the detection of individual-specific patterns of differently sized DNA restriction fragments containing different copy numbers of repeat sequences termed mini-satellites.

The identification of polymorphisms within genomes has been followed by many detailed studies which have greatly increased the understanding of their composition and function. Their identification also afforded the means by which specific genes could be identified and localised (e.g. by linkage analysis). The discovery of highly polymorphic repeat sequences in eukaryotic genomes has similarly led to the development of a new and valuable technology termed genetic fingerprinting. Developed in 1984 by Alec Jeffreys at Leicester, UK, genetic fingerprinting was initially based upon the discovery and use of repeat sequences within a non-coding region of the human myoglobin gene. Now termed mini-satellites because their sequence repetition locally distorts the ratio of AT/GC bases, probing of restriction enzyme-digested genomic DNAs with these repeat sequences unexpectedly revealed that the copy number of the repeats differed widely amongst individuals. Thus, the probe bound to a large number of DNA fragments of differing size, indicating that the repeat sequence was part of a large family of related repeat sequences (between 16 and 60 bp per repeat) scattered throughout the human genome. Crucially, the pattern of fragments detected was dramatically different between unrelated individuals, whilst the patterns obtained for offspring were also unique but contained some fragments identical to some of those present in the profiles obtained from both parents. The discovery and application of additional repeat sequences as probes, and the application of the polymerase chain reaction (PCR), has allowed genetic fingerprinting to become a widely adopted and accepted method in forensic and paternity investigations. It is also increasingly finding clinical applications (e.g. the detection of minimal residual disease or sample identification) and forms the basis of several gene mapping approaches utilised by genome mapping projects.

LEGAL ACCEPTANCE

Initially, acceptance of genetic fingerprinting in criminal and civil law was problematic. This was partly due to concerns over its accuracy, as the resulting from the submission of poorly controlled DNA fingerprinting data produced by a few biotechnology companies which at the time were inexperienced in dealing with the technically demanding nature of the technique. Tightening of procedural controls has however led to the widespread acceptance of genetic fingerprinting data as forensic evidence. Although the probability that two individuals may contain the same genetic fingerprint is very small, genetic fingerprinting is however predominantly used to eliminate suspects from an investigation, or to identify a victim from their remains, rather than to secure the conviction of a felon.

OAs: 78–88, 95, 739–56, 1145–53.
RAs: 260–3, 282, 307–8, 423–4.
SFR: 119–24, 126–8, 136–9, 166–8, 179–81, 225.

THE FINGERPRINTING PROCESS

The process of generating a genetic fingerprint is essentially simple, but nevertheless requires stringent control of the performance of individual steps and the nature of the probe used. The first step involves the isolation of high purity, high molecular weight DNA from the sample (Fig. 77.1). This is generally achieved using conventional proteinase K digestion followed by phenol/chloroform extraction. Although white blood cells or tissues are generally used as samples, the advent of the PCR has also allowed the direct analysis of single cells (e.g. sperm) or buccal scrape samples. The second step involves digestion of the DNA by restriction enzymes, such as *Hinf*I or *Hae*III, whose 4 bp recognition sequence occurs frequently in the genome but does not occur in the mini-satellite repeats. The ability to obtain high molecular weight DNA is crucial since this ensures that all informatively large DNA fragments will be represented following digestion.

The population of fragments produced, ranging from 1 kb to greater than 20 kb, are then size-separated by agarose gel electrophoresis. This cannot, however, resolve between alleles varying

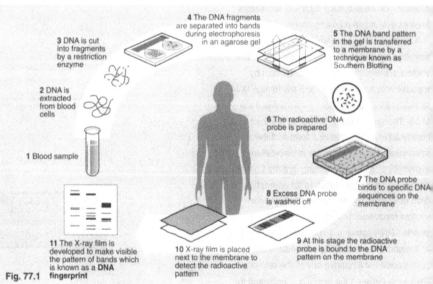

Fig. 77.1

1 Blood sample

2 DNA is extracted from blood cells

3 DNA is cut into fragments by a restriction enzyme

4 The DNA fragments are separated into bands during electrophoresis in an agarose gel

5 The DNA band pattern in the gel is transferred to a membrane by a technique known as Southern Blotting

6 The radioactive DNA probe is prepared

7 The DNA probe binds to specific DNA sequences on the membrane

8 Excess DNA probe is washed off

9 At this stage the radioactive probe is bound to the DNA pattern on the membrane

10 X-ray film is placed next to the membrane to detect the radioactive pattern

11 The X-ray film is developed to make visible the pattern of bands which is known as a **DNA fingerprint**

by a single repeat unit, which complicates subsequent interpretation. The electrophoresed bands are then transferred to a nylon membrane and sequentially incubated with several types of radiolabelled or non-isotopically (e.g. chemiluminescent) labelled single- (SLP) or multi-

locus (MLP) probes. Following autoradiography, the positions of the mini-satellite fragments can be determined within the final DNA profile or fingerprint by image analysis equipment (thereby eliminating human subjectivity).

INFORMATIVENESS OF FRAGMENTS

The genetic fingerprinting process relies upon the identification of the different lengths of DNA restriction fragments resulting from the presence of different numbers of repeat sequences. The most informative DNA fragments are those that represent regions or loci that have evolved the widest diversity of repeat sequence copy number. Consequently, a large pool of alleles are available which may be used to distinguish between individuals. Such fragments are often up to 20 kb in length, each containing 1000 or more repeats.

Single- and multi-locus probes

The first mini-satellite sequences used in genetic fingerprinting were termed multi-locus probes (MLPs) because the repeat sequences they detected are scattered throughout the genome at multiple locations. Latterly, several different single locus probes (SLPs) have been developed which are informative under appropriate conditions despite only being able to detect alleles at a single location. A range of highly repetitive DNA sequences (e.g. VNTRs or variable number of tandem repeats, and STRs or short tandem repeats) have now also been identified and are actively used in genetic fingerprinting both as SLPs and MLPs.

Genetic or DNA fingerprints can however, only be produced using MLPs, whilst the two-band pattern obtained using SLPs is referred to as a DNA profile (Fig. 77.2). DNA profiling using SLPs is insufficient to definitively establish the identity of an individual and is therefore not appropriate for use in parentage determinations. DNA profiling does however play a substantial role in forensic work, being involved in up to 80% of the casework. This is because DNA profiling requires as little as 20 ng of DNA (as much as can be obtained from a single hair root) rather than the 1 μg of DNA required for MLP-based testing.

The use of PCR in fingerprinting

Despite the large DNA fragments required, recent developments have allowed the coupling of PCR amplification to the detection of mini-satellite loci. Consequently, the discovery of polymorphisms within the repeating sequences of mini-satellites has led to the development of a PCR-based method known as mini-satellite variant repeat (MVR) analysis. This technique distinguishes an individual on the basis of the random distribution of repeat types along the length of the two alleles. This leads to a simple numerical coding of the detected repeat variation (and is hence also known as digital typing). This combines the advantages of PCR sensitivity and rapidity with the discriminating power of mini-satellite alleles.

Diversity of applications

DNA fingerprinting and profiling technologies have primarily been applied to human forensic and paternity investigations (e.g. to settle immigration issues). Since mini-satellites have been found in almost all eukaryotes, they have also proved valuable tools for the settling of paternity issues involving several animal and plant species. For example, they have been used to settle a paternity dispute in apples and to distinguish between exotic species of birds (e.g. parrots) bred legally in captivity and those illegally imported. Genetic fingerprinting is also increasingly being applied in biomedical reseach and therapy. For example, it has been used to monitor the remission of bone marrow transplants by specifically detecting patient-derived cancerous cells amongst normal donor cells (i.e. to detect minimal residual disease). It may also be used to definitively establish the identity or derivation of cell lines in both a clinical and research context.

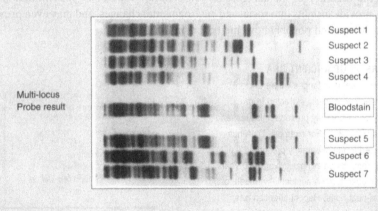

Multi-locus Probe result

Suspect 1
Suspect 2
Suspect 3
Suspect 4

Bloodstain

Suspect 5
Suspect 6
Suspect 7

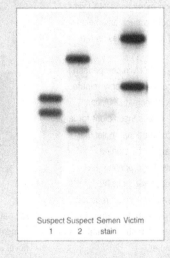

Single-locus Probe result

Suspect 1   Suspect 2   Semen stain   Victim

Fig. 77.2

78. ANALYSING ANCIENT DNA

The ability to isolate and analyse ancient/preserved DNA samples using the polymerase chain reaction is providing direct genetic evidence regarding recent historical, prehistoric, and evolutionary events.

Information derived from analysing genome structures and functions provides many potential benefits, especially in terms of understanding disease pathologies and developing diagnostic and/or therapeutic strategies. The analysis of contemporary nuclear and mitochondrial genomes is also providing several clues regarding the origins and evolution of many organisms. These studies are based upon calculating the rate at which mutations accumulate over time and using them as a form of molecular clock. Human prehistoric migrations have therefore been traced by the analysis of highly polymorphic alleles (e.g. some HLA genes of the major histocompatability complex of humans involved in immune system regulation), and their distribution in different geographic populations. Furthermore, analysis of polymorphisms within mitochondrial DNAs has been used to construct evolutionary trees which, for example, have identified the origins of a single ancestral human female (termed 'mitochondrial Eve'). Thanks to the remarkable stability of DNA, biologists, anthropologists and archaeologists are capable of using the analysis of ancient DNA samples as a more direct means of reaching back into time. Accordingly, the sensitivity of the polymerase chain reaction (PCR) has been exploited to *in vitro* amplify ancient DNA fragments obtained from many different organisms and sample types stretching back thousands or even millions of years into prehistory. Subsequent cloning, nucleotide sequencing and genetic fingerprint analysis are providing data which illuminate many aspects of human and world history. These data are helping to establish the accuracy of human lineages (e.g. relationships of Egyptian mummified remains), provide insights into social and environmental changes, and may even provide clues as to when, and possibly how, life began on our planet.

MUTATIONS MEASURE EVOLUTION?

It is now well accepted that mutations within the same gene in different species may be analysed to determine the degree to which the species are related on an evolutionary tree. Less well accepted is the suggestion by Allan Wilson at the University of California at Berkeley, USA, that the study of polymorphisms arising within a single species (e.g. within the mitochondrial DNA of humans) may be used to construct a molecular clock against which to measure evolution. Wilson calculated that mutations within mitochondrial DNA accumulate at the rate of 2–4 mutations (2–4% divergence) per million years. Analysing several modern human mitochondrial genes, he obtained a figure which on the molecular clock suggests that we evolved from a single common female African ancestor between 140 000 and 290 000 years ago. Since *Homo erectus* existed well before 400 000 years ago, this 'mitochondrial Eve' is unlikely to have been the only woman in existence at the time, yet only she carried the DNA from which modern DNAs are descended.

THE DIVERSITY OF ANCIENT DNA

There are now many instances where ancient DNA has been successfully extracted from accidentally or deliberately preserved tissues, and even bones (Fig. 78.1). For example, DNA has been extracted from deliberately preserved (i.e. mummified) remains dating back over 9000 years, and similarly aged tissues accidentally frozen, or preserved in peat bogs. Recent attention has also been focused upon extracting DNA from much more ancient sources such as insects accidentally fossilised within tree resins (i.e. entombed in amber).

Despite the stability of DNA in comparison to other biological macromolecules, only minute quantities of fragmented ancient DNA can however be obtained. This is insufficient for large-scale analysis using conventional molecular approaches. Their analysis, by cloning, sequencing and genetic fingerprinting, has consequently only been made possible by the application of the polymerase chain reaction which can accurately replicate (i.e. amplify) specific DNA fragments from such limited and degraded starting samples.

The purpose of cloning and analysing ancient DNA samples was not however the ability to recreate ancient or prehistoric organisms (evocatively expounded in Michael Crichton's novel *Jurassic Park*). These have instead been used to provide molecular evidence for historical and/or evolutionary lineages.

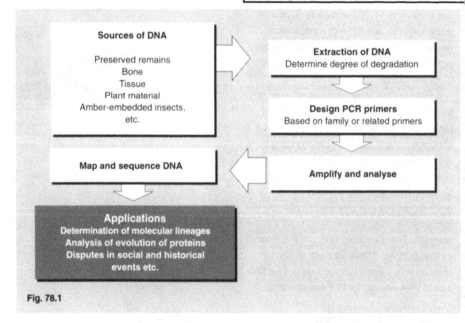

Fig. 78.1

OAs: 739–46, 756–70.
RAs: 264–71.
SFR: 36, 119–24, 136–9, 166–8, 222.

Tracing human history

The accessibility of ancient DNA to mutation analysis has been extremely useful in validating hypotheses regarding human evolution. The analysis of highly polymorphic HLA genes within ancient samples has been particularly important in directly validating calculations regarding our first female ancestor (i.e. mitochondrial Eve). These calculations are based upon measuring the occurrence and rate of accumulation of mutations in modern mitochondrial DNA samples. Such DNA analyses also have the potential to accurately place individuals on specific family trees. For example, genetic fingerprinting using highly polymorphic extragenic mini- or micro-satellite repeat sequences may be used to accurately determine the familial relationships between mummified Egyptian remains. Similarly, these approaches have also been used to identify more recent skeletal remains such as those of the last Tsar, his familiy and entourage (Fig. 78.2). They have also been used to identify other victims of political/'ethnic cleansing' programmes of mass murder. DNA-based investigations have also been used to trace human population migrations during prehistory. This potentially provides evidence which may be important in determining human evolutionary and social origins. Similarly, the analysis of non-human DNAs has also become an important tool. Accordingly, ancient plant DNA may be used to trace the original cultivation and dispersal of particular crop plants, thereby providing insights into environmental and social factors influencing key points in human history and prehistory.

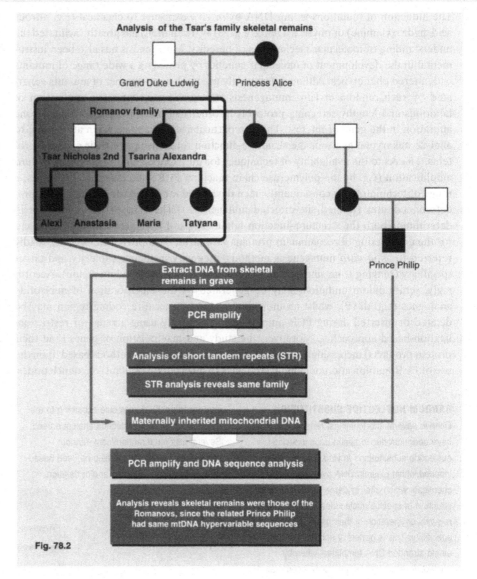

Analysis of the Tsar's family skeletal remains

Fig. 78.2

Tracing gene evolution

Combined with our understanding of the mechanisms by which mutations arise and are stably incorporated, the comparison of mutations within genes from different organisms has long been used to provide glimpses of the evolution of life on our planet. Many attempts have been made to construct phylogenetic trees based upon positioning different species on a branch according to the assumption that the amount of mutational differences indicates when they diverged during evolution (i.e. any mutational differences occurred after they diverged from a common ancestor). For the first time in our history, ancient DNAs including those extracted from insects embedded in amber, or microbes frozen during prehistory (e.g. recovered from ice cores taken in the frozen deserts of the Arctic or Antarctic) offer the potential means of validating, dismissing or modifying evolutionary theories. It has therefore been proposed that these studies may eventually define the origins of all life on our planet, for example, how modern genomes, with their great redundancy in terms of apparently unused/non-functional intronic or extragenic DNA, arose from a simple single cell organism. An intriguing possibility is that these investigations may identify ancient, but abandoned, genetic mechanisms or molecular precursors which may be studied, or even resurrected for modern exploitation. Perhaps the most forthcoming and directly relevant information to be derived will indicate when specific genes arose and when they functionally diverged. It is also very likely that we may be able to find out when some disease-generating genetic defects arose, and why they have been perpetuated within the population despite the disadvantage they currently confer (or whether in fact they conferred an advantage at other times in prehistory).

79. *IN VITRO* MUTAGENESIS METHODS

Random/semi-random *in vitro* mutagenesis can be achieved by several methods including the incorporation of nucleotide analogues, and enzyme-mediated restriction/ligation-based methods.

The induction of mutations within DNA by *in vivo* exposure to chemical (e.g. nitrous acid, hydroxylamine) or physical agents (e.g. X-ray irradiation) has greatly facilitated the understanding of molecular mechanisms of heredity. Mutagenesis has also been instrumental in the development of molecular genetics by providing a wide range of mutants with altered phenotypes. Although frequently used, the large number of mutants generated by such random *in vitro* mutagenesis strategies necessitates the application of laborious and lengthy screening protocols in order to isolate a mutant with only one alteration in the gene of interest. This is a particular disadvantage when attempting to analyse eukaryotic genes or the structure–function relationships of their encoded proteins. Thanks to the availability of techniques for the rapid cloning, sequencing, *in vitro* amplification (e.g. by the polymerase chain reaction, PCR) and expression of genes, a range of techniques has consequently been developed for introducing mutations at pre-determined sites (termed site-directed mutagenesis). These have been important in determining both the structure–function relationships of both genes and proteins, and for the engineering of recombinant proteins with novel or improved functions. Broadly referred to as *in vitro* mutagenesis methods, they vary in their complexity and target specificity (ranging from single or multiple codons to an entire coding region). Accordingly, semi-random mutations may be generated by the incorporation of nucleotide analogues (e.g. dITP), whilst an individual codon or an entire coding region may be deleted or inserted during PCR amplification or cloning using a range of restriction/ligation-based approaches. Moreover, the study and manipulation of genes (and their protein products) increasingly utilises site-directed mutagenesis methods based upon the use of PCR amplification or cloning strategies to incorporate 'mutant' oligonucleotides.

BIOENGINEERING A CLEANER WASH

The bacterial enzyme subtilisin is a broad specificity protease of considerable industrial importance, approximately 600 tonnes of which are used by the cleaning industry annually. Its widespread use in many brands of soap and washing powders is a direct consequence of the application of protein engineering approaches. These have allowed the development of recombinant forms of subtilisin which can be expressed in bulk *in vitro* and possess improved properties which enhance their cleaning performance. Accordingly, site-directed mutagenesis has been used to alter many of its enzymatic properties, including its specificity, pH profile and optimal working conditions (e.g. wash temperature and resistance to bleaching agents). Thus, thermostability has been increased by altering its sequence to resemble that of similar enzymes found in thermophilic (i.e. heat-tolerant) organisms, and resistance to oxidation improved by replacing methionine side chains.

RANDOM NUCLEOTIDE SUBSTITUTION

Despite several shortcomings, random mutations have been introduced based upon inducing nucleotide substitutions *in vitro*. These may be induced within plasmid DNA by exposure to chemicals which alter or damage the chemical structure of specific nucleotides, leading to their removal or alteration of their base-pairing specificity. This is generally achieved using single-stranded DNA templates, whereby nucleotide substitutions are stably incorporated during second strand synthesis using DNA polymerases which lack 3′ to 5′ proofreading activity (e.g. by *Taq* polymerase during PCR amplification). Similarly, random nucleotide susbstitutions can be induced via the enzymatic incorporation of nucleotide analogues, or enzymatic mis-incorporation using suboptimal reaction conditions (e.g. non-ideal buffer conditions or a deficiency of a specific nucleotide precursor). For example, nucleotide analogues such as deoxyinosine triphosphate (dITP) or 5-bromouracil (5bU) may be added to the PCR reaction mixture and subsequently incorporated during primer extension. Consequently, one, or more typically, several base substitutions are introduced per DNA molecule according to the amount and type of nucleotide analogue used. Several additional random/semi-random mutagenesis methods can also be used based upon substitution, insertion and/or deletion.

OAs: 109–11, 203–7, 465, 472, 613, 771–7, 1048–58. RAs: 272–6, 381–5, 401–2, 405. SFR: 58–60, 119–24, 126–8, 136–9, 169, 213.

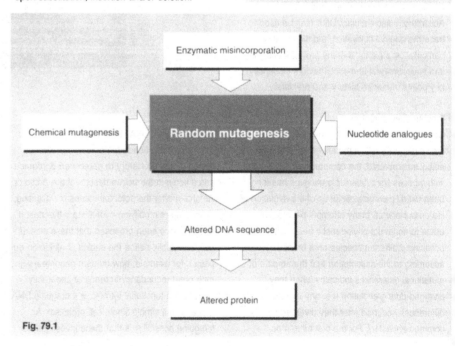

Fig. 79.1

4. Nucleic acid building blocks.
16. Types of mutation and their effects.
17. Forms of chemically altered DNA.
18. DNA repair mechanisms.
37. *In vivo* DNA replication.
42. Enzymes in molecular biology.
43. Restriction endonucleases.
54. Nucleotide sequencing methods.
55. Polymerase chain reaction.
59. Chemical synthesis of oligos.
60. Nucleotide analogues.
64. Inserting DNA into vectors.
66. Plasmid vector development.
71. Genomic/cDNA cloning strategies.
73. Gene mapping techniques.
74. Detecting DNA–protein interactions.
75. Detecting promoters and enhancers.
96. Protein engineering.

RESTRICTION/LIGATION METHODS

Restriction/ligation methods are based upon the modification of sequences in a DNA clone at the point where a restriction enzyme cleaves the DNA (see Fig. 79.2). Depending upon the desired location of the mutation (i.e. in a coding or non-coding region), such strategies can be used to investigate gene and/or protein structures and functions. Accordingly, cutting of a plasmid insert with a single restriction enzyme linearises the plasmid and creates two ends with 5′ (or 3′) overhangs. These can then be made blunt-ended by either removal of the overhangs (e.g. using S1 nuclease), or by the addition of bases using a DNA polymerase-mediated in-fill reaction. Religation of the blunt-ended plasmid therefore leads to a specific deletion or insertion of bases respectively (with the resultant loss of the original recognition site).

A variation of this technique, often termed linker insertion, utilises an oligonucleotide 'linker' to insert DNA sequences into the linearised DNA clone. This may be used to disrupt gene sequences, insert amino acids into a coding region, alter the reading frame, or insert a different restriction enzyme site to allow subsequent cutting of the linker (e.g. for mapping or the subsequent introduction of further mutations). A further variation on the method utilises exonucleases (e.g. *Bal*31 or exonuclease III) to remove one or both ends of the linearised plasmid prior to religation and recloning. By varying the digestion time and enzymes used, a series of 'nested' deletion mutants may be generated which vary in the amount of DNA removed from one or both ends of the linear fragment. Although such strategies have been useful in determining functional gene sequences, they are limited by the availability of appropriate restriction enzyme recognition sites.

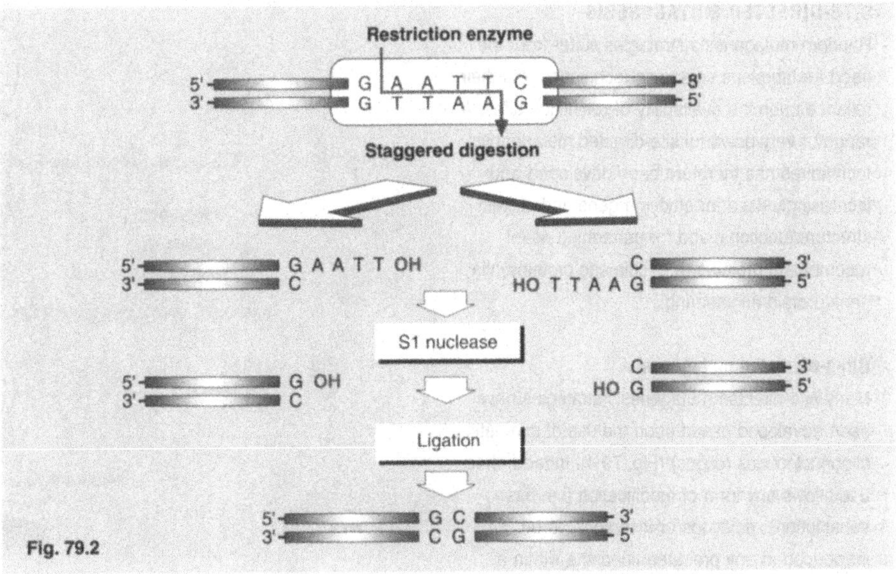

Fig. 79.2

Fig. 79.3

LINKER SCANNING MUTAGENESIS

Linker scanning mutagenesis is an alternative method which combines the use of linkers and restriction/ligation to allow the creation of deletions and base substitutions within defined areas of a cloned gene (Fig. 79.3). This is based upon the generation of two separate sets of unidirectional 3′ and 5′ deletion mutants using *Bal*31 exonuclease digestion. These mutants are each flanked by a synthetic 'linker' oligonucleotide sequence bearing the same restriction enzyme recognition sequence which consequently allows the religation of a 3′ deletion mutant to a 5′ deletion mutant. This generates a library of 'linker-scanning' mutants which contain limited deletions or nucleotide substitutions clustered at the linker site. DNA sequencing and functional analysis of the individual mutants (e.g. in reporter gene assays) can then be used to identify a particular base(s) important in determining a particular gene structure and/or function. This technique has therefore been widely used for the mapping and definition of transcriptional regulatory sequences (e.g. promoters and enhancers).

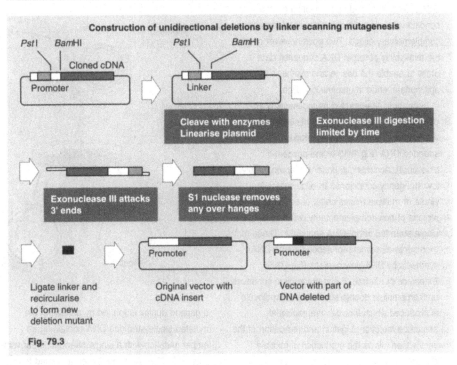

A range of strategies has been developed to enable site-directed introduction of substitutions, and insertions/deletions of varying length based upon the annealing of short or extremely large oligonucleotides.

SITE-DIRECTED MUTAGENESIS

Random mutagenesis strategies suffer from the need for laborious screening techniques and their reliance upon the availability of restriction sites. A range of very powerful site-directed mutagenesis techniques has therefore been developed and increasingly used for studying gene and protein structure/functions, and for generating novel recombinant proteins (e.g. chimeric proteins) via gene/protein engineering.

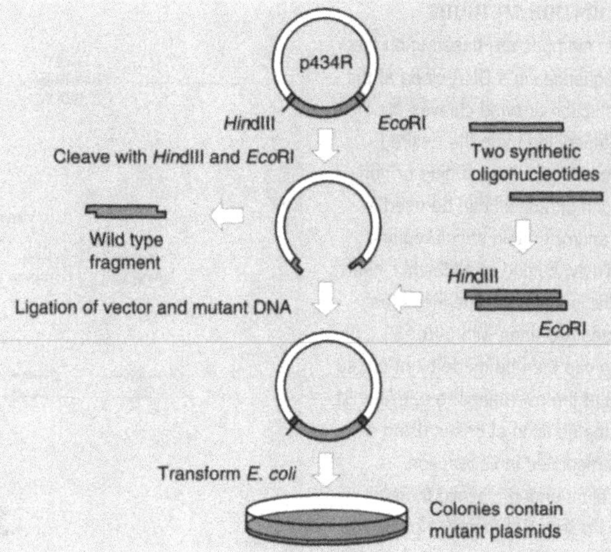

Oligo-mediated mutagenesis

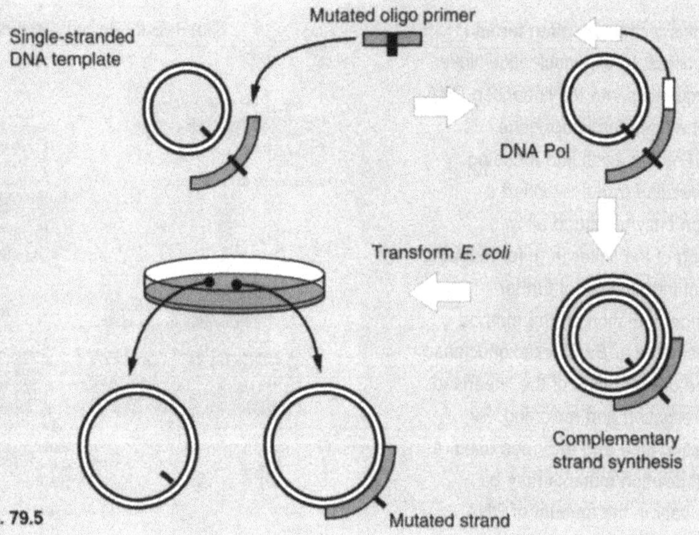

Many site-directed mutagenesis strategies have been developed based upon the use of synthetic oligonucleotides (oligos) (Fig. 79.4). Indeed, their use allows any form of modification (i.e. base substitutions, deletions or insertions) to be introduced at any predetermined site within a DNA molecule, based upon tailoring *in vitro* conditions to allow the hybridisation of imperfectly complementary oligos. This does however require the availability of target DNA sequence data in order to enable the design and synthesis of an appropriate, short 'mutagenising' oligo.

In general, site-directed mutagenesis is also performed on cloned DNA fragments contained within vectors which allow the production of single stranded DNA (e.g. M13 or the phagemid bluescript). Accordingly, under the appropriately low stringency conditions, an oligo containing single or multiple mismatches, or even extensive regions of non-complementarity, will anneal to the single-stranded target DNA sequence. These may then serve as primers for second strand DNA synthesis by DNA polymerases (Fig. 79.5). Extension of the free 3′ end of the oligo continues until a complete double-stranded DNA molecule is produced which includes the additional sequence mutation. Ligation and replication of the vector then allows the production of multiple copies, half of which contain the mutation, and half of which contain the original 'wild-type' sequence. Mutant clones can then be identified by high stringency hybridisation screening using the mutant oligo as the probe.

It is however also possible to enrich for mutant clones by inhibiting the growth of wild-type clones. For example, the gapped duplex approach employs an amber mutation which only allows growth in amber-suppressing hosts. Thus,

Fig. 79.4

Fig. 79.5

a gapped duplex is formed by annealing the mutated single-stranded DNA containing an amber mutation with a single-stranded linear wild-type (RF) DNA. An oligo is then annealed to the single-stranded region of the gapped duplex. DNA polymerase and DNA ligase then convert the heteroduplex to a double-stranded form which allows it to be used to transform an *E. coli* host strain lacking an amber supressor. This ensures that only mutant progeny are recovered, although the efficiency varies according to the target sequence. It does however mean that the mutant sequence will require recloning into another

vector before further rounds of mutation can be performed. A similar approach is based upon the use of host strains which lack or contain the enzyme uracil *N*-glycosylase, such that non-uracil-containing wild-type sequences are destroyed.

OAs: 520–6, 778–95, 828, 831–6, 880–1, 1099–1103. **RAs:** 272–6, 381–5, 401–2, 405. **SFR:** 58–60, 119–24, 126–8, 136–9, 169, 213.

The use of large oligos

A range of *in vitro* mutagenesis methods have now also been developed which use extremely large oligos for the introduction of mutations. For example, large double-stranded oligo cassettes may be chemically synthesised, generated by restriction digestion of cloned inserts, or by PCR amplification. A segment of vector insert may therefore be removed by digestion with a restriction enzyme cutting at two points, and the mutant oligo cassettes ligated in their place. Consequently, a range of mutants may be rapidly generated which contain single or multiple substitutions, or insertions/deletions spanning a few or many nucleotides.

A similar approach is employed in a method termed 'sticky feet-directed' mutagenesis, which allows large PCR-generated sequences to be inserted at specific sites (Fig. 79.6). This technique has become widely employed in gene and protein engineering as a simple and reliable means of creating chimeric genes/proteins (particularly for domain switching during antibody engineering). It essentially relies upon the ability of the two ends of a large single-stranded PCR product to anneal to complementary sites on an insert within a single-stranded DNA vector. Once annealed, the PCR product can be extended, ligated, and used to transform a suitable host strain. Mutants can then be selected based upon colony screening or the use of *E. coli* strains which allow discrimination (e.g. based upon the presence of uracil *N*-glycosylase).

The PCR mega-primer method

The PCR 'mega-primer' method is also a simple and versatile site-directed *in vitro* mutagenesis method (Fig. 79.7). This utilises three primers and two rounds of PCR amplification to generate large mutated DNA fragments from a cloned DNA template. An initial product is therefore generated by a first round PCR performed using a flanking primer and an internal mutant primer. This large product is then purified and used as one of the primers for a second round amplification of the target using a different flanking primer. This generates a very large PCR product defined by the two flanking primers, but which now contains the mutation introduced by the mutant internal primer. The final product can then be cloned into

an expression vector (by including restriction sites at the ends of the flanking primers), or used as the template for *in vitro* transcription and/or

translation (by including transcriptional promoter sequences on the 5′ end of primers).

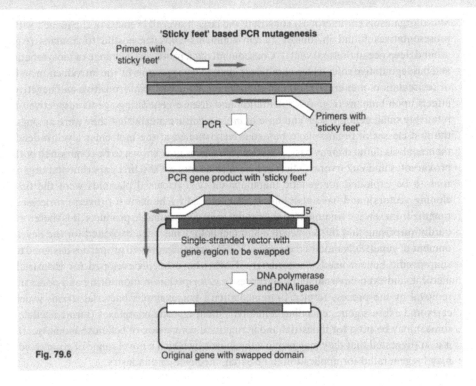

'Sticky feet' based PCR mutagenesis

Primers with 'sticky feet'

Primers with 'sticky feet'

PCR

PCR gene product with 'sticky feet'

Single-stranded vector with gene region to be swapped

DNA polymerase and DNA ligase

Fig. 79.6   Original gene with swapped domain

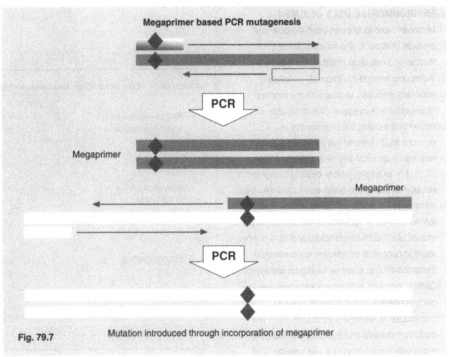

Megaprimer based PCR mutagenesis

PCR

Megaprimer

Megaprimer

PCR

Fig. 79.7   Mutation introduced through incorporation of megaprimer

80. GENETICALLY MODIFIED MICRO-ORGANISMS

Genetically modified microbes are increasingly being developed as bioreactors, biosensors and bioremediation agents for use in industry, medicine, agriculture, and to combat environmental pollution.

Micro-organisms (or microbes) constitute the largest overall biomass of the planet, with representatives found in almost all environments however hostile to humans (e.g. around deep ocean thermal vents). Consequently, microbes fulfill a range of biosynthetic and biodegradative roles which combine to maintain ecosystems by the introduction and/or recirculation of energy. Their enormous phenotypic variability, positive and negative effects upon humans (e.g. as fermentation and disease-generating agents respectively), relatively small genome size, and ease of *in vitro* culture meant that they were amongst the first classes of organisms to be extensively studied at the molecular level. Indeed, their analysis defined many basic genetic mechanisms now known to be common to both prokaryotes and eukaryotes. Microbes were also amongst the first experimental organisms to be exploited for genetic manipulation (e.g. bacterial plasmids were the first cloning vectors), and have also long been exploited by humans for diverse processes, ranging from sewage treatment to the production of consumable products. It is therefore hardly surprising that this knowledge and technology has been exploited for the development of genetically modified microbes (GMMs) with improved properties designed to suit specific human needs. Accordingly, GMMs have been developed for industrial, medical, and even environmental purposes (e.g. for pollution monitoring and pollutant removal by the process termed bioremediation). For example, bacterial strains which carry and express genes encoding cellulolytic multi-enzyme complexes (termed cellulosomes) may be used for industrial and agricultural conversion of cellulose biomasses. It is also suggested that they may provide the means by which a novel range of compounds may be generated for application in research, medicine and industry.

ENVIRONMENTAL USES OF GMMs

Microbes have long been used in water and sewage treatment. The levels and forms of human and industrial pollution are however increasing beyond the capacity of naturally occurring microbes to deal with the problem. Consequently a range of GMMs capable of performing aerobic transformations of environmental interest are being developed to remove recalcitrant chemical pollutants (Table 80.1). For example, GMMs bearing heavy metal resistance genes are being developed which can remove heavy metal pollutants contaminating industrial sites or released in industrial effluents. Indeed, such GMMs are envisaged as the most realistic approach to effective environmental bioremediation (i.e. act as biological remedies). GMMs, including bioluminescence-generating bacteriophages, are also being used as biosensors to sensitively detect, monitor, and quantify industrial and agricultural pollutants, especially those entering water supplies and rivers. Yet other GMMs are being developed for environmental release in order to combat a range of medical and agricultural problems. For example, GMMs are being developed to serve as live vaccines to protect humans, animals and crops from a range of microbial pathogens (e.g. by introducing resistance genes or stimulating innate immune protection mechanisms). GMMs have also been engineered to express insect toxins for the biological control of several insect pests.

ENVIRONMENTAL RELEASE OF GMMs

The potential health and safety risks posed by GMMs has led to the formulation and implementation of strict regulations governing their development and use. Adherence to these regulations is mandatory for all laboratories performing any type of genetic manipulation. These dictate that all proposed manipulations must be approved by appropriate scientific and governmental regulatory bodies, and are only performed in appropriately equipped and monitored containment facilities which minimise the possibility of inadvertant release. Other GMMs are however being developed for deliberate release into the environment, either as live vaccines or bioremediation agents. This depends upon the satisfaction of several legal requirements and technical challenges, for example, the design of appropriate promoters which ensure that the correct genes are expressed at the correct time, and in the correct amount, once the GMM is released into the environment.

Table 80.1

Applications of the production and release of genetically modified microbes

Bioremediation and waste management	• Biostimulation of groundwater • Strain improvement for aromatic degradation • Storage sites • Sludges from plant wood treatment • Composting of contaminated soil
Biological control and agriculture	• Crop nutrition using nitrogen fixers (*Rhizobium*) • Insect pest elimination (*Bacillus thuringiensis*) • Toxin or bacculovirus insecticides • Control of frost damage (*Pseudomonas syringae*) • Biomass degradation of cellulose
Food industries	• Food fermentations (yeast and lactic acid bacteria) • Bioluminescent phage detection of food pathogens
Oral vaccines	• Live attenuated *salmonellae* engineered to express other carbohydrate or protein antigens (combined vaccines)

OAs: 796–802, 962–4, 967.
RAs: 276–80, 332–8, 379.
SFR: 119–24, 126–8, 152, 169–72, 194, 227.

INDUSTRIAL USES OF GMMs

The biosynthetic and biodegradative capacity of microbes has long been exploited by humans for industrial processes (e.g. fermentation). Although an enormous diversity of microbes exist, their exploitation in industrial processes is still limited. For example, they may be too slow for industrial scale application, or they may produce unwanted by-products. Such problems may however be overcome through the genetic modification of microbes (although this may be limited by social acceptance). For example, many laboratory strains of genetically modified yeasts have been developed which improve the purity, yield and taste of beer. Their use in the fermentation industry has however been prevented by consumer resistance to 'genetically engineered beer'. There are many other potential applications for biosynthetic and biodegradative GMMs in industry (e.g. in cellulose processing).

Cellulose degradation by GMMs

Plant cellulose represents the most abundant source of carbon and renewable energy on the planet. Although its structural complexity and stability is exploited for the production of paper, cellulosic waste produced by agriculture and industry is also a major source of pollution. Despite the identification of a vast number of cellulolytic (i.e. cellulose-hydrolysing) bacteria and fungi, they have generally proved inappropriate for the industrial conversion of cellulose to useful products (e.g. animal feeds or fuels). However, biochemical and molecular genetic analysis of their multi-enzyme complexes (termed cellulosomes), and the enzyme systems of satellite microbes such as saccharolytic (i.e. sugar-hydrolysing) microbes vital for natural cellulose degradation, is opening up new avenues of exploitation. Accordingly, the composition of cellulosomes, and the activity of their subcomponents, have now been defined (Fig. 80.1). Essentially, each cellulosome is now known to be constructed from two subunits composed of several different domains: (i) a scaffoldin subunit comprised of cellulose-binding domains and a variable number of subunit-binding domains (termed cohesins) joined by linker sequences; and (ii) a catalytic subunit composed of catalytic domains, a docking domain

(termed a dockerin) and various linkers.

Armed with this knowledge, gene fusion techniques may be used to construct GMMs which combine different cellulosomal domains with a range of heterologous enzymes (e.g. ligninases, pectinases) cloned from other satellite microbes. This is enabling the construction of many different recombinant multi-enzyme complexes which are capable of efficiently degrading any type of cellulosic substrate. Gene engineering can also expand substrate specificities by constructing GMMs which contain exogenous cellulosomal domains which have undergone *in vitro* mutation, or which have been derived from other microbes.

GMMs and pharmaceuticals

Individual cellulosomal domains may also be of unlimited utility in many biotechnologies. For example, they may form the basis of reagents for a range of immobilization techniques and non-radioactive detection methods for many clinical diagnostic or industrial applications. Similarly, GMMs are increasingly playing a significant role in the development of a range of pharmacologically active compounds ranging from conventional drugs to recombinant antibodies or vaccines. They are also being used as *in vitro* bioreactors for the large-scale production of a range of compounds, including those destined for *in vitro* and *in vivo* clinical diagnosis and therapy.

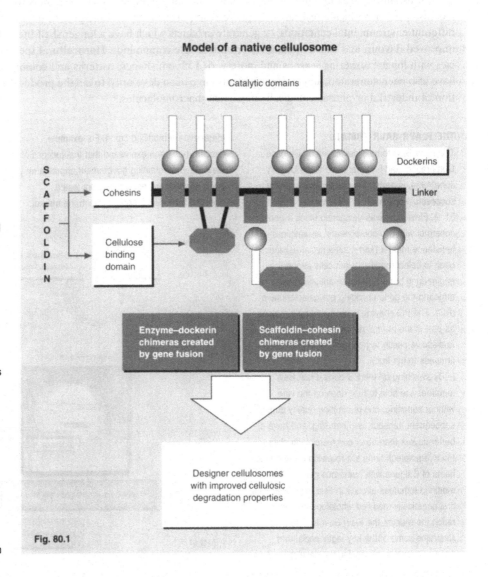

Model of a native cellulosome

Fig. 80.1

81. GENETICALLY ENGINEERED PLANTS

Transgenic and gene-targeting techniques are allowing the genetic modification of an enormous variety of plant species to improve their growth, resistance, and the nature or characteristics of their products.

Humankind has attempted to systematically manipulate plant characteristics for many hundreds of years by selective breeding approaches. Despite a greater understanding of plant genetics, and the availability of plant micro-propagation techniques, this conventional approach often requires many years of effort. Its application is also frequently restricted to sexually compatible plants, and may produce very variable results due to the fact that several genes are transferred in addition to the desired gene during crossing. The development of specialised vectors for cloning, introducing and tracking the fate of specific genes encoding desirable traits into plants is however overcoming these problems. Thus, many genes derived from different plant species, and other organisms, have now been cloned and subsequently transferred to generate so-called transgenic plants (i.e. containing transferred genes). Many other plants have also been genetically modified by targeting specific genes for alteration. Despite consumer concerns, the generation of new varieties of agricultural plants is now big business (often referred to as agro-biotechnology), with billions of dollars invested annually in research and development programmes. Indeed, over 2000 genetically engineered plant varieties are currently in field trials, and many thousands more are under development. Each of these aims to increase pest or herbicide resistance, improve crop performance, enable growth under different environmental conditions, or generate products which have a longer shelf life, improved flavour, and which contain fewer undesirable compounds. Horticultural species with longer flowering seasons and more varied bloom shapes, patterns and colour have also been generated, and transgenic plants have been developed to aid the production of industrial or pharmacologically important macromolecules.

THE REQUISITE DNA TECHNOLOGIES
Genetic modification of plants relies not only upon the understanding of plant genetics, and the role of specific genes and their protein products in biochemical and physiological processes, but on the development of appropriately tailored technologies. For example, techniques such as random amplification of polymorphic DNAs (RAPD) have been extremely useful in identifying and localising genes for subsequent use in crop improvement strategies. A vital aspect has been the development of suitable vector systems (principally based on Ti plamids or transposons) and methods for introducing genes into plant cell (e.g. by bolistic microprojectile bombardment, or electroporation of plant protoplasts). The regeneration of whole plants from single cells, and the tracking of transgenes (e.g. by RAPD or transposon tagging) has also been vital to the development and application of successful plant genetic modification strategies.

THE FLAVR SAVR TOMATO
The Flavr Savr tomato developed by Calgene Corporation, and first marketed in the USA in June 1994, is perhaps a classic example of the successful application of gene targeting (Fig. 81.1). Flavr Savr was developed using a gene construct with two components: an antibiotic resistance marker (*kan-r*, kanamycin resistance) used to select modified plant cells during the engineering process, and an antisense gene targeting the gene encoding polygalactonurase (PG). The PG enzyme is produced by tomatoes as part of the normal ripening process, and helps to dissolve pectin (a polysaccharide imparting firmness to the fruit).

By switching off the PG gene, Flavr Savr tomatoes are able to fully ripen on the vine without softening, can retain their quality during subsequent transport and handling, and have a better flavour than other commercial varieties. Many ag-biotech firms are following close on the heels of Calgene with numerous genetically modified tomatoes already in field trials. Being the first genetically modified wholefood product to reach the market, the Flavr Savr tomato also illustrates some of the key legal, social and economic factors influencing the development

of genetically modified crops. For example, concern has been expressed that the success of Flavr Savr in obtaining government approval may make it easier for less adequately tested genetically modified foods to reach the market.

OAs: 643–50, 660–6, 668, 725, 794–5, 803–22. RAs: 235–8, 253–4, 281–93, 406. SFR: 119–24, 126–8, 156–9, 169, 172, 209.

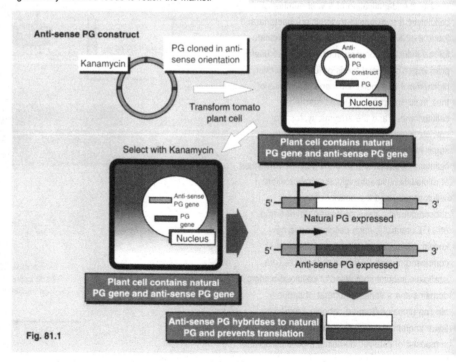

Fig. 81.1

DIVERSITY OF TRANSGENIC CROPS

Although more rapid than conventional plant breeding approaches, the development of a genetically engineered plant (or animal) still involves many years of experimentation and field trials before the product can be obtained and marketed. The considerable financial investment required means that the product must be able to hold an appreciable share of a large potential market if its development is to be feasible. For example, the development of the Flavr Savr tomato is estimated to exceed 10 million dollars, but is anticipated to capture a large proportion of the 200 billion dollar a year market for fresh tomatoes. Nevertheless, it is difficult to find a commercial plant species that has not undergone, or is undergoing, transgenic modification, primarily focused upon pest resistance, herbicide resistance or new crop products.

Pest resistance

Billions of dollars are spent annually to control crop pests using chemical pesticides, many of which also pose potential health hazards to humans and the environment. Several transgenic strategies have therefore now been developed to combat insect, fungal, bacterial or viral pests in crop plants as diverse as tobacco, cotton and cucumbers (Table 81.1). For example, the transfer of cloned viral genes encoding coat proteins has been shown to imbue the plant with resistance to the virus. Insect pests have also been deterred by generating transgenic plants which either express microbial (e.g. bacterial) toxins or natural plant proteins (e.g. serine proteases) which kill the insect or make the plant unpalatable to them.

Herbicide resistance

Weed control by chemical herbicides is still viewed as vital to maintain crop yields. Developing herbicides which are only active on weed plants is therefore a constant battle for organic chemists. To reduce the need for new herbicides, transgenic herbicide-resistant plants are being produced by increasing the production of the protein targeted by the herbicide. This is achieved by expressing mutant proteins unaffected by the herbicide, or by introducing a gene encoding a herbicide detoxifying enzyme.

New crop products

The ability to express modified transgenes in plants is also being exploited in numerous ways to yield novel products within traditional plant crops. Both transgenic plants and transgenic animals appear destined to be increasingly generated for the production of a range of compounds which have industrial and pharmacological application. For example, many commercial companies are developing transgenic oilseed plants which provide 'environmentally friendly' and renewable sources of fuel oils, or 'healthier' forms of edible oils. Thus, by introducing genes encoding specific enzymes used in lipid synthesis or metabolism, defined long chain oleochemicals can be produced which are difficult to produce naturally, or chemically using industrial processes. The ability to bulk produce useful compounds in plants is also now well recognised. Even transgenic plants expressing human or mouse antibodies have been produced. It is clear that the limits to transgenic plant exploitation have yet to be defined, by science as well as society.

CONSUMER CONCERNS

Genetic modification of food plants (and animals) has naturally aroused much concern over their ultimate safety, despite stringent legal and governmental safety requirements. Addressing such concerns is vital to future developments since this will ultimately determine their viability in the marketplace. Indeed, consumer polls suggest that there is considerable resistance to the concept of 'Frankenfood' (as it has often been described in the media).

Table 81.1 The diversity of transgenic crops

Plant/crop type	Plant modification or conferred trait
Apple	Resistance to insect attack
Canola (oilseed rape)	Seed oil modification, insect resistance
Maize	Insect and virus resistance, herbicide tolerance
Cucumber	Resistance to certain viruses
Melon	Resistance to certain viruses
Potato	Insect and virus resistance, starch increase
Rice	Modified seed storage protein, insect resistance
Soyabean	Modified seed protein storage, tolerance to herbicide
Sunflower	Modified seed storage protein
Tomato	Modified ripening, insect and virus resistance
Wheat	Expression of recombinant antibodies, human serum albumin

82. GENETICALLY ENGINEERED ANIMALS

The development of transgenic animals by altering or inserting genes within embryos is providing fundamental knowledge of gene functions, and is being exploited in agriculture, industry and medicine.

Molecular technologies originally developed for the genetic manipulation of single cell organisms (e.g. bacteria, viruses, and yeast) have since been applied to the transfer and modification of genes within multicellular eukaryotic organisms. Although the approaches involved are often very similar, their extension to mammalian (and plant) systems has required several additional molecular and cell biological technologies. For example, the development of specialised vectors, gene delivery and targeting systems have all been required in order to allow the engineering of specific genetic changes (e.g. by homologous recombination). Their application has also demanded advances in animal embryo and somatic cell propagation and transfer systems in order to generate viable mature organisms. Armed with these new technologies, the expression or function of specific genes or their protein products within living multicellular organisms can be altered. For instance, disruption (often termed 'gene knockout') and/or replacement of endogenous genes has been used to create strains of mice for the investigation of some human genetic diseases, and as models for the development of therapeutic approaches for use in humans. Alternatively, foreign genes have been introduced and stably incorporated into chromosomes to allow their subsequent transmission to progeny. Often referred to as 'transgenic' animals, these serve as models for studying the action of specific genes *in vivo*, and a convenient means of producing large amounts of specific gene products in an appropriate eukaryotic environment (i.e. as 'bio-reactors' to produce certain pharmaceuticals). With appropriate regulatory control of their production and use, such animals and the technologies involved offer many benefits to humans, not least the development of gene therapies for combatting several forms of human disease.

DECIDING ON NEW DEVELOPMENTS

All genetically modified organisms, be they microbes, plants or animals, require continual assessment and control to ensure they are safe for the environment and consumers. Although largely controlled by government agencies, the important issues of safety and desirability of any individual strain of genetically modified organism must ultimately reflect the views of all members of society. With the exception of cases involving genetically engineered foods where consumer resistance holds sway, this is clearly not the case. If, however, the mistakes and oversights of the past are not to be repeated, deciding upon the development of new forms of genetically modified organisms requires considerable discussion involving all members of society. This is only possible through the open disclosure of full and accurate information.

OAs: 659, 667–76, 679–80, 823–48, 880–1, 905–7. RAs: 235–40, 294–303, 320, 404–6. SFR: 119–24, 126–8, 156–9, 169, 172.

MAKING CHANGES IN ANIMAL GENES

The creation of genetically modified animals began with the development of transgenic mice as a means of studying gene function in higher organisms. This required the insertion of 'foreign' DNA (termed a transgene) into an appropriate vector (e.g. based on viral genomes) followed by its micro-injection into the nucleus of a developing embryo (Fig. 82.1a). This allowed the transgene to be stably integrated (e.g. by homologous recombination) into the embryo's chromosomes. Although a rather inefficient process which results in random numbers of genes being integrated at varying chromosomal locations, at least some of the embryos develop into viable offspring carrying the transgene following embryo transplantation into 'surrogate' mothers.

An alternative and now frequently used approach is to introduce the vector-borne DNA into cultured embryonic stem (ES) cells by transfection (e.g. electroporation or retroviral infection) (Fig. 82.1b). Correctly transfected ES cells can then be selected and introduced into developing embryos where they participate in normal tissue development. Homologous recombination in ES cell systems can also be used to target changes in specific tissues or

genes which alter either their expression or function. Thus, deletional mutations can be created which destroy the function of the gene or its protein product, or entire genes may be replaced (e.g. to correct mutant genes). Such gene targeting strategies are not only aiding the

study of human disease processes but are also providing new therapeutic strategies.

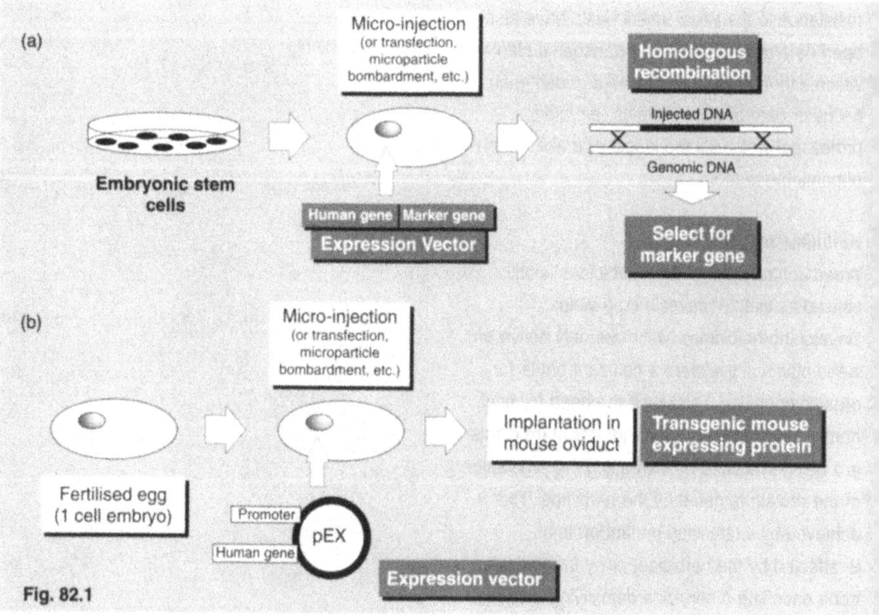

Fig. 82.1

Mice as disease models for humans

Several experimental strains of transgenic mice have now been produced by gene transfer in order to facilitate the investigation of human disease mechanisms (Table 82.1). Their generation has involved both the introduction of exogenous genes, and the disruption or replacement of endogenous genes For example, transgenic mice bearing human apolipoprotein (a) genes have been generated to facilitate the investigation of the genetic basis of atherosclerosis and heart disease. Similarly, strains of mice bearing disrupted dystrophin or CFTR genes have also been generated.

These transgenic strains have not only been valuable in elucidating the precise biochemical and genetic processes involved in disease progression, but are greatly facilitating the development and assessment of gene therapies for muscular dystrophy and cystic fibrosis respectively. Gene knockout techniques involving homologous recombination have also been used to create mice with severe combined immunodeficiency (so called SCID mice) or lacking expression of β_2-microglobulin. These are already proving invaluable experimental models for the investigation of immune mechanisms, and are being exploited for the development of novel diagnostic and therapeutic reagents (e.g. genetically engineered antibodies).

Transgenics in agriculture

In addition to the generation of transgenic mice strains, several types of transgenic farm animals have also been created. Similar to their transgenic plant strain counterparts, these have been generated to improve agricultural productivity, tolerance to environmental factors, and disease resistance. For example, transgenic cows and pigs bearing recombinant growth hormone genes were initially created in an attempt to improve growth rates and meat quality. Whilst the success of this particular approach was not significant, the creation of transgenic fish and animals bearing other growth-promoting genes has been much more successful. Furthermore, the introduction of viral coat protein-encoding genes capable of providing disease resistance has met with great success. This suggests that the development of such transgenic farm animals

Gene or region	Genetic alteration	Human equivalent
HPRT Hypoxanthine-guanine phosphoribosyl transferase	Inactivation of HPRT gene	HPRT deficiency
X chromosome	Mutation at X chromosome	Muscular dystrophy
α1-antitrypsin	Introduction of α1-AT gene mutant Z allele	Neonatel hepatitis
Amyloid precursor protein	Over expression of gene	Alzheimers disease
Tyrosine kinase	Constitutive expression of gene	Cardiac hypertrophy
HIV transactivator	Expression of HIV *tat* gene	Kaposi's sarcoma
c-*myc* oncogene	Expression of c-*myc* oncogene	Induction of tumourigenesis
Angiotensinogen	Expression of rat angiotensinogen gene	Hypertension
CET Protein Cholesterol ester transfer protein	Expression of CET gene	Atherosclerosis

Table 82.1

and fish is likely to increase as world agricultural needs rise.

Farming pharmaceuticals

Although transgenic technologies require further development to increase transgene expression rates, transgenic farm animals have also been developed for their potential as producers of certain pharmacological proteins. Accordingly, transgenic pigs, cows and sheep are allowing the production of a range of recombinant proteins which are directly secreted into the bloodstream or into milk. These proteins include human haemoglobin, insulin, the blood clotting factor IX, tissue plasminogen activator and interleukin 2. Such bioreactors may be of great commercial value, allowing recombinant proteins to be continually produced at high levels throughout the lifetime of the animal, and which can be readily harvested and easily recovered. Furthermore, they represent readily maintained mammalian expression systems for the production of correctly folded and post-translationally modified recombinant proteins.

83. MOLECULAR TECHNIQUES IN PRENATAL DIAGNOSIS

Molecular biology is increasing the accuracy and speed of prenatal diagnosis by both identifying disease genes and their mutations, and providing the means for their routine clinical detection.

Initially based upon the cytological detection of chromosomal abnormalities or the assay of biochemical markers (e.g. defective enzymes), prenatal diagnosis and carrier testing was consequently restricted to less than 50 disorders for which such suitable markers were available. Since their first application in 1978, molecular biological techniques have however revolutionised human clinical genetics in two principal ways. Firstly, they have been used to map, identify and clone disease genes, and provide polymorphic gene markers, often before the disease-associated gene itself was discovered. Over 200 single gene disorders are therefore now amenable to routine prenatal diagnosis and/or genetic counselling. Secondly, molecular biology has provided techniques and approaches which have increased the rapidity, reliability, accuracy, safety and timing of clinical diagnosis, thereby increasing the options available to affected/carrier parents. It is however important to realise that although the techniques used are the same, their diagnostic application is often quite different. For example, mapping using linkage analysis calculates the distance between restriction fragment length polymorphism (RFLP) markers and a gene based on the recombination frequency. In contrast, RFLPs are used diagnostically to 'track' genes through families, and the recombination frequency provides an indication of the error rate of the test. Three methods are now routinely used in prenatal diagnosis for mutation detection: (i) Southern blotting, (ii) pulse field gel electrophoresis, and (iii) the polymerase chain reaction (PCR). Of these, the PCR is probably having the most profound impact, by increasing the range of mutations amenable to detection, the rapidity with which diagnostic information can be obtained, and increasing permitting the analysis of smaller samples obtained more safely and at earlier times in gestation or embryo development.

INCREASING THE OPTIONS?

In only a few cases can prenatal diagnosis of affected fetuses provide therapeutic benefit (e.g. in the case of phenylketonuria where symptoms of mental retardation and restricted development can be avoided by the implementation of a phenylalanine-free diet from birth). Generally however, prenatal detection of an affected fetus has limited the options to choosing whether or not to abort the fetus. Molecular biological techniques, allied to advances in embryology, are altering this situation dramatically. Prenatal diagnosis can now be performed on *in vitro*-fertilised embryos obtained from a carrier thereby allowing the selection and implantation of only unaffected embryos. Whilst this has been argued to be a form of eugenics and potentially open to abuse, there is no doubting the value of this option to the individuals concerned. It is also possible that in the near future many detectable disease-associated gene defects could be corrected by somatic or germline gene therapies performed on fetuses or embryos.

OBTAINING SAMPLES FOR ANALYSIS

Conventional cytogenetic or biochemically based prenatal diagnosis has relied upon the analysis of samples of amniotic fluid containing fetal cells, urine and secreted/shed proteins obtained by the process of amniocentesis. Fetal cells are then harvested by centrifugation and their numbers expanded by *in vitro* culture (Fig. 83.1). Although this culture step is lengthy and often unsuccessful, it was routinely used to provide sufficient DNA for prenatal diagnoses using Southern blotting. A further significant drawback of amniocentesis is that it can only be performed successfully, and without significant risk to the fetus, at about the sixteenth week of gestation. The process of chorionic villus sampling (CVS), whereby a small amount (30 mg) of chorionic villi is removed by catheter, is not only safer but also allows samples to be taken between weeks 8 and 12 of gestation. Although the cells obtained by CVS may also be cultured, this step is often omitted in routine laboratories since sufficient DNA may be provided for direct analysis by both Southern and PCR-based methods. Furthermore, the use of the PCR has permitted prenatal diagnoses (including sex determinations) to be

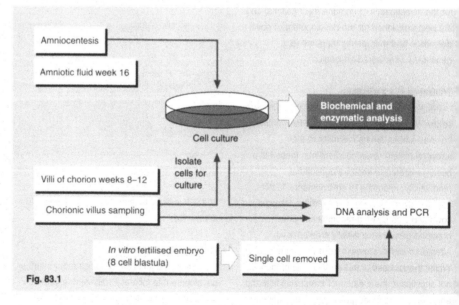

Fig. 83.1

performed on *in vitro*-fertilised embryos. Accordingly, sufficient DNA may be provided from a single cell removed at the 8 cell blastocyte stage.

OAs: 387–91, 464, 474–80, 491–3, 849–65, 868, 894–5, 892, 902. RAs: 203–6, 304–8, 427. SFR: 119–24, 136–9, 173–81, 227–9.

SOUTHERN BLOTTING

Until 1988, Southern blotting played a predominant role in prenatal diagnosis since it afforded a simple procedure whereby RFLPs could be detected and used in linkage analysis to establish linkage disequilibrium. This was particularly useful since many disease genes had not been cloned or their mutations determined, although RFLP markers were available which could be detected using labelled DNA probes. The availability of gene-, cDNA- and allele-specific probes has further aided Southern analysis allowing a range of diseases to be routinely detected by RFLP linkage. These include Duchenne muscular dystrophy (DMD), cystic fibrosis, sickle cell anaemia, β-thalassaemias, phenylketonuria, myotonic dystrophy and Huntington's disease. Although Southern analysis will undoubtedly continue to be used for the diagnosis of some disorders (e.g. fragile X syndrome) for many years to come, it is rapidly being replaced by PCR–based methods which are less labour intensive and do not require the 2–3 weeks often necessary before an autoradiographic result is produced.

Advantages of PCR amplification

Since 1988, the PCR has had an enormous impact on prenatal diagnosis and carrier determination. Indeed, many diseases such as cystic fibrosis, Duchenne and Becker's muscular dystrophy, and β-thalassaemia may now be detected solely using PCR-based methods (Fig. 83.2). Not only are significantly smaller amounts of fetal DNA used, but archival DNA (e.g. available on stored blood spots or fixed tissue samples) can also often be used to establish the carrier status of deceased members of a family. These data can be enormously useful in increasing the accuracy of linkage studies by increasing the informativeness of family trees.

It is however the rapidity, sensitivity and versatility of PCR-based methods which is having the greatest impact. Thus, the time taken for a single analysis may be reduced to a few days, and multiplex PCR using allele-specific oligonucleotides allows simultaneous detection of several different mutations in the same reaction tube. Consequently, in cases such as cystic fibrosis where over 200 different mutations have

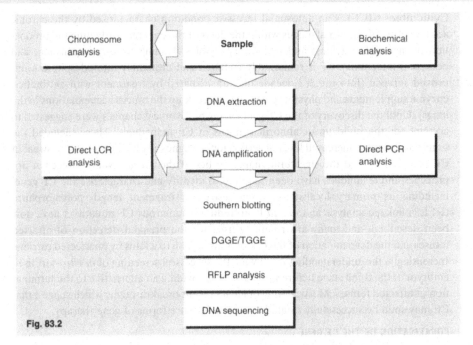

Fig. 83.2

been described, sequential analyses starting with the most frequently occurring mutations can be performed until a mutation is found or the possible existence of a mutation discounted. This saves both time and diagnostic laboratory costs by reducing the number of tests required.

The PCR is also capable of directly detecting a wide range of different mutations, including single base substitutions or deletion/insertions which are not amenable to detection by RFLP. Its sensitivity also means that single-copy genes are much more readily analysed. The ability of RT-PCR (reverse transcription-PCR) to amplify from mRNA transcripts also enables the PCR to be used to rapidly and sensitively detect mutations associated with diseases (e.g. DMD or chronic myeloid leukaemia) which are difficult to detect by Southern analysis.

Potential problems with the PCR

The PCR is limited to the analysis of genes for which sequences and mutations are known (i.e. to allow the design of appropriate primers). The rate of progress in gene mapping, especially as a result of the human genome mapping project, is however likely to diminish this limitation. Indeed, it is now often only a few months between the discovery of a disease gene or an associated

mutation, and the availability of PCR-based or even ligase chain reaction (LCR)-based diagnostic assays. Extreme care is however required for their performance and the interpretation of results in order to avoid the possibility that results have arisen due to cross-contamination, mispriming, or failure to amplify (or, where RFLP is used, the subsequent failure to digest the amplified PCR products).

Pulsed-field gel electrophoresis

Pulsed-field gel electrophoresis (PFGE) is a specialised technique often used for the analysis of large DNA fragments ranging betweeen 200 kb and 10 Mb. PFGE's technical complexity has however limited its application within prenatal diagnosis to the study of mutations within fragments exceeding the 40 kb limit amenable to study by Southern blotting. In particular, it has been extremely useful in analysing the large (2.3 Mb) dystrophin gene underpinning Duchenne and Becker's muscular dystrophy. PFGE allows unequivocal diagnosis by enabling the detection of deletions and duplications in a single experiment. A significant drawback is that the DNA for analysis has to be specially prepared to ensure it is not degraded (i.e. is of high molecular weight).

84. THE GENETICS OF CYSTIC FIBROSIS

The identification of the cystic fibrosis gene and its mutations has involved the combined use of a range of molecular biological techniques for gene mapping cloning and sequencing.

Cystic fibrosis (CF) is an autosomal recessive condition characterised by the chronic accumulation of abnormal mucus within the ducts of the pancreas (affecting the production of digestive enzymes) and respiratory airways (leading to lung impairment and persistent infections). CF is remarkably common and ultimately fatal, despite the improved survival (into the 3rd decade of life) mediated by treatment with antibiotics, enzyme supplements and physiotherapy (i.e. to break up the mucus accumulations in the lungs). Until the discovery of the CF gene, many biochemical changes were suggested to account for the build up of abnormal mucus in CF individuals. These included the suggestion that the increased level of certain ions (diagnostically detected in the sweat of CF patients) altered the structural characteristics of the mucus. Many molecular approaches and techniques have been required to identify and characterise the CF gene, including its primary localisation using restriction fragment length polymorphism (RFLP) linkage analysis and *in situ* hybridisation. Numerous CF mutations have now been described, and many are routinely used for the prenatal detection of affected fetuses and the determination of the risk of transmission to a fetus by unaffected carriers. Increasingly, this understanding is allowing the successful screening of *in vitro*-fertilised embryos at the 8 cell stage before implantation, providing an alternative to the termination of affected fetuses. Most exciting of all are recent breakthroughs which suggest that CF may soon be successfully treated using one or more forms of gene therapy.

LOCALISATION OF THE CF GENE

CF was the first condition where the disease gene was cloned based upon initial localisation by the linkage analysis of CF pedigrees. Thus, by 1985 *in situ* hybridisation and the demonstration of RFLP linkage disequilibrium with the *Met* oncogene (present in 90% of CF patients but 20% of normals) identified a 1.5 million base pair region within the long arm of chromosome 7 (band q31.3) as the location for the CF gene (see Fig. 84.1). Further linkage analysis of genomic DNA from many CF-affected families soon led to the discovery of further extragenic RFLPs (i.e. occurring outside the gene) which were more closely linked to the CF gene. This narrowed the CF gene location to a region of about 500 000 base pairs (i.e. 500 kb).

Without identifying the gene, and despite their different frequencies within populations, these RFLPs combined into haplotypes to provide early markers for prenatal diagnosis. Thus, the haplotype Xv-2c {-ve}, KM.19 {+}, MP6d-9 {+} had a high probability of being associated with CF in a study of 80 European CF families. Southern blot detection of these extragenic markers (now known to be located on the same side of the CF gene) has since been largely superseded by more rapid and direct polymerase chain reaction (PCR)-based mutation analyses and the use of intragenic markers based upon the presence of variable numbers of tandem repeats (VNTRs).

FREQUENCY OF CF

The frequency of CF varies widely amongst different ethnic groups, being remarkably commo (approx. 1 in 1000 affected, and 1 in 25 heterozygous carriers) in Caucasians and very rare in Negroid and Mongolian races. Different C haplotypes are also found which again vary in their frequency amongst different geographical populations. The identification of these haplotype has been, and remains, important for the development of molecular strategies for prenatal diagnosis. The frequency of CF has also influenced the course of research toward effectiv gene therapies for CF (i.e. determining its technical and economic feasibilty).

OAs: 683, 831–4, 849–50, 852, 855, 865–76. **RAs:** 248–9, 294, 304–9, 427. **SFR:** 119–24, 173–82, 184, 227–9.

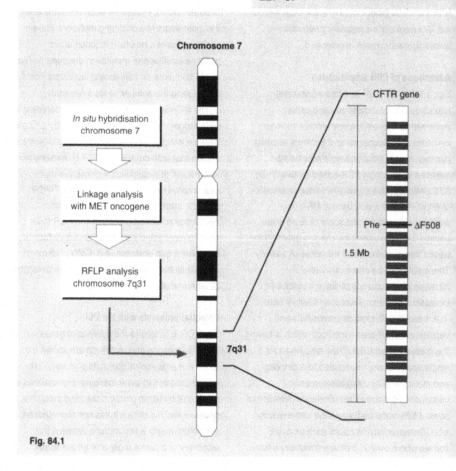

Chromosome 7

In situ hybridisation chromosome 7

Linkage analysis with MET oncogene

RFLP analysis chromosome 7q31

7q31

CFTR gene

Phe — ΔF508

1.5 Mb

Fig. 84.1

Isolating the CF gene

Having established several linked RFLP markers, genomic cloning strategies were used to create libraries employing phage and cosmid vectors. Consequently, clones were identified which allowed the sequencing of the putative CF gene-containing region starting from the upstream RFLP markers and moving towards the CF gene itself. Combined with cross-species DNA hybridisation studies, sequencing identified a clone containing a flanking sequence characteristic of a protein-encoding gene, and which detected a single mRNA transcript in northern blotting experiments. Part of this sequence then served as a probe for the detection and isolation of overlapping clones in a cDNA library generated from cultured, normal human sweat gland cells (likely to contain high levels of CF mRNA based upon previous biochemical clues). Sequencing of these overlapping cDNA clones revealed a complete mRNA transcript of 6.1 kb encoding a protein with the classical features of transmembrane ion channel-forming proteins (Fig. 84.2). Now termed the cystic fibrosis transmembrane conductance regulator (or CFTR), this has been shown to be encoded by 24 separate exons spanning 230 kb and encoding a 241 amino acid protein. Furthermore, various mutations have been described which alter its *in vivo* ion transport activity, consistent with early biochemical data.

Mutations within the CFTR gene

Genetic defects within the CFTR gene responsible for CF could not be detected by Southern or northern blotting experiments. Indeed, they revealed no apparent differences between CF and normal individuals. Instead, massive efforts were required which involved comparing the nucleotide sequences of cloned CFTR genes obtained from normal, CF carrier, and CF-affected individuals using newly developed YAC (yeast artificial chromosome) vectors, and sequence tagged sites (STSs) as markers. This eventually identified the first CF-causing mutation: a 3 bp deletion leading to the loss of a phenylalanine from the expressed protein at residue 508, and thus termed ΔF508 (i.e. Δ for deletion, F for phenylalanine, and 508 for the amino acid residue position). This

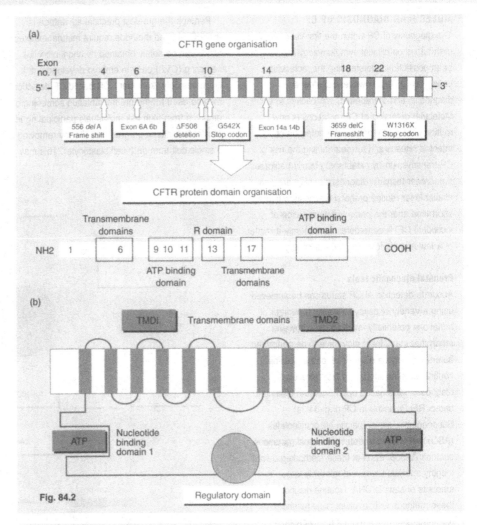

Fig. 84.2

Table 84.1 CFTR Mutations

Name	Mutations	Effect	Exon
A455E	C>A(1496)	Ala>*Glu*(455)	9
ΔF508	Deletion	Phe *del* (508)	10
Q439X	C>G(1609)	*Gln*>STOP	10
1717IG-A	G>A(1717−1)	Splice mutation	Intron10
ΔI507	Deletion	*Ile del* (507)	10
P574H	C>A(1853)	Pro>*His*(574)	12
R553X	C>T(1798)	Arg>STOP(553)	11
G551D	G>A(1784)	Gly>*Asp*(551)	11
R117H	G>A(482)	Arg>*His*(117)	4

mutation is remarkably common, accounting for an overall frequency of 68%, but varying in frequency amongst different geographical groups. Over 200 point substitution, insertional and deletional mutations within the CFTR gene exons or introns have since been identified (see Table 84.1). These result in mis-sense mutations, the creation of new stop codons, abnormal splicing of CFTR mRNA and frame-shift mutations. However, only 20 of these mutations commonly occur and these are therefore used as 'first line' diagnostic markers.

Molecular biological techniques are allowing accurate and rapid prenatal diagnosis of CF, and are enabling the development of effective (i.e. curative) germline and somatic cell-targeted gene therapies.

MOLECULAR DIAGNOSIS OF CF

The diagnosis of CF within the first year following birth based on clinical symptomology has long been possible. Understanding the molecular biology of CF has however transformed diagnostic, and therapeutic, approaches to CF. Molecular detection of CF mutations is now routinely available for prenatal detection of affected fetuses and for determining the risk of CF transmission by establishing family pedigrees (i.e. carrier testing). Diagnostic assays are now available for routine prenatal diagnosis within a short time after the primary determination of individual CFTR gene defects, often only a matter of a few months following their discovery.

Prenatal diagnostic tools

Accurate detection of CF status can be achieved using a variety of assays. The large number of mutations potentially involved does however mean that many tests may have to be performed before CF may be diagnosed, or the possibility confidently eliminated. The first molecular diagnoses were based upon Southern blotting to detect RFLPs linked to CF (Fig. 84.3a). Subsequently, allele-specific oligonucleotides (ASOs) were used to detect individual mutations such as ΔF508. Each of these methods requires lengthy analysis times and relatively large amounts of sample DNA. Routine diagnosis by these methods also involves large numbers of tests, many of which will not ultimately be required, thereby increasing the cost of individual diagnoses.

It is therefore not surprising that these methods are being superseded by the use of *in vitro* DNA amplification-based methods. Mutations are consequently most frequently detected by RFLP and ASO hybridisation performed on PCR amplification products, or the mutations directly detected by PCR using single or multiple (multiplex) ASO primers (Fig. 84.3b). These methods offer much greater accuracy and assay simplicity, short individual and overall analysis times (often days instead of weeks), cost effectiveness (by reducing the number of individual tests performed), and the ability to detect a wider range of genetic defects (e.g. based upon the use of highly polymorphic VNTR markers).

Perhaps the greatest practical advantage is that PCR-based methods require minute samples which can be safely obtained by chorionic villus sampling (CVS) early in embryo development. It is increasingly likely that PCR-based CF detection may be used for the pre-implantation screening of embryos from high risk individuals participating in *in vitro* fertilisation programmes (i.e. by removing a single cell from an 8 cell blastocyte). This may

even precede germline correction of the defects within embryos (e.g. using antigene strategies).

OAs: 683, 831–4, 849–50, 852, 855, 877–81. **RAs:** 248–9, 294, 304–9, 427. **SFR:** 119–24, 173–82, 184, 227–9.

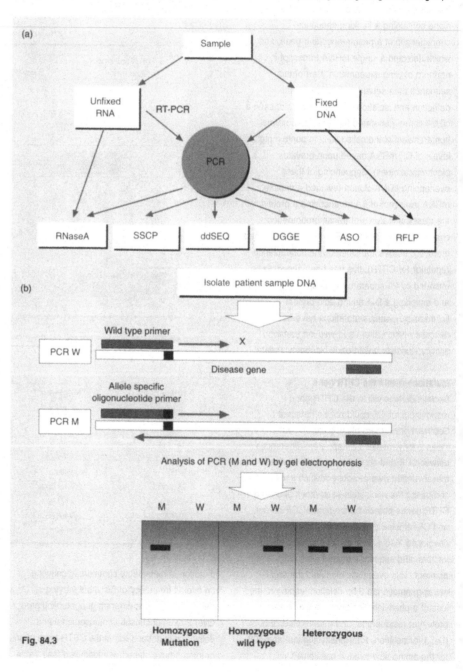

Fig. 84.3

STEPS TOWARDS GENE THERAPY

The high frequency of CF and the restricted number of CF haplotypes within several populations means that, purely on economic terms, CF represents a candidate for gene therapy. The many different CFTR gene mutations described to date does however favour the development of gene replacement or addition approaches. This avoids the need to develop several alternative forms of gene therapy (i.e. tailored to the defects within a particular group of CF patients). Several positive and crucial steps have already been taken towards the development and application of both somatic and germline gene therapies (Fig. 84.4).

(1) A full length CFTR gene has been assembled from three overlapping cDNA clones and inserted into vaccinia and/or retroviral vectors (each containing resistance markers to aid selection).

(2) Mammalian cells lines lacking the CFTR gene (CFTR⁻) have been created by chromosome-mediated gene transfer techniques or established from airway epithelial cells from a CF patient. Transfection of CFTR⁻ cells with the CFTR gene-containing vectors has demonstrated that the resultant cells express recombinant CFTR protein. Furthermore, the transfected cells exhibit the same functional ion efflux activity of normal tracheal cells. This demonstrated that somatic gene therapy may be possible using retroviral vectors to insert normal CFTR genes into the lung epithelial cells of CF patients (e.g. using nasal aerosols). Their expression would correct the abnormal ion flux of the cells, and would not require the direct targeting of the defective CFTR gene.

(3) Two research groups have now independently 'engineered' viable CF heterozygous (cf/+) and homozygous (cf/cf) transgenic mice by disrupting part of the protein-encoding region (exon 10) of the normal CFTR gene in embryonal stem cells. Both animal models expressed normal amounts of CFTR protein, however, both cf/cf strains expressed CFTR proteins with defective biochemical characteristics. They also varied in terms of CF pathological manifestations according to the types of mutations and how they were created.

These studies are providing vital clues to the molecular pathology of CF and pave the way towards *in vivo* models for each human mutation (e.g. ΔF508). Such *in vivo* models are required to both develop new, potentially curative, therapeutic approaches, and subsequently demonstrate that they can be safely used in humans. Recent decisions by scientific, medical and political agencies, and the apparent early success of human clinical trials, suggest that several forms of CF gene therapies may soon be available.

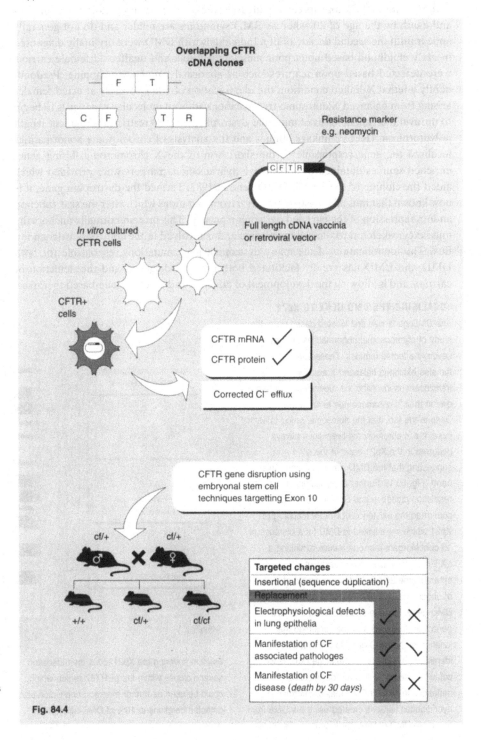

Fig. 84.4

Targeted changes		
Insertional (sequence duplication)		
Replacement		
Electrophysiological defects in lung epithelia	✓	✗
Manifestation of CF associated pathologes	✓	⟍
Manifestation of CF disease (*death by 30 days*)	✓	✗

85. THE DYSTROPHIN GENE AND MUSCULAR DYSTROPHIES

Localisation of the DMD gene by RFLP linkage analysis enabled the cloning and characterisation of a large (2.3 Mb) gene encoding the dystrophin protein, whose defective expression results in DMD/BMD.

The muscular dystrophies represent a group of genetically determined disorders, inherited in a sex-linked recessive mode, which affect approximately 1 in 3500 males. They are characterised by regressive alterations in muscle fibres which leads to progressive muscle weakening, atrophy, and loss of tendon reflexes. The two most common forms are Duchenne and Becker's muscular dystrophy (DMD and BMD) which vary in their severity. Thus, DMD appears early in life followed by the rapid progression of symptoms and death by the age of 20, whereas BMD symptoms are milder and do not generally appear until the second decade of life. Individuals with DMD were originally diagnosed in early childhood based upon poor muscle condition, and unaffected female carriers were detected based upon features such as abnormal lymphocyte capping. Predominantly a lethal X-linked condition, the identification of a few severely affected females arising from balanced X;autosome translocations allowed molecular geneticists to begin to unravel the genetic basis of muscular dystrophy. By 1983 restriction fragment length polymorphism (RFLP) linkage studies and the analysis of chromosome abnormalities localised the gene responsible to the short arm of the X chromosome. Using gene-enriched sources obtained by subtractive hybridisation, markers were provided which aided the cloning of the 2.3 Mb DMD gene in 1987. Termed the dystrophin gene, it is now known that muscular dystrophies arise from mutations which alter the size, function and/or expression of the encoded dystrophin protein. This protein normally anchors the muscle cytoskeleton to the cell membrane and is involved in the control of calcium ion flux. The identification of the many dystrophin gene mutations responsible for both DMD and BMD has greatly facilitated both prenatal diagnosis and the detection of carriers, and is allowing the development of effective and curative gene-based therapies.

LOCALISING THE DMD GENE TO Xp21

The DMD gene was first located based upon the study of chromosomal abnormalities in a few severely affected females. These affected females exhibited balanced X;autosome translocations involving the movement of a part of one of their X chromosomes to an autosome. Despite the fact that the autosome varied in each case, the X chromosome breakpoint always occurred in the Xp21 band of the short arm, suggesting that the DMD was localised to this band (Fig. 85.1). Further analyses using restriction digestion and Southern blotting confirmed the existence of RFLPs mapping to Xp21 which were linked to DMD (at a distance of 15 centiMorgans). In one female exhibiting a t(X;21) translocation, the breakpoint in chromosome 21 resulted in movement of a block of ribosomal RNA (rRNA) genes adjacent to the DMD gene. Linkage analysis using the rRNA gene sequences as a probe confirmed the location of the DMD gene to Xp21. It also identified an X chromosome clone (XJ1.1) which could be used as a marker for detecting mutations in 10% of DMD males. Subtractive hybridisation libraries created using the DNA from one male DMD patient which contained a large

deletion covering the Xp21 locus then identified several clones within the pERT87 region which could be used as further markers, and which also detected deletions in 10% of DMD-affected males.

BALANCED TRANSLOCATIONS IN DMD

Although an X-linked recessive condition usually affecting males, severely affected DMD females have been identified. These arise at an extremely low frequency by the combined action of a balanced translocation between regions of an autosome and an X chromosome, and X chromosome inactivation. Accordingly, equivalent regions of DNA are transferred leading to the disruption of the normal functioning of the DMD gene in one X chromosome. Under normal circumstances the DMD of the other X chromosome would remain active in most female cells since X chromosome inactivation is random. However, following an X;autosome balanced translocation, preferential inactivation of the normal X chromosome appears to occur. All cells therefore contain two active autosomes and an active X chromosome bearing the translocation. Consequently, only the translocated X chromosome containing the disrupted DMD gene remains active, generating a DMD-affected female.

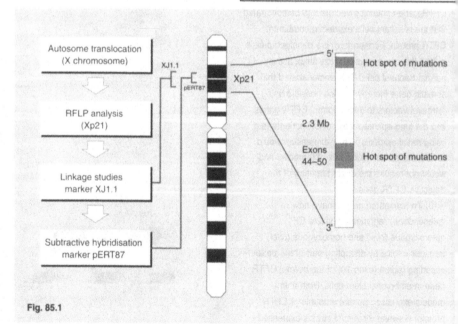

Fig. 85.1

OAs: 835–6, 857–62, 864–5, 882–9.
RAs: 250, 294, 304–8, 310–11, 410–11, 417. **SFR:** 119, 158–9, 173–83, 195, 220–1, 227–9.

Cloning the dystrophin gene

Study of a large number of DMD families using the XJ1.1 and pERT87 probes in RFLP linkage analysis soon revealed that the dystrophin gene was much larger than expected. Accordingly, these intragenic probes (i.e. binding within the gene) were calculated to be at least 5 cM (or 5 million bp) from the mutations involved in DMD (Fig. 85.2). The enormous size of the DMD gene-effectively prevented the use of genomic cloning strategies. Consequently, cDNA cloning strategies were employed since these clones would only represent the coding regions of the gene. Although the nature of the dystrophin gene-encoded protein remained unknown, the use of both pERT87 and XJ1.1 probes initially identified full length DMD protein-encoding cDNA clones in fetal and adult human skeletal muscle cDNA libraries. Cross-species hybridisations also demonstrated a high degree of conservation between human and other mammalian DMD genes. It is now known that the DMD gene is about 2.3 Mb in length, containing 79 exons of about 200 bp each, separated by introns up to 250 kb long. Northern blotting studies have also shown that the DMD gene is normally transcribed as a 14 kb mRNA, which is predicted to be translated into a membrane-associated rod-like protein of 3685 amino acids termed dystrophin.

The dystrophin protein

Antisera to dystrophin have revealed it to be a 427 kD ubiquitously expressed muscle protein representing 0.002% of the total muscle protein. As in many other instances, computerised sequence analyses have been invaluable in generating both structural and functional models of dystrophin. Computer database comparisons have also demonstrated the existence of four domains (denoted A, B, C and D) within the dystrophin molecule, and enabled putative functional roles to be ascribed to each domain based upon homologies with proteins of known function (Fig. 85.3). The central portion (domain B) is therefore predicted to form a triple helix of 2700 amino acids with homologies to two cytoskeletal proteins, α-actinin and spectrin. The 240 amino acid N-terminal A domain has homology with the actin filament-binding portion of α-actinin, and domain C has homology with the

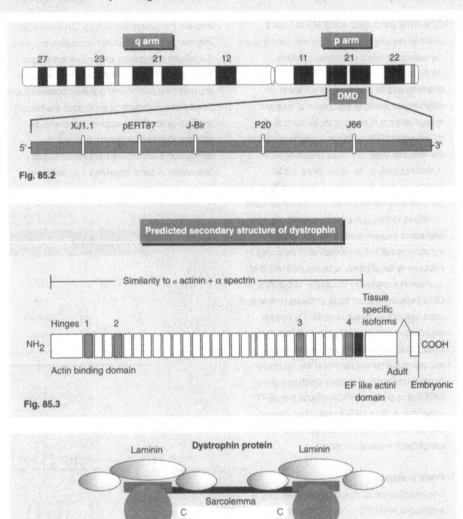

Fig. 85.2

Predicted secondary structure of dystrophin

Fig. 85.3

Fig. 85.4

C-terminal part of α-actinin. Domain D, located at the C-terminal region of dystrophin, bears no homology with known proteins. It is however probably involved in linking the dystrophin molecule to integral membrane proteins (e.g. dystrophin-associated glycoproteins or DAGs) which may control calcium ion flux across the muscle cell membrane. It is further speculated that two pairs of dystrophin molecules are arranged on the internal surface of muscle cells linking together three DAGs, with the A domains linked to F-actin molecules forming the cytoskeleton (Fig. 85.4). Mutations, particularly those leading to loss of the C-terminus of the dystrophin protein, consequently reduce normal muscle cell functions and lead to the pathophysiologies associated with DMD and BMD.

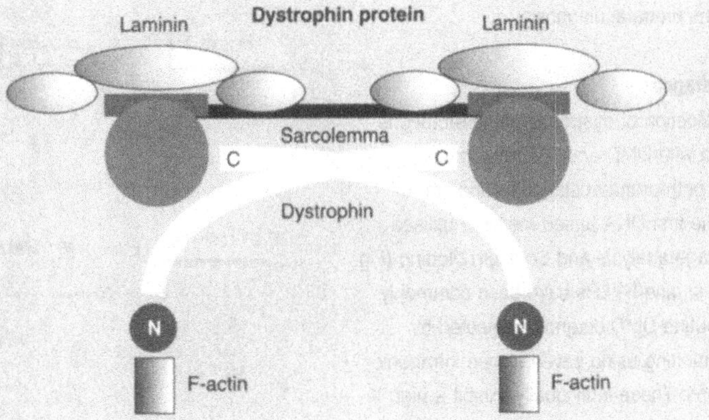

The cloning and identification of mutations within the dystrophin gene has revolutionised diagnosis of DMD/BMD, and is being exploited for the development of curative gene replacement-based therapies.

COMMON DMD AND BMD MUTATIONS

Disease severity appears to depend largely upon the amount of intact dystrophin produced. Accordingly, the less severe BMD is characterised by mutations which result in intermediate levels of expression of a larger, or truncated, form of dystrophin. In contrast, the more severe DMD condition is characterised by the very low level, or complete absence, of dystrophin protein. Analyses of the 2.3 Mb dystrophin gene in numerous DMD and BMD patients have demonstrated a range of different mutations including duplications, deletions and alterations in open reading frames. These variously result in the production of truncated or excessively large proteins, or reduced levels of dystrophin expression. Unusually, about 70% of DMD patients harbour large deletions (rather than point mutations), whereas about 7% contain duplications of parts of the gene. In fact, 50% of DMD patients have deletions confined to two regions in the central portion of the dystrophin gene, termed high frequency deletion regions (HDFRs), a proximal HFDR around the pERT locus, and a distal HFDR centred on exon 45. The remainder exhibit novel mutations, which complicates prenatal diagnosis.

Prenatal diagnosis

The identification of dystrophin gene mutations associated with DMD and BMD has greatly facilitated both prenatal diagnosis and carrier testing. The first DNA-based methods utilised RFLP linkage analysis and Southern blotting (Fig. 85.5). Thus, nine RFLPs have been commonly used in routine DMD diagnosis detected by Southern blotting using seven cloned intragenic DNA probes. These individually exhibit a high degree of recombination, ranging from 5 cM (XJ1.1 and pERT series) to 10 cM (754 and C7) recombination. In some instances this high degree of recombination can make diagnosis unreliable. Establishing haplotypes based on the use of two or more flanking RFLPs and allele markers can however be up to 99% accurate.

The diagnostic detection of RFLPs by genomic DNA probes has more recently been superseded by the use of cDNA probes in Southern blotting. This has enabled the direct detection of deletions and other abnormalities (e.g. duplications). This involves the generation of large DNA restriction fragments using 'rare-cutter' restriction enzymes (i.e. whose recognition sequences are 8 bp or more) which cut within introns. These are then separated by pulsed-field gel electrophoresis and deleted exons detected based upon the binding of cDNA probes. This approach can also be effective in detecting deletions in one X chromosome in carrier females based upon differences in band intensities (i.e. deletions appear as bands with half the normal intensity).

Ninety-eight per cent of the deletions detectable by cDNA analysis are now also routinely detected using the polymerase chain reaction (PCR). Most importantly, these methods reduce the amount of DNA required and, in the case of multiplex PCR, allow several potential mutations to be examined in a single test. The PCR is also used for accurate carrier detection based upon the use of a new class of polymorphic marker termed short tandem repeats or STRs (e.g. CA dinucleotide repeats). For example, a set of four STRs have been identified within the most deleted central region of the dystrophin gene.

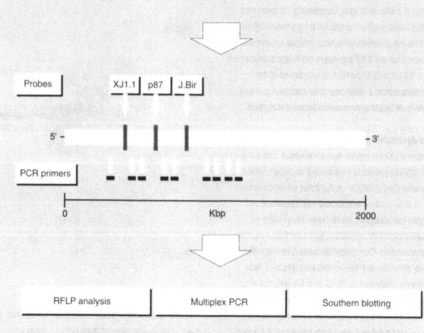

Probe	Restriction enyme	Alleles (Kbp)
pERT87.30	BgⅡ	30 Kbp/8 Kbp
pERT87.1	Xmnl	8.7 Kbp/7.5 Kbp
J.Bir	BamHI	21 Kbp/5 Kbp
XJ1≥1	Taql	3.8 K/5 Kbp
		3.8 Kbp/3.1 Kbp

Fig. 85.5

OAs: 835–6, 857–62, 864–5, 890–6, 1111–13. RAs: 294, 250, 304–8, 310–11, 410–11, 417. SFR: 119–20, 158–9, 173–835, 220–1, 227–9.

DEVELOPING DMD THERAPIES

Despite improvements in diagnosis, DMD and BMD remain incurable, and ultimately lethal conditions for which clinical treatment can only involve alleviating symptoms. However, the rapid advance in understanding the molecular basis of these conditions achieved by cloning the extremely large dystrophin gene, and the equally rapid development of recombinant DNA technologies for altering genes *in vivo*, suggest that a cure(s) may soon be available based upon supplying the missing dystrophin protein. Several such therapeutic approaches are currently being investigated which demonstrate that gene replacement or gene correction may be both feasible and practical.

Myoblast transfer therapy

Dystrophin replacement by myoblast transplantation does not involve recombinant DNA techniques. It exploits the ability of normal muscle stem cells, termed myoblasts, to fuse with the muscle fibres of DMD patients thereby allowing the synthesis of enough normal dystrophin to prevent progressive muscle degeneration (Fig. 85.6a). Although myoblast transfer decreases DMD progression by allowing the entire normal dystrophin gene and its regulatory sequences to be delivered to muscle fibres, several obstacles limit its effectiveness.

To be fully effective myoblasts must be injected into all muscles including key organs such as the heart and diaphragm (respiratory failure being the main cause of death). This is not only impractical but may lead to transplant rejection. Furthermore, it would not combat the problems of loss of brain function which are generally observed in DMD patients. Indeed, the early promise of myoblast transfer in producing normal levels of dystrophin synthesis in *mdx* mice (i.e. with dystrophin gene deletions similar to those in DMD) has not been seen in human trials.

Gene replacement therapy

The development of methods for replacing defective dystrophin genes is rapidly gaining momentum, despite several problems (including which type of gene constructs and vector delivery systems to use). Accordingly, the introduction (via an expression plasmid) of a full length mouse dystrophin cDNA gene into *mdx* mice (with the diaphragm weakening characteristic of human DMD) has been shown to allow dystrophin synthesis in transgenic animals (Fig. 85.6b). Although the level of dystrophin expression was greater than 50 times that in normal mice, these transgenic *mdx* mice showed marked improvement in the morphology of all muscles, and attained the normal force and power of their diaphragms. Based upon these experiments, dystrophin gene replacement appears to be feasible and will probably be the first approach to be available.

Alternative approaches

Additional gene therapy approaches are also being developed which could potentially compensate for the loss of dystrophin in DMD/BMD. For example, dystrophin-like genes (e.g. utrophin) which co-localise with DAGs in early development and are encoded by genes on autosomes may be artificially upregulated. It may also be possible to induce the alternative splicing of aberrant dystrophin RNAs to produce shorter, functional dystrophin molecules.

Alternative targets

The localisation, identification and cloning of the genes responsible for many other forms of muscular dystrophies (e.g. myotonic dystrophy) and several other neuromuscular diseases is also now possible. Cloning and identification of the genes and mutations involved will not only provide markers for prenatal diagnosis, but is also likely to be followed in many cases by the development of effective gene therapies shortly thereafter.

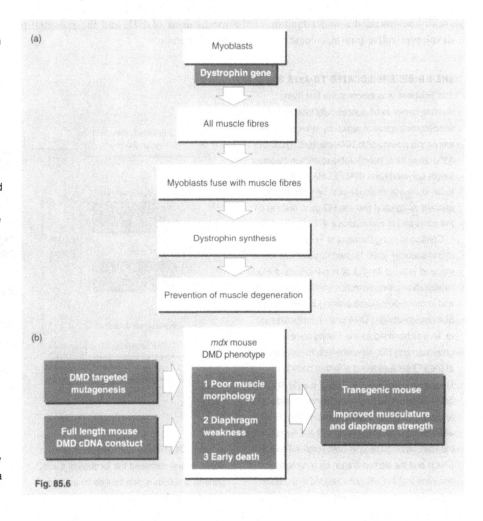

Fig. 85.6

86. IDENTIFYING THE GENE FOR HUNTINGTON'S DISEASE

A decade after linkage analysis localised the Huntington's disease gene to chromosome 4, extensive mapping and cloning has identified the gene (termed *IT15*) and its disease-associated mutations.

First described in 1872 by George Huntington of Ohio, USA, Huntington's disease (HD) is a progressive neurodegenerative disease affecting about 1 in 10 000 individuals. Inherited in an autosomal dominant fashion, HD is characterised by motor disturbance (giving rise to the distinctive choreic movement disorder), cognitive loss and psychiatric disturbances. These are now known to be caused by the loss of selected neurons within the brain. Generally a late onset disease appearing in the fourth and fifth decades of life and leading to death within 10 to 20 years, the occasional occurrence of a more severe form of HD in juveniles suggests that several factors may determine the age of onset. The biochemical basis for the selective neuronal loss characterising HD remains a mystery and prevents the development of effective treatments for delaying or preventing the onset or progression of HD. Nevertheless, molecular genetic studies have provided a major breakthrough in understanding the mechanisms of this disorder. In many ways, the case of the HD gene exemplifies the great effort required, and the technical difficulties encountered, when attempting to identify disease genes. Accordingly, the HD gene was first assigned to chromosome 4 (band 4p16.3) in 1983 by linkage analyses using polymorphic DNA markers. This immediately afforded the means for pre- and antenatal diagnosis of HD. However, the subsequent isolation of the HD gene (now termed *IT15*) bearing disease-associated mutations required a further decade of concerted study using many linkage mapping and cloning techniques. Although the function of the encoded protein (a 383 kD protein since termed Huntingtin) remains elusive, these studies have greatly advanced the understanding of the mechanisms of HD, and the potential to develop effective pharmacological and gene-based therapies.

FACTORS DICTATING AGE OF ONSET

Several factors governing the variability in the age of onset of HD have so far been suggested or implicated. These include the possession of an inherited but as yet unknown set of 'ageing genes', or maternally transmitted factors. It is also suggested that the age of onset is related to differences between individuals in the methylation of the HD gene region. Such differences may arise by a process termed genomic imprinting. Thus, the HD gene may become differentially modified (methylated) as it passes through the maternal or paternal germline. This then leads to earlier or higher levels of gene expression when, for example, transmitted by the father. Indeed, juvenile HD is known to be associated with paternal transmission of the disease allele. Genomic imprinting has also been suggested to be responsible for the early onset of spinocerebellar ataxia associated with paternal gene transmission, and the increased severity of neurofibromatosis associated with maternal gene transmission.

THE HD GENE IS LOCATED TO 4p16.3

The HD gene was amongst the first human disease genes to be successfully located to a specific chromosomal region by linkage analyses employing polymorphic DNA markers. Thus, by 1983 analysis of polymorphic restriction fragment length polymorphism (RFLP) DNA markers shown to be in linkage disequilibrium by recombination analysis suggested that the HD gene resided on the short arm of chromosome 4.

Confirmed using fluorescent *in situ* hybridisation (Fig. 86.1), the HD gene was further mapped to band 4p16.3 by non-fluorescent *in situ* hybridsation (using reflection contrast microscopy and immuno-peroxidase staining to reveal binding of a hapten-labelled DNA probe). Its localisation to 4p was heralded as the turning point in understanding HD. Nevertheless, the identification of the HD gene required a further decade of study involving the application of a wide range of detailed gene mapping and cloning techniques to analyse thousands of DNA samples from HD-affected families. This began with the formation of the Huntington's Disease Collaborative Research Group and the demonstration by recombination analyses that the HD gene resided in a 2.2 Mb region flanked by two polymorphic linkage

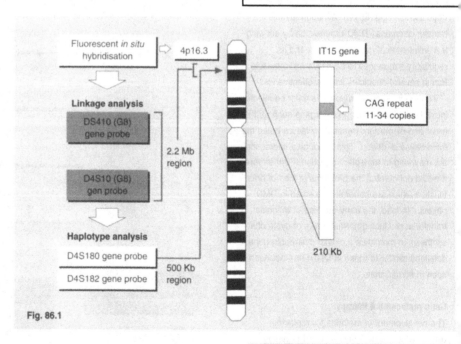

Fig. 86.1

markers, D4S10 (G8) and D4S98. Haplotype analysis then narrowed the location of the HD gene to a 500 kb region flanked by markers D4S180 and D4S182.

OAs: 192, 737–8, 863–4, 897–94.
RAs: 77–8, 304–8, 312–13, 427.
SFR: 53, 119–20, 162, 164–5, 173–82, 229.

Identifying the HD gene

The region flanked by G8 and D4S110 has become one of the most extensively studied segments of the human genome. Indeed, it has been subjected to many mapping techniques including positional DNA cloning, exon amplification and cDNA cloning in the search for the HD gene. Accordingly, following the analysis of a wide range of DNA markers (such as those developed by chromosome jumping from the G8 marker), cosmid and yeast artificial chromosome (YAC) clones spanning both the 2.2 Mb and 500 kb regions were constructed (Fig. 86.2). The techniques of exon amplification and exon trapping were then used to analyse the 16 cosmid contigs (i.e. contiguous, or overlapping, clones) spanning the 500 kb region. Both of these techniques have also been proven to be powerful and rapid methods for identifying coding regions (i.e. exons), based upon the ability of monkey COS cells to transcribe coding regions within random DNA fragments when introduced via appropriate vectors.

The subsequent detection of transcripts by the polymerase chain reaction then enabled four genes to be identified within this region: the α-adducin gene (*ADDA*), a potential transporter gene (*IT10C3*), a novel G-protein-coupled receptor kinase gene (*IT11*), and a gene denoted *IT15*. No genetic defects were however found in the *ADDA*, *IT10C3* and *IT11* genes which would implicate them in HD epidemiology.

Consequently, by isolating and sequencing overlapping cDNA clones representing the entire 210 kb of the *IT15* gene, it was found that this gene encoded a hitherto unknown protein (now termed Huntingtin) of 383 kD. Furthermore, the *IT15* gene contained repeat CAG motifs (potentially encoding seven amino acids) at the 5′ end which varied in number between unaffected and HD-affected European individuals. At least 17 *IT15* alleles have been found in normal individuals, each varying in size between 11 and 34 CAG copies, whereas HD-affected individuals have many more CAG copies, ranging from 42 to over 66. Although other mutations within the *IT15* gene locus have not been ruled out, the presence of an increased number of unstable CAG repeats is suggested to be responsible for the dominant HD phenotype (i.e. by altering either the

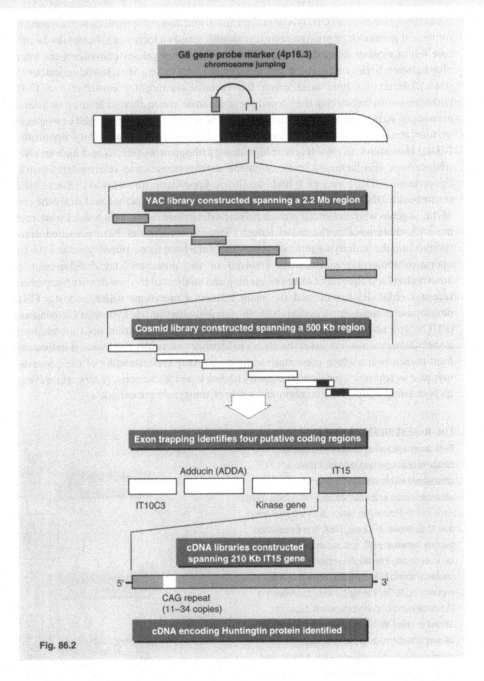

Fig. 86.2

properties of the mRNA transcripts or the protein). This information increases the ability to identify HD-affected individuals before disease manifestation (i.e. by DNA-based pre- and antenatal diagnostic methods). It may also allow the development of curative therapies (e.g. by gene replacement). Determining the function of the *IT15* gene-encoded protein, and the effects

of the HD-associated mutations, will also undoubtedly increase the understanding of the biochemical mechanisms involved in HD onset and progression. Such insights would be extremely valuable for the development of pharmacological agents capable of increasing the age of onset and/or decreasing the progression of HD.

87. LIPOPROTEIN GENES AND CORONARY HEART DISEASE

The analysis of lipoprotein transport genes within transgenic mice models is providing many insights into the molecular and genetic basis of atherosclerosis, and is aiding the search for new therapies.

Coronary heart disease (CHD) is of major clinical and economic significance, accounting for tens of thousands of premature deaths annually, and an increasing slice of the health-care bill of most developed countries. Although a complex disease involving the interplay between both genetic and environmental risk factors (e.g. diet, smoking, inactivity, stress), defects in lipid metabolism homeostasis significantly contribute to CHD pathogenesis by promoting the formation of atherosclerotic plaques leading to arterial narrowing. Accordingly, approximately 90% of heart attack patients exhibit one or more of four abnormal lipoprotein phenotypes: (i) high levels of low-density lipoprotein (LDL) cholesterol; (ii) low levels of high-density lipoproteins (HDL) and high levels of triglycerides; (iii) increased levels of chylomicron remnants and intermediate density lipoproteins (IDL); and (iv) high levels of lipoprotein (a) (Lp(a)). Since 1982, recombinant DNA techniques have increasingly been used to isolate and study the role of the 17 genes whose encoded protein products determine lipoprotein levels by interacting with cholesterol in the blood stream. These investigations have provided many insights into the underlying genetic basis of abnormal lipoprotein phenotypes and the co-operative molecular mechanisms involved in the initiation and development of atherosclerotic plaques. For example, cloning and analysis of the low density lipoprotein receptor (LDL-R) has defined the main classes of mutations which underlie CHD development amongst individuals with familial combined hypercholesterolaemia (FHC). The identification of defects and/or polymorphisms within lipid metabolism-associated genes also increases the ability to identify genetically predisposed individuals (and therefore introduce preventative measures). Our understanding of the processes involved in atherosclerosis, and the ability to develop new therapies, is however increasingly dependent upon the creation and study of transgenic mice models.

EXON SHUFFLING AND THE LDL-R

The LDL-R provides direct evidence for the modular construction of proteins and their evolution by the process of exon shuffling. The LDL-R gene is composed of 18 exons which may be grouped into six blocks, each encoding a protein domain of differing function. Four of these bear motifs common to many membrane proteins whereas the remaining two bear striking homologies to other proteins. Accordingly: (i) exon 1 encodes an N-terminal signal sequence required for cell surface expression; (ii) exon 16 encodes a transmembrane 'anchoring' domain; (iii) exon 15 encodes an oligosaccharide-binding region; (iv) exons 17 and 18 encode a cytoplasmic domain; (v) exons 2 to 6 encode the ligand-binding domain with remarkable homology to the complement component co-factor C9; and (vi) exons 7 to 14 are homologous to many proteins including epidermal growth factor, blood clotting factors IX, X and C, and the *Drosophila* notch protein.

LDL-R GENE DEFECTS AND FHC

FHC is an autosomal co-dominant disease, clinically characterised by raised plasma cholesterol levels resulting in premature atherosclerosis and heart attacks, often fatal, very early in life. Pioneering research by J.L. Goldstein and M.S. Brown in Texas, USA, first established the link between FHC and mutations within the gene encoding the cellular receptor for the cholesterol-rich LDL particles. Known as the LDL-receptor (LDL-R), it is responsible for delivering the cholesterol into the cell, where it can be stored or used as the substrate for the synthesis of new structures (e.g. the lipid bilayer of cell membranes) and energy-generating catabolism. Many different mutations (deletions and base substitutions) within the LDL-R gene have now been identified occurring throughout the gene (Fig. 87.1). These fall into four classes and are ultimately responsible for the defective uptake of LDL.

As with many other diseases (e.g. the muscular dystrophies), the existence of many different LDL-R mutations poses problems for the development of antigene therapies. Interestingly, the cloning and sequencing of LDL-R genes has provided direct evidence regarding the evolution of genes based upon exon shuffling events.

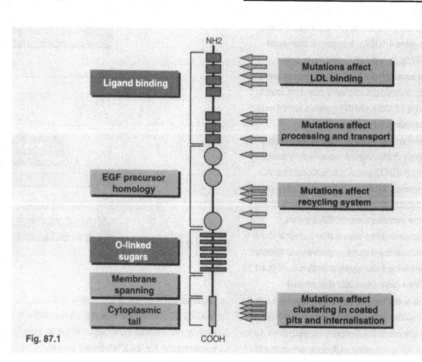

Fig. 87.1

OAs: 478, 837, 895–907, 1114.
RAs: 314–20, 410–11, 427.
SFR: 120, 158–9, 181–3, 186–7.

apo(a) polymorphisms and CHD

In recent years, cDNA and genomic cloning, sequencing, Southern blotting and transgenic mice studies have shown a causal link between size polymorphism of the human apolipoprotein (a) (apo(a)) gene and CHD. Structurally similar to plasminogen (a key enzyme in the regulation of blood clotting) the size of the apo(a) alleles appears to determine the susceptibility to CHD. The size of the alleles is determined by the number of reduplicated kringle IV domain-encoding gene segments, with smaller alleles conferring a higher risk of CHD by determining the level of serum apo(a)-bearing particles (i.e. Lp(a)). It is not precisely known how allele size determines apo(a), and hence Lp(a), serum levels although differences in messenger RNA synthesis and stability have been suggested. The fact that apo(a) genes are only expressed in liver cells initially posed many technical challenges. The application of new genetic analysis (see PCR) and recombinant DNA technologies (e.g. transgenics) is however overcoming such problems. Combined with cellular and biochemical studies, they may allow the early identification and treatment of individuals with high CHD susceptibility.

Transgenic studies

The ability to transfer and express human genes in animals has been particularly important in establishing the strong causal link between many lipoprotein transport genes (including Lp(a)) and the development of atherosclerosis. For example, mice do not naturally produce apo(a) or develop significant amounts of atherosclerosis, even when fed a high fat diet. In contrast, recently developed transgenic mice which have been transfected with cDNA clones encoding human apo(a) have been demonstrated to develop high numbers of morphologically large atherosclerotic plaques correlating with the expression of the transfected DNA (Fig. 87.2). Studies such as these are also providing the means by which to determine how apo(a) gene polymorphisms mediate their atherogenic potential. They may also provide the means to overcome the technical problems inhibiting the development and evaluation of new forms of antigene therapies (e.g. the liver tissue-specific synthesis of apo(a)). Similar benefits will

undoubtedly arise from the creation of transgenic mice bearing human *ApoA-I*, *ApoCIII* and/or *CII*, *Apo A-I*, *ApoCIII* and cholesterol ester transfer protein genes, or in which *ApoA-I* and *E* genes have been disrupted by gene 'knock-out' techniques.

New therapies

The ability to identify individuals at high risk of developing CHD should allow the instigation of effective disease avoidance or prevention strategies, thereby reducing the requirement for health-care resources. Many new therapies may also come from unravelling the complex and interactive genetic, cellular and biochemical mechanisms involved in atherosclerosis. Indeed, many studies are pointing the way to therapies that overcome problems such as the tissue-specific expression of particular genes. For example, experiments using the Watanabe strain

of hypercholesterolaemic rabbits have already demonstrated that the causal gene defects may be corrected by *in vitro* retroviral transfection of the animal's own hepatocytes by retroviral vectors containing the corrected gene. These are then successfully returned to the animal's liver to correct the animal's inherited hypercholesterolaemic condition. Strategies for directly correcting the genetic defects themselves (e.g. using gene replacement techniques) may, in themselves, only be effective before the onset of atherosclerotic plaque formation. However, having corrected a gene defect it may be possible to reverse existing arterial damage using pharmacological agents (including many cloned cellular factors such as recombinant cytokines). A further challenge remains to develop therapies effective for those who have already formed substantial numbers of atherosclerotic plaques and developed CHD.

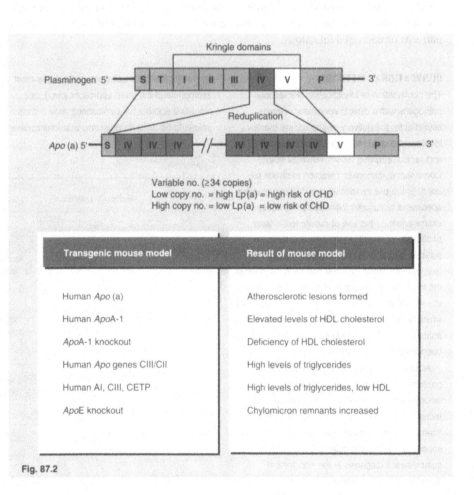

Variable no. (≥34 copies)
Low copy no. = high Lp(a) = high risk of CHD
High copy no. = low Lp(a) = low risk of CHD

Transgenic mouse model	Result of mouse model
Human *Apo* (a)	Atherosclerotic lesions formed
Human *ApoA-1*	Elevated levels of HDL cholesterol
ApoA-1 knockout	Deficiency of HDL cholesterol
Human *Apo* genes CIII/CII	High levels of triglycerides
Human AI, CIII, CETP	High levels of triglycerides, low HDL
ApoE knockout	Chylomicron remnants increased

Fig. 87.2

88. THE DETECTION OF MICROBIAL INFECTIONS

Nucleic acid probe hybridisation and *in vitro* amplification-based assays are increasingly being used for the clinical detection and strain typing of bacterial, viral, fungal, and protozoan infections.

Many microbes inhabiting the planet naturally provide beneficial effects (e.g. enterobacteria which assist food breakdown in the gut), or have been harnessed/exploited for a variety of purposes. Unfortunately, infection by an increasing number of microbes can have a profoundly deleterious effect upon human health (i.e. they are disease causing or pathogenic). The application of molecular biological techniques in basic and clinical research has identified a wide range of such microbial pathogens (e.g. the human immunodeficiency virus), and increased the understanding of their pathogenicity. They are also enhancing our ability to intervene therapeutically by developing effective, often novel, agents to target key stages of infection (e.g. anti-microbial drugs and vaccines). Nucleic acid-based techniques are also increasingly being used in modified formats as diagnostic tools for detecting and characterising (e.g. strain typing) microbial infections. These are complementing and/or extending the diagnostic capacity of conventional microbiological approaches (e.g. based upon staining, *in vitro* culture and immunological techniques). Thus, several forms of nucleic acid probe-based hybridisation assays, and *in vitro* amplification-based assays (e.g. utilising the polymerase chain reaction (PCR), ligase chain reaction, and Qβ-replicase), are increasingly used for the rapid detection of a range of bacterial, fungal and protozoan infections. Indeed, techniques such as reverse transcription-PCR (RT-PCR) are particularly useful in the diagnosis of infections caused by pathogenic viruses which, for a variety of reasons, are otherwise difficult to rapidly or sensitively detect. The rapidity and sensitivity of these methods therefore enables the initiation of effective treatments at appropriate times during infection, and the monitoring of residual infections. They may also be used as epidemiological tools for studying patterns of microbial infections.

CONVENTIONAL METHODS

The confirmation or identification of infectious pathogens within clinical specimens remains essential for the delivery of appropriate therapy. Ideal tests should be simple, reliable, specific, and rapid (i.e. giving results within 24 hours). Conventional diagnostic detection methods rely upon: (i) the use of specific dyes to stain relevant specimens or cultures followed by microscopic examination; (ii) the use of culture techniques based upon the ability to generate colonies in suitable media, or to detect infecting organisms (e.g. viruses) within cultured host cells by virtue of the induction of cytopathic changes (e.g. changes in morphology); and (iii) immunological techniques which detect the organism or specific microbial antigens, or pinpoint changes in pathogen-specific antibody responses.

Although conventional methods currently predominate in most routine diagnostic laboratories, they variously suffer from several technical shortcomings (see Table 88.1). For example, many organisms are difficult (or impossible) to grow in culture. Consequently, culture-based diagnosis is too slow for the detection of acute infections. Although immunological techniques are perhaps the most appropriate of the three approaches (e.g. being rapid and suitable for automation), they are often unable to be used for detecting viral infections or

ENGINEERING NEW THERAPIES

Many recombinant DNA techniques and approaches are currently assisting with the development of a wide range of conventional anti microbial drugs and vaccines. They are also being employed for the development of novel therapeutic approaches and reagents. For example, antisense oligonucleotides (e.g. peptide nucleic acids) have been developed for use as anti-viral agents. Furthermore, genetic engineering methods have been used to develop a novel anti-viral strategy (termed capsid-targeted viral inactivation) based upon the creation of capsid protein–nuclease fusion proteins. Shown experimentally to interfere with the replication of several types of virus including retroviruses, capsid–enzyme fusions may also be used to eliminate virions *in vivo*.

OAs: 410–11, 466, 491, 499, 510–11, 908–33. RAs: 191–4, 199–201, 204–6, 321, 333–5. SFR: 119–20, 136–41, 146–7, 179–81, 188–93.

for strain typing (e.g. due to lack of appropriate strain-specific protein markers and/or the inability to generate strain-specific antibody probes).

Table 88.1

	Direct visualisation	Culture techniques	Immuno-techniques
Type of analysis	Specific staining, light microscopy, EM	Growth on agar plates, liquid medium culture	Agglutination/ELISA/ fluorescence methods
Type of detection	Whole cell detection and morphology	Whole cell detection and cytopathic effects	Specific antigen detection
Time for assay	1–2 days minimum	1 week minimum (up to 6 weeks maximum)	1 day minimum
Problems	Unable to detect strain types	Some organisms difficult to culture	Unable to detect all strain types

236

DNA-based detection methods

The use of conventional approaches is now being challenged by a range of DNA/RNA-based assays. These are potentially capable of rapidly and sensitively detecting a broad range of microbial pathogens, including bacteria, viruses, fungi and protozoa (Table 88.2). Both radioactively and non-isotopically labelled nucleic acid (generally DNA) probes have become widely available and are beginning to be used for routine clinical microbial detection based upon their ability to hybridise with microbial DNA or RNA sequences. They have been used in many formats including Southern blotting, dot, spot and colony blotting, and *in situ*, sandwich, solution and capture hybridisation. Formats such as capture and solution hybridisation are increasingly being adopted since they are more rapid and are particularly well suited to automation.

By careful design of the probes, these assays may be capable of detecting not only species-specific sequences but also strain-specific sequence differences resulting from mutations or alternative splicing events. Some are also particularly suitable for target sequence quantitation and are therefore more applicable. For example, the ability to quantitate target sequence copy number in a specimen is particularly useful in distinguishing between active infections and latent or subclinical infections, particularly frequent in areas of endemic disease. A potential limitation of hybridisation assays is however their poor sensitivity, since relatively large samples are often required for the isolation of sufficient target nucleic acid. The sensitivity of these assays may however be increased by employing signal amplification labels, or using probes capable of recognising highly abundant microbial DNA or RNA sequences (e.g. highly repetitive sequences, multi-copy genes, or highly expressed genes). Alternatively, they may be used in conjunction with *in vitro* amplification assays such as the PCR. *In vitro* target or probe amplification techniques are however being developed for microbial detection in their own right. PCR-, ligase chain reaction-, and Qβ-replicase-based assays are now available for the detection of many microbial DNA and RNA target sequences. By careful primer design, these techniques can distinguish between species,

strains or individual clinical isolates. Their sensitivity and rapidity enables them to be used for the detection of much less abundant target copies (e.g. a single gene in a single microbe in a single specimen). Consequently, smaller samples are required and residual infections can be more readily detected. A balance must however be struck between sensitivity and informativeness. For example, extreme sensitivity may not be useful and can result in false positives (e.g. by accidental sample contamination by a single microbe).

Although suitable for automation and

quantitation, such assays are generally poor at discriminating between live and dead/attenuated microbes (since they all contain DNA/RNA). A promising solution currently under development for detecting bacterial infections employs phages engineered to contain the *lux* gene. Infection of live bacteria by such phages (which can be species- or strain-specific) results in the generation of a detectable and quantifiable bioluminescent signal. In contrast, attenuated or dead bacteria remain dark and are therefore not detected.

Table 88.2

Infectious agent	Standard detection	rDNA detection (Licensing Co.)
Gonorrhoea *Neisseria gonorrhoeae*	Growth on agar plates Immunofluorescence Immunoassay	PCR rRNA (Gen-probe)
Chlamydiae *Chlamydiae trachomatis*	Growth on agar plates Immunofluorescence Immunoassay	Gene probes (Roche diagnostics)
Legionella *Legionella pneumophila*	Growth on agar plates Serological tests	rRNA, PCR, PFGE (Gen-probe)
Listeria *Listeria monocytogenes*	Growth on agar plates Serological tests	Gene probes under tests
Mycobacteria *Mycobacteria tuberculosis*	Growth on agar plates (slow) Serological tests	Gene probes and PCR (Gene-probe and enzo)
Salmonellae	Growth on agar plates Serological tests	Gene probes and PCR (Amoco Gene-Track)
Cytomegalovirus	Viral cell culture (slow) Serological tests Immunological	Gene probes and PCR (Digene/ImClone/Roche)
Epstein Barr virus	Viral cell culture (slow) Serological tests Immunological	Gene probes and PCR (Digene/ImClone/Enzo)
HIV virus	Viral cell culture (slow) Serological tests Immunological	Gene probes and PCR (Digene/ImClone/Enzo/Roche)

89. MOLECULAR BIOLOGY OF HUMAN IMMUNODEFICIENCY VIRUS AND AIDS

AIDS is now believed to be caused by the infection of T helper cells (via surface CD4 receptors) with the RNA-containing HIV retrovirus, which replicates as a provirus following integration into the host genome.

First recognised in 1982 as a new disease, acquired immune deficiency syndrome or AIDS is associated with very high mortality rates. Initially described amongst homosexuals, AIDS is now known to be transmitted via the blood of affected individuals and may therefore potentially affect any members of the population, although several high-risk groups may be identified (e.g. needle-sharing intravenous drug users). Despite its extremely low incidence overall compared with other life-threatening diseases (e.g. cancer, heart disease and diabetes), initial fears and speculations concerning its mode of transmission catapulted AIDS into the media spotlight as a major world health problem of epidemic proportions. Consequently, this hitherto unknown human disease has already been extensively investigated, with billions of dollars invested in research into its cause, epidemiology and treatment (including 'preventative' advertising). By 1984 a virus now known as human immunodeficiency virus or HIV was identified as the probable causative agent. Whilst many aspects of AIDS, including the role of HIV and its origins, remain controversial subjects for debate, AIDS represents a remarkable example of the power of molecular biological techniques and approaches in rapidly elucidating disease mechanisms. Thus, recombinant DNA techniques are already unravelling the complexities of the HIV life cycle, and its epidemiology. They are also aiding the clinical diagnosis and management of HIV infections, and reducing HIV transmission via the screening of therapeutic blood-based products or the generation of recombinant proteins (e.g. blood factor VIII). Molecular technologies may also provide effective AIDS therapies (e.g. by aiding the development of recombinant-derived drugs targeting vulnerable points within the HIV life cycle), and the means for its ultimate eradication (e.g. via the development of effective HIV vaccines).

CONTROVERSIAL VIEWS ON AIDS

The appearance of AIDS has been linked to divine retribution for human immorality. Although this view is not shared by many, scientists have attempted to provide evidence supporting several more plausible arguments concerning the origins of AIDS. These arguments include the random mutation of an existing virus (originally derived from other animal species) to allow it to exploit a weakness in the immune system. It has also been suggested that HIV arose from naive experiments performed earlier this century using primate–human blood transfers as part of experiments to develop anti-malarial vaccines. Although feline and primate equivalents of HIV have been described, current evidence cannot yet adequately settle these disputes. The fact that HIV cannot be isolated from all AIDS patients has even led some scientists to suggest that HIV is not the cause of AIDS, whilst many question the epidemic status of AIDS. The social and scientific issues therefore appear as complex as the disease itself.

AIDS IS LINKED TO HIV INFECTION

AIDS was first recognised as a new sexually transmitted disease in 1982 by the Centre for Disease Control (USA) due to the increased incidence of the rare skin cancer Kaposi's sarcoma, and/or unusual infections amongst homosexuals (e.g. pnuemonia caused by *Pneumocystis carinii*) (Table 89.1). The term acquired immune deficiency syndrome was coined since infected individuals exhibited a severe immunodeficiency state characterised by a depletion of the T helper cells involved in combatting infections. Combined with evidence that AIDS was transmitted by the exchange of blood from an infected to an uninfected individual, this provided the first clue indicating that AIDS might be linked to some form of infection. The analysis of *in vitro*-cultured lymphocytes derived from patients at an early stage of disease quickly enabled the isolation of a retrovirus (termed human immunodeficiency virus or HIV) which infects and destroys human T helper cells. Many links have since been established between HIV and the development of AIDS. Accordingly, HIV is found in almost all AIDS patients and their numbers increase as the disease progresses.

Anti-HIV antibodies have also been found in asymptomatic individuals who eventually develop the disease. Furthermore, although it may be 10 years between HIV infection and the onset of AIDS, epidemiological studies have showed that AIDS only develops in areas where there has been prior HIV infection.

OAs: 411, 466, 499, 510–11, 910, 917–20, 934–43. RAs: 322–9, 337–8, 378, 413, 416. SFR: 120, 136–9, 191, 193–4, 207–10.

Table 89.1

The history of AIDS

1981	First reports of immune dysfunction leading to rare Kaposi sarcoma
1982	Disease recognised and named Aquired Immune Deficiency Syndrome (AIDS)
1983	Isolation by Montagnier in France of a Lymphadenopathy associated virus (LAV)
1983	Isolation by Gallo *et al.* in the USA of a human retrovirus named Human T cell leukaemia virus type III (HTLV-III)
1985	Further strain of virus dicovered in Africa
1985	Reports of AIDS virus entering T cells through CD4 receptor
1987	AIDS virus renamed HIV-I and HIV-II
1988	Research defines nucleotide sequence and functions of various genes
1990	Strategies designed against HIV enzyme functions
1992	Results of large scale trials of azidothymine (AZT) therapy reviewed

13. Organisation of genes within chromosomes.
23. Transcription.
24. Post-transcriptional processing of mRNA.
27. Transcription regulatory sequences.
30. *In vivo* translation.
32. Cell divison: mitosis and meiosis.
37. *In vivo* DNA replication.
41. Recombinant DNA technology.
42. Enzymes commonly used in molecular biology.
53. Converting mRNA into cDNA.
88. Molecular methods for detecting microbial infections.
91. Recombinant DNA techniques in vaccine development.
92. Receptors and cellular signalling.

THE HIV INFECTION CYCLE

Many molecular biological techniques and approaches have now been used to investigate the HIV life cycle. Consequently considerable data have been accumulated. HIV is now known to be a type of an unusual group of retroviruses, known as lentiviruses, which are capable of inducing cytopathic effects (e.g. host cell destruction).

HIV contains an unusually complex RNA genome which is replicated via a DNA intermediate following entry of the virus into host cells (Fig. 89.1). T helper cells are the primary targets for infection, although HIV may also reach the central nervous system via infection of migratory macrophages and dendrocytes. T helper cell entry is made possible because each HIV particle acquires a lipid membrane as it is released from an infected cell. This lipid membrane surrounds the protein core (containing the HIV RNA genome) and contains various viral coat proteins (e.g. glycoproteins gp120 and gp41) which act as ligands for the T helper cell CD4 receptor. Following interaction with the CD4 receptor, this lipid membrane fuses with the T helper cell membrane and releases the virus core into the host cell cytoplasm. Once inside, a viral reverse transcriptase synthesises a double-stranded DNA copy of the viral RNA genome which enters the nucleus and is integrated into the human genome as a provirus (Fig. 89.1b). This is then replicated with every cell division. Following integration, the provirus undergoes a series of complex transcriptional events which allow the formation of new lipid-coated viruses.

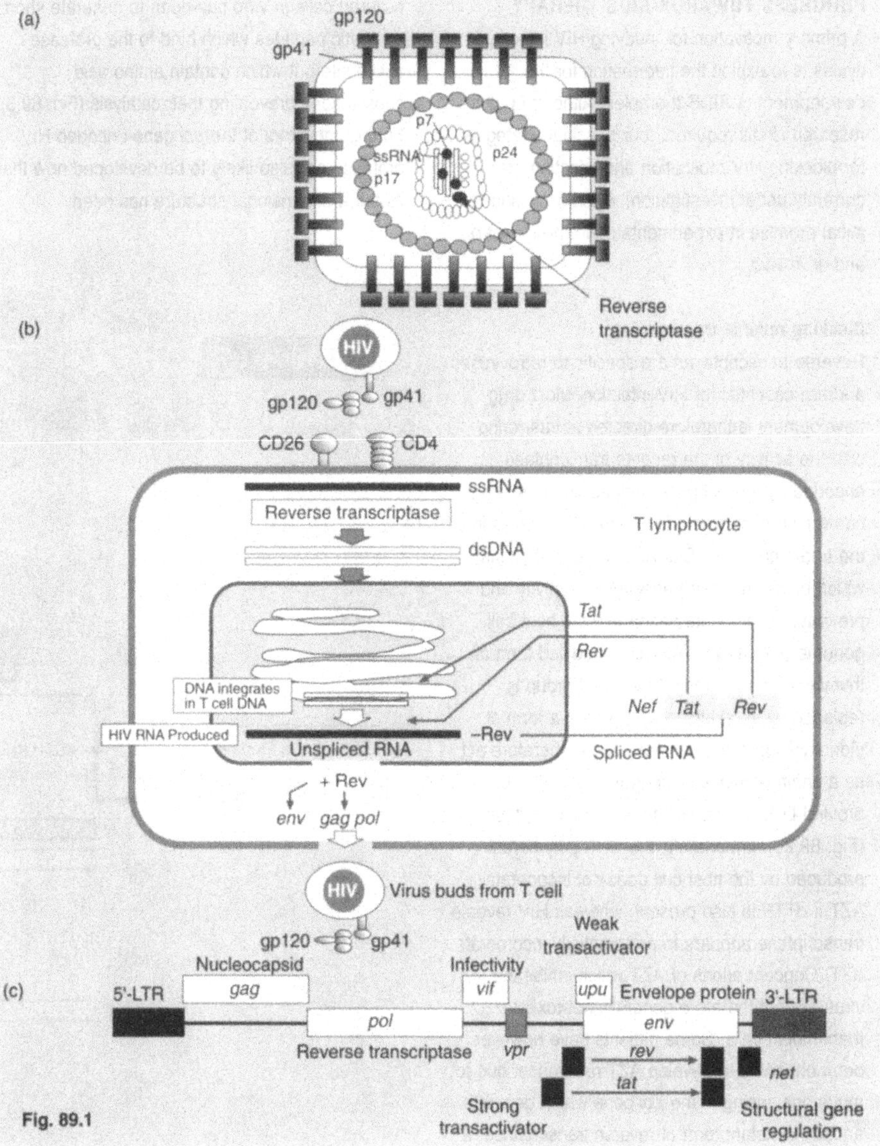

Fig. 89.1

Genes controlling HIV infection

The 9.2 kb RNA genome of HIV is arranged into nine separate genes (Fig. 89.1c). During reverse transcription within the host cell, these nine genes become flanked by identical structures termed long terminal repeats (or LTRs) at the 5′ (5′ LTR) and 3′ (3′ LTR) ends. These contain many regulatory (i.e. enhancer and promoter) elements and contain two unique stretches termed U3 and U5 at the 3′ and 5′ ends. Three of the genes are common to all retroviruses. They encode the capsid protein (*gag*), the envelope protein (*env*) and the reverse transcriptase and integrase (*pol*) required for provirus integration. All undergo an unusual form of alternative splicing. The remaining six small genes appear to primarily regulate viral gene expression. For example, two of the six genes, designated *rev* and *tat*, are now known to influence provirus transcription. Thus, at early stages of infection when there is an absence of Rev protein, only spliced viral RNA leaves the nucleus to be translated into Tat, Rev and Nef proteins. The Tat protein then acts upon the provirus to stimulate further transcription, and, once sufficient levels of Rev protein have accumulated, full length unspliced RNA transcripts leave the nucleus. These full length transcripts are destined to become new virus genomes. The function of the Nef protein remains to be determined but it appears to play a role in ensuring high levels of infection. The *env*, *gag* and *pol* genes are required to generate all the structural proteins, enzymes and coat proteins necessary for new virus particle formation at the cell membrane. The lipid layer containing the viral coat proteins is only added as the core particle emerges from the cell.

Knowledge of HIV infection cycles is enabling the development of recombinant AIDS vaccines and many therapies which may block HIV infectio and/or replication, or which destroy HIV-infected cells.

PROGRESS TOWARDS AIDS THERAPY

A primary motivation for studying HIV infection cycles is to exploit the information for the development of AIDS therapies. Although much research is still required, four principal strategies for blocking HIV replication and infection are currently under investigation, which have shown initial promise in experiments performed *in vitro* and/or *in vivo*.

Blocking reverse transcription

Reverse transcriptases are specific to retroviruses and are essential for HIV infection. Most drug development is therefore directed at interfering with the activity of the reverse transcriptase encoded by the *pol* gene. Indeed, the only drug currently proven to produce beneficial effects in the treatment of AIDS is AZT (or azidothymine) which blocks reverse transcriptase activity and prevents provirus integration into the host cell genome (Fig. 89.2a). AZT is a modified form of thymidine, in which the 3' hydroxyl group is replaced by an azido group (i.e. it is a form of dideoxynucleotide analogue). It can therefore act as a chain terminator during the synthesis of proviral DNA by the HIV reverse transcriptase (Fig. 89.2b). Importantly, the DNA polymerase produced by the host cell does not incoporate AZT if dTTP is also present, whereas HIV reverse transcriptase appears to preferentially incorporate AZT. Concentrations of AZT which inhibit the virus are not therefore generally not toxic to mammalian cells. Some patients have however been observed to develop AZT resistance, due to mutations arising in the *pol* gene which generate an AZT-resistant form of reverse transcriptase. It is however envisaged that additional dideoxynucleotides (e.g. based on dideoxyinosine) may be developed and used in alternating regimes to overcome the problem of AZT resistance.

Blocking HIV protease activity

Post-translational cleavage of the large polyproteins produced from the *gag* and *pol* genes by HIV protease is also vital to viral replication. Attention is therefore also focused towards developing drugs which will act as inhibitors of the HIV protease. One approach which successfully blocks HIV infection of cultured cells *in vitro* has been to generate short synthetic peptides which bind to the protease active site but which contain amino acid substitutions preventing their catalysis (Fig. 89.3). Further inhibitors of the *pol* gene-encoded HIV protease are also likely to be developed now that its three-dimensional structure has been determined by X-ray crystallography.

OAs: 944–56, 967, 1030, 1123. RAs: 324, 330–1, 337–8, 378, 413, 416. SFR: 120, 136–9, 191, 193–4, 207–10.

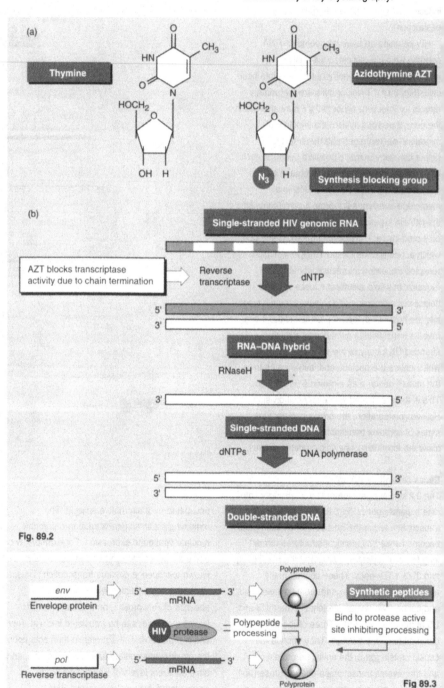

Fig. 89.2

Fig 89.3

Blocking CD4-mediated infection

Another vital part of the HIV infection cycle currently under investigation for therapeutic exploitation is the entry of the virus via the CD4 molecule expressed on the T helper cell surface (Fig. 89.4). Recombinant soluble forms of CD4 (sCD4) have now been generated which lack the membrane anchoring and cytoplasmic domain sequences. Initial experiments *in vitro* and in primate animal models have shown that sCD4 is able to prevent HIV infection by competing for binding to the virus surface-expressed gp120 protein. Unfortunately, such an approach may not be as effective in human patients and may not block the HIV infection of macrophages and dendrocytes. This would still allow HIV to infect the central nervous system. Recombinant CD4 molecules are however also being investigated as a potential means of blocking viral replication by binding to gp120 whilst inside infected cells. This would prevent gp120 expression on the cell surface. Preliminary experiments using *in vitro*-cultured cells, plasmid-borne sCD4 molecules, and gp120 suggest that this may also be an effective strategy for blocking HIV infection.

Targeting HIV-infected cells

Chimeric sCD4 molecules combining immunotoxins are also being investigated as a means to specifically destroy HIV-infected cells, similar to the use of antibody 'magic bullets' in the therapy of cancer. This strategy, and the use of antibody–toxin conjugates, is however of limited value since cells capable of being targeted (i.e. expressing surface gp120) are already releasing new infective viruses. It may however have value as a 'supporting' therapeutic approach (i.e. combined with other therapeutic strategies).

Progress towards AIDs vaccines

A feature still exploited to detect HIV infection is that the immune response makes anti-HIV antibodies (e.g. anti-gp120 neutralising antibodies). Unfortunately, this response is produced late in the course of infection and cannot eliminate virus-infected cells. The ability to clone and manipulate HIV coat protein genes is enabling the development of various potential AIDS vaccines based upon the production of recombinant proteins, peptides, or even genome-

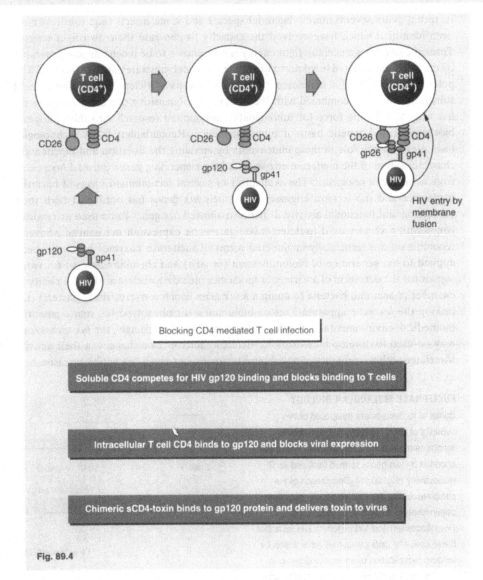

Fig. 89.4

devoid virion immunogens. For example, research has enabled the identification of a specific region within the gp120 viral coat protein termed the principal neutralising domain. This represents one of the dominant immunogenic sites recognised by the anti-HIV neutralising antibodies produced by AIDS patients (and immunised goats). This looped polypeptide from gp120, and a peptide from gp160, have already shown early promise in primate models. HIV coat protein- encoding genes have also been injected into human muscle cells where the expressed proteins serve as vaccination agents (i.e. immunogens).

DNA-based detection of HIV

Most assays for HIV detection have been based upon the detection of anti-HIV antibodies. The polymerase chain reaction is however now being used to diagnostically detect both HIV DNA and RNA, and to monitor HIV infections *in vivo* (e.g. during therapy). Since HIV has also been unwittingly transmitted via contaminated blood products, the PCR is also being used to screen human-derived biomedical products (e.g. factor VIII, blood transfusion materials, organs or bone marrow for transplants, etc.).

90. ENGINEERING MICROBIAL BIOLUMINESCENCE

Engineering of luciferase genes to produce bioluminescent phages and bacteria enables rapid and sensitive microbial detection and analysis for use in medical diagostics, and industrial and environmental monitoring.

In recent years several marine bacterial species and some insects (e.g. fireflies) have been identified which have evolved the capacity to generate their own light energy. Termed bioluminescence, this light energy is now known to be biochemically generated by the enzyme-mediated breakdown of specific chemical substrates, such as luminol. The potential to generate a luminescent signal as long as sufficient active enzyme and substrate is present, combined with the availability of sensitive detection instruments, has been the driving force for fundamental and applied research into the biological, biochemical and genetic basis of bioluminescence. Recombinant DNA technologies have played a key role in these endeavours by enabling the isolation and biochemical characterisation of the luciferase enzymes and their encoding genes (termed *lux* genes) from a variety of organisms. The availability of several recombinant forms of bacterial luciferases and the *in vitro* engineering of their *lux* genes has optimised both their expression and functional activity. It has also allowed *lux* genes to be used as reporter genes within vectors used to detect/investigate gene expression mechanisms. Several recombinant and genetically engineered forms of luciferase enzymes have also been applied to the generation of bioluminescent (*in vivo*) and chemiluminescent (*in vitro*) signals for the detection of a variety of molecules including nucleic acids. The ability to engineer phages and bacteria to contain a *lux* gene which converts 'dark' bacteria (i.e. lacking the *lux* gene apparatus) into a bioluminescent phenotype has many potential biomedical, environmental and industrial applications. Accordingly, the *lux* technology may be used to develop biosensors for detecting microbes or changes in their activity associated with the presence of pollutants and toxins, or to measure antibiotic resistance.

CHEMILUMINESCENT DNA DETECTION

The sustained chemical generation of light energy (termed enhanced chemiluminescence) has benefited from recombinant DNA studies of *in vivo* bioluminescence generation systems. It is perhaps therefore appropriate that chemiluminescence is increasingly being utilised by molecular genetic techniques as an alternative means of detecting the binding of gene probes (e.g. in hybridisation assays and some nucleotide sequencing methods). A variety of gene probe labelling strategies have consequently been developed which, combined with a range of substrate systems, may be used for the rapid and sensitive detection of specific DNA and RNA molecules via conventional autoradiography or the use of specialised photodetectors.

OAs: 323–30, 957–62. RAs: 191, 175, 332–5. SFR: 146–7, 181, 192–3.

LUCIFERASE MOLECULAR BIOLOGY

Bacterial luciferases are composed of two subunits of about 40 and 37 kDa molecular weight, termed α and β respectively. These are encoded by two genes termed *luxA* and *luxB* respectively (Fig. 90.1a). Comparison of the complete amino acid and nucleotide sequences obtained from bacterial *lux* genes of the *Photobacterium* and *Vibrio* genera suggests that these *luxA* and *luxB* genes may have arisen by tandem reduplication of an ancestral gene. *In vivo*, the *luxA* and *luxB* genes are located within a *lux* operon spanning about 9 kb, which contains additional structural genes (termed *luxC*, *luxD* and *luxE*) encoding the subunits of fatty acid reductase, and two regulatory genes (termed *luxR* and *luxI*). The organisation of these genes within the *lux* operon appears to be conserved amongst bioluminescent bacterial species. Although the luciferase active site is located within the α subunit, and the specific function of the β subunit remains unknown, both subunits are required for functional activity. Consequently, chimeric *luxA*–*luxB* genes have been constructed which are efficiently expressed as biologically active fusion proteins in several bacterial systems.

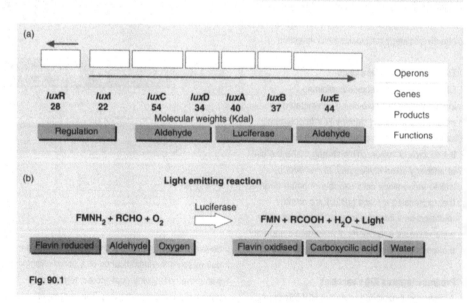

Fig. 90.1

Bioluminescence biochemistry

Bioluminescence is generated by the luciferase-catalysed oxidation of reduced flavin mononucleotide ($FMNH_2$) and a long chain aldehyde, generated by fatty acid reductase (Fig. 90.1b). Non-bioluminescent microbes (e.g. *E. coli*) lacking luciferase and fatty acid reductase genes, but containing an energy source and available molecular oxygen, can therefore be converted into a bioluminescent phenotype by transfection with the *lux* operon. Only the *luxA* and *luxB* genes, representing about 2 kb of DNA, are required since exogenous long chain aldehydes (e.g. dodecanal) can be supplied in the culture medium or sample buffer. Bioluminescence may therefore be generated in a wide range of prokaryotic organisms.

Phages and microbial detection

Because of their exquisite host cell specificity, bacteriophages have been used as a means of typing several forms of pathogenic bacteria for over 50 years. Such typing systems are however fraught with problems (e.g. lengthy assay times, the need for skilled operators and difficulties in maintaining laboratory stocks of phage and bacterial control cultures). The first meaningful application of the *lux* technology for microbial detection therefore began, in 1987, with the development of bacteriophages containing *lux* gene constructs. Following infection by the phage, the host bacterium consequently adopts a bioluminescent phenotype within minutes of the expression of the phage *lux* genes (Fig. 90.2). Given the wide range of bacteriophages capable of infecting only a single bacterial species or strain, and the ease and sensitivity of detection of a bioluminescent phenotype, this technology has much to commend it for the epidemiological analysis of pathogenic bacteria. Furthermore, it is possible to use this technology to monitor antibiotic resistance within the detected bacteria. Thus, antibiotic-resistant cells will maintain the bioluminescent phenotype in the presence of increasing levels of antibiotic in culture, whilst non-resistant cells will die and become 'dark' (i.e. stop producing bioluminescence).

Light to dark and back again

Any compound or environmental factor which impairs the intracellular biochemistry of the phage or bacterium also inhibits the generation of bioluminescence. *lux*-engineered phages and bacteria may therefore be used for the sensitive, rapid, and time-resolved monitoring of the concentrations of a range of substances which affect microbial biochemistry (based on the decrease in bioluminescence generated in standard cultures). For example, the efficacy of anti-microbial agents (e.g. disinfectants) or sterilisation procedures employed in clinical or industrial environments can be simply and rapidly determined. Similarly, this technology is applicable for simple and rapid detection and toxicity monitoring of industrial or environmental pollutants. *In vivo* induction of bioluminescence may also be used by research scientists to monitor prokaryote gene expression during/

Microbe detection by phage lux genes

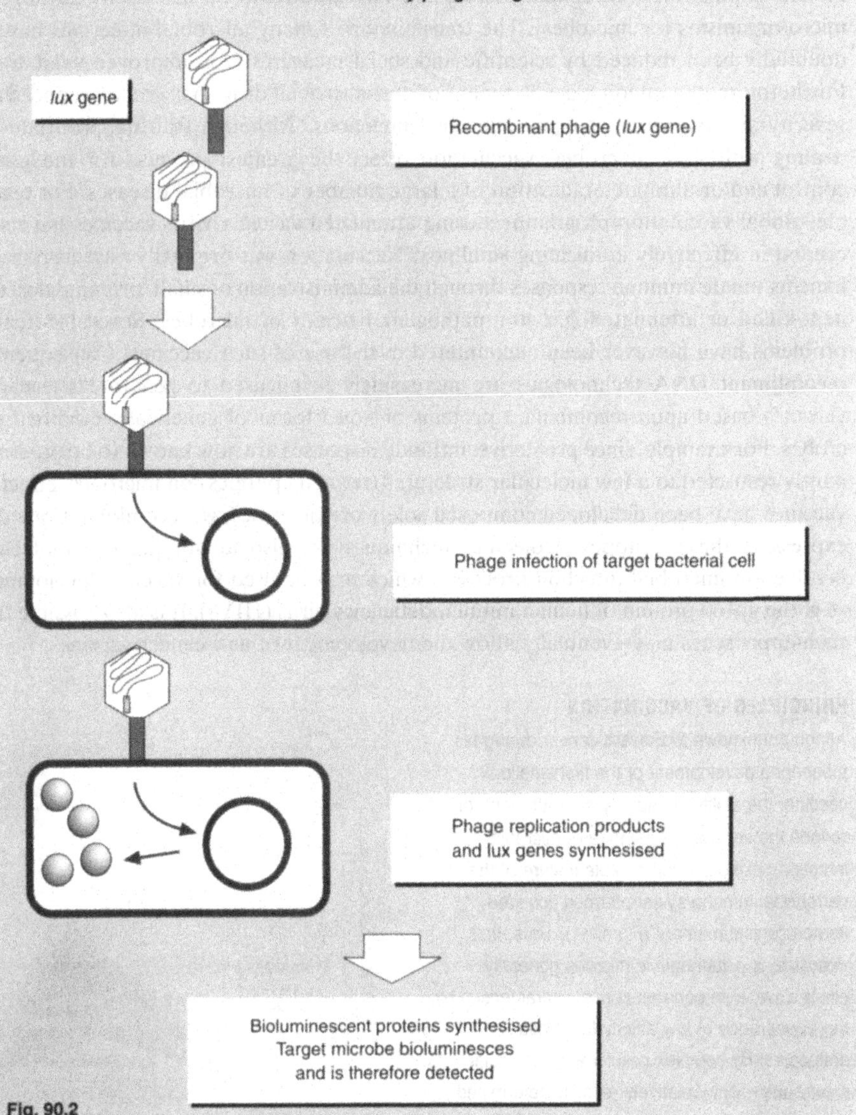

lux gene

Recombinant phage (*lux* gene)

Phage infection of target bacterial cell

Phage replication products and lux genes synthesised

Bioluminescent proteins synthesised
Target microbe bioluminesces and is therefore detected

Fig. 90.2

following environmental or toxic stress, or during cell development.

Just as bioluminescent-engineered microbes can be made 'dark' in response to external agents, external agents may also induce a bioluminescent phenotype in 'dark' microbes containing inactive *lux* genes. Of the several ways in which this may be achieved, the linkage of *lux* genes to genetic switches which are specifically activated by an external agent has many potential advantages. Such linking of *lux*

and genetic switches has already been successfully achieved for monitoring mercury and aromatic hydrocarbons. Thus, *lux* genes have been linked to promoters from the toluene plasmids of *Pseudomonas* (required for the use of toluene and xylene as substrates) and the *MerR* repressor gene (which confers mercury resistance upon bacteria). The presence of low levels of their respective substrates therefore results in the activation of the linked *lux* genes to produce a bioluminescent phenotype.

243

91. RECOMBINANT DNA METHODS IN VACCINE DEVELOPMENT

Cloning, *in vitro* mutagenesis and expression of recombinant microbial and immune proteins is greatly aiding the identification and isolation of antigens capable of being used as safe and effective vaccines.

Many of the epidemic diseases (e.g. cholera, typhoid, smallpox) which have devastated human populations throughout history are now known to be caused by pathogenic micro-organisms (or microbes). The transmission of many microbial infections has undoubtedly been reduced by scientific and social measures (e.g. improved sanitation). Furthermore, the ongoing development of anti-microbial drugs has greatly reduced the severity and mortality of many microbial infections. Although still inappropriate for dealing with some microbes, vaccination offers the greatest promise for the global control and/or ultimate eradication of a large number of microbial diseases. For example, global vaccination programmes using attenuated vaccinia virus vaccines have succeeded in effectively eradicating smallpox. Vaccination was originally based upon enhancing innate immune responses through the administration of small 'priming' doses of heat-killed or attenuated (i.e. non-pathogenic) strains of microbes. Several potential problems have however been encountered in the use of such vaccines. Consequently, recombinant DNA technologies are increasingly being used to generate alternative vaccines based upon recombinant proteins or novel forms of genetically modified microbes. For example, since protective antibody responses are now known to be predominantly restricted to a few molecular structures (termed epitopes) on microbes, effective vaccines have been developed composed solely of one, or a few, recombinant proteins expressing these epitopes. Molecular techniques are also identifying key molecules involved in microbial infection processes which may be used for vaccine development (e.g. the gp160 protein of human immunodeficiency virus (HIV)). It is also possible that such approaches may eventually allow the development of anti-cancer vaccines.

SOME POTENTIAL PROBLEMS

The use of attenuated or heat-killed microbes as the basis of vaccines has several potential drawbacks. Firstly, heat treament may lead to the loss of the normal structure (i.e. denaturation) of proteins vital for the recognition and recruitment of protective antibodies (a potential problem also faced by recombinant proteins produced in heterologous systems). Furthermore, microbial DNA may survive the heat-killing process and may be passively adsorbed by other non-pathogenic microbes (e.g. enterobacteria) causing their transformation into a pathogenic phenotype. The high rate of mutation observed in microbes also potentially limits the use of live attenuated microbes since they may also revert to a pathogenic phenotype. Indeed, such revertants have been detected in the stools from infants vaccinated with attenuated microbes. Some microbes (e.g. trypanosomes) also utilise mutational mechanisms to elude immune detection by altering surface protein expression.

PRINCIPLES OF VACCINATION

Although unknown to Edward Jenner during his pioneering development of the first smallpox vaccine, the ability of vaccination to prevent, or reduce the severity of, pathogenic microbial infections exploits a remarkable feature of the vertebrate immune system termed acquired immunological memory (Fig. 91.1). Thus, first exposure to a pathogenic microbe generally elicits a weak immune response, characterised by the appearance of low affinity IgM class antibodies. By complex cellular and molecular mechanisms only relatively recently determined and understood, the immune system retains a memory of this primary infection. Consequently, it is able to mount a very rapid and more effective antibody response (comprising higher levels of high affinity antibodies) upon second exposure to the microbe.

Vaccination involves the induction of immunological memory to ensure the mounting of a protective secondary response to any microbe. This was originally achieved by immunisation with attenuated or heat-killed microbes bearing the molecular structures (epitopes or antigenic determinants) recognised by immune cells and antibodies. More recently, recombinant DNA

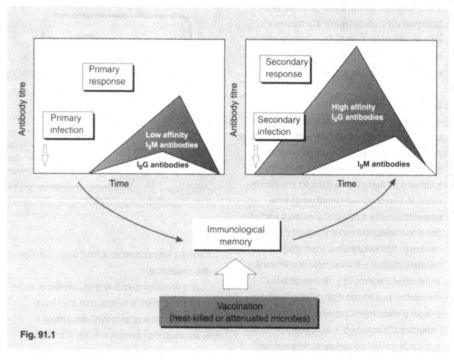

Fig. 91.1

technologies were used to identify individual microbial proteins with vaccine potential. Their application is also enabling the development of safer, and more effective, forms of vaccine.

OAs: 955–6, 963–74.
RAs: 336–8.
SFR: 119–20, 181, 194.

Recombinant DNA approaches

The ability to clone, manipulate and express microbial and immune proteins (e.g. antibodies) has provided many insights into the cellular, molecular and genetic events underpininng microbial pathogenesis and immune responses. These technologies are proving particularly useful for the investigation of microbes capable of avoiding innate immune responses, and/or those which are technically difficult to study. This has allowed the identification and production of many of the proteins involved in infection or immune processes (Fig. 91.2). The resultant recombinant proteins may then be used for structural and functional analysis, and in many cases may be used directly as the basis of a vaccine. For example, recombinant forms of the HIV surface protein gp160 are currently being investigated for their potential as an effective AIDS vaccine (e.g. by eliciting protective anti-gp160 neutralising antibodies which may block HIV infection). *In vitro* mutagenesis has also been used for epitope mapping (i.e. identification of individual protein structures recognised by antibodies) by creating site-directed mutants. This approach is crucial to the development of 'immuno-dominant' epitope-expressing recombinant and/or synthetic peptides for use as vaccines.

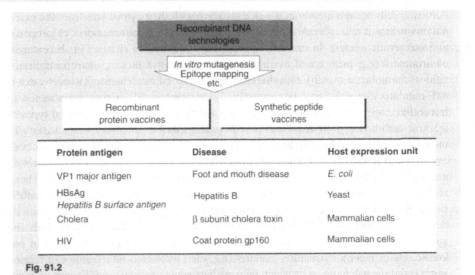

Fig. 91.2

Protein antigen	Disease	Host expression unit
VP1 major antigen	Foot and mouth disease	E. coli
HBsAg Hepatitis B surface antigen	Hepatitis B	Yeast
Cholera	β subunit cholera toxin	Mammalian cells
HIV	Coat protein gp160	Mammalian cells

'Engineered' vaccines

The attenuated vaccinia virus used as a vaccine for the eradication of smallpox has been widely used as a vector for the introduction of genes encoding the major antigens of especially virulent pathogens (e.g. hepatitis B, HIV). Accordingly, the vaccinia virus has been engineered to contain several dozens of foreign genes by insertion into a site flanking the vaccinia promoter (Fig. 91.3). Non-infectious microbial particles have also been engineered which overcome the problems associated with the use of heat-killed or attenuated microbes. Their efficacy may also be enhanced by increasing the expression of endogenous or exogenous surface proteins bearing desirable epitopes. Similarly, the efficacy of recombinant protein- and peptide-based vaccines may be enhanced by creating chimeric genes expressing fusion proteins composed of the microbial protein/peptide fused to an immuno-stimulatory molecule (e.g. a cytokine or antibody

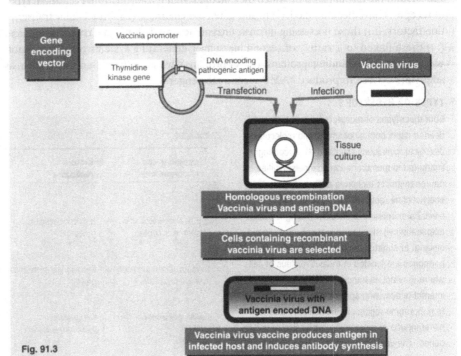

Fig. 91.3

fragment).

Creating anti-cancer vaccines?

The vast majority of vaccines already developed, or currently under development, are directed towards combatting microbial infections. It has however been suggested that vaccines may be developed which afford protection against several forms of cancer. Such anti-cancer vaccines comprising recombinant tumour cell-specific surface proteins may therefore be developed which elicit immune surveillance and elimination mechanisms in the same way as microbial surface proteins. Unfortunately, the difficulty in identifying tumour-associated surface proteins specific for a particular tumour phenotype currently impedes anti-cancer development.

92. RECEPTORS AND CELLULAR SIGNALLING MECHANISMS

Recombinant DNA technologies have defined four basic types of receptor, which (with the exception of steroid hormone receptors) are anchored within the cell membrane by similar transmembrane regions.

All living cells possess membranes which separate both their cytoplasms from the external environment and individual internal compartments (e.g. mitochondria, chloroplasts and eukaryotic nuclei). In order to appropriately respond to changes in the external environment (e.g. presence of nutrients, hormones, growth factors, neurotransmitters, light or chemotactic stimuli), cells therefore require specific mechanisms whereby external regulatory signals may be transmitted (or transduced) across membranes to intracellular sites of action. This is accomplished by the possession of several types of integral membrane proteins. These act as ligand-specific receptors for a variety of molecules whose binding leads to the transduction of several forms of signals across both external and internal membranes. The importance of cellular receptors in maintaining normal cellular functions, and their role/association with various diseases, has long been realised. Receptors have therefore become major targets for basic and applied scientific research (e.g. for investigating disease pathologies, and for drug development). Initially detected and functionally characterised using biochemical and/or bioassay methods, the application of molecular biological techniques has led to substantial advances in our knowledge of receptor structure, genetics (e.g. their evolution into families of related genes), and the mechanisms by which they mediate signal transduction. Four main types of receptors have been defined: (i) ligand-gated ion channels, (ii) intracellular transcription factors, (iii) those possessing intrinsic enzymatic activity for signal transduction, and (iv) those linked to a variety of second messenger-generating systems by effector molecules (e.g. GTP-binding proteins (or G-proteins) which activate enzymes such as adenylate cyclase to produce cAMP second messengers).

TYPES OF RECEPTORS

Four main types of receptors have now been defined using both biochemical and molecular biological techniques (Table 92.1). Accordingly, ligand-gated membrane ion channels, typified by neurotransmitter receptors such as the nicotinic acetylcholine receptor, trigger changes in the electrical properties of cells by altering ion efflux across the cell membrane in response to ligand binding. In contrast, receptors for steroid hormones are located in the cell nucleus of cells where they act as transcription factors which interact directly with specific genomic regions (e.g. hormone regulatory elements; HREs) to either induce or repress the expression of certain genes. The vast majority of cellular receptors are however involved in transducing ligand-mediated signals via a variety of second messengers (e.g. cAMP, cGMP, calcium ions, phospholipids). These receptors fall into two principal classes: (i) the G protein-linked receptors which activate second messenger-generating systems (e.g. adenylate cyclase) by coupling to specific GTP-binding proteins (or G-proteins); and (ii) those such as the T lymphocyte CD4 protein (which also acts as a receptor for entry of the AIDS virus) which are linked to enzymic proteins associated with the cytoplasmic face of the cell membrane, or those such as the growth factor receptors which possess intrinsic enzymic activity (e.g. protein kinase activity leading to signal transduction by protein phosphorylation).

RECEPTORS AND DISEASE

The pivotal role of receptors in determining cellular functions means that receptor defects underpin many disease pathologies. For example polymorphic defects within the ligand-binding domains of low-density lipoprotein receptors account for the manifestations of familial combined hypercholesterolaemia. Similarly, the unregulated proliferation of some cancer cells, and the dysfunctional neuronal activity associated with many diseases of the central nervous system, stem from the expression of functionally defective cellular receptors. These include several oncoproteins which represent truncated and therefore constitutively activated growth factor receptors. Receptors may also serve as portals for the entry of pathogenic microbes into specific cells (e.g. the infection of T cells by the human immunodeficiency virus).

Table 92.1

Ligand gated ion channels	Steroid receptors	Second messenger	Protein kinase/ phosphates
E.g. Nicotinic acetyl choline receptor	E.g. Glucocorticoid receptor	E.g. Rhodopsin	E.g. PDGF receptor
Multisubunit membrane spanning prptein	Single transmembrane protein	Single transmembrane protein	Single transmembrane protein
Cell depolarization ion efflux	Cell regulation by external stimuli	Cell regulation by external stimuli	Cell regulation by external stimuli
Has an extracellular-transmembrane action	Results in movement of external signal to nucleus (TF like)	Results in activation of G protein cAMP systems	Has inherent protein kinase/phosphatase activity

OAs: 702, 975–81, 1005–10, 1031, 1044–5. RAs: 316–17, 328, 339–43, 356–61, 374, 380, 389–90. SFR: 119–20, 181, 195–7, 199–202.

BASIC RECEPTOR STRUCTURES

Despite their often high copy number per cell, the isolation of sufficient membrane receptor proteins for bio-physical analysis has proved difficult. The biochemical analysis of small amounts of highly purified receptor proteins isolated by affinity chromatography did however reveal the first insights into their structural diversity. Current understanding of receptor structure has therefore largely been derived from the analysis of receptor-encoding genomic and cDNA clones. Although recombinant receptor proteins remain technically difficult subjects for three-dimensional structural analysis (e.g. due to alterations in post-translational modifications), sequence analysis of various cloned receptor proteins has revealed that most are structurally and evolutionarily related. Membrane receptor proteins can therefore be grouped into a few families. Furthermore, each member of a family is comprised of different combinations of a limted number of protein domains, each of which determines their ligand binding specificity and signal transduction mechanisms. Each different family member is therefore believed to have arisen during evolution by gene duplication and gene shuffling. Similarly, receptors show enormous size variation, ranging from small single polypeptides to those composed of many different polypeptide subunits encoded by separate genes (also believed to have evolved by the process of gene duplication). A common feature of membrane receptor proteins is however the presence of one or more membrane-spanning (or transmembrane) regions.

Transmembrane regions

Common to all receptors except steroid hormone receptors, transmembrane regions are α-helical structures composed of hydrophobic amino acids inserted perpendicularly (or at a slight angle) into the membrane lipid bilayer. These serve to anchor the receptor protein in the membrane lipid bilayer, or they may contribute to the formation of channels through the membrane (depending on the nature of the receptor). The number of transmembrane regions varies between receptor types and classes from one to many per receptor or polypeptide subunit. For example, many polypeptide growth factor receptors have a single transmembrane helix linking the extracellular ligand-binding domain to a cytoplasmic domain with intrinsic enzymic activity. In contrast, G-protein-linked receptors (e.g. bacteriorhodopsin and adrenergic receptors) all contain seven transmembrane α-helical regions (Fig. 92.1). These are linked by hydrophilic peptide loops exposed on the external and cytoplasmic faces of the cell membrane which are responsible for ligand binding and coupling to G-proteins. Ligand-gated ion channel receptors (e.g. the acetylcholine receptor), however, contain four transmembrane helices per subunit.

Although first observed by electron microscopy of receptor protein crystals, their location within many receptors can now be quite accurately predicted from deduced amino acid sequences based upon the creation of hydropathy plots (in which each amino acid is assigned a hydrophobicity index value). Transmembrane helices within ion channel-forming receptors may however be less accurately predicted since these probably contain charged residues which normally line the inside of the ion channel.

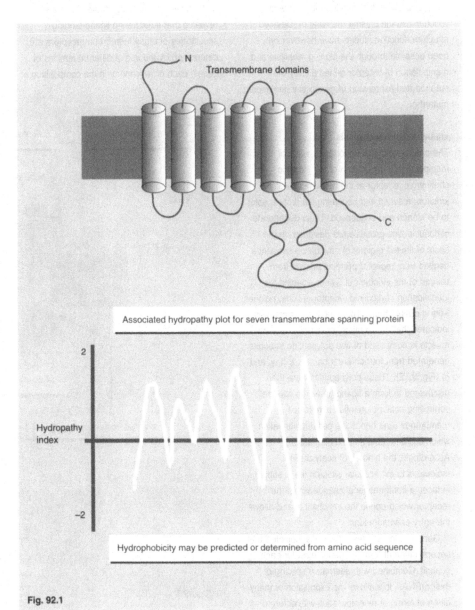

Fig. 92.1

The cloning, *in vitro* mutation and expression of receptor genes has identified the key receptor structures and the diversity of mechanisms employed for transducing signals across cell membranes.

LIGAND BINDING AND SIGNALLING

In order to fulfil their role in transducing cellular signals, all types of receptors contain specific regions (or individual protein domains) responsible for determining ligand binding specificity and signal transduction respectively. Much of the initial knowledge regarding the structural and functional diversity of these receptor domains has been provided by biochemical studies. These involved proteolytic digestion, antibody binding, chemical modification, and reconstitution experiments (in which purified receptor proteins are inserted into lipid vesicles which mimic the plasma membrane). Detailed structure–function studies have however only been possible through the cloning, analysis and manipulation of receptor genes (i.e. by the creation and expression of genetically engineered mutants).

Ligand-gated ion channels

The pharmacological importance, ease of manipulation, and location of the nicotinic cholinergic receptor at the neuromuscular junction, resulted in it becoming the first receptor to be cloned and sequenced. Thus, degenerate oligonucleotide probes were designed on the basis of limited regions of amino acid sequence derived from receptor proteins purified from tissues of the electric eel. These enabled the identification of nicotinic receptor-encoding clones from a cDNA library. Their subsequent analysis indicated that the nicotinic receptor in vertebrate muscle is composed of five polypeptide subunits generated from four different genes (α, β, γ, and δ) (Fig. 92.2). These polypeptides were then discovered to form a ligand-gated ion channel controlling sodium ion influx across cell membranes (and hence the cell depolarisation which initiates rapid muscle contraction). Accordingly, the binding of acetylcholine molecules to extracellular sites on the α subunits induces a conformational change within the receptor which opens the ion channel and allows the entry of sodium ions.

Gene cloning studies have since indicated that several different forms of genes exist for each subunit. Combined with alternative splicing of their mRNAs, this allows the expression of many different forms of receptor, each with different kinetic properties and ligand specificities. Sequence comparisons between different ligand-gated ion channel receptor clones have indicated that both ion selectivity and pore size are defined by the nature of the amino acids within the transmembrane regions of one or more subunits. Although sequence comparisons have contributed to the identification and modelling of ligand-binding sites on the α subunits, their precise nature and localisation has also been determined/ confirmed by the study of cells expressing recombinant receptors bearing point substitutions and/or deletions. Cloning studies have also revealed that the differing ligand binding specificities of many membrane receptors are determined by the use of different families of genes, each of which vary in the composition of their extracellular domains. In contrast to ligand-gated ion channel receptors which transmit a signal directly across the membrane, ligand binding to many families of receptors results in signal transduction via the activation of intrinsic or associated enzymic domains (e.g. by mediating protein phosphorylation). Alternatively, signal transduction involves the activation of second messenger-generating systems via linkage to GTP-binding proteins (G-proteins).

OAs: 777, 895–7, 982–92, 1072, 1077. **RAs:** 344–54, 356–61, 374, 380, 389–90. **SFR:** 119–20, 181, 195–7, 199–202.

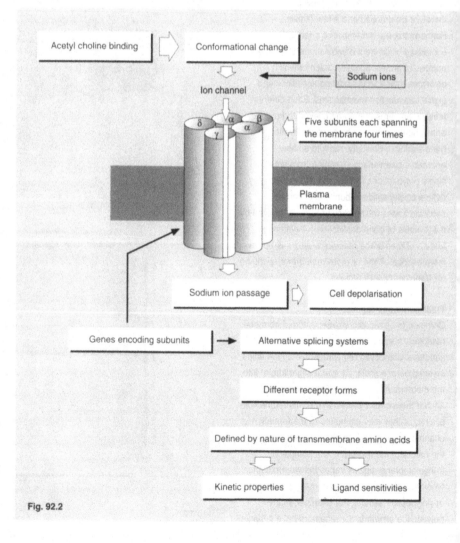

Fig. 92.2

Second messenger systems

Growth hormone receptors are fundamental for maintaining normal cell growth (as evidenced by the effect of certain oncogene products upon cell proliferation). Cloning and expression studies involving *in vitro*-mutagenised receptor genes have established that many peptide receptors include both an extracellular ligand-binding domain and an intrinsic cytoplasmic enzymic domain linked via a single transmembrane helix (Fig. 92.3). For example, the epidermal growth factor (EGF) receptor possesses two extracellular EGF-binding domains linked to a protein tyrosine kinase domain by a single transmembrane helix. EGF binding therefore activates the kinase which adds phosphate groups to (i.e. phosphorylates) the tyrosines of its own polypeptides or those of other proteins. Indeed, protein phosphorylation signalling via protein kinases appears to be a widely used mechanism for controlling cellular processes including cell cycling.

A range of enzymic domains and second messenger systems have now been identified, including protein kinases and/or guanylate cyclase. Some receptors contain phosphatase domains which can counteract phosphorylation-mediated signals, whilst others (e.g. the CD4 protein) lack intrinsic enzymic activity but associate with, and activate, an intracellular enzymic polypeptide.

A further large family of receptors, including those activated by many hormones, neurotransmitters and visual signals (e.g. rhodopsin), have been identified which are linked to second messenger systems by a family of GTP-binding proteins (or G-proteins). These G-proteins possess GTPase activity and are now known to act as molecular switches and signal amplification systems which give rise to a cascade of any of a number of common second messengers (Fig. 92.4). For example, hormone binding can activate adenylate cyclase to generate cAMP. This cAMP then activates cAMP-dependent protein kinases, or activates phospholipases which hydrolyse inositol phospholipids to produce two forms of second messengers which affect several different protein kinases. The definitive role of G-proteins in signal transduction was established by the construction of chimeric receptor genes derived from cDNA

clones of different subtypes, and their functional analysis following expression in *Xenopus* oocytes (i.e. by injection of *in vitro*-transcribed mRNAs). In addition to their role in signal transduction, G-proteins are now known to be involved in protein synthesis, intracellular transport and exocytosis. Indeed, cloning and homology screening has identified numerous G-protein families within a variety of plants, animals and even single cell

organisms such as yeast. Most G-proteins are now known to be heterotrimers consisting of one copy of an α, β and γ domain encoded by sets of related genes. Their full genetic, structural and functional diversity has yet to be elucidated, however their restricted receptor specificity has made them potential targets against which specific pharmacological agents may be developed and therapeutically directed.

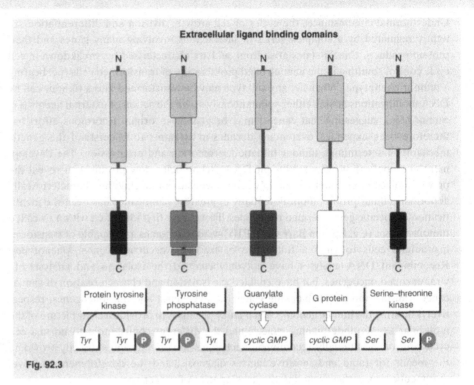

Extracellular ligand binding domains

| Protein tyrosine kinase | Tyrosine phosphatase | Guanylate cyclase | G protein | Serine–threonine kinase |

Fig. 92.3

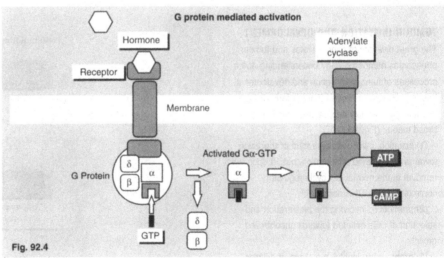

G protein mediated activation

Fig. 92.4

93. ONCOGENES AND THE MOLECULAR BASIS OF CANCER

Recombinant DNA analysis of the events occurring during viral transformation of mammalian cells has revealed the existence of cancer-causing oncogenes derived from normal proto-oncogenes.

Under normal circumstances the cycles of cell growth, division and differentiation are tightly regulated by a complex series of interactions involving many genes and their protein products. Cancer states arise from, and are characterised by, breakdown in cell cycle control resulting in the uncontrolled proliferation of transformed cells (i.e. bearing a tumour phenotype). Virtually any cell type may be transformed into a tumour cell by DNA modifications arising either spontaneously or via the action of external agents (i.e. carcinogens), suggesting that cancers may be genetic in origin. Enormous effort has therefore been expended over many decades in attempts to understand the genetic mechanisms determining tumour initiation, promotion and progression. The development and application of recombinant DNA technologies now promises to reveal the precise molecular mechanisms and key genes involved in inducing this characteristically abnormal cellular proliferation. Potentially of greatest significance has been the identification of specific genes, termed oncogenes. These were first identified within so-called tumour viruses (e.g. Epstein Barr virus; EBV) whose genomes are capable of transforming animal cells to yield a full range of naturally occurring cancer phenotypes. Recombinant DNA analyses have not only enabled the isolation and analysis of viral-encoded oncogenes, but have enabled the isolation and characterisation of similar cancer-causing and cancer-blocking genes (i.e oncogenes and anti-oncogenes, respectively) littering the human genome. Their study promises to provide a clear picture of the molecular events underpinning cancer, and the different contributors to normal cell cycle control. Additionally, recombinant techniques are directly and indirectly providing the means for rapid and sensitive cancer diagnosis, and the development of novel therapeutic approaches including the development of cancer vaccines.

CARCINOGENS AND DNA DAMAGE

Many forms of cancer are directly caused by avoidable physical or chemical agents, termed carcinogens. For example, lung cancers can be caused by carcinogenic compounds in tobacco smoke, skin cancers may arise from excessive exposure to solar ultraviolet radiation (i.e. during sunbathing), and forms of leukaemia have been linked to exposure to gamma radiation emanating from nuclear power stations, and 'fall out' from accidental or deliberate atomic explosions (e.g. Chernobyl). Many carcinogens are also mutagens capable of inducing several forms of chemical damage to DNA. The outcome of this damage depends upon which genes are affected and when the damage occurs. Thus, cancers may arise when the damage affects genes involved with cell signalling and/or cell cycle control, and birth defects may arise when developmental genes are damaged during early embryogenesis.

TUMOUR INITIATION AND DEVELOPMENT

The great diversity of cancer states and tumour phenotypes hampered early understanding of the processes of tumour formation and development. It is now however believed that the formation of any tumour is a multistep process involving three broad stages (Fig. 93.1):

(1) initiation, involving some form of triggering towards unregulated proliferation, based upon the removal of the normal cell cycle regulatory controls (e.g. via DNA damage);

(2) promotion, involving the proliferation and selection of cells initiated towards uncontrolled growth;

(3) progression, involving a range of cellular changes which allow the formation of altered blood supplies (i.e. tumour vascularisation) and the adoption of a malignant and/or metastatic state (i.e. the formation of small clumps of tumour cells termed metastases which may migrate to other sites to form new tumour growths).

The initial breakthrough in understanding the genetic basis of tumour initiation was provided in 1911 by experiments performed by Peyton Rous. These demonstrated that the small genomes of certain viruses could be taken up by cultured

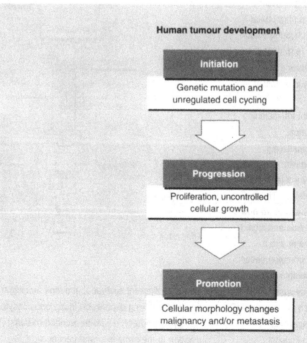

Human tumour development

Initiation

Genetic mutation and unregulated cell cycling

Progression

Proliferation, uncontrolled cellular growth

Promotion

Cellular morphology changes malignancy and/or metastasis

Fig. 93.1

animal cells and cause them to develop tumour phenotypes, a principle still exploited for the formation of lymphoblastoid cell lines (e.g. secreting human monoclonal antibodies).

OAs: 112, 526, 839, 993–8. RAs: 35–6, 38, 300, 355–9. SFR: 49, 56–7 61, 120, 181, 197–204.

TUMOUR VIRUSES AND CANCER

Many different viruses have the ability to enter (i.e. infect) animal cells, and several have been directly or indirectly linked with tumour initiation. For example, cervical cancer has been linked to infection by members of the human papilloma virus (HPV) group, and some types of viruses (e.g. human T cell leukaemia virus (HTLV)) are thought to play a role in the induction of some leukaemias. Although the precise mode of action is unclear, the transforming activity of tumour viruses is known to be due to the stable integration of all or part of the viral genome into the host cell genome by recombination (a feature which has since been exploited for the development of viral vectors for animal cell transfection). Virally transformed mammalian cancer cells consequently owe their uncontrolled proliferation to the insertion of a viral sequence(s) which alters the normal level of expression of regulatory genes (e.g. resulting in inappropriate levels of a protein or enzyme). Indeed, it is the analysis of virally transformed mammalian cells which provided the first evidence for the existence of specific cancer-causing genes (i.e. oncogenes).

Oncogenes and proto-oncogenes

The first oncogenes to be identified were those of tumour viruses and are accordingly termed viral oncogenes (designated v-*onc*). It was however soon discovered that this relatively small set of viral genes was also represented in normal mammalian cells (Fig. 93.2a). Accordingly, using viral oncogene sequences as probes, Southern blot hybridisation studies of mammalian cell genomic DNAs revealed the existence of corresponding genes, termed cellular proto-oncogenes (designated c-*onc*). The analysis of c-*onc* and v-*onc* genes led to the realisation that proto-oncogenes were actually normal genes. Furthermore, these proto-oncogenes encoded proteins that were involved in signal transduction mechanisms employed to maintain normal cell growth and development. Further molecular biological studies also established that all cancer cells contained one or more oncogenes which had arisen from these proto-oncogenes. In other words, the proliferation of cancer cells was due to the conversion of proto-oncogenes into

oncogenes. Indeed, v-*onc* genes may have arisen during evolution by the integration of a copy of a normal c-*onc* gene into the viral genome. This is supported by the fact that viral oncogenes found in RNA-containing retroviruses lack the introns found in cellular oncogenes.

Oncogenes are now known to arise from proto-oncogenes in a variety of ways (Fig. 93.2b). Following viral transformation, proto-oncogenes may become overtranscribed by their association with inserted viral regulatory sequences, or the proto-oncogene may mutate resulting in an abnormally functioning protein product. Similarly, in the absence of virus infection many types of mutations may arise spontaneously within cellular proto-oncogenes to generate an oncogene. For example, base substitutions may alter the functional activity of the proto-oncogene-encoded protein, and chromosome breakage and/or rearrangement may bring a proto-oncogene under the influence of a new regulatory element. Alternatively, the proto-oncogene may be duplicated or amplified, thereby altering its transcriptional rate. Many different oncogenes and proto-oncogenes have now been identified in organisms as diverse as yeasts and humans.

(a)

Oncogene	Function	Chromosome	Tumour type
abl	Tyr Kinase	9q34.1	B cell tumours
bcl-2	Inhibit apoptosis	18q21.3	Lymphoma
erb-B2	EGF receptor	17q11.2	Breast cancer
fos	Trans factor	14q21	B cell tumour
jun	Trans factor	1p32	?
myc	Trans factor	8q24+	Burkitts lymphoma
N-myc	Trans factor	2p24	Neuroblastoma
Ha-ras	GTPase	11p15	Colon and lung cancer
sis	PDGF like	23q12	Breast cancer
src	Tyr kinase	20q12	Colon and stomach cancer

(b)
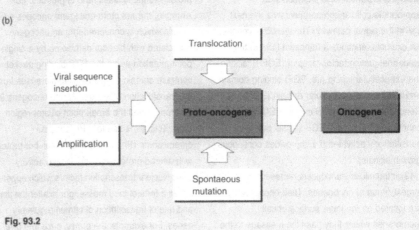

Fig. 93.2

251

Cancers are now known to arise from multiple mutations within proto-oncogenes encoding proteins involved in signal transduction and/or cell cycle control, leading to the production of oncoproteins.

PROTO-ONCOGENE FUNCTIONS

The application of recombinant DNA technologies has allowed the identification, isolation and characterisation of greater than 60 oncogenes and their corresponding proto-oncogenes. Most of these proto-oncogenes encode proteins involved in the signalling pathway whereby external signals are delivered to the intracellular regulatory machinery controlling cellular growth, cell division and development. Conversion of a proto-oncogene therefore often results in the production of an oncoprotein which short-circuits the normal signal transduction pathway. Mutations either alter the function of the proto-oncogene protein, or alter the timing or rate of its transcription. This results in the continuous transmission of growth/division signals to the cell in the absence of any external messages.

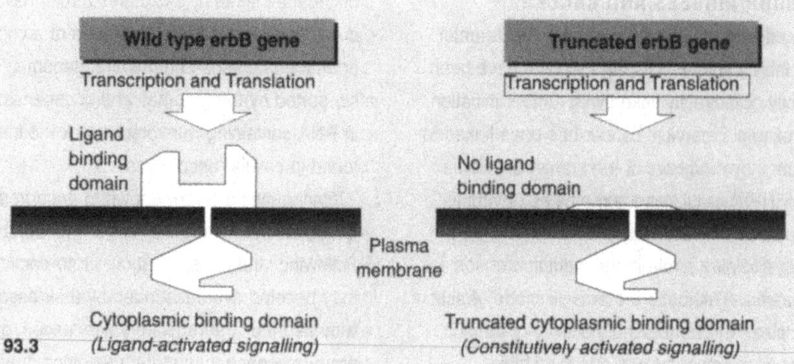

Types of oncogenes

Oncogenes may be divided into four broad classes based upon the function of their encoded oncoprotein.

(1) Growth factors are responsible for the delivery of external signals which stimulate cell growth, proliferation and differentiation. Accordingly, the *sis* oncogene encodes a type of platelet-derived growth factor (PDGF) which is secreted from, and stimulates the growth of, the *sis*-containing cell (i.e. it acts as an autocrine mitogenic signal to promote cell growth and division).

(2) Growth factor receptors are located on cell membranes where they receive the external regulatory signals provided by soluble growth factors and transmit this information to appropriate cellular compartments via the next link in the signal pathway. The v-*erbB* oncogene, for example, encodes a truncated form of the epidermal growth factor receptor (EGFR) lacking the extracellular ligand (i.e. EGF)-binding domain and a truncated intracellular domain (Fig. 93.3). Consequently, the v-*erbB*-encoded EGFR protein cannot react to normal EGF signals but is constitutively activated (i.e. it provides continuous growth signals).

(3) Intracellular transducers represent the largest group of oncogenes. Their oncoproteins are located on the inner surface of cell membranes where they pass the message to the

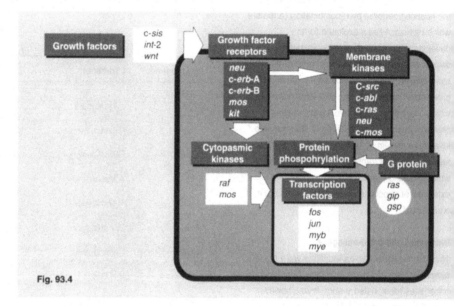

Fig. 93.3   (Ligand-activated signalling) / (Constitutively activated signalling)

Fig. 93.4

nuclear compartment (see Fig. 93.4). These oncogenes generally encode altered or translocated forms of protein–tyrosine and protein–serine kinases, and G-proteins. For example, the *ras* proto-oncogene encodes a G-protein which is converted into an oncogene associated with bladder carcinoma by a single point mutation within the GTP-binding pocket. In contrast, chronic myeloid leukaemia arises from the translocation of the c-*abl* proto-oncogene into an area termed the break point cluster region (*bcr*). This gives rise to a Philadelphia chromosome (Ph') and a large c-abl–bcr protein with altered protein–tyrosine kinase activity.

(4) Nuclear transcription factors which appear to act as a form of third messenger to alter the timing and rate of transcription of certain regulatory genes. For example, the c-*myc*, c-*fos* and c-*rel*

proto-oncogenes are normally expressed at low levels in quiescent cells. Their expression is however rapidly increased in response to certain mitogenic signals, with the consequent transcription of genes involved in cell growth and division. For example, gene translocation and rearrangement often causes the conversion of the c-*myc* proto-oncogene into the *myc* oncogene resulting in an increase in c-*myc* gene expression. The consequent accumulation of c-*myc*-encoded transcription factors therefore alters the expression of regulatory genes normally controlled by c-*myc* transactivation. This leads to the unregulated proliferation of the transformed cell.

OAs: 112, 526, 839, 999–101.
RAs: 35–6, 38, 355–67. **SFR:** 49, 56–7, 61, 120, 181, 197–204.

Amplified oncogenes and HSRs

Some cancer cells have also been described which contain karyotypic abnormalities characterised by the presence of a 'double minute' chromosome or the possession of homogeneously stained regions (HSRs). These are generated by the amplification of cellular proto-oncogenes resulting in their activation (i.e. they then act as oncogenes). Both double minute chromosomes and HSRs appear to represent interchangeable forms of the same amplification events (Table 93.1). They consequently vary in size and number depending upon the nature of the amplification events which have occurred in particular tumours. For example, they were first described in human neuroblastomas where hybridisation studies revealed that the HSRs were predominantly composed of amplified *myc* sequences.

Anti-oncogenes are protective

Recombinant DNA investigations have also identified a group of proto-oncogenes whose encoded proteins appear to block cell division. These are consequently often referred to as anti-oncogenes. For example, the p53 protein (so called because it is a 53 000 Da phospho-protein) is an anti-oncoprotein which may normally play a protective role in cells. Accordingly, the expression of the p53 gene in normal cells results in the suppression of cell growth by arresting the cell in G1. This is thought to be achieved by the accumulation and binding of p53 proteins to several sites on DNA (Fig. 93.5). Its accumulation following DNA damage may therefore allow time for the cell to repair the damage before commencing the next cell division cycle, or its participation in programmed cell death (i.e. apoptosis). The analysis of many solid tumours has indicated that point mutations and deletions within the p53 gene (e.g. at certain residues in exon 7) lead either to lack of p53 expression, or the production of altered p53 proteins which are unable to suppress cell growth. Inheritance of a damaged p53 gene is therefore associated with elevated susceptibility to a variety of cancers. Anti-oncoproteins are therefore believed to be the products of normal tumour-suppressor genes. They are consequently the focus of considerable attention, particularly since they may be

Table 93.1

Oncogene	Tumour Type	Amplification
c-*myc*	Small cell lung carcinoma	5–30 X
N-*myc*	Neuroblastoma	5–1000 X
c-*myb*	Acute myeloid leukaemia	5–10 X
n-*ras*	Mammary cancer cell line	5–10 X
c-*erb*-B	Epidermal carinoma	30 X
K *ras* 2	Lung, colon, bladder cancer	4–20 X

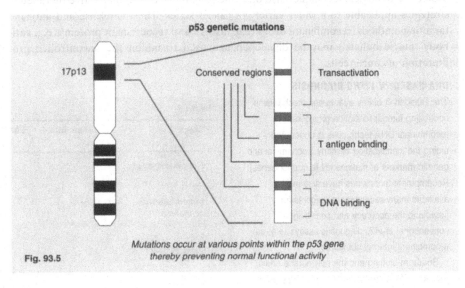

p53 genetic mutations

17p13 — Conserved regions — Transactivation — T antigen binding — DNA binding

Mutations occur at various points within the p53 gene thereby preventing normal functional activity

Fig. 93.5

developed as therapeutic agents for certain tumours.

Cancer requires multiple events

Certain oncogenes are common to many tumours and their detection greatly facilitates cancer diagnosis. It is however clear that the mutation of a single proto-oncogene is generally insufficient to transform a cell into a malignant tumour phenotype. Thus, the multi-step process of tumour initiation, promotion and progression involves the accumulation of several mutations within the same or multiple proto-oncogenes.

Tumour progression may also require additional mutations leading to the production of soluble secreted factors such as angiogenesis factors which direct blood flow into solid tumour masses in order to sustain their growth. This requirement for an accumulation of mutations makes it difficult to unravel the complexity of steps determining when and how cancers develop. Nevertheless, continued research is likely to increase our understanding of tumorigenesis and provide data which may prove invaluable for cancer diagnosis and therapy.

94. MOLECULAR DIAGNOSIS AND THERAPY OF CANCERS

Cancer diagnosis and therapy is being aided by the direct and indirect use of molecular biological techniques, which allow the *in vitro* and/or *in vivo* detection, and targeting of tumour cells

The increasing incidence, high level of associated mortality, and enormous diversity of human cancer phenotypes requires numerous strategies for their clinical diagnosis and therapy. Conventionally, diagnosis relied upon the detection of: (i) tumour-specific secreted biochemical protein markers (e.g. ectopic hormones or tumour-associated antigens); (ii) tumour-associated changes in the histological appearance (i.e. morphology) and/or surface-expressed markers of malignancy (e.g. certain cluster designation antigens) on biopsied cells; or (iii) solid tumour masses *in vivo* (e.g. by X-ray). The realisation that cancers represent a form of genetic disease and the investigation of the mechanisms of tumour initiation and development using molecular biological techniques have allowed the identification of many new tumour markers. Although many of these are also detectable by conventional diagnostic methods, the identification of genetic malignancy markers has resulted in molecular biological techniques becoming an essential component of the approaches routinely used for diagnosis and therapy monitoring. Recombinant DNA techniques are also indirectly aiding cancer diagnosis via the development of improved diagnostic reagents (e.g. recombinant antibodies for *in vivo* tumour imaging). This increases the ability to detect cancer cells at an earlier stage of disease and increases the success of conventional radiation- and drug-based (chemo-) therapies. Molecular biological techniques also offer the promise of new and effective therapeutic strategies applicable to a wider variety of cancer states. Thus, antisense and antigene targeting methods, recombinant drugs (e.g. taxols), and recombinant proteins (e.g. antibody 'magic bullets') may be developed and used to inhibit the uncontrolled proliferation of cancer cells.

ALTERED GLYCOSYLATION PATTERNS

Tumour cells have long been known to vary in their expression of specific proteins, and/or altered forms of the proteins. Indeed, many of these have been used as tumour-specific markers for cancer diagnosis. In recent years the changes in glycosylation profile of molecules expressed on cancer cell surfaces have been the subjects of detailed analysis. In particular, the polymorphic high molecular weight O-linked glycoproteins encoded by the *MUC1* gene and expressed on epithelial cells and epithelial carcinoma cells (termed polymorphic epithelial mucins or PEMs) have been extensively studied. Identification of these altered glycosylation patterns suggests that they may be exploited for *in vitro* and *in vivo* diagnosis (e.g. based on PEM-specific antibody probes). This information may also be used to develop novel cancer immuno-therapies (e.g. based upon the use of PEMs as immunogens to elicit tumour-specific cytotoxic antibodies).

DNA-BASED *IN VITRO* DIAGNOSIS

The elucidation of key events and mechanisms controlling tumour formation provided by recombinant DNA techniques is undoubtedly aiding the identification of many biochemical and genetic markers of malignancy (e.g. oncogenes). Recombinant techniques have also made available many recombinant reagents for improving the accuracy and sensitivity of conventional *in vitro* diagnostic assays (e.g. as recombinant internal standards).

Southern blotting and the polymerase chain reaction (PCR) are however increasingly being used directly to diagnostically detect tumour-associated DNA mutations (e.g. point mutations, amplifications, deletions, and chromosomal rearrangements). For example, Southern blotting is often used to detect c-*abl*–*bcr* translocations associated with chronic myeloid leukaemias. *In situ* hybridisation is also being used to detect tumour cell-specific transcription patterns in biopsied materials. The rapidity and sensitivity of PCR-based methods also allows the early detection and diagnosis of solid and diffuse tumours (e.g. haematological malignancies) based upon the analysis of blood cell or needle biopsy samples (Table 94.1). The ability to detect cancer cells early in disease development is

particularly important in determining the ultimate success of many cancer therapies, many of which are maximally effective if instigated at the correct stage in tumour development (e.g. before metastases form).

Table 94.1

Stage	Chromosomal site	Alteration	Gene	Diagnosis (Molecular cytogenetic histology)
Normal epithelium				
Hypoproliferative epithehum	5q	Loss	APC	
Early adenoma				
Intermediate adenoma	12p	Activation	K-ras	
Late adenoma	18q	Loss	DCC	
Carcinoma	5q	Loss	p53	
Metastasis				

OAs: 433, 749–52, 974, 1015–27, 1089–91, 1098, 1115, 1120–2, 1125. **RAs:** 212, 367–72, 398, 406–7, 414, 427. **SFR:** 56–7, 120, 180–1, 201–10, 219–20.

AIDING *IN VIVO* TUMOUR IMAGING

The ability to detect solid tumours *in vivo* (e.g. by X-ray) remains an important aspect of cancer diagnosis and therapy. Such conventional approaches are however less than satisfactory for the early detection of tumours since only large, electron-dense tumours may be revealed. Tumour imaging using labelled (e.g. radioactive) tumour-specific monoclonal antibodies (Mabs) has been evaluated as an alternative approach. They have however generally proved inadequate due to poor *in vivo* half-lives, their high non-specific uptake (e.g. via antibody Fc receptors on normal cells), and their poor tissue penetration (i.e. inability to reach the tumour). Recombinant/engineered Mabs are however overcoming such problems. Accordingly, a range of recombinant tumour-specific Mabs fragments (e.g. representing single antigen-binding sites) have been generated, and directly labelled. These readily penetrate tissues and, since they lack Fc regions, have been demonstrated to deliver detectable labels in a highly tumour-specific manner. The same approaches are also being used to create antibody-based 'magic bullets' for therapy.

Anitbody 'magic bullets'

Cancer therapies have hitherto been primarily based upon the use of gamma radiation, and/or chemotherapy. The success of these methods is however limited by their toxic effects on normal as well as cancer cells (which may prevent administration of sufficiently high and effective doses). A range of recombinant antibody reagents, often referred to as 'magic bullets', have therefore been developed. For example, chimeric molecules linking a tumour antigen-specific antibody-binding domain to a cytotoxic drug may enable delivery of high doses of drugs or immunotoxins to the tumour without excessive damage to normal tissues (Fig. 94.1). Another strategy depends upon the use of an antibody–enzyme chimeric molecule which activates an inactive pro-drug previously localised to the tumour site. Alternatively, chimeric antibody reagents may increase the destruction of tumour cells by recruiting the patient's own cytotoxic cells. This involves the use of fragments containing one tumour antigen-specific domain, and another which binds and activates receptors on cytotoxic T lymphocytes. Originally of rodent origin, strategies for antibody humanisation have also overcome the initially encountered problem of their rapid removal *in vivo* (arising from the elicitation of a human anti-mouse antibody (or HAMA) response).

Recombinant anti-cancer drugs

Recombinant DNA technologies are also enabling the development, and large-scale production, of natural and/or modified anti-cancer reagents (see Fig. 94.1). For example, recombinant forms of immuno-regulators (e.g. interleukin-2 and granulocyte–macrophage colony-stimulating factors) have been developed and used to potentiate elimination of some types of cancer cells by innate immune responses. The bulk production of the Yew tree-derived anti-cancer drug taxol is also being facilitated by the application of transgenic methods.

Anti-sense/antigene therapies

Current research is also focused upon the development of recombinant DNA-based methods which may be directly applied as potential cancer therapies (see Fig. 94.1). For example, the development of antisense or gene-targeting approaches may potentially be exploited to switch off activated oncogenes within tumours, or block the synthesis of the encoded oncoprotein. An outstanding problem, however, is how to specifically target tumour cells without affecting normal cells.

Monitoring therapy efficacy

The PCR and genetic fingerprinting techniques are also increasingly being used to monitor the effectiveness of therapies (including bone marrow transplants) and maximise their potential. For example, the sensitivity of the PCR is greatly aiding the detection of minimal residual disease, i.e. the existence of small numbers of cancer cells which may have evaded or become tolerant to therapy. If undetected and untreated, these may allow the formation of a new and more resistant tumours.

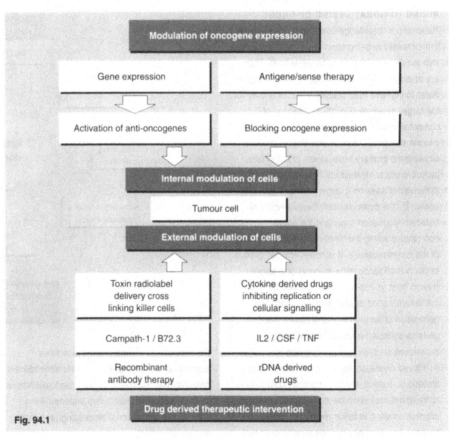

Fig. 94.1

255

95. DRUG DEVELOPMENT USING RECOMBINANT DNA APPROACHES

The discovery and evaluation of peptide- and oligonucleotide-based drugs through rational design or random screening increasingly depend upon recombinant DNA-based approaches and reagents.

The development of pharmacologically active chemical compounds (i.e. drugs) currently underpins the treatment of most genetic and non-genetic diseases. Indeed, the development of several antigene and antisense therapies also depends upon the generation of novel compounds such as peptide nucleic acids (PNAs) which can modulate the expression of specific genes. Conventional strategies for the development of pharmacologically active compounds have relied upon the random or progressive synthesis of panels of chemical compounds (e.g. by the successive alteration of chemical groups). This is followed by the screening for pharmacological activity, and assessment of their toxicity (i.e. using a range of appropriately designed *in vitro* and *in vivo* assays/bioassays). In recent years, however, recombinant DNA approaches have aided the identification of targets for potential drug development (i.e. by defining disease-determining molecular events and interactions). They are also increasingly being employed directly as tools for the development and assessment of new compounds, including several forms of recombinant proteins. For example, the availability of recombinant cellular receptors and their associated effector protein complexes (e.g. G-proteins) is enabling the rational design of receptor agonists and antagonists (i.e. compounds capable of inducing or blocking receptor activity) by providing the three-dimensional structural data necessary for computer-assisted molecular modelling. Furthermore, prokaryotic surface display and differential display libraries, and transfectant animal cell lines harbouring recombinant receptors, are increasingly being used in drug screening. Genetic engineering methods are also increasing the ability to generate novel recombinant proteins with pharmacological potential (e.g. antibody mimetics).

ANTIBODY-BASED DRUG DESIGN

The genetic manipulation of antibodies is creating a wide range of reagents for *in vivo* diagnostic and therapeutic use. Studies of antigen-combining site structures within antibodies is also providing valuable lessons for drug design. Accordingly, synthetic peptides representing antibody-binding determinants (termed complementarity-determining region or CDR peptides), and synthetic peptide mimetics representing a conformationally restricted organic ring mimicking the structure of a CDR loop, have been produced which retain the capacity to bind antigen. It has been suggested that the design of such small, bioactive peptides based upon antigen-combining sites may serve as models for the design of other bioactive peptides with pharmacological potential.

AIDING RATIONAL DESIGN OF DRUGS

Rational (i.e. knowledge-based) design principles are increasingly being used to increase the speed with which new drugs may be developed. They are also being used to redesign existing drugs in order to improve their specificity and/or affinity for the target molecule (Fig. 95.1). The use of computer-assisted molecular modelling is an important aspect of the rational design process, allowing the primary assessment of potentially bioactive drugs for their ability to spatially interact with desired sites on a target molecule (i.e. act as ligands). This depends upon the availability of detailed information regarding the atomic coordinates within the three-dimensional structure of the target molecule. It is however often difficult to purify sufficiently large amounts of a target protein from *in vivo*- or *in vitro*-derived cells for the determination of atomic coordinates. The generation of large amounts of recombinant proteins suitable for structural analysis using techniques such as nuclear magnetic resonance and X-ray crystallography is overcoming these limitations. Indeed, the target protein complexed to its ligand may often be studied directly, and the information used to refine molecular modelling programs. It must however be remembered that

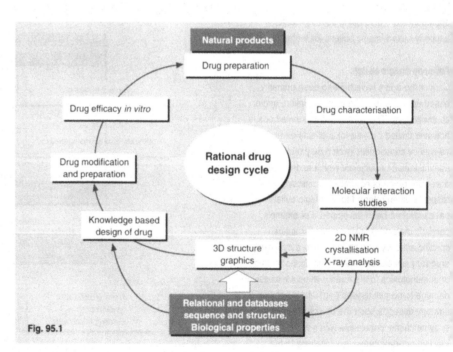

Fig. 95.1

the accuracy of rational design processes depends upon the expression of the recombinant proteins in appropriate (i.e. homologous) systems to ensure that the protein has undergone the correct post-translational processing (e.g. folding and glycosylation).

OAs: 55, 595–600, 796–802, 944–54, 1025, 1028–41, 1049. **RAs:** 280, 303, 354, 370–9, 381–5, 398–401. **SFR:** 120, 169–70, 208–212.

Aiding random drug screening

Random screening methods presently remain the most widely used and productive approaches for the discovery of new drugs of natural or synthetic origin. Based initially upon high-throughput screening principles to assess their binding to purified or cell-borne potential target molecules, such random screening processes are being revolutionised by the application of emergent recombinant DNA technologies (Fig. 95.2). For example, the ability to produce so-called 'epitope libraries' composed of millions of different random peptide sequences expressed on the surface of bacteriophages has radically facilitated the screening for compounds with bioactive potential. Phages bearing such peptides may therefore be selected by 'panning' over immobilised antibodies, receptors or other appropriate ligand-binding molecules. The affinity of the phage-borne peptide for the immobilised protein allows the desired phages to be recovered, propagated, and the DNA encoding the peptide may be isolated for analysis (i.e. sequencing). The selected peptide may also be expressed and assessed for bioactivity. This may be followed by further random or rational design cycles (e.g. involving *in vitro* mutagenesis steps).

Similar approaches are being adopted to identify oligonucleotides (termed aptamers) with potential as therapeutic agents (e.g. as anti-viral agents, or for antisense and antigene therapies). Random oligonucleotide libraries containing 10^{13} DNA molecules have therefore been generated and screened against a range of potential target proteins or enzymes which do not normally interact with oligonucleotides. Any bound oligonucleotides are eluted, *in vitro* amplified by the polymerase chain reaction, rescreened, and sequenced.

Aiding drug evaluation

An essential part of drug development is the evaluation of compounds for their bioactivity and toxicity. Such screening approaches include whole cell assays, membrane receptor assays, transcriptional assays and *in vivo* lethal dose determinations (i.e. LD_{50} doses). Each of these are benefiting from the application of recombinant DNA techniques. For example, the pharmaceutical industry is using cloned receptor

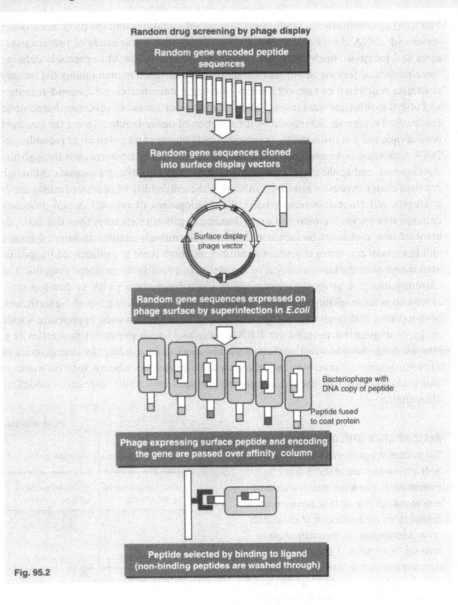

Fig. 95.2

genes expressed on microbial and mammalian cell lines containing few endogenous receptors to more definitively screen panels of synthetic chemicals or natural products. Panels of such cell lines, bearing receptors representing the natural molecular diversity of related receptors, are also being used as alternatives to conventional bioassays. These may be used for both the reassessment of the specificity of existing drugs and assessment of compounds produced by random or rational design. Transcriptional screening using cultured cells transfected with a

transcriptional unit linked to a reporter system (e.g. an enzyme) is also being used to assess the effect of potential nucleic acid or protein modulatory compounds by monitoring the level of reporter molecules induced. Techniques such as northern blotting, *in situ* hybridisation, and ribonuclease protection assays are also being employed to measure transcriptional effects *in vitro* and *in vivo*. As the sophistication of these approaches increases, toxicity tests involving LD_{50} determinations may become redundant, or the numbers of tests required will be reduced.

96. PROTEIN ENGINEERING

The ability to predictably engineer proteins depends upon a fundamental understanding of protein structure, and the availability of structural databases and accurate molecular modelling methods.

The strategy of introducing random or site-directed mutations into the protein-encoding regions of cDNA clones has commonly been employed for the study of protein structures and functions. Such strategies have been particularly useful in precisely defining those molecular regions and/or specific amino acids involved in maintaining the tertiary structures required for a range of protein–protein, protein–nucleic acid, ligand–receptor, and enzyme–substrate interactions. Simple random (or iterative) processes based upon randomised mutagenesis followed by the selection of desired mutants from the resultant pool of molecules are still widely exploited. The full power and potential of recombinant DNA technologies to engineer proteins will however only be harnessed through the development and application of knowledge-based (rational) design processes. Although iterative design processes have benefited from the availability of improved mutagenesis strategies and clone selection systems, the development of rational design processes depends more upon an improved understanding of protein structures than the development of new molecular biological techniques. Accordingly, understanding the nature and functional correlates of protein structures, and how these are influenced by specific amino acid substitutions, is critical to the ability to predictably engineer proteins. The current paucity of protein three-dimensional structures upon which to develop sufficiently accurate modelling systems is therefore the greatest limiting factor. Nevertheless, both iterative and rational design processes have already been used to generate a wide range of engineered proteins (and RNA molecules) with improved properties (e.g. increased stability and functional activity). These are in turn aiding the development of sufficiently accurate predictive models as each new product yields structural information which may be used to expand computer databases and refine molecular modelling algorithms.

REDESIGNING RIBOZYMES

Protein engineering principles have also been used to redesign ribozymes (the unusual RNA molecules which catalyse many biological process with efficiencies approaching those of traditional protein enzymes). Accordingly, both rational and iterative design processes have been used to create ribozymes with altered substrate specificities and kinetic rates which may be used as molecular catalysts, or for the correction of genetic defects in vivo. For example, new variants of self-cleaving hammerhead RNAs (termed aptamers) have been selected from a library of randomly amplified and mutagenised RNAs. These exhibit enhanced RNA ligation and cleavage properties or enhanced catalytic turnover rates. Rational redesign of the internal guide site of the self-splicing group I intron of Tetrahymena pre-rRNA has also altered its substrate specificity.

BASIC PROTEIN STRUCTURES

The function of a protein ultimately depends upon its final three-dimensional structure (or conformation). Four levels of protein structure may be defined (Fig. 96.1): (i) primary structure, defined by the linear sequence of amino acids within a polypeptide; (ii) secondary structure, involving the formation of local folded conformations such as α helices or β strands; (iii) tertiary structure, resulting from the packaging of secondary structural elements into compact domains; (iv) quaternary structure, involving the final arrangement of the functional protein (often involving the close association of several different polypeptides). The formation of tertiary and quaternary structures is extremely important in determining functional regions within proteins (i.e. active sites or binding sites) by bringing together amino acids which may be far apart within the (primary) polypeptide sequence.

Many regular features may be discerned within protein secondary structures. Protein folding is primarily driven by the need to protect hydrophobic amino acid side chains from the aqueous environment, resulting in the formation

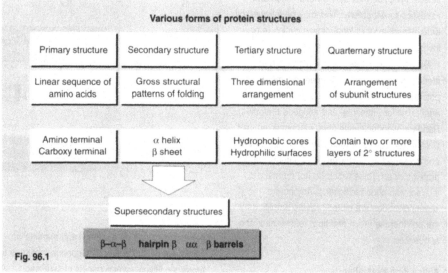

Fig. 96.1

of globular structures containing a hybrophobic core and a hydrophilic surface. This is firstly achieved by the folding of the polypeptide into two forms of regular core structures (α helices or β strands) separated by loops of polypeptide which often contribute to the active or binding site of the functional protein. Combination and arrangement of such core secondary structures

subsequently generates several different structural motifs (e.g. helix-turn-helix motifs).

OAs: 519–26, 595–600, 771–93, 802, 1042–6. **RAs:** 272–5, 375–7, 380, 383, 395–7. **SFR:** 120, 169–70, 194, 205–6, 211–13.

8. Types of protein–DNA interaction.
11. Types of RNA molecules.
14. Anatomy of eukaryote genes.
23. Transcription.
25. tRNA/rRNA processing events.
27. Transcription regulatory sequences.
29. Transcription factors.
30. *In vivo* translation.
41. Recombinant DNA technology.
55. Polymerase chain reaction.
62. Fundamental principles of cloning.
68. Phagemid vectors.
71. Genomic/cDNA cloning strategies.
74. Detecting protein–DNA interactions.
75. Detecting promoters and enhancers.
79. *In vitro* mutagenesis.
92. Receptors and cellular signalling.
98. Engineering antibody genes.

PROTEIN STRUCTURE PREDICTION

One of the main problems limiting the rational design of new protein molecules is the ability to predict protein structures (see Fig. 96.2). Although no universal method exists, several predictive methods have been developed based upon identifying the constraints imposed upon protein secondary and tertiary structures by their amino acid sequence.

Homology-based methods

Proteins with homologous amino acid sequences usually have similar three-dimensional structures. This is often due to the possession of conserved structural core regions containing elements used to construct protein scaffolds of similar three-dimensional structure. Database comparisons involving homologous sequences of proteins whose tertiary structure have already been determined (e.g. by crystallisation studies) may therefore indicate the presence of similar scaffold structures. Similarly, the approximate structure of the various loop regions connecting scaffold regions may be indicated by database comparisons with loop regions from known protein structures. Graphic modelling of the data from such alignments using sophisticated computer software and energy minimisation calculations can then be used to propose the conformation of the whole protein. Although these analyses have proved useful (e.g. in modelling the hypervariable loops of antibodies involved in antigen binding), their accuracy depends upon the availability of sufficient solved structures against which comparisons may be made.

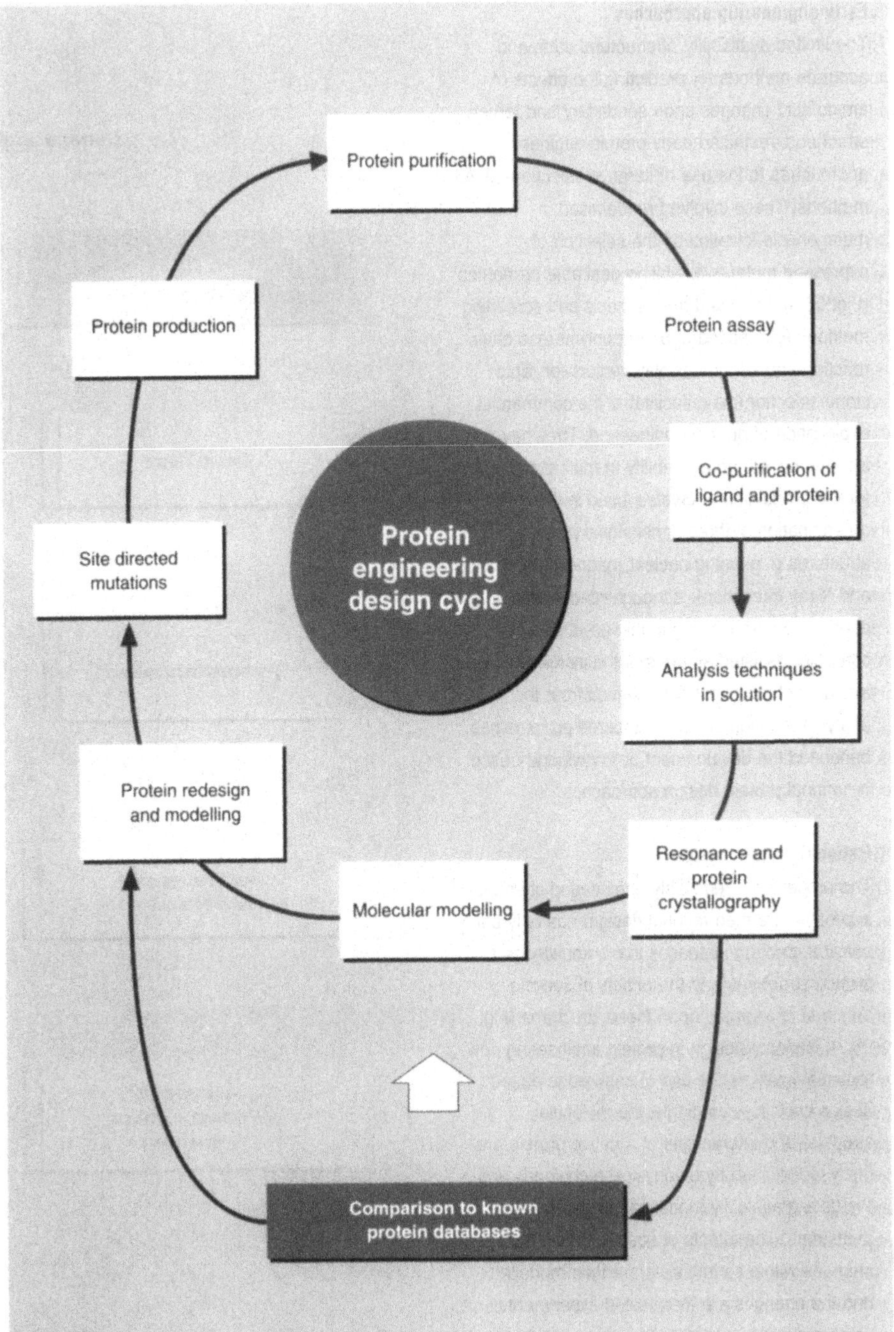

Fig. 96.2

Algorithmic methods

Several methods for predicting secondary structures have also been developed based on statistical algorithms which assume that the local sequence determines the local immediate structure. For example, the Chou and Fasman method is based upon the statistical analysis of the frequency of occurrence of a given amino acid within an α helix, β strand or turn within known X-ray structures to predict the likelihood of a given sequence to adopt regions of each particular structure. Accordingly, a list of average probability values is assigned which take into account neighbouring amino acids. Consecutive amino acids are therefore initially analysed for the occurrence of proposed secondary structures until a loop region or irregular loop coil is encountered which has a low probability of existing as an α helix or β strand. Such predictive algorithms are often able to predict an α helix, β strand or loop. The success and accuracy with which individual structures may be predicted is however generally very poor when used in the absence of additional physical data.

Protein engineering has evolved from random iterative selection to the use of knowledge-based design cycles to alter the conformation, stability, and functional activity of a range of important proteins.

Early engineering approaches

The limited availability of structural data, and accurate methods for predicting the effects of amino acid changes upon secondary and tertiary structures, restricted early protein engineering approaches to the use of iterative selection-based methods. These involved randomised mutagenesis followed by the selection of expressed mutants exhibiting desirable properties (Fig. 96.3). Improved mutagenesis and screening methods (e.g. based upon the polymerase chain reaction and surface display vectors for rapid clone selection) have facilitated the commercial exploitation of protein engineering. They have also increased the availability of mutant proteins for the production of crystals used in the determination of three-dimensional protein structures (e.g. using nuclear magnetic resonance and X-ray diffraction). Although randomised approaches are still commonly used, the computer-assisted analysis of the increased amounts of structural data derived from the analysis of recombinant/engineered proteins has enhanced the development of 'knowledge-based' or rational protein design approaches.

Rational protein design

The process of predictably engineering changes in proteins (termed rational design) has only been possible through increases in our knowledge of protein structures and the effects of specific amino acid changes upon these structures (Fig. 96.3). Rational design in protein engineering now routinely involves the use of knowledge-based design cycles. Accordingly, the molecular structure and interactions of a native protein are firstly determined by biophysical techniques and the data graphically modelled using computer algorithms. The effects of specific amino acid changes within certain sites are then modelled, and the changes are then tested experimentally using molecular biological techniques such as site-directed mutagenesis. The changes in the function and three-dimensional structure of the expressed mutant protein resulting from the amino acid substitutions are then determined. These data are then used to begin another design cycle. Each completed cycle increases the computer database of known protein structures, therefore the accuracy of predictions based upon

modelling algorithms increases with each cycle. Consequently, rational design cycles are increasingly being used to generate a range of proteins which exhibit altered conformational stability and functional activity.

General engineering approaches

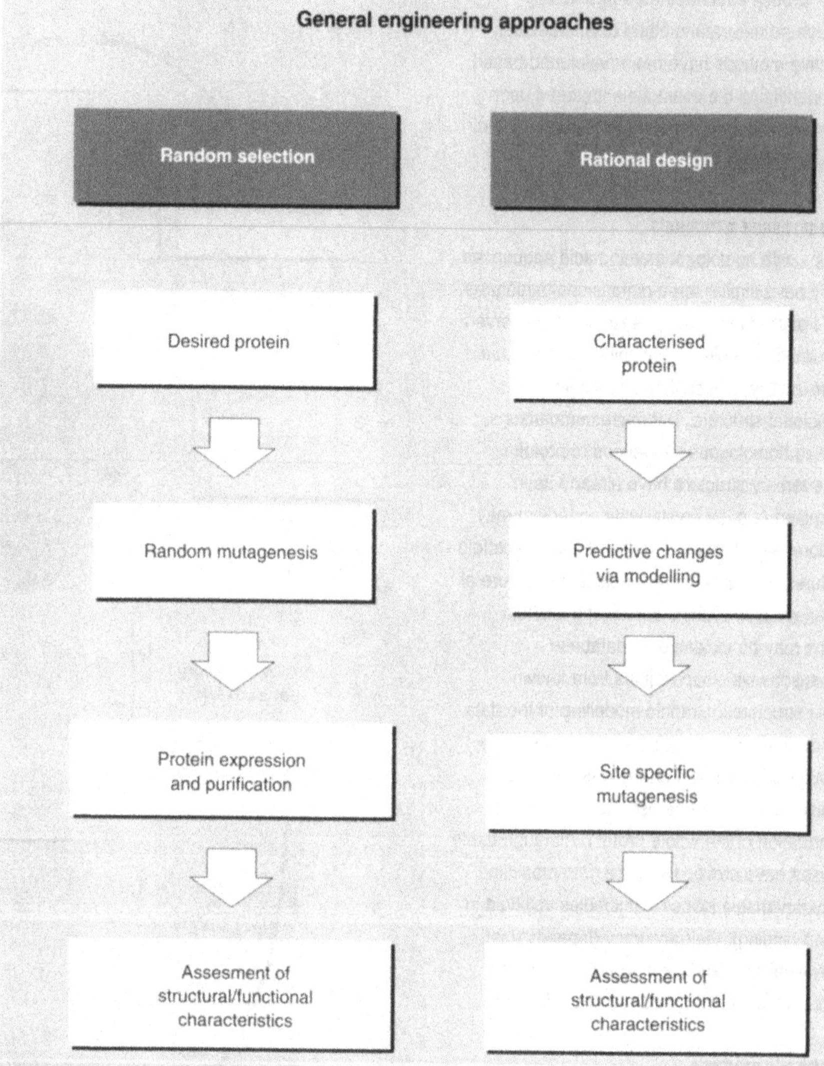

Fig. 96.3

OAs: 519–26, 595–600, 771–93, 802, 1035–41. **RAs:** 272–5, 375–7, 381–5, 295–7. **SFR:** 120, 169–70, 194, 205–6, 211–13.

Improving protein stability

Stability is an important factor which determines the manipulation, production, storage and application of proteins. Strategies for increasing protein stability via genetic engineering have often relied upon the use of three-dimensional models of thermostable proteins to direct the introduction of multiple amino acid substitutions (Fig. 96.4a). The introduction of metal-binding sites is however also proving to be successful as an approach to increase conformational stability, and reduce proteolytic degradation. Accordingly, naturally occurring metallo-proteins often contain important motifs within α helices and β strands which bind to metal ions (e.g. His-X(1-3)-His, where X represents any amino acid). These are known to enhance their thermodynamic stability, possibly by maintaining the protein in a folded state. The strategic introduction of as few as two such motifs near the surface of many proteins can therefore increase their conformational stability. For example, the introduction of a single His at position 58 in cytochrome c creates a chelating site with an existing His at position 39, such that chelation of Cu^{2+} iminodiacetate results in enhanced conformational stability. Chelation of metal ions that exchange their ligands slowly (e.g. the substitution of inert metals Co^{3+}, Ru^{3+}), has also been found to increase stability.

Local unfolding events which render specific sites more susceptible to proteolytic degradation can also be reduced by introducing metal ion-binding sites known to increase conformational stability. For example, the grafting of a Ca^{2+} binding loop from a thermostable homologue (thermitase) onto one of the auto-proteolysis sites of the serine protease subtilisin (widely used in industry) reduces its susceptibility to proteolysis. The engineering of metal chelation sites also appears to be an effective and widely applicable approach for increasing protein thermostability. Thus, the commonality of α helices and β sheets allows the insertion of simple metal-binding sites onto the surfaces of many proteins with little disruption of their normal function or structure. These metal ion-binding sites may also facilitate metal affinity purification processes. It is also possible that they may act as immobilisation and/or nucleation sites to enhance protein crystallisation (thereby aiding structural analyses).

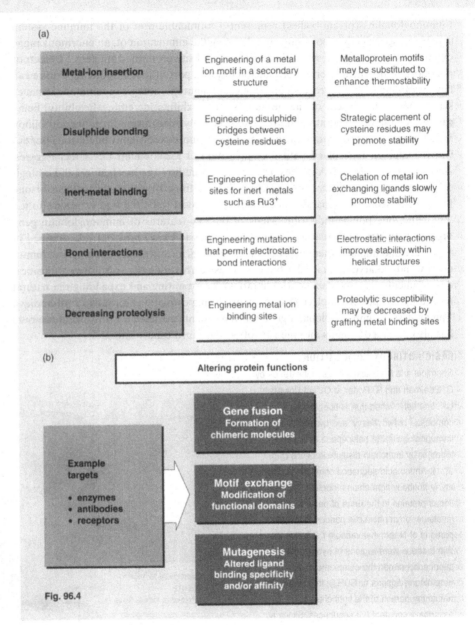

Fig. 96.4

Altering protein functions

The function of all proteins depends upon their ability to specifically bind to other molecules. Alteration of the structure of such binding sites therefore has profound effects upon their functional activity. Enzymes, antibodies and receptors are of great importance in biology and biotechnology. Consequently, many well-defined structures have been obtained for these classes of molecules, and their binding/active sites have been primary targets for protein engineering. Accordingly, amino acid substitutions have been introduced within enzyme active sites to enhance their affinity for substrate binding, or alter their catalytic/biosynthetic activity (Fig. 96.4b). The ligand-binding specificity of antibodies and receptors has also been altered, and the antigen-binding sites of antibodies have been altered to give them intrinsic enzymic activity (i.e. to generate catalytic antibodies or abzymes).

97. IMMUNOGLOBULIN GENETICS

The diversity of antibody structures available in nature is partly determined by the existence of multiple families of genes, each encoding a different portion of the light or heavy chain polypeptides.

Immunoglobulins (or antibodies) represent a formidable arm of the immune system capable of specifically recognising, and dictating the elimination of, an enormous range of naturally occurring and synthetic substances (collectively termed antigens). Defects or polymorphisms which alter the production of just a part of the antibody repertoire can have profound biological consequences (e.g. autoimmune, allergic or immunodeficiency syndromes). This aspect of the immune system exhibits incredible flexibility, being capable of rapidly generating antibodies specifically recognising an estimated billion different molecular structures on antigens. This attribute posed many intellectual puzzles which remained unresolved until the application of recombinant DNA technologies. Accordingly, in 1976 DNA studies first demonstrated how the biochemical and biological attributes of antibodies were genetically determined by the construction of one polypeptide from two separate genes. This also proved to be the first exception to the 'one gene, one polypeptide' rule. Many of the key features of immunoglobulin gene organisation, structure, rearrangements and expression have now been determined in both mice and humans, thereby providing precedents for several new concepts in immunology and eukaryotic molecular biology. These discoveries have also driven technical developments (e.g. recombinatorial libraries) for exploiting and expanding the natural repertoire of antibody molecules (e.g. by producing chimeric and catalytic antibodies). The analysis of immunoglobulin gene rearrangements associated with several tumours has also provided insights into the role of oncogenes in cancer initiation.

BASIC ANTIBODY COMPOSITION

Chemical and biological studies in the 1960s by G. Edelman and R. Porter at Oxford University, UK, first established that antibodies were composed of two 'heavy' and two 'light' immunoglobulin (Ig) polypeptide chains held together by interchain disulphide bonds (Fig. 97.1). Amino acid sequence comparisons of many antibody light chains (secreted as Bence Jones proteins in the urine of patients with myeloma, a lymphoid cell cancer) highlighted portions of N-terminal variable (V) sequence. Within these were regions of hypervariable sequence (termed the complementarity determining regions or CDRs). In contrast, the remaining portion of the light chain had essentially constant (C) sequences. Similarly, separate regions (or domains) have been confirmed within the heavy chains.

Each domain exhibits a characteristic structure comprising two anti-parallel β-pleated sheets, joined by an intrachain disulphide bond, with loops at either end. It is now known that VL and VH domains are responsible for the interaction with antigen. Differences in antigen specificity are therefore largely determined by their CDR sequences which form the amino distal polypeptide loops involved in antigen interaction. The C domains are responsible for interaction with a variety of soluble and cellular molecules

THE ANTIBODY DOMAIN BLUEPRINT

The characteristically repeating domain structure of antibody polypeptides suggests that they evolved by a process of gene duplication. A large number of immunologically related receptor molecules mediating cellular cross-talk, adhesion, effector function activation and immune regulation (including 'self' and 'non-self' discrimination) have since been shown to contain similar antibody domain structures. These molecules are therefore referred to as members of the immunoglobulin supergene family. Their similarity in molecular construction and utilisation of similar genetic mechanisms reinforces the concept of evolutionary duplication from an ancestral domain-encoding gene. It also enhances the understanding of mechanisms controlling immune functions which may provide ways to therapeutically modulate *in vivo* immune responses.

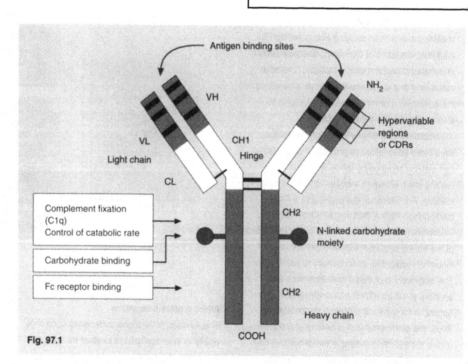

Fig. 97.1

which effect antigen elimination (i.e. effector molecules). Several types of heavy chain and two types of light chain polypeptides have been described, with only one type of light chain and heavy chain used to construct a single antibody molecule.

OAs: 93–4, 276–80, 1000–2, 1065–70. RAs: 25, 32, 35–6, 131, 380, 386–90. SFR: 37–9, 56–7, 120, 177–8, 181, 214–17.

Antibody polypeptide diversity

Five classes (or isotypes) of antibody, designated IgM, IgG, IgD, IgA, and IgE, have been characterised in humans, distinguished by their heavy chains (designated μ, γ, δ, α, and ε respectively). The variation in constant region sequence, number of domains, interchain interactions and glycosylation sites between the heavy chains determine the overall composition and differing effector functions of the different antibody classes. Four subclasses of γ and two subclasses of α chains have also been defined based on sequence and structural differences. For example, IgG3 molecules contain an extended hinge region comprising multiple inter-heavy chain disulphide bonds located between the first and second constant domains. Two classes of light chain, designated kappa (κ) and lambda (λ), have also been defined which differ in constant region sequence, with each existing in several minor isoforms termed allotypes.

Gene diversity and organisation

The human heavy and light chain polypeptides are now known to each be encoded by multiple gene segments present within the genome at three discrete chromosomal locations (or loci).

(1) *The heavy chain locus.* Located on human chromosome 14, it contains several clusters of gene segments, each of which encodes a different portion of the heavy chain polypeptide. Thus, the heavy chain variable region of the polypeptide is encoded by a V_H, J_H (joining) and D_H (diversity) gene segment. Each of these is present as a cluster or family of related gene segments separated by introns. Each V_H gene segment is also preceded by an upstream leader sequence separated by an intron. The actual number of human V_H, D_H and J_H gene segments is estimated to be of the order of 250–1000, 20, and 6, 3 pseudogenes respectively, although recent studies suggest that the number and variety of gene segments may vary between individuals. In contrast, the constant regions of each isotype and subclass are encoded by a single C gene segment composed of several exons, clustered downstream of the V-D-J segments.

(2) *The kappa light chain locus.* Located on human chromosome 2, this also contains discrete V and J gene families which combine to encode the V_L domain, and a group of C genes each encoding the various κ chain allotypes.

(3) *The lambda light chain locus.* Located on human chromosome 22, the lambda light chain locus is very different in organisation to the kappa chain locus, although the precise number of V, J and C gene segments may vary between individuals.

The different V-D-J and C gene segments utilised by the human immune system are believed to have arisen from an ancestral gene through gene duplication events (supported by the existence of several pseudogenes). Their rearrangement is the primary mechanism involved in the creation of antibody diversity.

Regulatory sequences

A number of transcriptional promoters have also been identified located proximal to the gene segments. These include the TATA box at position −25, an octanucleotide element (ATGCAAT) at −70 and a heptamer element (CTAATGA). An enhancer is also located in the heavy chain locus upstream of the first Cμ exon. This accounts for the high transcriptional rates observed following rearrangement of V-D-J segments which brings the complex into range of the enhancer. Specialised switch elements are also located upstream of the heavy chain C gene segments. These are involved in class switching during B cell development.

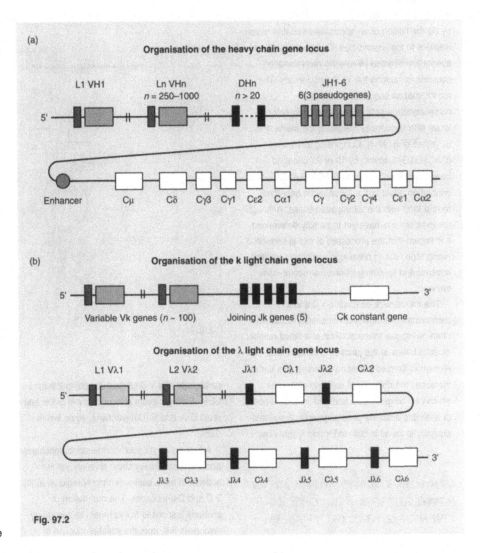

Fig. 97.2

(a) Organisation of the heavy chain gene locus

(b) Organisation of the k light chain gene locus / Organisation of the λ light chain gene locus

97. IMMUNOGLOBULIN GENETICS/Contd

Functional antibodies are generated by the highly regulated combination of several processes, including gene segment rearrangement, imprecise recombination, the insertion of nucleotides, and somatic mutation.

GENERATING ANTIBODY DIVERSITY

Unlike other protein-encoding genes, immunoglobulin genes are uniquely able to undergo events which lead to germline rearrangements prior to the synthesis of a heteronuclear RNA transcript. The relatively large number of available germline V, J and D gene elements makes a substantial contribution to antibody diversity by allowing different combinations of these gene elements to be formed. This is often termed combinatorial diversification.

Confined to one rearrangement of light chain genes, two rearrangement events occur amongst the heavy chain genes (Fig. 97.3). Accordingly, specific heavy chain variable region exons are assembled by two recombination events: (i) the linking of a diversity to a joining element, followed by (ii) the linking of an appropriate variable region element to the recombined (D-J) element. These events are directed by specific recombination sequences flanking the V, D and J exons. The recombination sequences represent complementary sets of bases which may form loops with sequences preceding the element to be joined (Fig. 97.4). Comprising a heptamer (CACAGTG) followed by 12 or 23 unpaired bases, and then a nonamer (ACAAAAACC), their orientation ensures the 12 unpaired bases will form a loop with the 23 unpaired bases. Although the mechanisms have yet to be fully determined, it is known that the processes of signal sequence recognition, strand breakage, nucleotide addition/ removal and re-joining involves a recombinase enzyme.

The mechanism controlling V-J and V-D-J recombination is however not entirely accurate, often giving rise to the addition of a small number of extra bases at the junction between these elements. Termed junctional diversity, this further increases the diversity of antibody structures which can be generated from the limited number of available antibody gene segments. Junctional diversity is found in both the κ and λ light chains,

OAs: 93–4, 276–80, 1000–2, 1071–7.
RAs: 25, 32, 35–6, 131, 380, 386–90.
SFR: 37–9, 56–7, 120, 177–8, 181, 211–12, 214–17.

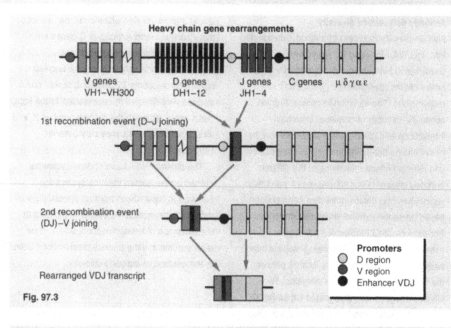

Fig. 97.3

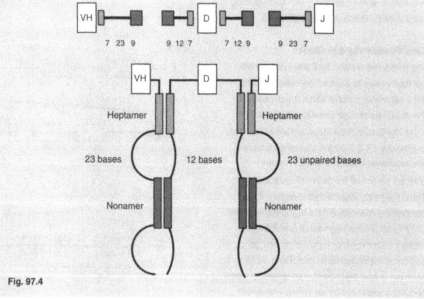

Fig. 97.4

particularly at a V-J junction encoding the third CDR. A similar event also occurs within the heavy chain D-J and V-(DJ) junctions, again within CDR3.

A further, as yet poorly understood, mechanism also increases heavy chain diversity via the addition of extra bases, termed N regions, at the V-D and D-J junctions. The maturation of antibody responses from primary to secondary responses (i.e. from the initial production of

predominantly IgM molecules with low affinity for antigen to predominantly IgG molecules with high affinity for antigen) also involves another poorly characterised mechanism which leads to the introduction of somatic mutations within the variable regions. These mutations may not be entirely randomly introduced since they appear to occur most frequently around, or within, the CDRs.

Regulation of gene expression

Antibody gene rearrangement and expression
follow a defined sequence during the different
stages of B cell differentiation. These stages in
the production and secretion of a functional
antibody are controlled by both genetic
mechanisms and the cross communication
between lymphoid cells mediated by soluble
factors (e.g. lymphokines).

Producing a functional antibody

Functional antibodies are only secreted from
terminally differentiated B lymphocytes termed
plasma cells. Each plasma cell secretes a single
type of antibody (i.e. mono-specific) despite being
diploid in nature. Crucial to the development of
monoclonal antibodies, this arises from a unique,
and as yet undefined, mechanism termed allelic
exclusion which occurs during B cell
differentiation (see Fig. 97.5). Accordingly, D-J
recombination in pro-B cells (the earliest cells
shown to undergo DNA rearrangements) occurs
on both chromosomes. This is followed by the
joining of V region genes and the final
recombination of V-D-J with a μ constant region
gene segment at the pre-B cell stage. It is at this
stage that one of the two alleles is completely
switched off.

Following rearrangement and expression of a
functional heavy chain, light chain gene
rearrangements begin at the κ loci. The λ chain
genes however only undergo rearrangement if
both κ loci fail to produce a functional gene. This
finally results in the production of a functional IgM
bound to the membrane of unstimulated Bμ cells
via a transmembrane anchor (encoded by two
small mini-exons located immediately
downstream of the Cμ exons). An RNA
processing step then gives rise to a Bμ + δ cell
expressing both IgM and IgD on the surface.

Differential RNA splicing is also involved in the
switch from the synthesis of membrane IgM to
secreted IgM following the first interaction with
antigen (possibly triggered by a mechanism
involving interleukins 4 and 5). Differential RNA
splicing is also involved in the switch from IgM to
IgD. Switching from IgM production to any other
heavy chain isotype is a specialised process
involving recombination with special elements
(termed switch elements). These are found 2–

3 kb upstream of each constant region gene and
organised into multiple repeats of the sequence
GAGCT-GGGGT spanning up to 10 kb.
Transcripts arising from this switching event

remain under the control of the heavy chain
enhancer in the same relative position to the
transcriptional promoter.

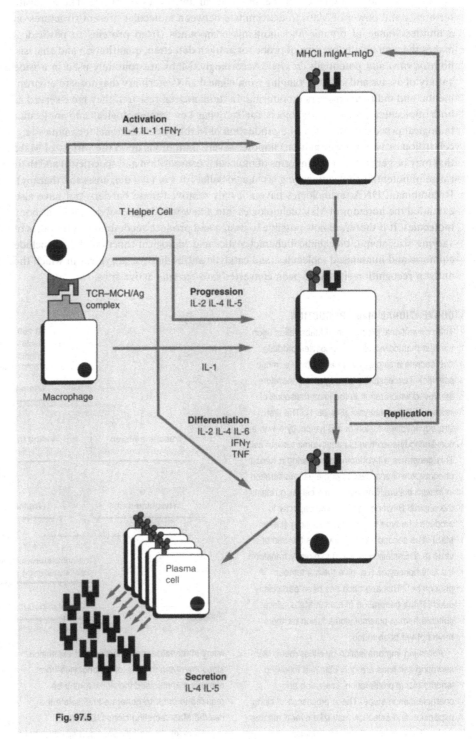

Fig. 97.5

98. GENETIC ENGINEERING OF RECOMBINANT ANTIBODIES

Conventional monoclonal antibody production limitations have been overcome by recombinant DNA technologies, including the PCR, repertoire cloning and immune reconstitution of SCID mice.

Beginning in 1975 with the formation of hybridomas by Cesar Milstein and Georges Kohler at Cambridge, UK, the development of techniques for the immortalisation of B lymphocytes revolutionised immunology. In particular, they had a significant impact upon many areas of basic and applied life sciences by providing continuous supplies of highly specific antibody reagents (i.e. monoclonal antibodies or Mabs). The exquisite sensitivity and power of Mabs to discriminate between molecular present structures on a limitless range of organic and inorganic compounds (from proteins to pesticides) makes them an invaluable type of probe for antigen detection, quantitation and analysis, both *in vitro* and potentially *in vivo*. Accordingly, Mabs are routinely used in a wide variety of assays and settings, ranging from clinical and veterinary diagnosis to environmental and industrial process monitoring. In fundamental research they have served as both molecular probes and subjects for studying key immunological and molecular biological processes. However, the production of Mabs by conventional techniques (e.g. cell fusion or viral transformation) imposes severe limitations upon the variety of Mabs that may be generated (both in terms of molecular composition and specificity) and their range of potential applications (e.g. as 'magic bullets' for *in vivo* diagnosis and therapy). Recombinant DNA technologies have not only removed these barriers, but have also extended the horizons of Mab technologies into a new world of novel designer antibody molecules. It is therefore now possible to design and produce recombinant antibodies of varying size, shape, biochemical characteristics and biological function. These include chimeric and humanised molecules, and catalytic antibodies (or abzymes) in which the antigen recognition sites have been converted into enzyme active sites.

THE Mabs REVOLUTION

In addition to providing constant supplies of high specific probes for the detection and quantitation of an enormous range of substances, conventional Mab production has provided many additional benefits. The hybridoma or transforme B cell lines have themselves been vital as subjects used to define the cellular, molecular and genetic mechanisms maintaining antibody-mediated immune protection, and the defects leading to immune dysfunction. For example, current knowledge concerning the role of somati hypermutation in generating antibody diversity has been determined using hybridoma cell lines Similarly, knowledge regarding the three-dimensional structure of antibodies has largely been derived from X-ray crystallographic studies of purified Mabs. Conventional Mab production technologies have therefore underpinned the development of recombinant DNA-based antiboc technologies.

CONVENTIONAL Mab PRODUCTION

The conventional production of Mabs relies upon the fact that individual B lymphocytes produce and secrete a single type of antibody (i.e. mono-specific). Two essential strategies are therefore employed which result in the immortalisation of individual B lymphocytes (Fig. 98.1). The first strategy involves fusion of a B lymphocyte with a non-antibody-secreting plasmacytoma tumour cell. This generates a hybridoma cell bearing a tumour phenotype which allows its perpetual proliferation in *in vitro* culture. The hybridoma cell also retains the original B lymphocyte-derived capacity to produce one type of antibody specificity (i.e. an Mab). The second strategy involves the use of a virus (e.g. Epstein Barr virus) to directly transform the B lymphocytes (i.e. give them a tumour phenotype). This approach has been particularly used for the generation of human Mabs, since suitable human plasmacytoma fusion partners have not yet been found.

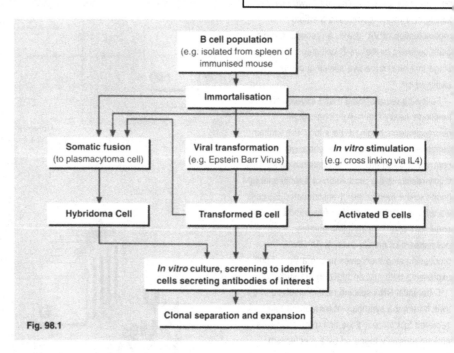

Fig. 98.1

Following immortalisation by either route, Mab-secreting cell lines are only obtained following lengthy clonal proliferation, screening and characterisation steps. These processes of clonal expansion and selection may take over 6 months and suffer from low efficiencies. Consequently, many immortalisation, selection and expansion steps, involving the use of lymphocytes from numerous immunised individuals, may be required in order to generate and isolate the desired Mab-secreting clones. Despite improvements in the efficiency and specificity of the fusion process (e.g. via the use of electrofusion methods, and the generation of improved plasmacytoma lines), several additiona factors limit the range of Mabs that can be generated by conventional approaches.

Technical limitations

The primary requirement for Mab production by hybridoma techniques is the availability of sufficient numbers of suitably activated B lymphocytes (i.e. producing antibodies of the desired antigen specificity). This generally requires repeated immunisation (often via several routes) and recovery of the appropriately activated B cells (generally from the spleen). Consequently, most Mabs have been of rodent origin. This however restricts the range of Mabs that can be obtained to the repertoire of antibodies naturally produced by the rodent immune system. For example, the production of murine Mabs specific for certain human red blood cell antigens (e.g. for use in blood group typing) has not been possible using this technology since mice do not produce an immune response to these molecules. The production of human Mabs is similarly limited by the availability of sufficiently large numbers of activated human B cells. Conventional human Mab production therefore requires extremely large amounts of blood or excised lymphoid nodes obtained from several naturally immunised (e.g. vaccinated) human donors. Furthermore, most human Mab-secreting cell lines produce only very low levels of antibody and are generally unstable, often losing their capacity to divide and/or secrete antibody during the course of long-term *in vitro* culture.

Recombinant DNA approaches

Thanks to a greater understanding of the natural process determining antibody diversity, the limitations associated with conventional methods of Mab production have been overcome by the development and application of three basic techniques/approaches.

(1) The polymerase chain reaction has allowed the *in vitro* amplification of antibody heavy and light chain polypeptide-encoding DNA segments from mRNA transcripts to be isolated from limited numbers of B cells, or even a single B cell. These DNA fragments can then be cloned into appropriate vectors to generate recombinant

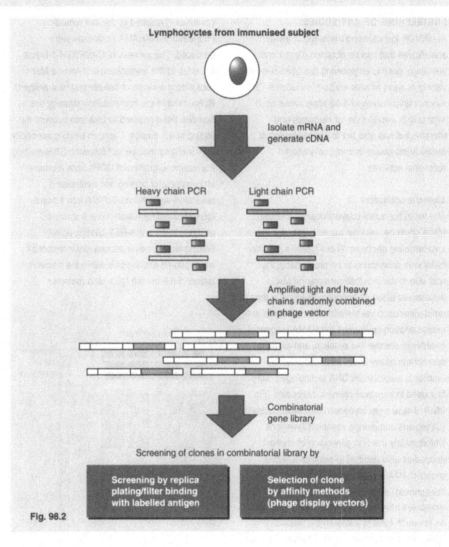

Lymphocytes from immunised subject

Isolate mRNA and generate cDNA

Heavy chain PCR   Light chain PCR

Amplified light and heavy chains randomly combined in phage vector

Combinatorial gene library

Screening of clones in combinatorial library by

Screening by replica plating/filter binding with labelled antigen

Selection of clone by affinity methods (phage display vectors)

Fig. 98.2

human/rodent Mabs (Fig. 98.2).

(2) Repertoire cloning has allowed the random insertion of heavy and light chain-encoding DNA fragments into specialised expression vectors. This enables the artificial recombination of light and heavy chain-encoding DNA segments to create a limitless diversity of antibody Fab fragments with 'unnatural' antigen-binding specificities. The creation of these so-called combinatorial or hierarchical libraries effectively overcomes the restrictions imposed by normal *in vivo* immune responses upon the range of antibody specificities which can be obtained (Fig. 98.2). The development and use of surface display phagemid vectors has also vastly increased the ease with which clones harbouring desirable heavy and light chain combinations (i.e.

with the desired antibody specificities) can be recovered. Desirable clones can therefore be isolated or selected using affinity chromatography columns rather than lengthy and difficult conventional screening techniques (e.g. filter hybridisation).

(3) Immune system reconstitution of mice lacking a normal immune system with human B lymphocytes has overcome difficulties in obtaining sufficient quantities of appropriately activated human B cells. Accordingly, human B lymphocytes introduced into severe combined immune deficiency (or SCID) mice may be activated by immunisation with appropriate antigens. These B lymphocytes may then be isolated and used as a source of mRNA for PCR amplification and repertoire cloning.

Genetic engineering has generated many forms of novel chimeric or humanised Mabs, subfragments and conjugates, and catalytic Mabs (or abzymes), for many types of *in vitro* and *in vivo* applications.

ENGINEERING OF ANTIBODIES

In addition to increasing the range of antigen specificities that can be obtained (i.e. by chain shuffling), genetic engineering has been used in a variety of ways to alter antibody molecules. This has not only increased their applications as *in vitro* and *in vivo* probes for diagnosis and therapy, but has also led to the generation of antibody molecules with radically altered functional activities.

Chimeric antibodies

The initial hope that conventionally generated Mabs could be used as 'magic bullets' for the *in vivo* targeting of abnormal cell types (e.g. tumour cells) was soon found to be problematic. This was due to the fact that such murine Mabs themselves acted as antigens, resulting in their rapid elimination via the elicitation of human anti-mouse antibodies (termed the HAMA response). In order to combat this problem, and to recruit appropriate heavy chain-mediated effector functions, recombinant DNA technologies were first used to construct chimeric antibodies (Fig. 96.3). These were composed of human constant (C) regions and murine variable (V) regions. Unfortunately this first generation of chimeric antibodies also resulted in significant, albeit reduced, HAMA responses due to the 'foreignness' of the murine V regions. Such molecules have however been widely employed as research tools to establish the precise molecular structures involved in effector function activation.

Humanisation of antibodies

HAMA responses have since been considerably reduced by adopting strategies variously referred to as humanisation or antibody reshaping. This originally involved the 'grafting' of murine Mab sequences encoding just the complementarity-determining regions (CDRs) of the heavy and light chain variable regions into the respective positions within a human Mab (Fig. 98.3). The first such humanised antibody (called CAMPATH-1-H) utilised the CDRs of a rat Mab specific for a human monocyte/lymphocyte surface antigen grafted into a human IgG1 Mab. *In vivo* administration of this reshaped Mab in patients with non-Hodgkin's lymphoma (a type of lymphoid

cell tumour) resulted in disease remission. Furthermore, no HAMA responses were produced. The success of CAMPATH-1-H has since led to the humanisation of murine Mabs recognising a range of cellular and viral antigens. Refinement of the humanisation strategy has however been required to take into account the finding that the original antigen binding specificity and/or affinity may be lost following CDR grafting. For example, grafting of CDRs from a murine Mab capable of binding and neutralising respiratory syncitial virus (RSV) into a human IgG1 Mab initially resulted in a humanised antibody with no anti-RSV binding activity. Binding was however subsequently restored by altering three amino acids within the framework region of the human IgG1 Mab (between

positions 91 and 94). Three general strategies are therefore currently used for antibody humanisation/reshaping: (i) the use of human V regions whose framework sequences most closely correspond to those from which the murine CDRs are derived; (ii) the 'veneering' of murine CDR surface residues into V regions to resemble the existing human profile; (iii) the additional introduction of minimal 'strategic' changes in human Mab framework sequences to complement the CDR grafting process.

OAs: 128–9, 479, 595, 600, 1086–98, 1021–6. **RAs:** 199–204, 299, 369–70, 372–3, 391–403. **SFR:** 120–4, 126–8, 169, 179, 181, 205, 218–19.

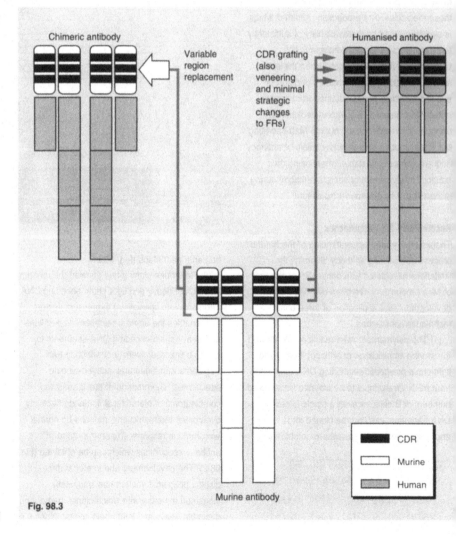

Fig. 98.3

Fragments and conjugates

Despite the defined antigen specificity and effector functions of reshaped Mabs, two additional factors have been found to significantly limit their utility as *in vivo* 'magic bullets' for diagnosis or therapy: (i) their non-specific uptake and accumulation in certain organs or tissues (e.g. by virtue of heavy chain (Fc) interactions with cellular receptors); and (ii) the poor tissue penetration of the relatively large (150 kD) antibody molecules which consequently prevents them from reaching the intended target (e.g. a solid tumour). Recombinant Mab subfragments, ranging from Fab molecules to 'minimal recognition units' (MRUs) representing a single CDR, have therefore been generated (Fig. 98.4).

Of these, Fv subfragments representing only heavy and light chain variable domains are likely to be maximally effective *in vivo* since they are small enough to penetrate all tissues. Furthermore, they do not possess significant sites for their interaction with cellular receptors, and are therefore unlikely to evoke HAMA responses. Such Fv fragments may be generated either by the natural association of the light chain and heavy chain V region fragments, or their physical linkage during cloning by means of an artificial linker molecule.

Genetic manipulations have also been used to generate a range of Mab and subfragments with additional functionalities (e.g. an enzyme or immunotoxin). Such conjugates have many potential *in vitro* and *in vivo* applications (e.g. for directly detecting analytes, or labelling and destroying tumour cells). It remains to be seen how far this process of molecular trimming and paring can be taken with beneficial results.

Catalytic antibodies (abzymes)

The binding sites of antibodies and enzymes are remarkably similar in that both are involved in specifically recognising and binding an antigen or substrate respectively. The range of naturally occurring enzyme active sites is, however, very restricted in comparison to the possible diversity of antigen binding specificities of antibodies. The development of repertoire cloning involving randomised antibody chain shuffling, and the availability of protein engineering techniques (e.g. *in vitro* mutagenesis strategies), has therefore

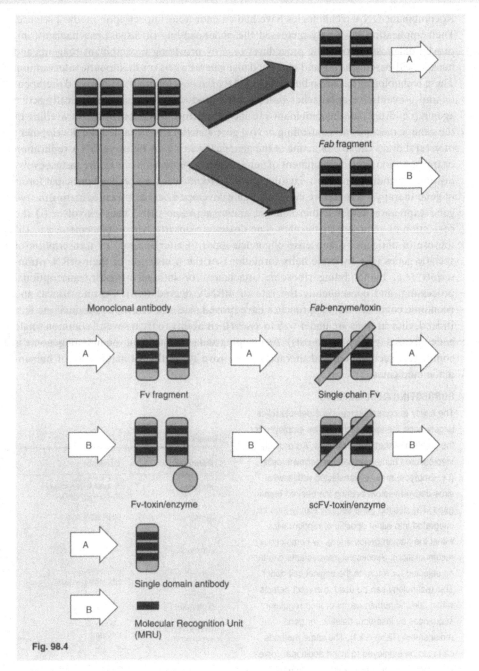

Fig. 98.4

enabled the creation of antibodies which are capable of not only binding an antigen (or any molecule) but also mediating its catalysis. Termed catalytic antibodies or abzymes, these are envisaged to be more rapidly and easily generated than enzyme-derived counterparts generated by conventional protein engineering. They are consequently likely to find many

applications, including their use as *in vivo* therapeutic agents for the destruction of aberrant cells or pathogenic microbes by chemically degrading surface molecules, or within industrial syntheses (e.g. for the generation of pharmaceuticals, chemicals and novel synthetic materials).

99. CURRENT APPROACHES TO GENE THERAPY

Two main gene therapy strategies exist based upon: (i) directly altering gene defects in somatic or germline cells, and (ii) modulating gene expression using antigene and antisense oligonucleotides.

Rocombinant DNA technologies have had an enormous impact upon medical science. Their application has vastly increased the understanding of disease mechanisms, improved diagnostic/prognostic procedures (e.g. by providing recombinant reagents and nucleic acid-based assays), and identified many new targets for therapeutic intervention. These technologies have also been exploited for the development of improved therapeutic and preventative approaches based on the production of pharmacologically active agents (i.e. drugs) and recombinant vaccines respectively. Although the unravelling of the genetic mechanisms controlling *in vivo* gene functions has enabled the development of several drugs capable of acting at the genetic level (e.g. by blocking DNA replication or transcription), the development of gene therapies represents one of the fastest evolving and expanding, and most exciting areas of biomedical science. Two principal forms of gene therapy are therefore currently being developed based upon modulating *in vivo* gene expression either within somatic and/or germline cells. These involve: (i) the correction of gene defects through gene disruption, insertion, or replacement, and (ii) the use of antigene or antisense oligonucleotides to alter the rate of transcription of specific genes (e.g. by triple helix formation), or the translation of their mRNA transcripts (e.g. by inhibiting ribosome attachment or influencing post-transcriptional processing, and consequently the rate of mRNA degradation). Despite ethical and economic concerns, both approaches have proved safe and effective in animal models. Indeed, clinical trials are under way to assess their ability to treat several common single gene diseases (e.g. cystic fibrosis). As mapping and sequencing of the human genome is completed, such gene-based therapies may prove to be the ultimate form of human-made medicines.

ETHICAL AND ECONOMIC CONCERNS

The ability to correct genetic defects at the germline level using antigene therapies has raised several ethical concerns. The most significant of these is how to avoid potential abuses of the technology. For example, they may be used as part of eugenics programmes to improve the genetic constitution of specific individuals or entire social groups (i.e. as a sophisticated form of 'social cleansing'). The high costs involved also effectively restrict the development of antigene therapies directed towards those diseases which are sufficiently common. Concern has also been expressed that the high costs of gene therapies would restrict their availability to affluent individuals within developed countries in order to assure reasonable returns on the investments required for their development.

OAs: 536–42, 667–76, 823–39, 877–81, 1099–128. RAs: 212, 223–4, 294, 299, 404–17. SFR: 120, 142–5, 158–9, 181, 207–10, 220–1.

CORRECTING GENE DEFECTS

The ability to correct human gene defects relies largely upon the same technologies employed for the creation of transgenic aminals. Accordingly, *in vitro*-cultured human somatic or germline cells (i.e. embryos) may be transfected with animal virus-derived vectors bearing the desired human gene. The desired gene sequence may then be integrated into either specific or random sites within the human genome (e.g. by homologous recombination). Successful transfectants can then be selected for return to the original cell donor. This technology can be used to correct defects within specific human genes or their regulatory sequences by insertion, deletion, or gene replacement (Table 99.1). The same methods can also be employed to insert additional copies of normal functional genes within a human genome (e.g. to compensate for inactive genes).

Their application for human gene therapy depends primarily upon the identification of disease-associated gene defects, and the availability of cloned or engineered genes (or gene segments) representing the normal gene equivalent. However, it also depends upon the development of appropriate viral vectors which can: (i) accommodate the desired gene sequence (particularly if entire genes or large segments are involved); (ii) be used to infect defined human cells; (iii) target specific genes or genomic regions without affecting the functional activity of normal genes; and (iv) be safely used in humans (e.g. lack an origin of replication and are therefore incapable of independent replication). Although not yet possible, the construction of stable mammalian artificial chromosomes (MACs) may revolutionise such approaches to human gene therapy.

Table 99.1

Deficiency	Disorder	Human gene	Target cell
Immune	Adenine deaminase deficiency (ADA)	Adenosine deaminase	T, B cell, ADA-fibroblasts
Immune	PNP deficiency	Purine nucleoside phosphorylase (PNP)	PNP-fibroblasts
Emphysema	Deficiency (α1AT)	α1-antitrypsin	Human liver cells
Gaucher disease	GC deficiency Storage disorder	Glucocer-ebrosidase (GC)	GC-fibroblasts
Haemo-globinopathies	Thalassemia	β-globin	Fibroblasts
Lesch–Nyham syndrome	Metabolic deiciency	Hypoxanthine guanine phosphoribosyl transferase (HGPRT)	HPRT-cells
Urea cycle	Metabolic deficiency	Ornithine transcarbamylase	OTC-cells
Amino acid	Metabolic deficiency	Phenylketonuria	Hepatoma cells

5 and 48. Formation of DNA duplexes.
23. Transcription.
24. Processing of mRNA transcripts.
27. Transcription regulatory sequences.
30. *In vivo* translation.
34. Genetic recombination mechanisms.
37. *In vivo* DNA replication.
59. Chemical synthesis of oligos.
67. Yeast-derived vectors.
69. Plant and animal cell vectors.
82. Genetically engineered animals.
83. Prenatal diagnosis methods.
84. Genetics of cystic fibrosis.
89. HIV and AIDS.
91. Recombinant vaccine development.
94. Cancer diagnosis and therapy.
95. Recombinant drug development.
100. Human genome mapping project.

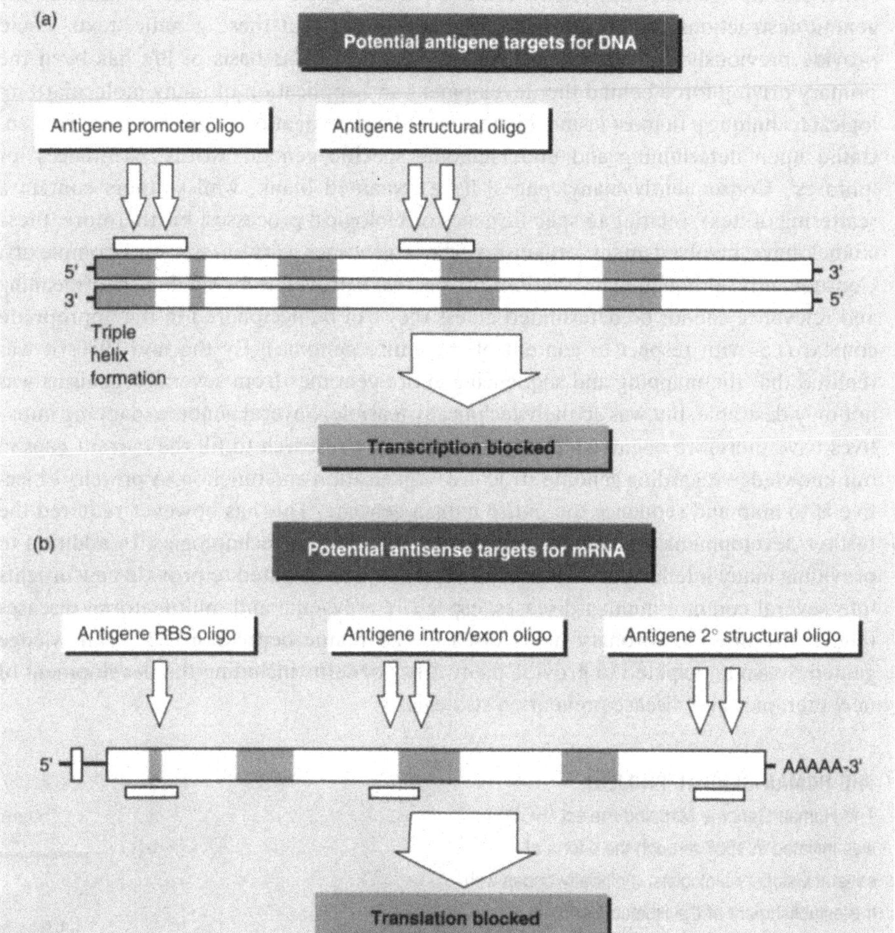

Fig. 99.1

Oligonucleotide-based therapy

Antigene or antisense oligonucleotides (oligos) which bind to DNA or mRNA sequences respectively may also be used to therapeutically modulate gene expression. These approaches exploit the ability of short oligonucleotides (e.g.11-15mers) to interact with complementary DNA or mRNA sequences with extreme specificity; capable of discriminating between sequences varying by a single base.

Antigene oligos may accordingly inhibit the expression of a defective gene by the formation of a triple helix by binding (Fig. 99.1a) within the major groove of a specific region of double-stranded genomic DNA (via Hoogsteen interactions). This may inhibit transcription, and consequently reduce the production of a defective protein if binding occurs within the coding regions or competes with transcription factors for promoter binding sites. Similarly, their binding to other regulatory sequences (e.g. steroid responsive elements) may prevent abnormal responses to transciptional regulators (e.g. some hormones).

In contrast, antisense oligos bind to single stranded mRNA (Fig. 99.1b). Depending upon the region targeted, this may induce changes in mRNA secondary structure which block translation or increase their susceptibility to degradation. Antisense oligos directed to the 'Shine-Dalgarno' site of ribosome attachment may therefore prevent translation or inhibit splicing reactions, or provide double-stranded regions accessible to degradation by endogenous ribonucleases (e.g. RNase H).

Factors limiting oligo use

Several factors currently need to be addressed before therapeutic oligos can be widely exploited. Adequate methods are still required for their delivery to specific human target cells *in vivo* and to mediate their pharmacokinetically favourable uptake into the cytoplasm or nucleus. These delivery systems may possibly be based upon the linkage of oligos to liposomes, proteins and other polymeric compounds. Although antisense oligos are inherently more stable than their antigene equivalents, their structural instability and susceptibility to nuclease degradation also potentially limit their utility *in vivo*. Peptide nucleic acids (PNAs) in which the sugar–phosphate backbone of the DNA molecule is replaced by a structure resembling those in a peptide link (e.g. 2-aminoethylglycine) may however provide a solution. Indeed, the recent development of oligo or PNA clamps in which two PNA strands are linked together by base analogues may further increase the stability of triple helix structures. Although still relatively expensive to synthesise, concerns have also been expressed over their use. This stems from their potential long-term toxicity (e.g. antigenicity, mutagenicity), the unknown fate of degraded oligos, and their potential for genomic incorporation. Despite these concerns, antisense oligos have already been used to successfully inhibit the expression of oncogenes, cellular genes and viral genes both *in vitro*, and in limited *in vivo* studies.

ANTI-VIRAL OLIGO THERAPY

Many potential diseases are anticipated to be amenable to antisense oligo-based therapy. Of these, viral diseases represent excellent targets since many viral genes have no cellular counterpart. Consequently, the technique is likely to be highly selective, targeting only viral cells. Considerable effort is therefore directed towards the development of antisense oligos as potential anti-viral agents. For example, many groups are attempting to develop antisense oligo-based AIDS therapies based upon targeting specific mRNA transcripts produced by infective human immunodeficiency viruses (HIV), especially those transcribed from the HIV *Tat* gene.

100. THE HUMAN GENOME MAPPING PROJECT

Mapping of the entire human genome requires the use of a range of genetic and physical techniques, particularly those based on the generation and analysis of huge overlapping DNA clones (or contigs).

An organism's genome has often been likened to a book whose pages contain all the genetic instructions required for life. The realisation that these genetic 'texts' could provide previously unimagined insights into the molecular basis of life has been the primary driving force behind the development and application of many molecular biological techniques. In most instances, however, the investigation of genomes has concentrated upon determining and understanding specific genetic 'words', 'sentences' or 'phrases'. Consequently many 'pages' have remained blank, whilst others contain a scattering of 'text' relating to specific genes or biological processes. Furthermore, these studies have involved many organisms whose genomes vary in size and complexity. Consequently, although many sections of 'text' have proven to be similar, their meaning and relevance cannot be determined unless they can be deciphered in the appropriate context (i.e. with respect to content of the entire genome). By the mid-1980s it was realised that the mapping and sequencing entire genomes from several organisms was not only desirable, but was actually technically feasible. Several genome mapping initiatives have therefore begun which aim to coordinate research to fill the current gaps in our knowledge regarding genome structure, organisation and function. A primary objective is to map and sequence the entire human genome. This has however required the further development of mapping, sequencing and analysis technologies. In addition to providing many intellectual answers, this initiative is anticipated to provide new insights into several common human diseases, especially polygenic and multifactorial diseases (e.g. schizophrenia, coronary heart disease, and manic depression). The knowledge gained is also anticipated to provide many other benefits including the development of new therapies and disease prevention strategies.

OTHER GENOME MAPPING TARGETS

In addition to the Human Genome Mapping Project, projects are also currently under way aimed at mapping and sequencing the entire genomes of several 'reference' organisms. These include the already extensively investigated genomes of some bacteria (*E. coli*), yeast (*Saccharomyces*), fruit flies (*Drosophila melonogaster*) and mice (*Mus musculus*). Another principal target is the small and structurally simple genome of the roundworm *Caenorhabditis elegans*. This genome has already been particularly useful in answering questions concerning the genetic mechanisms involved in the development of multicellular organisms. The small haploid genome (about 10^8 bp) of the flowering plant *Arabidopsis thaliana* is also a major target since many of its physical and genetic attributes make it an excellent plant model for investigation.

THE HUMAN GENOME PROJECT

The Human Genome Mapping Project (HGMP) was initiated in 1988 through the efforts of several visionary biologists. It officially began with the establishment of the Human Genome Organisation (HUGO), whose remit is to administer and collate the data derived from research centres around the world. The aim of the HGMP is to completely map and sequence the human genome by the year 2005. This will be achieved by: (i) constructing genetic maps of all chromosomes using arrays of standardised probes; (ii) constructing physical maps showing lengths and distances (in base pairs); and (iii) nucleotide sequencing Fig. 100.1. These aims are essentially being achieved simultaneously with some 2% of human genes already mapped to specific chromosomal locations and sequenced. Extensive organisation, funding and technological innovation are however primary factors determining the success of the HGMP. Funding is provided from several sources, particularly government agencies and private sponsorships/investments. For example, most funding in the USA comes from the Department of Energy, National Institute of Health, National Science Foundation, and the Howard Hughes Medical Institute. Consequently, many of the technological innovations in cloning, mapping and sequencing required to cost-effectively complete the task by the 2005 deadline have now been developed and are being applied.

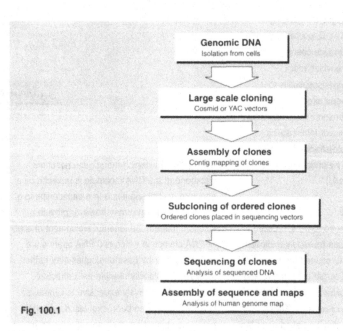

Fig. 100.1

Genomic DNA
Isolation from cells

Large scale cloning
Cosmid or YAC vectors

Assembly of clones
Contig mapping of clones

Subcloning of ordered clones
Ordered clones placed in sequencing vectors

Sequencing of clones
Analysis of sequenced DNA

Assembly of sequence and maps
Analysis of human genome map

OAs: 451–4, 625–6, 688–9, 1129–54. **RAs:** 219–24, 244, 247–9, 418–25. **SFR:** 43–4, 119–24, 126–30, 160–5, 168, 222–5, 227–9.

Mapping the human genome

Mapping of the human genome began with the application of several conventional techniques based upon calculating gene recombination frequencies, or the localisation of specific genes by automated chromosome sorting and/or *in situ* hybridisation using labelled DNA probes. However, the recent production of a map covering 90% of the human genome testifies to the development and application of a new generation of genetic and physical mapping techniques (Fig. 100.2). Many of these would not however have been so productive without the development and application of numerous supporting technologies. Accordingly, given the enormous size of the human genome, the ability to generate libraries of extremely large (> 3 kb), overlapping genomic fragments in cosmids and yeast artificial chromosome (YAC) vectors has been of particular importance. Termed 'contigs', these overlapping clones (generated by partial digestion with certain specific restriction enzymes prior to insertion in the vector) have played, and continue to play, a crucial role in the HGMP. A physical map can therefore be initially constructed by aligning the contigs based upon the sequence of their overlapping ends.

Nevertheless, ordering and mapping of contigs requires the use of several genetic and physical mapping strategies. For example, physical mapping has been dependent upon techniques such as restriction fragment fingerprinting and inter-*Alu*–polymerase chain reaction (PCR) hybridisation (Fig. 100.3). Genetic mapping has also increasingly been based upon the use of sequence-tagged sites (STSs). These have been of particular importance since they may be used as markers on both physical and genetic maps for each chromosome. They also serve as common reference points for different research groups each utilising separate approaches.

Regardless of the nature of the mapping strategies employed, several technologies have played significant and/or key supporting roles. Southern blotting and probe hybridisation assays have consequently been crucial for contig alignment. Similarly, the ability to separate and/or isolate the large DNA fragments present in YAC clones has required sophisticated electrophoretic techniques (e.g. pulsed-field gel electrophoresis).

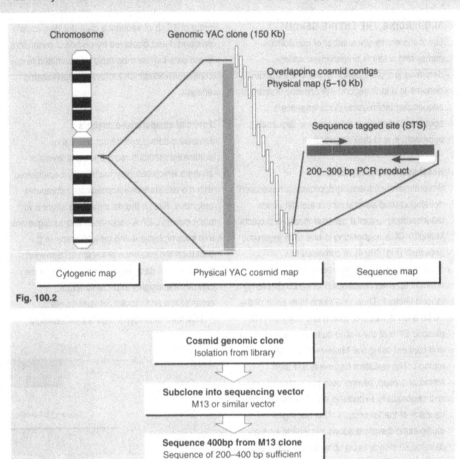

Fig. 100.2

Fig. 100.3

This has been a vital step in both the mapping process and the subsequent sequencing of individual contigs.

Speeding up the process

Mechanisation, automatisation and computerisation have all played a significant role in enabling the HGMP to achieve its aims in a cost-effective and timely manner. Thus, many aspects of mapping and sequencing have only been accomplished through massive investments in machinery which can robotically perform some of the simple and complex procedures involved. These range from simple liquid handling/dispensing devices and automated Southern/dot blotting machines, to PCR amplification workstations and automated sequencers whose data output is linked to powerful computers which enable rapid map construction and sophisticated sequence analyses to be performed.

273

Sequencing of the entire human genome by large-scale and automated techniques, and subsequent computer-assisted analysis, is providing insights into its structure, organisation, function and evolution.

SEQUENCING THE ENTIRE GENOME

Given the enormous number of nucleotides comprising a single human chromosome, determining the sequence of the entire human genome is a task beyond the capacity of manual sequencing techniques. Such large-scale sequencing therefore requires new sequencing approaches, and their automation.

Novel sequencing approaches

Several novel sequencing approaches have been developed (and several others are still under development) to assist genome mapping projects. Multiplex DNA sequencing is one such ingenious approach (Fig. 100.4). It combines the sequencing of clones from as many as 20 different libraries constructed using 20 differently tagged vectors. Thus, one clone from each of the 20 libraries is isolated, mixed and cultured. The plasmid DNA of the mixed culture is then isolated, and cleaved using the Maxam–Gilbert sequencing method. The resultant fragments are then electrophoresed, blotted onto nylon membranes and sequentially hybridised with probes specific for each of the 20 vectors. This pooling or multiplexing therefore allows nucleotide sequence data for 20 clones to be obtained per gel track, using a single sequencing reaction, a single gel and single nylon filter. A similar approach to large-scale sequencing, termed sequencing by hybridisation (SbH), involves hybridising the DNA of unknown sequence with an enormous array of short oligonucleotides. Overlapping oligonucleotides which form perfect duplexes with the DNA of interest are subsequently identified and analysed. This permits the reconstruction of the original target DNA sequence. Combined with developments in sequencing micro-chips containing immobilised oligonucleotides, SbH may also play a major role in future large-scale sequencing projects.

Automated sequencing

It is, however, the development of automated sequencing apparatus based upon the use of fluorescently labelled dideoxynucleotides which is currently making the greatest contribution to genome sequencing initiatives. Linked to PCR-based isolation and sequencing techniques, each of these machines is capable of generating of the order of 10 kb of sequence data per day. Initially processed and displayed by on-board computers, these data can then be rapidly transmitted to any computer workstation for further sophisticated analysis.

Forms of computerised analysis

Sequence data derived from manual or automated methods require several levels of analysis which can only realistically be achieved with the assistance of sophisticated computer programs. Firstly, the data must be aligned with other genomic DNA sequences and amalgamated into known physical and genetic maps (e.g. based on the presence of known STS markers). Secondly, the data are only really useful if they can provide insights into the structure, organisation and location of specific genes.

This form of analysis requires increasingly sophisticated and rapid software which allow: (i) the searching of vast databases for sequence homologies in order to define and locate consensus sequences; (ii) the identification of gene-specific sequence motifs, such as those characteristic of functional protein-encoding genes (e.g. CpG islands, exon–intron boundaries); (iii) the identification of open reading frames and the performance of translations (i.e. the determination of amino acid sequences of encoded polypeptides); and (iv) the analysis of sequences using algorithms for predicting nucleic acid and protein folding motifs.

OAs: 451–4, 625–6, 68–9, 1155–63. **RAs:** 31, 198, 219–24, 244, 247–9, 426–8. **SFR:** 43–4, 119–24, 126–30, 160–5, 168, 222–5, 227–9.

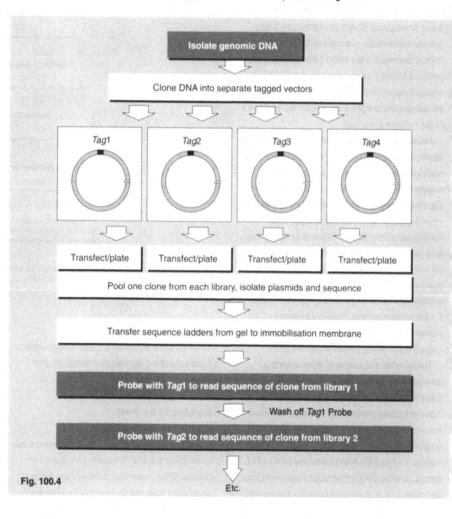

Fig. 100.4

Some interesting early results

Mapping and sequencing of the human genome (as part of the HGMP, or via independent studies focused upon specific chromosomal regions) has already provided many insights into the structure, organisation, evolution and function of both genes and their intergenic regions (Fig. 100.5). It is now evident that genes vary widely in size, often due to the number of the introns they contain. Thus, the 800 bp α-globin gene has only three introns, whilst the > 2 Mb dystrophin gene has some 60–70 introns (although its mRNA corresponds to less than 1% of the DNA).

Sequencing has also revealed that several genes (e.g. the *NF-1* von Recklinghausen gene encoding a growth regulator implicated in neurofibromatosis-1, factor VIII and the steroid 21 hydroxylase gene *P450c21*) contain other genes embedded within their introns which are read in the opposite direction to the main gene. Albeit infrequently, such 'genes-within-genes' now appear to be present within the genomes of many eukaryotes, ranging from *Drosophila* to humans. It is also evident that genes with a common ancestry frequently occur in defined clusters paralleling their expression in early development. However, there does not appear to be any clustering of genes whose encoded proteins are organ-specific or which perform related functions (e.g. enzymes involved in successive steps of a metabolic pathway). Indeed, genes encoding subunits of hetero-multimeric proteins, cytosolic and mitochondrial forms of the same enzyme, and receptors and their corresponding ligands, are usually located on different chromosomes. For example, the genes for collagen α1 and α2 are located to 17q and 7q respectively, the hormone EGF and the EGF receptor genes have been mapped to 4q and 7p respectively, and the insulin and insulin receptior genes have been mapped to 11p and 19p respectively.

Mapping and sequencing of the yeast genome has also recently suggested that most genes may be located to GC-rich regions. It has therefore been suggested that the 10% of the human genome remaining to be mapped due to the problems encountered by their high GC content may actually contain up to 50% of the genes comprising the genome.

The availability of such information will

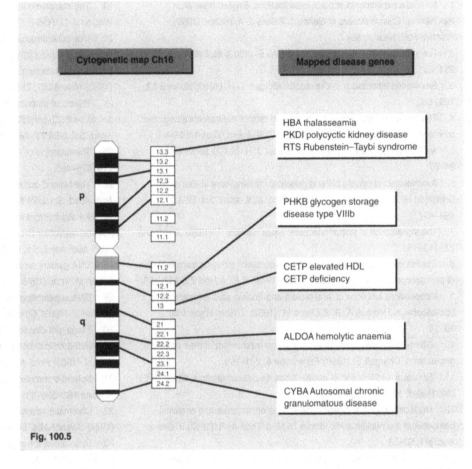

Fig. 100.5

undoubtedly aid the search for disease genes, and the development/application of many diagnostic and therapeutic techniques.

Balancing the benefits and deficits

It has been argued that the great effort and cost involved in achieving the aims of the HGMP cannot be justified in terms of the limited benefits and potential problems that will arise. For example, it has been argued that a very substantial period of time will be required following its completion before the hope of identifying many disease genes will be realised. Furthermore, some argue that the money currently directed to the HGMP is detracting from the financing of research more likely to identify disease genes by conventional means. However valid these objections may or may not be, society as a whole must address concerns regarding the potential socio-economic impacts the HGMP may

have. For example, the ability to identify disease susceptibility genes or genes determining certain behavioural patterns may be utilised in gene screening programmes. This may prevent some individuals from obtaining life or medical insurance cover, and even employment. Similarly, prenatal genetic screening may inevitably encourage the adoption of eugenic attitudes, whereby fetuses deemed 'imperfect' based upon the possession of genetic markers are routinely terminated. The potential benefits anticipated to accrue from the completion of the HGMP must therefore be carefully considered. Informed judgements must then be made which ensure the drafting and implementation of sufficiently stringent legislative measures to protect both society and the rights of the individual (e.g. by limiting the access to gene data, and limiting the ways this may be used).

BIBLIOGRAPHY 1:

ORIGINAL ARTICLES

1. Mendel's experiments in plant hybridisations. English translation reprinted in: *Classic Papers in Genetics*. Peters J. A. (editor) (1959) Prentice-Hall, New Jersey.

2. The chromosomes in heredity. Sutton W. S. (1903) *Biol. Bull.* **4**; 231–251.

3. Sex-limited inheritance in *Drosophila*. Morgan T. H. (1910) *Science* **32**; 120–122.

4. The linear arrangement of six sex-linked factors in *Drosophila* as shown by mode of association. Sturtevant A. H. (1913) *J. Exp. Zool.* **14**; 39–45.

5. Artificial transmutation of the gene. Muller J. H. (1927) *Science* **46**; 84–87.

6. A correlation of cytological and genetical crossing-over in *Zea mays*. Creighton H. B. & McClintock B. (1931) *Proc. Natl. Acad. Sci. USA* **17**; 492–497.

7. The significance of pneumonococcal types. Griffith, F. (1928) *J. Hygeine* **27**; 113–159.

8. Studies on the chemical nature of the substance inducing transformation of pneumococcal types. Avery O. T. *et al.* (1944) *J. Exp. Med.* **79**; 137–158.

9. Independent functions of viral protein and nucleic acid in growth of bacteriophage. Hershey A. D. & Chase M. (1952) *J. Gen. Physiol.* **36**; 39–56.

10. Chemical specificity of nucleic acids and mechanism of their enzymatic degradation. Chargaff E. (1950) *Experientia* **6**; 201–209.

11. Structure and function of nucleic acids as cell constituents. Chargaff E. (1951) *Fed. Proc.* **10**; 654–659.

12. Nucleotides, part X. Some observations on structure and chemical behaviour of the nucleic acids. Brown D. M. & Todd A. R. (1952) *J. Chem. Soc.* **pt1**; 52–58.

13. Nucleotides part XIX. Pyrimidine deoxyribonucleoside diphosphates. Dekker C. A. *et al.* (1953) *J. Chem. Soc.* **pt1**; 947–951.

14. The bases of the nucleic acids of some bacterial and animal viruses: the occurrence of 5-hydroxymethylcytosine. Wyatt G. R. & Cohen S. S. (1953) *Biochem. J.* **55**; 774–782.

15. Linkage of polynucleotides through phosphodiester bonds by an enzyme from *E. coli*. Olivera B. M. & Lehman I. R. (1967) *Proc. Natl. Acad. Sci. USA* **57**; 1426–1433.

16. Electron microscopy of sodium desoxyribonucleate by use of a new freeze-drying method. Williams R. C. (1952) *Biochim. Biophys. Acta* **9**; 237–239.

17. Molecular configuration in sodium thymonucleate. Franklin R. E. & Gosling R. G. (1953) *Nature* **171**; 740–741.

18. Molecular structure of nucleic acids: a structure for deoxyribonucleic acid. Watson J. D. & Crick F. H. C. (1953) *Nature* **171**; 737–738.

19. Genetical implications of the structure of deoxyribonucleic acid. Watson J. D. & Crick F. H. C. (1953) *Nature* **171**; 964–967.

20. Molecular structure of deoxypentose nucleic acids. Wilkins M. H. F. *et al.* (1953) *Nature* **171**; 738–740.

21. The complementary structure of deoxyribonucleic acid. Crick F. H. C. & Watson J. D. (1954) *Proc. Roy. Soc.* **223**; 80–96.

22. Molecular structure of a left-handed DNA fragment at atomic resolution. Wang A. H. *et al.* (1979) *Nature* **282**; 680–686.

23. Crystal structure analysis of a complete turn of B-DNA. Wing R. M. *et al.* (1980) *Nature* **287**; 755–758.

24. Effects of methylation on a synthetic polynucleotide: the B-Z transition in poly(dG-m5dC).poly(dG-m5dC). Behe M. & Felsenfeld G. (1981) *Proc. Natl. Acad. Sci. USA* **78**; 1619–1623.

25. The anatomy of A-, B-, and Z-DNA. Dickerson R. E. *et al.* (1982) *Science* **216**; 475–485.

26. The twisted circular form of polyoma DNA. Vinograd J. *et al.* (1965) *Proc. Natl. Acad. Sci. USA* **53**; 1104–1111.

27. An activity from mammalian cells that untwists superhelical DNA – a possible swivel for DNA replication. Champoux J. J. & Dulbecco R. (1972) *Proc. Natl. Acad. Sci. USA* **69**; 143–146.

28. DNA gyrase: an enzyme that introduces superhelical turns into DNA. Gellert M. *et al.* (1976) *Proc. Natl. Acad. Sci. USA* **73**; 3872–3876.

29. DNA supercoiling and its effects upon DNA structure and function. Wang J. C. *et al.* (1983) *Cold Spring Harbor Symp. Quant. Biol.* **47**; 251–257.

30. Flipping of cloned d(pCpG)n.d(pCpG)n DNA sequences from a right- to a left-handed helical structure by salt, Co(III), or negative supercoiling. Peck L. J. *et al.* (1982) *Proc. Natl. Acad. Sci. USA* **79**; 4560–4564.

31. Spheroid chromatin units (v bodies). Olins A. L. & Olins D. E. (1974) *Science* **183**; 330–332.

32. Chromatin structure: a repeating unit of histones and DNA. Kornberg R. (1974) *Science* **184**; 868–871.

33. Structure of the nucleosome core particles at 7Å resolution. Richmond T. J. *et al.* (1984) *Nature* **311**; 532–537.

34. Crystallographic structure of the octameric histone core of the nucleosome at a resolution of 3.3 Å. Burlinghame R. W. *et al.* (1985) *Science* **228**; 546–553.

35. Structure of the 30nm chromatin fibre. Felsenfeld G. & McGlee J. D. (1986) *Cell* **44**; 375–377.

36. The bacterial chromosome and its manner of replication as seen by autoradiography. Cairns J. (1963) *J. Mol. Biol.* **4**; 407–409.

37. Shape and fine structure of nucleoids observed on sections of ultra-rapidly frozen and cryosubstituted bacteria. Hobot J. A. *et al.* (1985) *J. Bacteriol.* **162**; 960–971.

38. Multiple molecular species of circular R-factor DNA isolated from *Escherichia coli*. Cohen S. N. & Miller C. A. (1969) *Nature* **224**; 1273–1277.

39. Experimental puffs in salivary gland chromosomes of *Drosophila hydei*. Berendes H. D. *et al.* (1965) *Chromosoma* **16**; 35–46.

40. The structure of histone depleted metaphase chromosomes. Paulson J. & Laemmli U. (1977) *Cell* **12**; 817–828.

41. Sequence of centromere separation: role of centromeric heterochromatin. Vig B. (1982) *Genetics* **102**; 795–806.

42. Histone H5 in the control of DNA synthesis and cell proliferation. Sun J.-M. et al. (1989) Science **245**; 68–71.

43. Instability of a 550-base pair DNA segment and abnormal methylation in fragile X syndrome. Oberle I. et al. (1991) Science **252**; 1097–1102.

44. Homology among DNA-binding proteins suggests use of a conserved super-secondary structure. Sauer R. T. et al. (1982) Nature **298**; 447–451.

45. The molecular basis of DNA-protein recognition inferred from the structure of the Cro repressor. Ohlendorf D. H. et al. (1982) Nature **298**; 718–723.

46. The helix-turn-helix DNA-binding motif. Brennan R. G. & Matthews B. W. (1989) J. Mol. Biol. **264**; 1903–1906.

47. Interactions between heterologous helix-loop-helix proteins generate complexes that bind specifically to a common DNA sequence. Murre C. et al. (1989) Cell **58**; 537–544.

48. Crystal structure of an engrailed homeodomain–DNA complex at 2.8 Å resolution: a framework for understanding homeodomain–DNA interactions. Kissinger C. R. et al. (1990) Cell **63**; 579–590.

49. Proposed structure for the zinc-binding domains from transcriptional factor IIIA and related proteins. Berg J. M. (1988) Proc. Natl. Acad. Sci. USA **85**; 99–102.

50. Three-dimensional solution structure of a single zinc finger DNA-binding domain. Lee M. S. et al. (1989) Science **245**; 635–637.

51. The leucine zipper; a hypothetical structure common to a new class of DNA-binding proteins. Landschulz W. H. et al. (1988) Science **240**; 1759–1764.

52. Scissors-grip model for DNA recognition by a family of leucine zipper proteins. Vinson C. R. et al. (1989) Science **246**; 911–916.

53. Evidence that the leucine zipper is a coiled coil. O'Shea E. K. et al. (1989) Science **243**; 538–542.

54. Solution structure of the glucocorticoid receptor DNA-binding domain. Hard T. et al. (1990) Science **249**; 157–160.

55. Design of DNA-binding peptides based on the leucine zipper motif. O'Neil K. T. et al. (1990) Science **249**; 774–778.

56. Chromosome differentiation and pairing behaviour of polyploids: an assessment on preferential metaphase I associations in colchicine-induced autotetraploid hybrids within the genus Secale. Benavente E. & Orellana J. (1991) Genetics **128**; 433–442.

57. Sequence and organisation of the human mitochondrial genome. Anderson S. et al. (1981) Nature **290**; 457–465.

58. Sequence and gene organisation of mouse mitochondrial DNA. Bibb M. J. et al. (1981) Cell **26**; 167–180.

59. Complete sequence of bovine mitochondrial DNA. Anderson S. et al. (1982) J. Mol. Biol. **156**; 683–717.

60. Structure of a ribonucleic acid. Holley R. W. et al. (1965) Science **147**; 1462–1465.

61. A complete mapping of the proteins in the small ribosomal subunit of E. coli. Capel M. S. et al. (1987) Science **238**; 1403–1406.

62. RNA–protein interactions in 30S ribosomal subunits: Folding and function of 16S rRNA. Stern S. et al. (1989) Science **244**; 783–790.

63. Three dimensional tertiary structure of yeast phenylalanine transfer RNA. Kim S. H. et al. (1974) Science **185**; 435–440.

64. Structure of yeast phenylalanine tRNA at 3Å resolution. Robertus J. D. et al. (1974) Nature **250**; 546–551.

65. Three dimensional structure of transfer RNA. Rich A. & Kim S. (1978) Scientific American **238**; 52–62.

66. Structural basis of anticodon loop recognition by glutaminyl-tRNA synthetase. Rould M. A. et al. (1991) Nature **352**; 213–218.

67. Acceptor end binding domain interactions ensure correct aminoacylation of transfer RNA. Weygand-Durasevic I. et al. (1993) Proc. Natl. Acad. Sci. USA **90**; 2010–2014.

68. Rules that govern tRNA identity in protein synthesis. McClain W. H. (1993) J. Mol. Biol. **234**; 257–280.

69. Variable and constant components of chromosomes. Mirsky A.E. & Ris H. (1949) Nature **163**; 666–667.

70. The centromeric region in the scanning electron microscope. Heneen W. (1982) Hereditas **97**; 311–314.

71. Yeast centromere DNA is in a unique and highly ordered structure in chromosomes and small circular minichromosomes. Bloom K. S. & Carbon J. (1982) Cell **29**; 305–317.

72. A highly conserved repetitive DNA sequence, (TTAGGG)n, present at the telomeres of human chromosomes. Moyzis R. K. et al. (1988) Proc. Natl. Acad. Sci. USA **85**; 6622–6626.

73. Gene within a gene: nested Drosophila genes encode unrelated proteins on opposite DNA strands. Henikoff S. et al. (1986) Cell **44**; 33–42.

74. A novel member of the thyroid/steroid hormone receptor family encoded by the opposite strand of the rat c-erbAα transcriptional unit. Lazar M. A. et al. (1989) Mol. Cell. Biol. **9**; 1128–1136.

75. Transcript encoded on the opposite strand of the human steroid 21-hydroxylase/complement C4 gene locus. Morel Y. et al. (1989) Proc. Natl Acad. Sci. USA **86**; 6582–6586.

76. Identification and characterisation of transcripts from the neurofibromatosis 1 region: the sequence and genomic structure of EV12 and mapping of other transcripts. Cawthorn R. M. et al. (1990) Genomics **7**; 555–565.

77. A transcribed gene in an intron of human Factor VIII gene. Levinson B. et al. (1990) Genomics **7**; 7–11.

78. Repeated sequences in DNA. Britten R. J. & Kohne D. E. (1968) Science **161**; 529–540.

79. Ubiquitous interspersed repeated DNA sequences in mammalian genomes. Jelinek W. R. et al. (1980) Proc. Natl. Acad. Sci. USA **77**; 1398–1402.

80. An abundant cytoplasmic 7S RNA is partially complementary to the dominant interspersed middle repetitive DNA sequence family in the human genome. Weiner A. (1980) Cell **22**; 209–218.

81. The Alu family of dispersed repetitive sequences. Schmid C. W. & Jelenik W. R. (1982) Science **216**; 1065–1070.

82. Alu sequences are processed 7SL RNA genes. Ullu E. & Tschudi C. (1984) Nature **312**; 171–172.

83. A fundamental division in the *Alu* family of repeated sequences. Jurka J. & Smith T. (1988) *Proc. Natl. Acad. Sci. USA* **85**; 4775–4778.

84. LINE-1: a mammalian transposable element. Fanning T. G. & Singer M. F. (1988) *Biochim. Biophys. Acta* **910**; 203–212.

85. Human genome organisation: *Alu*, LINES, and the molecular structure of metaphase chromosome bands. Korenberg J. R. & Rykowski M. C. (1988) *Cell* **53**; 391–400.

86. Hypervariable 'minisatellite' regions in human DNA. Jeffreys A. J. *et al.* (1985) *Nature* **314**; 67–73.

87. Abundant class of DNA polymorphisms which can be typed using the PCR. Weber J. L. & May P. E. (1989) *Am. J. Hum. Genet.* **44**; 488–496.

88. Repeat unit sequence variation in minisatellites: a novel source of DNA polymorphism for studying variation and mutation by single molecule analysis. Jeffreys A. J. *et al.* (1990) *Cell* **60**; 473–485.

89. Molecular cloning and characterisation of the human α-like globin gene cluster. Fritsch E. F. *et al.* (1980) *Cell* **19**; 959–972.

90. The gross anatomy of a tRNA gene cluster at region 42Å of the *D. melanogaster* chromosome. Yen P. H. & Davidson N. (1980) *Cell* **22**; 137–148.

91. The organisation and expression of histone gene families. Hentschel C. C. & Birnstiel M. L. (1981) *Cell* **25**; 301–305.

92. Human ribosomal RNA genes: orientation of the tandem array and conservation of the 5′ end. Worton R. G. *et al.* (1988) *Science* **239**; 64–68.

93. Organisation of the constant region gene family of the mouse immunoglobulin heavy chain. Shimizu A. *et al.* (1982) *Cell* **28**; 499–506.

94. Clusters of genes encoding mouse transplantation antigens. Steinmetz M. *et al.* (1982) *Cell* **28**; 489–498.

95. A highly polymorphic locus in human DNA. Wyman A. *et al.* (1980) *Proc. Natl. Acad. Sci. USA* **77**; 6754–6758.

96. The structure of a human α-globin pseudogene and its relationship to α-globin gene duplication. Proudfoot N. J. & Maniatis T. (1980) *Cell* **21**; 537–544.

97. Split genes and RNA splicing. Crick F. H. C. (1979) *Science* **204**; 264–271.

98. An amazing sequence arrangement at the 5′ ends of adenovirus 2 messenger RNA. Chow L. T. *et al.* (1977) *Cell* **12**; 1–8.

99. Ovalbumin gene: evidence for a leader sequence in mRNA and DNA sequences at the exon–intron boundaries. Breathnach R. *et al.* (1978) *Proc. Natl Acad. Sci. USA* **75**; 4853–4857.

100. Generation of authentic 3′ termini of an H2A mRNA *in vivo* is dependent upon a short inverted DNA repeat and on spacer sequences. Birchmeier C. *et al.* (1982) *Cell* **28**; 739–745.

101. The sequence 5′-AAUAAA-3′ forms part of the recognition site for polyadenylation of late SV40 mRNAs. Fitzgerald M. & Shenk T. (1981) *Cell* **24**; 251–260.

102. Spliced segments at the 5′ termini of adenovirus–2 late mRNA. Berget S. M. *et al.* (1977) *Proc. Natl Acad. Sci. USA* **74**; 3171–3175.

103. Evidence for an intron-contained sequence required for the splicing of yeast RNA polymerase II transcripts. Langford C. J. & Gallwitz D. (1983) *Cell* **33**; 519–527.

104. The 3′-terminal sequence of *E. coli* 16S ribosomal RNA: complementarity to nonsense triplets and ribosome binding sites. Shine J. & Delgarno L. (1974) *Proc. Natl. Acad. Sci. USA* **71**; 1342–1346.

105. High levels of *de novo* methylation and altered chromatin structure at CpG islands in cell lines. Antequera F. *et al.* (1990) *Cell* **62**; 503–514.

106. Chromosome disorders associated with mental retardation. Brewster T. & Gerald P. (1978) *Pediatr. Ann.* **7**; 82–89.

107. Origin of chi46,XX/46,XY chimerism in a human true hermaphrodite. Dewald G. *et al.* (1980) *Science* **207**; 321–323.

108. X-chromosome inactivation may explain the difference in viability of XO humans and mice. Ashworth A. *et al.* (1991) *Nature* **351**; 406–408.

109. Mutations in barley induced by X-rays and radium. Stadler L. J. (1928) *Science* **68**; 186–187.

110. Hydroxylamine mutagenesis of HSV DNA and DNA fragments: introduction of mutations into selected regions of the viral genome. Chu C. T. *et al.* (1979) *Virology* **98**; 168–181.

111. Mechanisms of spontaneous and induced frameshift mutation in bacteriophage T4. Streisinger G. & Owen J. (1985) *Genetics* **109**; 633–659.

112. Identifying environmental chemicals causing mutations and cancer. Ames B. (1979) *Science* **204**; 587–593.

113. Distribution of 5-methylcytosine in pyrimidine sequences of deoxyribonucleic acids. Doskocil J. & Sorm F. (1962) *Biochim. Biophys. Acta* **55**; 953–959.

114. Modified nucleosides and bizarre 5′-termini in mouse myeloma mRNA. Adams J. M. & Cory S. (1975) *Nature* **255**; 28–33.

115. Molecular models for DNA damaged by photoreaction. Pearlman D. A. *et al.* (1985) *Science* **227**; 1304–1308.

116. Enzymatic breakage and joining of deoxyribonucleic acid, I. Repair of single-strand breaks in DNA by an enzyme system from *E. coli* infected with T4 bacteriophage. Weiss B. & Richardson C. C. (1967) *Proc. Natl. Acad. Sci. USA* **57**; 1021–1028.

117. Methyl-directed repair of DNA base-pair mismatches *in vitro*. Lu A. L. *et al.* (1983) *Proc. Natl. Acad. Sci. USA* **80**; 4639–4643.

118. Different base/base mismatches are corrected with different efficiencies by the methyl-directed DNA mismatch repair system in *E. coli*. Kramer B. *et al.* (1984) *Cell* **38**; 879–888.

119. Heterogeneous DNA damage and repair in the mammalian genome. Bohr V. A. *et al.* (1987) *Cancer Res.* **47**; 6426–6436.

120. DNA mismatch correction in a defined system. Lahue R. S. *et al.* (1989) *Science* **245**; 160–164.

121. Genetic dissection of the biochemical activities of the RecBCD enzyme. Amundsen S. K. *et al.* (1990) *Genetics* **126**; 25–40.

122. Formation and resolution of recombination intermediates by *E. coli* RecA and RuvC proteins. Dunderdale H. J. *et al.* (1991) *Nature* **354**; 506–510.

123. Requirement for the replication protein SSB in human DNA excision repair. Coverley D. *et al.* (1991) *Nature* **349**; 538–541.

124. Human nucleotide excision nuclease removes thymine dimers from DNA by incising the 22nd phosphodiester bond 5′ and the 6th phosphodiester bond 3′ to the photodimer. Huang J.-C. *et al.* (1992) *Proc. Natl. Acad. Sci. USA* **89**; 3664–3668.

125. Molecular cloning of a mouse DNA-repair gene that complements the defect of group A xeroderma pigmentosum. Tanaka K. *et al.* (1989) *Proc. Natl. Acad. Sci. USA* **86**; 5512–5516.

126. Complementation of a DNA repair defect in xeroderma pigmentosum cells by transfer of human chromosome 9. Kaur G. P. & Athwal R. S. (1989) *Proc. Natl. Acad. Sci. USA* **86**; 8872–8876.

127. Analysis of a human DNA excision repair gene involved in group A xeroderma pigmentosum and containing a zinc-finger domain. Tanaka K. *et al.* (1990) *Nature* **348**; 73–76.

128. The mouse mutation severe combined immunodeficiency (SCID) is on chromosome 16. Bosma G. C. *et al.* (1989) *Immunogenetics* **29**; 54–57.

129. The *scid* mutation in mice causes a general defect in DNA repair. Fulop G. M. & Phillips R. A. (1990) Nature **347**; 479–482.

130. Sequential test for the detection of linkage. Morton N. E. (1955) *Am. J. Hum. Genet.* **7**; 277–318.

131. Estimation of recombination fraction in human pedigrees: efficient computation of the likelihood for human linkage studies. Ott J. (1974) *Am. J. Hum. Genet.* **26**; 588–597.

132. Construction of a genetic linkage map in man using restriction fragment length polymorphisms. Botstein D. *et al.* (1980) *Am. J. Hum. Genet.* **32**; 314–331.

133. Genetic linkage: interpreting Lod scores. Risch N. (1992) *Science* **255**; 803–804.

134. Inborn errors of metabolism. Garrod A. E. (1908) *Lancet* **2**; 1–7, 73–79, 142–148, 214–220.

135. Genetic control of biochemical reactions in *Neurospora*. Beadle G. W. & Tatum E. L. (1941) *Proc. Natl. Acad. Sci. USA* **27**; 499–506.

136. Sickle cell anaemia: a molecular disease. Pauling L. *et al.* (1949) *Science* **110**; 543–548.

137. Gene mutations in human haemoglobin: the chemical difference between normal and sickle cell haemoglobin. Ingram V. M. (1957) *Nature* **180**; 326–328.

138. On the colinearity of gene structure and protein structure. Yanofsky C. *et al.* (1964) *Proc. Natl Acad. Sci. USA* **51**; 266–272.

139. General nature of the genetic code for proteins. Crick F. H. C. *et al.* (1961) *Nature* **192**; 1227–1232.

140. RNA code words and protein synthesis II: nucleotide sequence of a valine code word. Leder P. & Nirenberg M. W. (1964) *Proc. Natl. Acad. Sci. USA* **52**; 420–427.

141. The *in vitro* synthesis of a copolypeptide containing two amino acids in alternating sequence dependent upon a DNA-like polymer containing two nucleotides in alternating sequence. Nishimura S. *et al.* (1965) *J. Mol. Biol.* **13**; 302–324.

142. Codon–anticodon pairing: the wobble hypothesis. Crick F. H. C. (1966) *J. Mol. Biol.* **19**; 548–555.

143. Different pattern of codon recognition by mammalian mitochondrial tRNAs. Barrell B. G. *et al.* (1980) *Proc. Natl Acad. Sci. USA* **77**; 3164–3166.

144. Codon recognition rules in yeast mitochondria. Bonitz S. G. *et al.* (1980) *Proc. Natl Acad. Sci. USA* **77**; 3167–3170.

145. Diverged genetic codes in protozoans and bacterium. Fox T. D. (1985) *Nature* **314**; 132–133.

146. Deciphering divergent codes. Grivell L. A. (1986) *Nature* **324**; 109–110.

147. Central dogma of molecular biology. Crick F. H. C. (1970) *Nature* **227**; 561–563.

148. The dependence of cell-free synthesis in *E. coli* upon naturally occurring or synthetic polyribonucleotides. Nirenberg M. W. & Matthaei J. H. (1961) *Proc. Natl. Acad. Sci. USA* **47**; 1588–1602.

149. A soluble ribonucleic acid intermediate in protein synthesis. Hoagland M. B. *et al.* (1958) *J. Biol. Chem.* **231**; 241–257.

150. An unstable intermediate carrying information from genes to ribosomes for protein synthesis. Brenner S. *et al.* (1961) *Nature* **190**; 576–581.

151. Unstable ribonucleic acid revealed by pulse labelling of *E. coli*. Gros F. *et al.* (1961) *Nature* **190**; 581–585.

152. The enzymatic incorporation of ribonucleotides into polyribonucleotides and the effect of DNA. Hurwitz J. *et al.* (1960) *Biochem. Biophys. Res. Commun.* **3**; 15–19.

153. Incorporation of the adenine ribonucleotide into RNA by cell fractions from *E. coli* B. Steven A. (1960) *Biochem. Biophys. Res. Commun.* **3**; 92–96.

154. Enzymatic incorporation of ribonucleotide triphosphates into the interpolynucleotide linkages of ribonucleic acid. Weiss S. B. (1960) *Proc. Natl. Acad. Sci. USA* **46**; 1020–1030.

155. Cyclic re-use of the RNA polymerase sigma factor. Travers A. A. & Burgess R. R. (1969) *Nature* **222**; 537–540.

156. A study of the unwinding of DNA and shielding of the DNA grooves by RNA polymerase by using methylation with dimethylsulphate. Melnikova A. F. *et al.* (1978) *Eur. J. Biochem.* **83**; 301–309.

157. Unwinding of the DNA helix by *E. coli* RNA polymerase. Wang J. C. *et al.* (1977) *Nucleic Acids Res.* **4**; 1225–1241.

158. Size of the unwound region of DNA in *E. coli* RNA polymerase and calf thymus RNA polymerase II ternary complexes. Gamper H. B. & Hearst J. E. (1983) *Cold Spring Harbor Symp. Quant. Biol.* **47**; 447–453.

159. Physiochemical studies on interactions between DNA and RNA polymerase. Unwinding of the DNA helix by *E. coli* RNA polymerase. Wang J. C. *et al.* (1977) *Nucleic Acids Res.* **4**; 1225–1241.

160. Rate limiting steps in RNA chain initiation. McClure W. R. (1980) *Proc. Natl. Acad. Sci. USA* **77**; 5634–5638.

161. Interaction between RNA polymerase II, factors, and template leading to accurate transcription. Fire A. *et al.* (1984) *J. Biol. Chem.* **259**; 2509–2521.

162. Five intermediate complexes in transcription initiation by RNA polymerase II. Buratowski S. *et al.* (1989) *Cell* **56**; 549–561.

163. A model for transcription termination suggested by studies on the trp attenuator *in vitro* using base analogs. Farnham P. J. & Platt T. (1980) *Cell* **20**; 739–748.

164. Rho-independent termination; dyad symmetry in DNA causes RNA polymerase to pause during transcription *in vitro*. Farnham P. J. & Platt T. (1981) *Nucleic Acids Res.* **9**; 563–577.

165. Transcription termination factor Rho is an RNA–DNA helicase. Brennan C. A. *et al.* (1987) *Cell* **48**; 945–952.

166. Adenine-rich polymer associated with rabbit reticulocyte messenger RNA. Lim L. & Canellakis E. S. (1970) *Nature* **227**; 710–712.

167. An adenylic acid-rich sequence in messenger RNA of HeLa cells and its possible relationship to reiterated sites in DNA. Darnell J. E. *et al.* (1971) *Proc. Natl. Acad. Sci. USA* **68**; 1321–1325.

168. Polyadenylic acid sequences in the heterogeneous nuclear RNA and rapidly-labelled polyribosomal RNA of HeLa cells: possible evidence for a precursor relationship. Edmonds M. *et al.* (1971) *Proc. Natl. Acad. Sci. USA* **68**; 1336–1340.

169. A polynucleotide segment rich in adenylic acid in the rapidly-labelled polyribosomal RNA component of mouse sarcoma 180 ascites cells. Lee S. Y. *et al.* (1971) *Proc. Natl. Acad. Sci. USA* **68**; 1331–1335.

170. Polyadenylic acid sequences in yeast messenger ribonucleic acid. McLaughlin C. S. *et al.* (1973) *J. Biol. Chem.* **248**; 1466–1471.

171. Lariat RNAs as intermediates and products in the splicing of messenger RNA precursors. Padgett R. A. *et al.* (1984) *Science* **225**; 898–903.

172. Stepwise assembly of a pre-mRNA splicing complex requires U-snRNPs and specific intron sequences. Frendewey D. & Keller W. (1985) *Cell* **42**; 355–367.

173. The 5′ terminus of the RNA moiety of U1 small nuclear ribonucleoprotein particles is required for the splicing of messenger RNA precursors. Kramer A. *et al.* (1984) *Cell* **38**; 299–307.

174. A factor, U2AF, is required for U2 snRNP binding and splicing complex assembly. Ruskin B. *et al.* (1988) *Cell* **52**; 207–219.

175. Base pairing between U2 and U6 snRNAs is necessary for splicing of a mammalian pre-mRNA. Wu. J. & Manley J. L. (1991) *Nature* **352**; 818–821.

176. The RNA moiety of ribonuclease P is the catalytic subunit of the enzyme. Guerrier-Takada C. *et al.* (1983) *Cell* **35**; 849–857.

177. Self-splicing RNA: autoexcision and autocyclization of the ribosomal RNA intervening sequence of *Tetrahymena*. Kruger K. *et al.* (1982) *Cell* **31**; 147–157.

178. The *Tetrahymena* ribozyme acts like an RNA restriction endonuclease. Zaug A. J. *et al.* (1986) *Nature* **324**; 429–433.

179. RNA processing errors in patients with beta-thalassemias. Ley T. J. *et al.* (1982) *Proc. Natl. Acad. Sci. USA* **79**; 4775–4779.

180. Specific transcription and RNA splicing defects in five cloned β-thalassaemia genes. Treisman R. *et al.* (1983) *Nature* **302**; 591–596.

181. Two RNAs can be produced from a single immunoglobulin μ gene by alternative RNA processing pathways. Early P. *et al.* (1980) *Cell* **20**; 313–319.

182. Alternative RNA processing in calcitonin gene expression. Amara S. G. *et al.* (1982) *Nature* **298**; 240–244.

183. Splice commitment dictates neuron-specific alternative RNA processing in calcitonin/CGRP gene expression. Leff S. E. *et al.* (1987) *Cell* **48**; 517–524.

184. Three novel brain tropomyosin isoforms are expressed from the rat α-tropomyosin gene through the use of alternative promoters and alternative RNA processing. Lees-Miller J. P. *et al.* (1990) *Mol. Cell. Biol.* **10**; 1729–1742.

185. A conserved AU sequence from the 3′ untranslated region of GM-CSF mRNA mediates selective mRNA degradation. Shaw G. & Kamen R. (1986) *Cell* **46**; 659–667.

186. Removal of poly(A) and consequent degradation of c-*fos* mRNA facilitated by 3′ AU-rich sequences. Wilson T. & Treisman R. (1988) *Nature* **336**; 396–399.

187. Processing of 45S nucleolar RNA. Weinberg R. A. & Penman S. (1970) *J. Mol. Biol.* **47**; 169–178.

188. Enzymatic modification of transfer RNA. Soll D. (1971) *Science* **173**; 293–299.

189. Feedback regulation of ribosomal protein gene expression in *E. coli*: structural homology of ribosomal RNA and ribosomal protein mRNA. Nomura M. *et al.* (1980) *Proc. Natl. Acad. Sci. USA* **77**; 7084–7088.

190. Correlation between the abundance of *E. coli* transfer RNAs and the occurrence of the respective codons in its protein genes. Ikemura T. (1981) *J. Mol. Biol.* **146**; 1–21.

191. DNA modification mechanisms and gene activity during development. Holliday R. & Pugh J. E. (1975) *Science* **187**; 226–232.

192. X inactivation, differentiation, and DNA methylation. Riggs A. D. (1975) *Cytogenet. Cell. Genet.* **14**; 9–25.

193. Selective silencing of eukaryotic DNA. Sager R. & Kitchen R. (1975) *Science* **189**; 426–433.

194. Identification of a promoter component involved in positioning the 5′ termini of simian virus 40 early mRNAs. Ghosh P. K. *et al.* (1981) *Proc. Natl. Acad. Sci. USA* **78**; 100–104.

195. Sequences upstream from the TATA box are required *in vivo* and *in vitro* for efficient transcription from adenovirus serotype 2 major late promoter. Hen R. *et al.* (1982) *Proc. Natl. Acad. Sci. USA* **79**; 7132–7136.

196. Enhancers and transcription factors in the control of gene expression. Wasylyck B. (1988) *Biochim. Biophys. Acta* **951**; 17–35.

197. Bacteriophage T7 early promoters: nucleotide sequences of two RNA polymerase binding sites. Pribnow D. (1975) *J. Mol. Biol.* **99**; 419–443.

198. Transcriptional control signals of a eukaryotic protein-coding gene. McKnight S. L. & Kingsburg R. (1982) *Science* **217**; 316–324.

199. Sequence determinants of promoter activity. Youderian P. *et al.* (1982) *Cell* **30**; 843–853.

200. Requirement for an upstream element for optimal transcription of a bacterial tRNA gene. Lamond A. I. & Travers A. A. (1983) *Nature* **304**; 248–250.

201. The repeated GC-rich motifs upstream from the TATA box are important elements of the SV40 early promoter. Everret R. D. *et al.* (1983) *Nucleic Acids Res.* **11**; 2447–2451.

202. Simian virus 40 early- and late-region promoter functions are enhanced by the 72 base pair repeat inserted at distant locations and inverted orientations. Fromm M. & Berg P. (1983) *Mol. Cell. Biol.* **3**; 991–998.

203. *In vivo* expression of *lac* promoter variants with altered −10, −35, and spacer sequences. Ackerson J. W. & Gralla J. D. (1983) *Cold Spring Harbor Symp. Quant. Biol.* **47**; 473–476.

204. Multiple nuclear factors interact with the immunoglobulin enhancer sequences. Sen R. & Baltimore D. (1986) *Cell* **47**; 705–716.

205. Two distinct promoter elements in the human ribosomal RNA gene identified by linker scanning mutagenesis. Haltiner M. M. *et al.* (1986) *Mol. Cell. Biol.* **6**; 227–239.

206. Fine structure genetic analysis of a β-globin promoter. Myers R. M. *et al.* (1986) *Science* **232**; 613–618.

207. Fine-structure mapping of the three mouse α-fetoprotein gene enhancers. Godbout R. *et al.* (1988) *Mol. Cell. Biol.* **8**; 1169–1176.

208. The SV40 enhancer contains two distinct levels of organisation. Ondek B. *et al.* (1988) *Nature* **333**; 40–45.

209. Genetic regulatory mechanisms in the synthesis of proteins. Jacob F. & Monod J. (1961) *J. Mol. Biol.* **3**; 318–356.

210. Transposition of the *lac* region in *E. coli*. I. Inversion of the *lac* operon and transduction of *lac* by f80. Beckwith J. R. & Signer W. R. (1966) *J. Mol. Biol.* **19**; 254–265.

211. Isolation of the *lac* repressor. Gilbert W. & Muller-Hill B. (1966) *Proc. Natl. Acad. Sci. USA* **56**; 1891–1898.

212. Isolation of the λ phage repressor. Ptashne M. (1967) *Proc. Natl. Acad. Sci. USA* **57**; 306–313.

213. The operators controlled by the λ phage repressor. Ptashne M. & Hopkins N. (1968) *Proc. Natl. Acad. Sci. USA* **60**; 1282–1287.

214. Isolation of pure *lac* operon DNA. Shapiro J. *et al.* (1969) *Nature* **224**; 768–774.

215. Mechanism of activation of catabolite-sensitive genes: a positive control system. Zubay G. *et al.* (1970) *Proc. Natl. Acad. Sci. USA* **66**; 104–110.

216. Measurements of unwinding of *lac* operator by repressor. Wang J. *et al.* (1974) *Nature* **251**; 247–249.

217. New features of the regulation of the tryptophan operon. Bertrand K. *et al.* (1975) *Science* **189**; 22–26.

218. Attenuation in the *E. coli* tryptophan operon: the role of RNA secondary structure involving the Trp codon region. Oxender D. L. *et al.* (1979) *Proc. Natl. Acad. Sci. USA* **76**; 5524–5528.

219. Interaction of the cAMP receptor protein with the *lac* promoter. Simpson R. B. (1980) *Nucleic Acids Res.* **8**; 759–766.

220. Structure of the *cro* repressor from bacteriophage λ and its interaction with DNA. Anderson W. F. *et al.* (1981) *Nature* **290**; 754–758.

221. The operator-binding domain of λ repressor: structure and DNA recognition. Pabo C. O. and Lewis M. (1982) *Nature* **298**; 443–447.

222. Structural similarity in the DNA-binding domains of catabolite gene activator and cro repressor proteins. Steitz T. A. *et al.* (1982) *Proc. Natl. Acad. Sci. USA* **79**; 3097–3100.

223. Structure of the catabolite gene activator protein at 2.9Å resolution: incorporation of amino acid sequence and interactions with cyclic-AMP. McKay D. B. *et al.* (1982) *J. Biol. Chem.* **257**; 9518–9524.

224. Cro repressor protein and its interaction with DNA. Matthews B. W. *et al.* (1983) *Cold Spring Harbor Symp. Quant. Biol.* **47**; 427–433.

225. Isolation of transcription factors that discriminate between different promoters recgnised by RNA polymerase II. Dynan W. S. & Tjian R. (1983) *Cell* **32**; 669–680.

226. Affinity purification of sequence-specific DNA binding proteins. Kadonaga J. T. & Tjian R. (1986) *Proc. Natl. Acad. Sci. USA* **83**; 5889–5893.

227. Cell-type-specific protein binding to the enhancer of SV40 in nuclear extracts. Davidson I. *et al.* (1986) *Nature* **323**; 544–548.

228. Isolation of cDNA encoding transcription factor Sp1 and functional analysis of the DNA-binding domain. Kadonaga J. T. *et al.* (1987) *Cell* **51**; 1079–1090.

229. A cellular DNA-binding protein that activates eukaryotic transcription and DNA replication. Jones K. A. *et al.* (1987) *Cell* **48**; 79–91.

230. Molecular cloning of an enhancer binding protein: isolation by screening a library with a recognition site DNA. Singh H. *et al.* (1988) *Cell* **52**; 415–423.

231. Binding of transcription factor TFIID to the major late promoter during *in vitro* assembly potentiates subsequent initiation by RNA polymerase II. Workman J. L. & Roeder R. (1987) *Cell* **51**; 613–619.

232. Factors involved in specific transcription by mammalian RNA polymerase II: functional analysis of initiation factors TFIIA and TFIID and identification of a new factor operating at sequences downstream of the initiation site. Reinberg D. *et al.* (1987) *J. Biol. Chem.* **262**; 3322–3330.

233. Factors involved in specific transcription by mammalian RNA polymerase II: purification and analysis of initiation factors TFIIB and TFIIE. Reinberg D. & Roeder R. G. (1987) *J. Biol. Chem.* **262**; 3310–3321.

234. Factors involved in specific transcription by mammalian RNA polymerase II: transcription factor IIS stimulates elongation of RNA chains. Reinberg D. & Roeder R. G. (1987) *J. Biol. Chem.* **262**; 3331–3342.

235. Repetitive zinc-binding domains in the protein transcription factor IIIA from *Xenopus* oocytes. Miller J. *et al.* (1985) *EMBO J.* **4**; 1609–1614.

236. Assembly of the peptide chain of haemoglobin. Dintzis H. M. (1961) *Proc. Natl. Acad. Sci. USA* **47**; 247–261.

237. Nascent polypeptide chains emerge from the exit domain of the large ribosomal subunit: immune mapping of nascent chains. Bernabeu C. & Lake J. A. (1982) *Proc. Natl. Acad. Sci. USA* **79**; 3111–3115.

238. Interaction of tRNA with 23S rRNA in the ribosmal A, P, and E sites. Moazed D. & Noller H. F. (1991) *Nature* **352**; 36–42.

239. Initiation of protein synthesis in bacteria at a translational start codon of mammalian DNA: effects of the preceding nucleotide sequence. Change A. C. *et al.* (1980) *Proc. Natl. Acad. Sci. USA* **77**; 1442–1446.

240. Characterisation of translational initiation sites in *E. coli*. Stormo G. D. *et al.* (1982) *Nucleic Acids Res.* **10**; 2971–2996.

241. N-formylmethionyl-tRNA as the initiator of protein synthesis. Adams J. M. & Capecchi M. R. (1966) *Proc. Natl. Acad. Sci. USA* **55**; 147–155.

242. Sequence of initiation factor IF2 gene: unusual protein features and homologies with elongation factors. Sacredot C. *et al.* (1984) *Proc. Natl. Acad. Sci. USA* **81**; 7787–7791.

243. Molecular cloning and sequence of the *Bacillus stearothermophilus* translational initiation factor IF2 gene. Brombach M. *et al.* (1986) *Mol. Gen. Genet.* **205**; 97–102.

244. Eukaryotic signal sequence transports insulin antigen in *Escherichia coli*. Talmadge K. *et al.* (1980) *Proc. Natl. Acad. Sci. USA* **77**; 5230–5233.

245. The OmpA signal peptide directed secretion of staphylococcal nuclease A by *Escherichia coli*. Takahara M. *et al.* (1985) *J. Biol. Chem.* **260**; 2670–2674.

246. The *secE* gene encodes an integral membrane protein required for protein export in *Escherichia coli*. Schatz J. P. *et al.* (1989) *Gene. Devel.* **3**; 1035–1044.

247. Signal peptide for protein secretion directing glycophospholipid membrane anchor attachment. Caras I. W. & Weddell G. N. (1989) *Science* **243**; 1196–1198.

248. Protein translocation across membranes. Verner K. & Schatz G. (1988) *Science* **140**; 17–26.

249. Effects of signal peptide changes on the secretion of bovine somatotropin (bsT) from *Escherichia coli*. Klein B. K. *et al.* (1992) *Protein Eng.* **5**; 511–517.

250. Increasing the efficiency of protein export in *Escherichia coli*. Perez-Perez J. *et al.* (1994) *BioTechnology* **12**; 178–180.

251. Control of folding of proteins secreted by a high expression secretion vector, pIN–111-ompA: 16-fold increase in production of active subtilisin E in *Escherichia coli*. Takagi H. *et al.* (1988) *BioTechnology* **6**; 948–950.

252. Protein folding intermediates and inclusion body formation. Mitraki A. & King J. (1989) *BioTechnology* **7**; 690–697.

253. Chaperonin-mediated protein folding at the surface of GroEL through a 'molten globule'-like intermediate. Martin J. *et al.* (1991) *Nature* **352**; 36–42.

254. Protein folding in the cell. Gething M.-J. & Sambrook J. (1992) *Nature* **355**; 33–45.

255. The organisation and duplication of chromosomes as revealed by autoradiographic studies using tritium-labelled thymidine. Taylor J. H. *et al.* (1957) *Proc. Natl. Acad. Sci. USA* **43**; 122–128.

256. Genetic homology and crossing over in the X and Y chromosomes of mammals. Burgoyne P. S. (1982) *Hum. Genet.* **61**; 85–90.

257. A view of interphase chromosomes. Manuelidis L. (1990) *Science* **250** 1533–1540.

258. The Barr body is a looped X chromosome formed by telomere association. Walker C. L. *et al.* (1991) *Proc. Natl. Acad. Sci. USA* **88**; 6191–6195.

259. Two different microtubule-based motor activities with opposite polarities in kinetochores. Hyman A. A. & Mitchison T. J. (1991) *Nature* **351**; 206–211.

260. Chromosome length controls chromosome segregation in yeast. Murray A. W. *et al.* (1986) *Cell* **45**; 529–536.

261. Genetic control of the cell cycle in yeast: V. Genetic analysis of *cdc* mutants. Hartwell L. H. *et al.* (1973) *Genetics* **74**; 267–286.

262. Gene required for G1 for committment to cell cycle and in G2 for contr of mitosis in fission yeast. Nurse P. & Bisset Y. (1981) *Nature* **292**; 558–560.

263. The *Xenopus* cdc2 protein is a component of MPF, a cytoplasmic regulator of mitosis. Dunphy W. G. *et al.* (1988) *Cell* **54**; 423–431.

264. Purified maturation-promoting factor contains the product of a *Xenopus* homolog of the fission yeast cell cycle control gene *cdc2+*. Gautier J. *et al.* (1988) *Cell* **54**; 433–439.

265. Activation of cdc2 protein kinase during mitosis in human cells: cell cycle-dependent phosphorylation and subunit rearrangment. Draetta G. & Beach D. (1988) *Cell* **54**; 17–26.

266. Cyclin: a protein specificed by maternal mRNA in sea urchin eggs that is destroyed at each cleavage division. Evans T. *et al.* (1983) *Cell* **33**; 389–396.

267. The clam embryo protein Cyclin A induces entry into M phase and the resumption of meiosis in *Xenopus* oocytes. Swenson K. I. *et al.* (1986) *Cell* **47**; 861–870.

268. A family of cyclin homologs that control the G1 phase in yeast. Hadwiger J. A. *et al.* (1989) *Proc. Natl. Acad. Sci. USA* **86**; 6255–6259.

269. An essential G1 function for the cyclin-like proteins in yeast. Richardson H. E. *et al.* (1989) *Cell* **59**; 1127–1133.

270. cdc2 protein kinase is complexed with both Cyclin A and B: evidence for proteolytic inactivation of MPF. Draetta G. *et al.* (1989) *Cell* **56**; 829–838.

271. Cyclin synthesis drives the early embryonic cell cycle. Murray A. W. & Kirschner M. W. (1989) *Nature* **339**; 275–280.

272. The role of cyclin synthesis and degradation in the control of maturation promoting factor activity. Murray A. W. *et al.* (1989) *Nature* **339**; 280–286.

273. Apoptosis: a basic biologic phenomenon with wide-ranging implications in tissue kinetics. Kerr J. F. R. *et al.* (1972) *Br. J. Cancer* **26**; 239–257.

274. Chromosome breakage accompanying genetic reconstruction in bacteriophage. Meselson M. & Weigle J. J. (1961) *Proc. Natl. Acad. Sci. USA* **B**; 857–868.

275. A mechanism for gene conversion in fungi. Holliday R. (1964) *Genet. Res.* **5**; 282–304.

276. Genetic recombination: the nature of the crossed strand-exchange between two homologous DNA molecules. Sigal N. & Alberts B. (1972) *J. Mol. Biol.* **71**; 789–793.

277. A general model for genetic recombination. Meselson M. & Radding C. M. (1975) *Proc. Natl. Acad. Sci. USA* **72**; 358–361.

278. On the mechanism of recombination: electron microscopic observation of recombination intermediates. Potter H. & Dressler D. (1976) *Proc. Natl. Acad. Sci. USA* **73**; 3000–3004.

279. Making Holliday junctions. Leach D. (1987) *Nature* **329**; 290–291.

280. Scanning tunneling microscopy of recA–DNA complexes coated with a conducting film. Amrein M. *et al.* (1988) *Science* **240**; 514–516.

281. Genetic exchange in *Salmonella*. Zinder N. D. & Lederberg J. (1952) *J. Bacteriol.* **64**; 679–699.

282. Studies on transformation of *Haemophilus influenzae*. Goodgal S. H. (1961) *J. Gen. Physiol.* **45**; 205–228.

283. Genetic mapping in *Bacillus subtilis*. Goldthwaite C. *et al.* (1967) *J. Mol. Biol.* **27**; 163–185.

284. Specialized transduction of tryptophan markers in *Escherichia coli* K12 by bacteriophage f80. Matsushiro A. (1963) *Virology* **19**; 475–482.

285. Transposable mating-type genes in *Saccharomyces cerevisiae*. Hicks J. B. *et al.* (1979) *Nature* **282**; 478–483.

286. Genetic organisation of transposon Tn10. Foster T. J. *et al.* (1981) *Cell* **23**; 201–213.

287. Transposon-mediated site-specific recombination *in vitro*: DNA cleavage and protein–DNA linkage at the recombination site. Reed R. R. & Grindley N. D. F. (1981) *Cell* **25**; 721–728.

288. Genetic evidence that Tn10 transposes by a nonreplicative mechanism. Bender J. & Kleckner N. (1986) *Cell* **45**; 801–815.

289. Tissue specificity of *Drosophila* P element transposition is regulated at the level of mRNA splicing. Laski F. A. *et al.* (1986) *Cell* **44**; 7–19.

290. The organisation and expression of the yeast transposon, Ty. Fulton A. M. *et al.* (1987) *Microbiol. Sci.* **4**; 180–185.

291. Ty elements transpose through an RNA intermediate. Boeke J. *et al.* (1985) *Cell* **40**; 491–500.

292. Ty element transposition: reverse transcriptase and virus-like particles. Garfinkel D. *et al.* (1985) *Cell* **42**; 507–517.

293. Activation of trypanosome surface glycoprotein genes involves a duplication-transposition leading to an altered 3′ end. Bernards A. *et al.* (1981) *Cell* **27**; 497–505.

294. Typanosome variable surface glycoproteins: composite genes and order of expression. Thon G. *et al.* (1990) *Genes Devel.* **9**; 1374–1383.

295. P-element mediated enhancer detection: an efficient method for isolating and characterising developmentally regulated genes in *Drosophila*. Wilson C. *et al.* (1989) *Genes Devel.* **3**; 1301–1313.

296. The replication of DNA in *Escherichia coli*. Meselson M. & Stahl F. W. (1958) *Proc. Natl. Acad. Sci. USA* **44**; 671–682.

297. Biological synthesis of deoxyribonucleic acid. Kornberg A. (1960) *Science* **131**; 1503–1508.

298. Mechanisms of DNA chain growth. Okazaki T. & Okazaki R. (1969) *Proc. Natl. Acad. Sci. USA* **64**; 1242–1248.

299. DNA replication. Kornberg A. (1988) *J. Biol. Chem.* **263**; 1–4.

300. Sequential initiation of lagging and leading strand synthesis by two different polymerase complexes at the SV40 DNA replication origin. Tsurimoto T. *et al.* (1990) *Nature* **346**; 534–539.

301. Enzymatic synthesis of deoxyribonucleic acid. XXXVI. A proofreading function for the 3′ to 5′ exonuclease activity in deoxyribonucleic acid polymerases. Brutlag D. & Kornberg A. (1972) *J. Biol. Chem.* **247**; 241–248.

302. Genetic control of cell division patterns in the *Drosophila* embryo. Edgar B. A. & O'Farrell P. H. (1989) *Cell* **57**; 177–187.

303. Mutations affecting segment number and polarity in *Drosophila*. Nusslein-Volhard C. & Weischaus E. (1980) *Nature* **287**; 795–801.

304. Molecular genetics of the bithorax complex in *Drosophila melanogaster*. Bender W. *et al.* (1983) *Science* **221**; 23–29.

305. Genomic and cDNA clones of the homeotic locus *Antennapedia* in *Drosophila*. Garber R. L. *et al.* (1983) *EMBO J.* **2**; 2027–2036.

306. The molecular organisation of the *Antennapedia* locus of *Drosophila*. Scott M. P. *et al.* (1983) *Cell* **35**; 763–776.

307. A conserved DNA sequence found in homeotic genes of the *Drosophila* Antennapedia and bithorax complexes. McGinnis W. *et al.* (1984) *Nature* **308**; 428–433.

308. Spatial distribution of transcripts from the segmentation gene *fushi tarazu* during *Drosophila* embryonic development. Hafen E. *et al.* (1984) *Cell* **37**; 833–841.

309. Transcription pattern of the *Drosophila* segmentation gene *hairy*. Ingham P. W. *et al.* (1985) *Nature* **318**; 439–445.

310. Correlative changes in homeotic and segmentation genes expression in *Kruppel* mutant embryos of *Drosophila*. Ingham P. W. *et al.* (1986) *EMBO J.* **5**; 1659–1665.

311. Regulation and function of the *Drosophila* segmentation gene *fushi tarazu*. Hiromi Y. & Gehring W. J. (1987) *Cell* **50**; 963–974.

312. The gradient morphogen *bicoid* is a concentration-dependent transcriptional activator. Struhl G. *et al.* (1989) *Cell* **57**; 1259–1273.

313. Gradients of *Kruppel* and *knirps* gene products direct pair-rule gene stripe patterning in the posterior region of the *Drosophila* embryo. Pankratz M. J. *et al.* (1990) *Cell* **61**; 309–317.

314. The dorsal morphogen is a sequence-specific DNA-binding protein that interacts with a long-range repression element in *Drosophila*. Ip Y. T. *et al.* (1991) *Cell* **64**; 439–446.

315. The murine and *Drosophila* homeobox gene complexes have common features of organisation and expression. Graham A. *et al.* (1989) *Cell* **57**; 367–347.

316. Human *Hox4.2* and *Drosophila Deformed* encode similar regulatory specificities in *Drosophila* embryos and larvae. McGinnis N. *et al.* (1990) *Cell* **63**; 969–976.

317. The protein encoded by the *Arabidopsis* homeotic gene *agamous* resembles transcription factors. Yanofsky M. F. *et al.* (1990) *Nature* **346**; 35–39.

318. The developmental gene *knotted–1* is a member of a maize homeobox gene family. Vollbrecht E. *et al.* (1991) *Nature* **350**; 241–243.

319. A cluster of Antennapedia-class homeobox genes in a nonsegmented animal. Kenyon C. & Wang B. (1991) *Science* **253**; 516–517.

320. Homeobox proteins as sequence-specific transcription factors. Levine M. & Hoey T. (1988) *Cell* **55**; 537–540.

321. The structure and function of the homeodomain. Scott M. P. *et al.* (1989) *Biochim. Biophys. Acta* **989**; 25–49.

322. Transgenes as probes for active chromosomal domains in mouse development. Allen N. D. *et al.* (1988) *Nature* **333**; 852–855.

323. The growth of bacteriophage. Ellis E. L. & Belbruck M. (1939) *J. Gen. Physiol.* **22**; 365–383.

324. The growth of bacteriophage and lysis of the host. Delbruck M. (1940) *J. Gen. Physiol.* **23**; 643–660.

325. The intracellular growth of bacteriophage. Doermann A. H. (1952) *J. Gen. Physiol.* **35**; 645–656.

326. Fine structure of a genetic region in bacteriophage. Benzer S. (1955) *Proc. Natl. Acad. Sci. USA* **41**; 344–354.

327. Nucleotide sequence of bacteriophage fX174 DNA. Sanger F. *et al.* (1977) *Nature* **265**; 687–695.

328. Nucleotide sequence of bacteriophage λ DNA. Sanger F. *et al.* (1982) *J. Mol. Biol.* **162**; 729–773.

329. Features of bacteriophage lambda: analysis of the complete nucelotide sequence. Daniels D. *et al.* (1983) *Cold Spring Harbor Symp. Quant. Biol.* **47**; 1009–1024.

330. The complete nucleotide sequence of bacteriophage T7 DNA, and the locations of T7 genetic elements. Dunn J. J. & Studier F. W. (1983) *J. Mol. Biol.* **166**; 477–535.

331. Viral RNA-dependent DNA polymerase. Baltimore D. (1970) *Nature* **226**; 1209–1211.

332. Viral RNA-dependent DNA polymerase. Temin H. M. & Mizutani S. (1970) *Nature* **226**; 1211–1213.

333. Molecular cloning of the DNA ligase gene from bacteriophage T4. I. Characterisation of the recombinants. Wilson G. G. & Murray N. E. (1979) *J. Mol. Biol.* **132**; 471–491.

334. Substrate and sequence specificity of eukaryotic DNA methylase. Gruenbaum Y. *et al.* (1982) *Nature* **295**; 62–622.

335. Cloning and expression of the bacteriophage T3 RNA-polymerase gene. Morris C. E. *et al.* (1986) *Gene* **41**; 193–200.

336. Proteolytic cleavage of native DNA polymerase into two different catalytic fragments. Klenow H. *et al.* (1971) *Eur. J. Biochem.* **22**; 371–381.

337. Deoxyribonucleic acid polymerase from the extreme thermophile *Thermus aquaticus.* Chien A. *et al.* (1976) *J. Bacteriol.* **127**; 1550–1557.

338. High fidelity DNA synthesis by the *Thermus aquaticus* DNA polymerase Eckert K. A. & Kunkel T. A. (1990) *Nucleic Acids Res.* **18**; 3739–3744.

339. Bacteriophage SP6-specific RNA polymerase. I. Isolation and characteristaion of the enzyme. Butler E. T. & Chamberlain M. J. (1982) *J. Biol. Chem.* **257**; 5772–5778.

340. Cloning and expression of the gene for bacteriophage T7 RNA polymerase. Davanloo P. *et al.* (1984) *Proc. Natl. Acad. Sci. USA* **81**; 2035–2039.

341. DNA restriction-modification enzymes of phage P1 and plasmid p15B. Subunit functions and structural homologies. Hadi S. M. *et al.* (1982) *J. Mol. Biol.* **165**; 19–34.

342. Host specificity of DNA produced by *E. coli.* X. *In vitro* restriction of phage fd replicative form. Linn S. & Arber W. (1968) *Proc. Natl. Acad. Sci. USA* **59**; 1300–1306.

343. DNA restriction enzyme from *E. coli.* Meselson M. & Yuan R. (1968) *Nature* **217**; 1110–1114.

344. A restriction enzyme from *Haemophilus influenzae.* I. Purification and general chracterisation. Smith H. O. & Wilcox K. W. (1970) *J. Mol. Biol.* **51**; 379–391.

345. A restriction enzyme form *Haemophilus influenzae.* II. Base sequence the recognition site. Kelly T. J., Jr. & Smith H. O. (1970) *J. Mol. Biol.* **51**; 393–409.

346. Specific cleavage of simian virus 40 DNA by restriction endonuclease *Haemophilus influenzae.* Danna K. & Nathans D. (1971) *Proc. Natl. Acad. Sci USA* **68**; 2913–2917.

347. Cleavage of DNA by RI restriction endonuclease generates cohesive ends. Mertz J. E. & Davis R. W. (1972) *Proc. Natl. Acad. Sci. USA* **69**; 3370–3374.

348. A suggested nomenclature for bacterial host modification and restrictio systems and their enzymes. Smith H. O. & Nathans D. (1973) *J. Mol. Biol.* **81** 419–423.

349. Nucleotide sequence specificity of restriction endonucleases. Smith H. O. (1979) *Science* **205**; 455–462.

350. Site-specific cleavage of DNA at 8- and 10-base pair sequences. McClelland M. *et al.* (1984) *Proc. Natl. Acad. Sci. USA* **81**; 983–987.

351. Isolation of high molecular weight DNA from mammalian cells. Gross-Bellard M. *et al.* (1973) *Eur. J. Biochem.* **36**; 32–38.

352. Preparative fractionation of DNA by reversed phase column chromatography. Hardies S. C. & Wells R. D. (1976) *Proc. Natl. Acad. Sci. USA* **73**; 3117–3121.

353. Rapid isolation of high molecular weight plant DNA. Murray M. & Thompson W. (1980) *Nucleic Acids Res.* **8**; 4321–4325.

354. Isolation of DNA from biological specimens without extraction with phenol. Buffone G. & Darlington G. (1985) *Clin. Chem.* **31**; 164–165.

355. Purification of DNA from formaldehyde fixed and paraffin embedded human tissue. Goelz S. E. *et al.* (1985) *Biochem. Biophys. Res. Commun.* **130**; 118–126.

356. Extraction of cellular DNA from human cells and tissues fixed in ethanol. Smith L. J. *et al.* (1987) *Anal. Biochem.* **160**; 135–138.

357. Purification and analysis of RNA from paraffin embedded tissues. Rupp G. M. & Locker J. (1988) *BioTechniques* **6**; 56–60.

358. Ribonuclease inhibitor from human placenta. Blackburn P. *et al.* (1977) *J. Biol. Chem.* **252**; 5904–5910.

359. Inhibition of intractable nucleases with ribonucleoside–vanadyl complexes: isolation of messenger ribonucleic acid from resting lymphocytes. Berger S. & Birkenmeir C. (1979) *Biochemistry* **18**; 5143–5149.

360. Isolation of biologically active ribonucleic acid from sources enriched in ribonuclease. Chirgwin J. *et al.* (1979) *Biochemistry* **18**; 5290–5294.

361. Rapid extraction of high molecular weight RNA from cultured cells and granulocytes for northern analysis. Birnboim H. C. (1988) *Nucleic Acids Res.* **16**; 1487–1497.

362. A procedure for the isolation of mammalian messenger ribonucleic acid. Brawerman G. *et al.* (1972) *Biochemistry* **11**; 637–640.

363. Ribonucleic acid isolated by cesium chloride centrifugation. Glisin V. *et al.* (1974) *Biochemistry* **13**; 2633–2637.

364. Purification of biologically active globin messenger RNA by affinity chromatography on oligo-thymidylic acid-cellulose. Aviv H. & Leder P. (1972) *Proc. Natl. Acad. Sci. USA* **69**; 1408–1412.

365. Single-step method of RNA isolation by acid guanidinium thiocyanate-phenol-chloroform extraction. Chomczynski P. & Sacchi N. (1987) *Anal. Biochem.* **162**; 156–162.

366. Precipitation of nucleic acids. Wallace D. M. (1987) *Meth. Enzymol.* **152**; 41–46.

367. Recovery of DNA from gels. Smith H. O. (1980) *Meth. Enzymol.* **65**; 371–388.

368. A reliable method for the recovery of DNA fragments from agarose and acrylamide gels. Dretzen G. *et al.* (1981) *Anal. Biochem.* **112**; 295–300.

369. An optimized freeze-squeeze method for the recovery of DNA fragments from agarose gels. Tautz D. & Renz M. (1983) *Anal. Biochem.* **132**; 14–19.

370. Electron microscope heteroduplex methods for mapping regions of base sequence homology. Davis R. W. *et al.* (1971) *Meth. Enzymol.* **21**; 413–430.

371. Electron microscopy and restriction mapping reveal additional intervening sequences in the chicken ovalbumin split gene. Garapin A. C. *et al.* (1978) *Cell* **14**; 629–639.

372. Electron microscopic visualisation of nucleic acids and of their complexes with proteins. Fischer H. W. & Williams R. C. (1979) *Annu. Rev. Biochem.* **48**; 649–660.

373. Direct observation of native DNA structures with the scanning tunnelling microscope. Beebe T. P. *et al.* (1989) *Science* **243**; 370–372.

374. Scanning tunnelling microscopy of Z-DNA. Arscott P. G. *et al.* (1989) *Nature* **339**; 484–486.

375. Atomic-scale imaging of DNA using scanning tunnelling microscopy. Driscoll R. J. *et al.* (1990) *Nature* **346**; 294–296.

376. Protein–DNA contacts in the structure of a homeodomain–DNA complex determined by nuclear magnetic resonance spectroscopy in solution. Otting G. *et al.* (1990) *EMBO J.* **9**; 3085–3092.

377. High resolution structure of an HIV zinc finger-like domain via a new NMR-based distance geometry approach. Summers M. F. *et al.* (1990) *Biochemistry* **29**; 329–340.

378. The gel electrophoresis of DNA. Aaji C. & Borst P. (1972) *Biochim. Biophys. Acta* **269**; 192–200.

379. Chain length determination of small double- and single-stranded DNA molecules by polyacrylamide gel electrophoresis. Maniatis T. *et al.* (1975) *Biochemistry* **14**; 3787–3794.

380. Sizing and mapping of early adenovirus mRNAs by gel electrophoresis of S1 endonuclease-digested hybrids. Berk A. J. & Sharp P. A. (1977) *Cell* **12**; 721–732.

381. Measurement of DNA length by gel electrophoresis. Southern E. M. (1979) *Anal. Biochem.* **100**; 319–323.

382. Gel electrophoresis of restriction fragments. Southern E. M. (1979) *Meth. Enzymol.* **68**; 152–176.

383. Separation of chromosomal DNA molecules from yeast by orthoganol-field-alteration gel electrophoresis. Carle G. R. & Olsen M. V. (1986) *Nucleic Acids Res.* **12**; 5647–5664.

384. Separation of yeast chromosome-sized DNAs by pulsed field gradient gel electrophoresis. Schwartz D. C. & Cantor C. R. (1984) *Cell* **37**; 67–75.

385. Electrophoretic separations of large DNA molecules by periodic inversion of the electric field. Carle G. R. *et al.* (1986) *Science* **232**; 65–68.

386. A model for the separation of large DNA molecules by crossed field gel electrophoresis. Southern E. M. *et al.* (1987) *Nucleic Acids Res.* **15**; 5925–5943.

387. Nearly all single base substitutions in DNA fragments joined to a GC-clamp can be detected by denaturing gradient gel electrophoresis. Myers R. M. *et al.* (1985) *Nucleic Acids Res.* **13**; 3131–3145.

388. Detection of single base substitutions by ribonuclease cleavage at mismatches in RNA:DNA duplexes. Myers R. M. *et al.* (1985) *Science* **230**; 1242–1246.

389. Detection of single base substitutions in total genomic DNA. Myers R. M. *et al.* (1985) *Nature* **313**; 495–497.

390. Comprehensive detection of single base changes in human genomic DNA using denaturing gradient gel electrophoresis and a GC clamp. Abrams E. S. *et al.* (1990) *Genomics* **7**; 463–475.

391. Detection of polymorphisms of human DNA by gel electrophoresis as single-strand conformation polymorphisms. Orita M. *et al.* (1989) *Proc. Natl. Acad. Sci. USA* **86**; 2766–2770.

392. DNA electrophoresis in microlithographic arrays. Volkmuth W. D. & Austin R. H. (1992) *Nature* **358**; 600–602.

393. Strand separation and specific recombination in deoxyribonucleic acids: physical chemical studies. P. Doty *et al.* (1960) *Proc. Natl. Acad. Sci. USA* **46**; 461–476.

394. Strand separation and specific recombination in deoxyribonucleic acids: biological studies. Marmur J. & Lane L. (1960) *Proc. Natl. Acad. Sci. USA* **46**; 453–461.

395. Sequence complementarity of T2-DNA and T2-specific RNA. Hall B. D. & Spiegelman S. (1961) *Proc. Natl. Acad. Sci. USA* **47**; 137–146.

396. The formation of hybrid DNA molecules, and their use in studies of DNA homologies. Schildkraut C. L. *et al.* (1961) *J. Mol. Biol.* **3**; 595–617.

397. The effect of electrolytes on the stability of the deoxyribonucleate helix. Hamaguchi K. & Geiduschek E. P. (1962) *J. Am. Chem. Soc.* **84**; 1329–1338.

398. Kinetics of renaturation of DNA. Wetmur J. G. & Davidson N. (1968) *J. Mol. Biol.* **31**; 349–370.

399. Room temperature method for increasing the rate of DNA reassociation by many thousand fold: the phenol emulsion reassociation technique. Kohne D. E. *et al.* (1977) *Biochemistry* **16**; 5329–5341.

400. Acceleration of nucleic acid hybridisation rate by polyethylene glycol. Amasino R. M. (1986) *Anal. Biochem.* **152**; 304–407.

401. Predicting DNA duplex stability from the base sequence. Breslauer K. J. *et al.* (1986) *Proc. Natl. Acad. Sci. USA* **83**; 3746–3750.

402. Homology probing: identification of cDNA clones encoding members of the protein-serine kinase family. Hanks S. (1987) *Proc. Natl. Acad. Sci. USA* **84**; 388–392.

403. A quantitative assay for DNA–RNA hybrids with DNA immobilised on a membrane. Gillespie D. & Spiegelman S. (1965) *J. Mol. Biol.* **12**; 829–842.

404. A quantitative assay for DNA–RNA hybrids with DNA immobilised on a membrane. Gillespie D. *et al.* (1965) *J. Mol. Biol.* **12**; 829–842.

405. A membrane-filter technique for the detection of complementary DNA. Denhardt D. T. (1966) *Biochem. Biophys. Res. Commun.* **23**; 641–646.

406. DNA–DNA hybridisation on nitrocellulose filters. I. General considerations and non-ideal kinetics. Flavell R. A. *et al.* (1974) *Eur. J. Biochem.* **47**; 535–543.

407. Determination of nucleic acid sequence homologies and relative concentrations by a dot hybridisation procedure. Kafatos F. C. *et al.* (1979) *Nucleic Acids Res.* **7**; 1541–1552.

408. Hybridisation of denatured RNA and small DNA fragments transferred to nitrocellulose. Thomas P. S. (1980) *Proc. Natl. Acad. Sci. USA* **77**; 5201–5205.

409. Hybridisation properties of immobilised nucleic acids. Gingeras T. R. *et al.* (1987) *Nucleic Acids Res.* **15**; 5373–5390.

410. A solution hybridisation assay for ribosomal RNA from bacteria using biotinylated probes and enzyme-labelled antibody to DNA:RNA. Yehle C. O. *et al.* (1987) *Mol. Cell. Probes* **1**; 177–193.

411. A sensitive solution hybridisation technique for detecting RNA in cells: application to HIV in blood cells. Pelligrino M. G. *et al.* (1987) *BioTechniques* **5**; 452–459.

412. Sandwich hybridisation as a convenient method for detection of nucleic acids in crude samples. Ranki M. *et al.* (1983) *Gene* **21**; 77–85.

413. Use of DNA immobilised on plastic and agarose supports to detect DNA by sandwich hybridisation. Polsky-Cynkin R. *et al.* (1985) *Clin. Chem.* **31**; 1438–1443.

414. Sensitive detection of genes by sandwich hybridisation and time-resolved fluorimetry. Dahlen P. *et al.* (1987) *Mol. Cell. Probes* **1**; 159–168.

415. Nucleic acid hybridisation assays employing dA-tailed capture probes. Multiple capture methods. Morrissey D. V. *et al.* (1989) *Anal. Biochem.* **81**; 345–359.

416. Fast quantification of nucleic acid hybrids by affinity-based hybrid collection. Syvanen A-C. *et al.* (1986) *Nucleic Acids Res.* **14**; 5037–5048.

417. Quantitation of polymerase chain reaction products by affinity-based hybrid collection. Syvanen A-C. I (1988) *Nucleic Acids Res.* **16**; 11327–11337.

418. A homogeneous nucleic acid hybridisation assay based on strand displacement. Vary C. P. H. (1987) *Nucleic Acids Res.* **15**; 6883–6897.

419. Simplified format for DNA probe-based tests. Bains W. (1991) *Clin. Chem.* **37**; 248–253.

420. Detection of specific sequences among DNA fragments separated by gel electrophoresis. Southern E. M. (1975) *J. Mol. Biol.* **98**; 503–517.

421. Efficient transfer of large DNA fragments from agarose gels to diazobenzyloxymethyl-paper and rapid hybridisation by using dextran sulphate. Wahl G. M. *et al.* (1979) *Proc. Natl. Acad. Sci. USA* **76**; 3683–3687.

422. Rapid transfer of DNA from agarose gels to nylon membranes. Reed K. C. & Mann D. A. (1985) *Nucleic Acids Res.* **13**; 7207–7221.

423. Quantitative molecular hybridisation on nylon membranes. Cannon G. *et al.* (1985) *Anal. Biochem.* **149**; 229–237.

424. Reduction of background problems in nonradioactive Northern and Southern blot analyses enables higher sensitivity than P-based hybridisations. Engler-Blum G. *et al.* (1993) *Anal. Biochem.* **210**; 235–244.

425. Rapid and sensitive colorimetric method for visualising biotin-labelled DNA probes hybridised to DNA or RNA immobilised on nitrocellulose: bio-blots. Leary J. J. *et al.* (1983) *Proc. Natl. Acad. Sci. USA* **80**; 4045.

426. Efficiency of *in situ* hybridisation as a function of probe size and fixation technique. Moench T. *et al.* (1985) *J. Virol. Meth.* **11**; 119–130.

427. Rapid, high-resolution *in situ* hybridisation histochemistry with radioiodinated synthetic oligonucleotides. Lewis M. *et al.* (1986) *J. Neurosci. Res.* **16**; 117–124.

428. Novel non-isotopic *in situ* hybridisation technique detects small (1 kb) unique sequences in routinely G-banded human chromosomes: fine mapping of N-myc and β-NGF genes. Garson J. A. *et al.* (1987) *Nucleic Acids Res.* **15**; 4761–4769.

429. Mapping small DNA sequences by fluorescence *in situ* hybridisation directly on banded metaphase chromosomes. Fan Y.-S. *et al.* (1990) *Proc. Natl. Acad. Sci. USA* **87**; 6223–6227.

430. Method for detection of specific mRNAs in agarose gels by transfer to diazobenzyloxymethyl-paper and hybridisation with DNA probes. Alwine J. C. *et al.* (1977) *Proc. Natl. Acad. Sci. USA* **74**; 5350–5354.

431. Quantitative analysis of *in situ* hybridisation methods for the detection of actin gene expression. Lawrence J. & Singer R. (1985) *Nucleic Acids Res.* **13**; 1777–1799.

432. Visualisation of mRNA transcription of specific genes in human cells and tissues using *in situ* hybridisation. Lum J. (1986) *BioTechniques* **4**; 30–32.

433. ErbB2 amplification in breast cancer analyzed by flourescent *in situ* hybridisation. Kallioniemi O. P. *et al.* (1992) *Proc. Natl. Acad. Sci. USA* **89**; 5321–5325.

434. Quantitation of mRNA by the polymerase chain reaction. Wang A. M. *et al.* (1989) *Proc. Natl. Acad. Sci. USA* **86**; 9717–9721.

435. Absolute mRNA quantification using the polymerase chain reaction (PCR). A novel approach by PCR aided transcript titration assay (PATTY). Becker-Andre M. & Hahlbrock K. (1989) *Nucleic Acids Res.* **17**; 9437–9446.

436. Analysis of cytokine mRNA and DNA: detection and quantitation by competitive polymerase chain reaction. Gilliland G. *et al.* (1990) *Proc. Natl. Acad. Sci. USA* **87**; 2725–2729.

437. Polymerase chain reaction-aided analysis of gene expression in frozen tissue sections. Luqmani Y. A. *et al.* (1992) *Anal. Biochem.* **200**; 291–295.

438. Bacterial cloning of plasmids carrying copies of rabbit globin messenger RNA. Rabbitts T. H. (1976) *Nature* **260**; 221–225.

439. Synthesis of full length cDNAs from four partially purified oviduct mRNAs. Buell G. N. *et al.* (1978) *J. Biol. Chem.* **253**; 2471–2482.

440. Synthesis of double-stranded DNA complementary to lysozyme, ovomucoid and ovalbumin mRNAs. Wickens M. P. *et al.* (1978) *J. Biol. Chem.* **253**; 2483–2491.

441. Enzymatic synthesis of deoxyribonucleic acid by the avian retrovirus reverse transciptase *in vitro*: optimum conditions required for transcription of large ribonucleic acid templates. Retzel E. F. *et al.* (1980) *Biochemistry* **19**; 513–524.

442. 5′ terminal sequences of eukaryotic mRNA can be cloned with high efficiency. Land H. *et al.* (1981) *Nucleic Acids Res.* **9**; 2251–2266.

443. High efficiency cloning of full length cDNA. Okayama H. & Berg P. (1982) *Mol. Cell. Biol.* **2**; 55–58.

444. A simple and very effective method for generating cDNA libraries. Gubler U. & Hoffman B. J. (1983) *Gene* **25**; 263–269.

445. Rapid production of full length cDNAs from rare transcripts: amplification using a single gene specific oligonucleotide primer. Frohman M. A. *et al.* (1988) *Proc. Natl. Acad. Sci. USA* **85**; 8998–9002.

446. A rapid method for determining sequences in DNA by primed synthesis with DNA polymerase. Sanger F. & Coulson A. R. (1975) *J. Mol. Biol.* **94**; 444–448.

447. A new method for sequencing DNA. Maxam A. M. & Gilbert W. (1977) *Proc. Natl. Acad. Sci. USA* **74**; 560.

448. DNA sequencing with chain termination inhibitors. Sanger F. *et al.* (1977) *Proc. Natl. Acad. Sci. USA* **74**; 5463–5467.

449. DNA sequencing with *Thermus aquaticus* DNA polymerase and direct sequencing of polymerase chain reaction amplified DNA. Innis M. A. *et al.* (1988) *Proc Natl. Acad. Sci. USA* **85**; 9436–9440.

450. Affinity generation of single-stranded DNA for dideoxy sequencing following the polymerase chain reaction. Michell L. G. & Merril C. R. (1989) *Anal. Biochem.* **178**; 239–242.

451. Fluorescence detection in automated DNA sequence analysis. Smith L. M. *et al.* (1986) *Nature* **321**; 674–679.

452. A system for rapid DNA sequencing with fluorescent chain-terminating dideoxynucleotides. Prober J. M. *et al.* (1987) *Science* **238**; 336–341.

453. Development of an automated procedure for fluorescent DNA sequencing. Wilson R. K. *et al.* (1990) *Genomics* **6**; 626–634.

454. Capillary gel electrophoresis for rapid, high resolution DNA sequencing. Swerdlow H. & Gesteland R. (1990) *Nucleic Acids Res.* **18**; 1415–1419.

455. DNA sequence analysis with a modified bacteriophage T7 DNA polymerase. Tabor S. & Richardson C. C. (1987) *Proc. Natl. Acad. Sci. USA* **84**; 4767–4771.

456. A short primer for sequencing DNA cloned in the single-stranded Phage vector M13mp2. Anderson S. *et al.* (1985) *Nucleic Acids Res.* **8**; 1731–1743.

457. Enzymatic amplification of β-globin sequences and restriction site analysis for diagnosis of sickle-cell anemia. Saiki R. K. *et al.* (1985) *Science* **230**; 1350–1354.

458. Direct cloning and sequence analysis of enzymatically amplified genomic sequences. Scharf S. J. *et al.* (1986) *Science* **233**; 1076–1078.

459. Primer-directed enzymatic amplification of DNA with thermostable DNA polymerase. Saiki R. K. *et al.* (1988) *Science* **239**; 487–491.

460. PCR amplification from paraffin-embedded tissues: recommendation on fixatives for long-term storage and prospective studies. Greer C. E. *et al.* (1991) *PCR Methods Applic.* **1**; 46–50.

461. Molecular genetic analysis of DNA obtained from fixed, air-dried, or paraffin embedded sources. Grunewald K. *et al.* (1991) *Ann. Haematol.* **62**; 108–114.

462. Utility of PCR for DNA analysis from dried blood spots on filter paper blotters. McCabe E. R. B. (1991) *PCR Methods Applic.* **1**; 99–106.

463. Simple non-invasive method to obtain DNA for gene analysis. Lench N. *et al.* (1988) *Lancet* **11**; 1356.

464. Direct electrophoretic detection of the allelic state of a single DNA molecule in human sperm using the polymerase chain reaction. Li H. *et al.* (1990) *Proc. Natl. Acad. Sci. USA* **87**; 4580–4584.

465. Optimization of the polymerase chain reaction with regard to fidelity: modified T7, *Taq*, and Vent DNA polymerases. Ling L. L. *et al.* (1991) *PCR Methods Applic.* **1**; 63–69.

466. Effects of primer–template mismatches on the polymerase chain reaction: human immunodeficiency virus type–1 model studies. Kwok S. *et al.* (1990) *Nucleic Acids Res.* **18**; 99–1005.

467. Antibodies as thermolabile switches: high temperature triggering for the polymerase chain reaction. Sharkey D. J. *et al.* (1994) *BioTechnology* **12**; 506–509.

468. The effect of temperature and oligonucleotide primer length on specificity and efficiency of amplification by the polymerase chain reaction. Wu D. Y. *et al.* (1991) *DNA Cell Biol.* **10**; 233–238.

469. A computer program for selection of oligonucleotide primers for polymerase chain reaction. Lowe T. *et al.* (1990) *Nucleic Acids Res.* **18**; 1757–1761.

470. OLIGOSCAN: a computer program to assist in the design of PCR primers homologous to multiple DNA sequences. Montpetit M. L. *et al.* (1992) *J. Virol. Methods* **36**; 119–128.

471. HyperPCR: A Macintosh Hypercard program for determination of optimal PCR annealing temperature. Osborne B. I. (1992) *CABIOS* **8**; 33.

472. Synthesis of oligodeoxyribonucleotides containing degenerate bases and their use as primers in the polymerase chain reaction. Lin P. K. T. & Brown D. M. (1992) *Nucleic Acids Res.* **19**; 5149–5152.

473. Attachment of a 40 base pair G+C rich sequence (GC-clamp) to genomic DNA fragments by polymerase chain reaction results in improved detection of single base changes. Sheffiled D. C. *et al.* (1989) *Proc. Natl. Acad. Sci. USA* **86**; 232–236.

474. Analysis of any point mutation in DNA. The amplification refractory mutation system. Newton C. R. *et al.* (1989) *Nucleic Acids Res.* **17**; 2503–2516.

475. Rapid and sensitive detection of point mutations and DNA polymorphisms using the polymerase chain reaction. Orita M. *et al.* (1989) *Genomics* **5**; 874–879.

476. PCR-SSCP: a simple and sensitive method for detection of mutations in the genomic DNA. Hayashi K. (1991) *PCR Method Applic.* **1**; 34–38.

477. Protocols for an improved detection of point mutations by SSCP. Spinardi L. *et al.* (1991) *Nucleic Acids Res.* **19**; 4009.

478. Evaluation of running conditions for SSCP analysis: application of SSCP for detection of point mutations in the LDL receptor gene. Leren T. P. *et al.* (1993) *PCR Methods Applic.* **3**; 159–162.

479. Rapid amplfication of complementary DNA from small amounts of unfractionated RNA. Doherty P. J. *et al.* (1989) *Anal. Biochem.* **177**; 7–10.

480. Rapid non-radioactive detection of mutations in the human genome by allele-specific amplification. Okayama H. *et al.* (1989) *J. Lab. Clin. Med.* **114**; 105–112.

481. Maximising sensitivity and specificity of PCR by preamplification heating. D'Aquila R. T. *et al.* (1991) *Nucleic Acids Res.* **19**; 3749.

488. Single-stranded DNA binding protein facilitates amplification of genomic sequences by PCR. Oshima R. G. (1992) *BioTechniques* **13**; 188.

489. Formamide can dramatically improve the specificity of PCR. Sarkar G. *et al.* (1990) *Nucleic Acids Res.* **18**; 7465.

490. Use of uracil DNA glycosylase to control carry-over contamination in polymerase chain reactions. Longo M. C. *et al.* (1990) *Gene* **93**; 125–128.

491. New targets for nucleic acid amplification. Quirus E. & Gonzalez L. (1993) *Med. Clin.* **101**; 141–143.

492. A ligase-mediated gene detection technique. Landegren U. *et al.* (1988) *Science* **241**; 1077–1080.

493. Genetic disease detection and DNA amplification using cloned thermostable ligase. Barany F. (1991) *Proc. Natl. Acad. Sci. USA* **88**; 189–193.

494. Ligase chain reaction – overview and application. Czajka J. *et al.* (199) *PCR Meth. Applic.* **3**; 551–564.

495. Isothermal *in vitro* amplification of DNA by a restriction enzyme/DNA polymerase system. Walker G. T. *et al.* (1992) *Proc. Natl. Acad. Sci. USA* **89** 392–396.

496. Nucleic acid sequence-based amplification (NASBA). Compton J. (199) *Nature* **350**; 91–92.

497. NASBA. Malek L. *et al.* (1994) *Methods Mol. Biol.* **28**; 253–260.

498. Isothermal *in vitro* amplification of nucleic acids by a multi-enzyme reaction modelled after viral replication. Guatelli J. C. *et al.* (1990) *Proc. Natl. Acad. Sci. USA* **87**; 1874–1878.

499. Transcription-based amplification system and detection of amplified human immunodeficiency virus type I with a bead-based sandwich hybridisation assay format. Kwoh D. Y. *et al.* (1989) *Proc. Natl. Acad. Sci. USA* **86**; 1173–1177.

500. Self-sustained sequence replication (3SR): an isothermal transcription-based amplification system alternative to PCR. Fahy E. *et al.* (1991) *PCR Methods Applic.* **1**; 25–33.

501. Issues of variability, carryover prevention and detection in 3SR based assays. Fahy E. *et al.* (1994) *PCR Meth. Applic.* **3**; 883–894.

502. Empirical aspects of SDA. Walker G. T. (1993) *PCR Meth. Applic.* **3**; 1–6.

503. Quantitative assays based upon the use of replicatable hybridisation probes. Lomeli H. *et al.* (1989) *Clin. Chem.* **35**; 1826–1831.

504. Exponential amplification of recombinant-RNA hybridisation probes. Lazardi P. M. *et al.* (1989) *Biotechnology* **6**; 1197–1202.

505. Synthesis of an amplifiable reporter RNA for bioassays. Chu B. C. F. *et al.* (1986) *Nucleic Acids Res.* **14**; 559–603.

506. Exponential amplification of recombinant-RNA hybridisation probes. Lizardi P. M. *et al.* (1988) *BioTechnology* **6**; 119–202.

507. The cloning of a self-replicating RNA molecule. Levisohn R. & Spiegelman S. (1968) *Proc. Natl. Acad. Sci. USA* **60**; 866–872.

508. Localisation of the Qβ replicase recognition site in MDV–1 RNA. Nishihara T. *et al.* (1983) *J. Biochem.* **93**; 669–674.

509. Amplifying DNA probe signals: a 'Christmas tree' approach. Fahrlander P. D. (1988) *BioTechnology* **6**; 1165–1168.

510. Synthesis and characterisation of branched DNA (bDNA) for the direct and quantitative detection of CMV, HBV, HCV, and HIV. Urdea M. S. (1993) *Clin. Chem.* **39**; 725–726.

511. Direct and quantitative detection of HIV–1 RNA in human plasma with branched DNA (bDNA) signal amplification assay. Urdea M. S. *et al.* (1993) *AIDS* **7**; S11–S14.

512. Efficient translation of tobacco mosaic virus RNA and rabbit globin 9S RNA in a cell-free system from commercial wheat germ. Roberts B. E. & Patterson B. M. (1973) *Proc. Natl. Acad. Sci. USA* **70**; 2330–2334.

513. An efficient mRNA-dependent translation system from reticulocyte lysates. Pelham H. R. B. & Jackson R. J. (1976) *Eur. J. Biochem.* **67**; 247–256.

514. Vectors for selective expression of cloned DNAs by T7 RNA polymerase. Rosenberg A. H. *et al.* (1987) *Gene* **56**; 125–135.

515. Purification and mapping of specific mRNAs by hybridisation selection and cell free translation. Ricciardi R. P. *et al.* (1979) *Proc. Natl. Acad. Sci. USA* **76**; 4927–4931.

516. Universal promoter for gene expression without cloning: expression-PCR. Kain K. C. *et al.* (1991) *BioTechniques* **10**; 366–374.

517. Access to a messenger RNA sequence or its protein product is not limited by tissue or species specificity. Sarkar G. & Sommer S. S. (1989) *Science* **244**; 331–334.

518. Protein truncation test (PTT) for rapid detection of translation-terminating mutations. Roest P. A. M. *et al.* (1993) *Hum. Mol. Genet.* **2**; 1719–1721.

519. Enzymatic incorporation of a new base pair into DNA and RNA extends the genetic alphabet. Piccirilli J. A. *et al.* (1990) *Nature* **343**; 33–37.

520. A general method for site-specific incorporation of unnatural amino acids into proteins. Noren C. J. *et al.* (1989) *Science* **244**; 182–188.

521. Site-specific incorporation of non-natural residues during *in vitro* protein biosynthesis with semi-synthetic aminoacyl-tRNAs. Bain J. D. *et al.* (1991) *Biochemistry* **30**; 5411–5421.

522. Selenocysteine: the 21st amino acid. Bock A. *et al.* (1991) *Mol. Microbiol.* **5**; 515–520.

523. Ribosome-mediated incorporation of a non-standard amino acid into a peptide through expansion of the genetic code. Bain J. D. *et al.* (1992) *Nature* **356**; 537–539.

524. Protein biosynthesis with conformationally restricted amino acids. Mendel D. *et al.* (1993) *Am. J. Chem. Soc.* **115**; 4359–4360.

525. Probing protein stability with unnatural amino acids. Mendel D. *et al.* (1992) *Science* **256**; 1789–1802.

526. Probing the structure and mechanism of ras protein with an expanded genetic code. Chung H. H. *et al.* (1993) *Science* **259**; 806–809.

527. Enzymatic synthesis of oligodeoxynucleotides. Chang L. M. S. & Bollum F. J. (1971) *Biochemistry* **10**; 536–542.

528. Efficient *in vitro* synthesis of biologically active RNA and DNA hybridisation probes from plasmids containing bacteriophage SP6 promoter. Melton D. A. *et al.* (1984) *Nucleic Acids Res.* **12**; 7035–7056.

529. Rapid production of vector-free biotinylated probes using the polymerase chain reaction. Lo Y-M. D. *et al.* (1988) *Nucleic Acids Res.* **16**; 8719.

530. Rapid isolation of DNA probes within specific chromosome regions by interspersed repetitive sequence polymerase chain reaction. Ledbetter S. A. *et al.* (1990) *Genomics* **6**; 475–481.

531. Rapid synthesis of oligodeoxyribonucleotides: a new solid phase method. Gait M. J. & Sheppard R. C. (1977) *Nucleic Acids Res.* **4**; 1135–1158.

532. Total synthesis of a gene. Khorana H. G. (1979) *Science* **203**; 614–625.

533. Expression in *E. coli* of a chemically synthesised gene for the hormone somatostatin. Itakura K. *et al.* (1977) *Science* **198**; 1056–1063.

534. Total synthesis and cloning of a gene coding for the ribonuclease S proteins. Nambiar K. P. *et al.* (1984) *Science* **223**; 1299–1301.

535. Total synthesis of a gene for bovine rhodopsin. Ferreti L. *et al.* (1986) *Proc. Natl. Acad. Sci. USA* **83**; 599–603.

536. Chemical synthesis of deoxyribonucleotides by the phosphoramidate method. Caruthers M. H. *et al.* (1987) *Methods Enzymol.* **154**; 287–313.

537. Anti-sense and anti-gene properties of peptide nucleic acids. Hanvey J. C. *et al.* (1992) *Science* **258**; 1481–1484.

538. The synthesis of oligonucleotides containing an aliphatic amino group at the 5'-terminus: synthesis of fluorescent DNA primers for use in DNA sequence analysis. Smith L. M. *et al.* (1985) *Nucleic Acids Res.* **13**; 2399–2412.

539. Biotin-labelled synthetic oligodeoxyribonucleotides: chemical synthesis and uses as hybridisation probes. Chollet A. & Kawashima E. H. (1985) *Nucleic Acids Res.* **13**; 1529–1541.

540. The synthesis of protected 5'-amino–2',5'-dideoxyribonucleoside–3'-O phosphoramidates: applications of 5'-amino oligodeoxyribonucleotides. Spoat B. S. *et al.* (1987) *Nucleic Acids Res.* **15**; 6181–6196.

541. Ribozymes as potential anti-HIV–1 therapeutic agents. Sarver N. *et al.* (1990) *Science* **247**; 1222–1225.

542. Ribozyme-mediated repair of defective mRNA by targeted trans-splicing. Sullenger B. A. & Cech T. R. (1994) *Nature* **371**; 619–622.

543. Derivatization of unprotected polynucleotides. Chu B. C. F. *et al.* (1983) *Nucleic Acid Res.* **11**; 6513–6529.

544. 7-Deaza–2'deoxyguanosine–5'-triphosphate: enhanced resolution in M13 dideoxy sequencing. Barr P. J. *et al.* (1986) *BioTechniques* **4**; 428–432.

545. Structure independent DNA amplification by PCR using 7-deaza–2'-deoxyguanosine. McConlogue L. *et al.* (1988) *Nucleic Acids Res.* **16**; 9869.

546. DNA containing the base analogue 2-aminoadenine: preparation, use as hybridisation probes and cleavage by restriction endonucleases. Choillet A. & Kawashima E. (1988) *Nucleic Acids Res.* **16**; 305–317.

547. Novel biotinylated nucleotide analogs for labelling and colorimetric detection of DNA. Gebeychu G. *et al.* (1987) *Nucleic Acids Res.* **15**; 4513–4534.

548. Inosine incorporation in GC rich RNA probes increases hybridisation sequence specificity. Varshney U. *et al.* (1988) *Nucleic Acids Res.* **16**; 4162.

549. A colorimetric method for DNA hybridisation. Renz M. & Kurz C. (1984) *Nucleic Acids Res.* **12**; 3435–3444.

550. Chemically modified nucleic acids as immunodetectable probes in hybridisation experiments. Tchen P. *et al.* (1984) *Proc. Natl. Acad. Sci. USA* **81**; 3466–3470.

551. Sensitive non-radioactive dot-blot hybridisation using DNA probes labelled with chelate group substituted psoralen and quantitative detection by europium ion fluorescence. Oser A. *et al.* (1988) *Nucleic Acids Res.* **16**; 1181–1196.

552. Chemical and enzymatic biotin-labelling of oligodeoxyribo-nucleotides. Kempe T. *et al.* (1985) *Nucleic Acids Res.* **13**; 45–57.

553. Detection of biotinylated nucleic acid hybrids by antibody-coated gold colloid. Tomlinson S. *et al.* (1988) *Anal. Biochem.* **171**; 217–222.

553. Iodination of nucleic acids *in vitro.* Commerford S. L. (1971) *Biochemistry* **10**; 1993–1999.

554. Efficient synthesis of high specific activity 35S-labelled human β-globin pre-mRNA. Johnson M. & Johnson B. (1984) *BioTechniques* **2**; 156–162.

556. Synthesis of highly radioactively labelled RNA hybridisation probes with synthetic single-stranded DNA oligonucleotides. Wolfl S. *et al.* (1987) *Nucleic Acids Res.* **15**; 858.

557. Enzymatic synthesis of biotin-labelled polynucleotides: novel nucleic acid affinity probes. Langer P. R. *et al.* (1981) *Proc. Natl. Acad. Sci. USA* **78**; 6633–6637.

558. A technique for radiolabelling DNA restriction endonuclease fragments to high specific activity. Feinberg A. P. & Vogelstein B. (1983) *Anal. Biochem.* **132**; 6–13. (Addendum: (1984) *Anal. Biochem.* **137**; 266–267.)

559. Labelling deoxyribonucleic acid to high specific activity *in vitro* by nick translation with DNA polymerase I. Rigby P. W. J. *et al.* (1977) *J. Mol. Biol.* **113**; 237–251.

560. Improved hybridisation assays employing tailed oligonucleotide probes: a direct comparison with 5′ end-labelled oligonucleotide probes and nick-translated plasmid probes. Collins M. & Hunsaker W. (1985) *Anal. Biochem.* **151**; 211–224.

561. Introduction of 5′-terminal functional groups into synthetic oligonucleotides for selective immobilisation. Bischoff R. *et al.* (1987) *Anal. Biochem.* **164**; 336–344.

562. Specific labelling of 3′-termini of RNA with T4 ligase. England T. E. *et al.* (1980) *Methods Enzymol.* **65**; 65–74.

563. Terminal labelling and addition of homopolymer tracts to duplex DNA fragments by terminal deoxynucleotidyl transferase. Roychoudhury R. *et al.* (1976) *Nucleic Acids Res.* **3**; 863–877.

564. The generation of radiolabelled DNA and RNA probes with the polymerase chain reaction. Schowalter D. B. & Sommer S. S. (1989) *Anal. Biochem.* **177**; 90–94.

565. Production of discrete high specific activity DNA probes using polymerase chain reaction. Jansen R. & Ledley F. D. (1989) *Gene Anal. Tech.* **6**; 79–83.

566. Reagents suitable for the crosslinking of nucleic acids to proteins. Fink G. *et al.* (1980) *Anal. Biochem.* **108**; 394–401.

567. Non-radioactive hybridisation probes prepared by the chemical labelling of DNA and RNA with a novel reagent, photobiotin. Forster A. C. *et al.* (1985) *Nucleic Acids Res.* **13**; 745–761.

568. Preparation of oligodeoxynucleotide–alkaline phosphatase conjugates and their use as hybridisation probes. Jablonski E. *et al.* (1986) *Nucleic Acids Res.* **14**; 6115–6128.

569. A chemical method for introducing haptens onto DNA probes. Keller G. H. *et al.* (1988) *Anal. Biochem.* **170**; 441–450.

570. Monoadduct forming photochemical reagents for labelling nucleic acids for hybridisation. Albarella J. P. *et al.* (1989) *Nucleic Acids Res.* **17**; 4293–4308.

571. Labelling of DNA probes with a photoactivatable hapten. Keller G. H. (1989) *Anal. Biochem.* **177**; 392–395.

572. Enzymatic joining of DNA strands: a novel reaction of diphosphopyridine nucleotide. Zimmerman S. B. *et al.* (1967) *Proc. Natl. Acad. Sci. USA* **57**; 1841–1848.

573. Enzymatic joining of polynucleotides. V. A DNA adenylate intermediate in the polynucleotide joining reaction. Olivera B. M. *et al.* (1968) *Proc. Natl. Acad. Sci. USA* **61**; 237–244.

574. Biochemical method for inserting new genetic information into DNA of simian virus 40: circular SV40 DNA molecules containing lambda phage genes and the galactose operon of *E. coli.* Jackson D. *et al.* (1972) *Proc. Natl. Acad. Sci. USA* **69**; 2904–2909.

575. Construction of biologically functional bacterial plasmids *in vitro.* Cohen S. *et al.* (1973) *Proc. Natl. Acad. Sci. USA* **70**; 3240–3244.

576. Enzymatic end-to-end joining of DNA molecules. Lobban P. & Kaiser A. D. (1973) *J. Mol. Biol.* **79**; 453–471.

577. Ligation of *Eco* RI endonuclease-generated DNA fragments into linear and circular structures. Dugaiczyk A. *et al.* (1975) *J. Mol. Biol.* **96**; 171–184.

578. Temperature dependence of the joining by T4 DNA ligase of termini produced by type II restriction endonucleases. Ferretti L. & Sgaramella V. (1981) *Nucleic Acids Res.* **9**; 85–93.

579. Macromolecular crowding allows blunt end ligation by DNA ligases from rat liver or *Escherichia coli.* Zimmerman S. B. & Pheiffer B. (1983) *Proc. Natl. Acad. Sci. USA* **80**; 5852–5856.

580. Chemical synthesis of restriction enzyme recognition sites useful for cloning. Scheller R. *et al.* (1977) *Science* **196**; 177–180.

581. Synthetic adaptors for cloning DNA. Rothstein R. J. *et al.* (1979) *Methods Enzymol.* **68**, 98–109.

582. Addition of homopolymers to the 3′ ends of duplex DNA with terminal transferase. Nelson T. & Brutlag D. (1979) *Methods Enzymol.* **68**; 41–50.

583. An improved method for utilizing terminal transferase to add homopolymers to the 3′ termini of DNA. Deng G. & Wu R. (1981) *Nucleic Acids Res.* **9**; 4173–4188.

584. Characterisation of the homopolymer tailing reaction catalyzed by terminal transferase. Implications for the cloning of cDNA. Michelson A. M. & Orkin S. H. (1982) *J. Biol. Chem.* **256**; 1473–1482.

585. Viable molecular hybrids of bacteriophage lambda and eukaryotic DNA. Thomas M. *et al.* (1974) *Proc. Natl. Acad. Sci. USA* **71**; 4579–4583.

586. Phage lambda receptor chromosomes for DNA fragments made with restriction endonuclease III of *Haemophilus influenzae* and restriction endonuclease I of *Escherichia coli*. Murray K. & Murray N. E. (1975) *J. Mol. Biol.* **98**; 551–564.

587. Charon phages: safer derivatives of bacteriophage lambda for DNA cloning. Blattner F. R. *et al.* (1977) *Science* **196**; 161–169.

588. EK2 derivatives of bacteriophage lambda useful in the cloning of DNA from higher organisms: the λgt WES system. Leder P. *et al.* (1977) *Science* **196**; 175–177.

589. Lambdoid phages that simplify recovery of *in vitro* recombinants. Murray N. E. *et al.* (1977) *Mol. Gen. Genet.* **150**; 53–61.

590. A new host-vector system allowing selection for foreign DNA inserts in bacteriophage λgt WES. Davison J. *et al.* (1979) *Gene* **8**; 69–80.

591. Lambda replacement vectors carrying polylinker sequences. Frischauf A. M. *et al.* (1983) *J. Mol. Biol.* **179**; 827–842.

592. Improved M13 cloning phage cloning vectors and host strains: nucleotide sequences of the M13mp18 and pUC19 vectors. Yanisch-Peron C. *et al.* (1985) *Gene* **33**; 103–119.

593. Filamentous coli phage M13 as a cloning vehicle: insertion of a *Hind*II fragment of the *lac* regulatory region in M13 replicative form *in vitro*. Messing J. *et al.* (1977) *Proc. Natl. Acad. Sci. USA* **74**; 3642–3646.

594. Cloning in single-stranded bacteriophage as an aid to rapid DNA sequencing. Sanger F. *et al.* (1980) *J. Mol. Biol.* **143**; 161–178.

595. Filamentous fusion phage: novel expression vectors that display cloned antigens on the virion surface. Smith G. P. (1985) *Science* **228**; 1315–1317.

596. Selecting high affinity binding proteins by monovalent phage display. Lowman H. B. *et al.* (1991) *Biochemistry* **30**; 10832–10838.

597. Linkage of recognition and replication functions by assembling combinatorial antibody Fab libraries along phage surfaces. Kang A. S. *et al.* (1991) *Proc. Natl. Acad. Sci. USA* **88**; 4363–4366.

598. Assembly of combinatorial libraries on phage surfaces (Phabs): the gene III site. Barbas C. F. *et al.* (1991) *Proc. Natl. Acad. Sci. USA* **88**; 7978–7982.

599. Display of biologically active proteins on the surface of filamentous phages: cDNA cloning system for selection of functional gene products linked to the genetic information responsible for their production. Crameri R. & Suter M. (1993) *Gene* **137**; 69–75.

600. Cloning in bacteriophage lambda vector for the display of binding proteins on filamentous phage. Hogrefe H. H. *et al.* (1993) *Gene* **137**; 85–91.

601. A dye-bouyant-density method for the detection and isolation of closed circular duplex DNA: the closed circular DNA in HeLa cells. Radloff R. *et al.* (1967) *Proc. Natl. Acad. Sci. USA* **57**; 1514–1521.

602. A rapid method for the identification of plasmid deoxyribonucleic acid in bacteria. Eckhardt T. (1978) *Plasmid* **1**; 584–588.

603. A rapid alkaline extraction procedure for screening recombinant plasmid DNA. Birnboim H. C. & Doly J. (1979) *Nucleic Acids Res.* **7**; 1513–1523.

604. Methods for isolating large bacterial plasmids. Gowland P. C. & Hardmann D. J. (1986) *Microbiol. Sci.* **3**; 252–254.

605. Construction and characterisation of new cloning vehicles. I. Ampicillin resistant derivatives of the plasmid pMB9. Bolivar F. *et al.* (1977) *Gene* **2**; 75–93.

606. Construction and characterisation of new cloning vectors. II. A multipurpose cloning system. Bolivar F. *et al.* (1977) *Gene* **2**; 95–113.

607. Construction and characterisation of new cloning vehicles. III. Derivatives of plasmid pBR322 carrying unique *Eco* RI sites for selection of *Eco* RI generated recombinant DNA molecules. Bolivar F. (1978) *Gene* **4**; 121–136.

608. Complete nucleotide sequence of the *E. coli* plasmid pBR322. Sutcliffe G. (1979) *Cold Spring Harbor Symp. Quant. Biol.* **43**; 77–90.

609. *In vitro* gene fusions that join an enzymatically active beta-galactosidase segment to amino-terminal fragments of exogenous proteins: *Escherichia coli* plasmid vectors for the detection and cloning of translational initiation signals. Casadaban M. J. *et al.* (1980) *J. Bacteriol.* **143**; 971–980.

610. A plasmid cloning vector for the direct selection of strains carrying recombinant clones. Dean D. (1981) *Gene* **15**; 99–102.

611. Versatile low-copy-number plasmid vectors for cloning in *Escherichia coli*. Stoker N. G. *et al.* (1982) *Gene* **18**; 335–341.

612. A novel cloning vector for the direct selection of recombinant DNA in *E. coli*. Hennecke H. *et al.* (1982) *Gene* **19**; 231–234.

613. The pUC plasmids: an M13mp7-derived system for insertion mutagenesis and sequencing with synthetic universal primers. Vieira J. & Messing J. (1982) *Gene* **19**; 259–268.

614. pEMBL: a new family of single-stranded plasmids. Dente L. *et al.* (1983) *Nucleic Acids Res.* **11**; 1645–1655.

615. New runaway-replication-plasmid cloning vectors and suppression of runaway replication by novobiocin. Uhlin B. E. *et al.* (1983) *Gene* **22**; 255–265.

616. A plasmid expression vector that permits the stabilisation of both mRNAs and proteins encoded by the cloned genes. Duvoisin R. M. *et al.* (1986) *Gene* **45**; 193–201.

617. Positive selection vectors based on xylose ultilisation suppression. Stevis P. E. & Ho N. W. Y. (1987) *Gene* **55**; 67–74.

618. Production of single-stranded plasmid DNA. Vieira J. & Messing J. (1987) *Methods Enzymol.* **153**; 3–11.

619. Cloning and stable maintenance of 300 kilobase pair fragments of human DNA in *Escherichia coli* using an F-factor-based vector. Shizuya H. *et al.* (1992) *Proc. Natl. Acad. Sci. USA* **89**; 8794–8797.

620. Transformation of yeast by a replicating hybrid plasmid. Beggs J. D. (1978) *Nature* **275**; 104–109.

621. High frequency transformation of yeast: autonomous replication of hybrid DNA molecules. Struhl K. *et al.* (1979) *Proc. Natl. Acad. Sci. USA* **76**; 1035–1039.

622. Isolation of a yeast centromere and construction of functional small circular chromosomes. Clarke L. & Carbon J. (1980) *Nature* **287**; 504–509.

623. Characterisation of a yeast replicative origin (ars2) and construction of stable minichromosomes containing cloned yeast centromere DNA (CEN3). Hsiao C. L. & Carbon J. (1981) *Gene* **15**; 157–166.

624. Cloning yeast telomeres on linear plasmid vectors. Szostak J. & Blackburn E. H. (1982) *Cell* **29**; 245–255.

625. Construction of artificial chromosomes in yeast. Murray A. W. & Szostak J. W. (1983) *Nature* **305**; 189–193.

626. Cloning of large segments of exogenous DNA into yeast by means of artificial chromosome vectors. Burke D. T. *et al.* (1987) *Science* **236**; 806–813.

627. Bending the rules: the 2μ plasmid of yeast. Murray J. A. H. (1987) *Mol. Microbiol.* **1**; 1–14.

628. A system of shuttle vectors and yeast host strains designed for efficient manipulation of DNA in *Saccharomyces cerevisiae*. Sikorski R. S. & Hieter P. (1989) *Genetics* **122**; 190–127.

629. Extrachromosomal maintenance and amplification of yeast artificial chromosome DNA in mouse cells. Featherstone T. & Huxley C. (1993) *Genomics* **17**; 267–278.

630. Isolation of genes by complementation in yeast: molecular cloning of a cell cycle control gene. Nasmyth K. A. & Reed S. I. (1980) *Proc. Natl. Acad. Sci. USA* **77**; 2119–2123.

631. Isolation of a gene from *Drosophila* by complementation in yeast. Henikoff S. *et al.* (1981) *Nature* **289**; 33–37.

632. Complementation used to clone a human homologue of the fission yeast cell cycle control gene cdc2. Lee M. G. & Nurse P. (1987) *Nature* **327**; 31–35.

633. Construction of plasmids carrying the cl gene of bacteriophage λ. Backman K. *et al.* (1976) *Proc. Natl. Acad. Sci. USA* **73**; 4174–4178.

634. Phasmids: hybrids between Col E1 plasmids and *E. coli* bacteriophage lambda. Brenner S. *et al.* (1982) *Gene* **17**; 27–44.

635. The tac promoter: a functional hybrid derived from the trp and lac promoters. DeBoer H. A. *et al.* (1983) *Proc. Natl Acad. Sci. USA* **80**; 21–25.

636. M13 vectors with T7 polymerase promoters: transcription limited by oligonucleotides. Eperon I. C. (1986) *Nucleic Acids Res.* **14**; 2830.

637. λZAP: a bacteriophage lambda expression vector with *in vivo* excision properties. Short J. M. *et al.* (1988) *Nucleic Acids Res.* **16**; 7583–7600.

638. A broad-range-host shuttle system for gene insertion into the chromosomes of Gram-negative bacteria. Barry G. F. (1988) *Gene* **71**; 75–84.

639. Cosmids: a type of plasmid gene cloning vector that is packageable *in vitro* in bacteriophage heads. Collins J. & Hohn B. (1978) *Proc. Natl. Acad. Sci. USA* **75**; 4242–4246.

640. A small cosmid for efficient cloning of large DNA fragments. Hohn B. & Collins J. (1980) *Gene* **11**; 291–298.

641. Rapid and efficient cosmid cloning. Ish-Horowicz D. & Burke J. F. (1981) *Nucleic Acids Res.* **9**; 2989–2998.

642. Double cos site vectors: simplified cosmid cloning. Bates P. F. & Swift R. A. (1983) *Gene* **26**; 137–146.

643. Transfection and transformation of *Agrobacterium tumefaciens*. Holsters M. *et al.* (1978) *Mol. Gen. Genet.* **163**; 181–187.

644. A binary plant vector strategy based upon separation of the Vir- and T-region of *Agrobacterium tumefaciens* Ti plasmid. Hoekema A. *et al.* (1983) *Nature* **303**; 179–180.

645. Binary *Agrobacterium* vectors for plant transformation. Bevan M. (1984) *Nucleic Acids Res.* **12**; 8711–8721.

646. *Agrobacterium*-mediated transformation of germinating seeds of *Arabidopsis thaliana*: a non-tissue culture approach. Feldmann K. A. & Marks M. D. (1987) *Mol. Gen. Genet.* **208**; 1–9.

647. *Agrobacterium*-mediated transformation of rice (*Oryza sativa* L.). Raineri D. M. *et al.* (1990) *BioTechnology* **8**; 33–38.

648. A vector for introducing new genes into plants. Chilton M. D. (1983) *Scientific American* **248**; 50–59.

649. Expression of a bacterial gene in plants using a viral vector. Brisson N. *et al.* (1984) *Nature* **310**; 511–514.

650. Gene amplification and expression in plants by a replicating geminivirus vector. Hayes R. J. *et al.* (1988) *Nature* **334**; 179–182.

651. The genome of simian virus 40. Reddy V. B. *et al.* (1978) *Science* **200**; 494–502.

652. Complete nucleotide sequence of SV40 DNA. Fiers W. *et al.* (1978) *Nature* **273**; 113–120.

653. Construction of hybrid viruses containing SV40 and λ phage DNA segments and their propagation in cultured monkey cells. Goff S. P. & Berg P. (1976) *Cell* **9**; 695–705.

654. Construction and applications of a highly transmissible murine retrovirus shuttle vector. Cepko C. L. *et al.* (1984) *Cell* **37**; 1053–1062.

655. Lineage analysis in the vertebrate nervous system by retrovirus-mediated gene transfer. Price J. *et al.* (1987) *Proc. Natl. Acad. Sci. USA* **84**; 156–160.

656. Packaging recombinant DNA molecules into bacteriophage particles *in vitro*. Hohn B. & Murray K. (1977) *Proc. Natl. Acad. Sci. USA* **74**; 3259–3263.

657. *In vitro* packaging of lambda and cosmid DNA. Hohn B. (1979) *Methods Enzymol.* **68**; 299–309.

658. Competence for DNA transfer of ouabain resistance and thymidine kinase: clonal variation in mouse L-cell recipients. Corsaro C. M. & Pearson M. L. (1981) *Somatic Cell Genet.* **7**; 601–616.

659. High-fidelity transcription of 5S DNA injected into *Xenopus* oocytes. Brown D. D. & Gurdon J. B. (1977) *Proc. Natl. Acad. Sci. USA* **74**; 2064–2068.

660. Fertile transgenic rice plants regenerated from transformed protoplasts. Shimamoto K. *et al.* (1989) *Nature* **338**; 274–276.

661. Genetically transformed maize plants from protoplasts. Rhodes C. A. *et al.* (1988) *Science* **240**; 204–207.

662. Transgenic plants of tall fescue (*Festula arundinacea* Screb.) obtained by direct gene transfer to protoplasts. Wang Z. *et al.* (1992) *BioTechnology* **10**; 691–696.

663. Direct gene transfer to cells of a graminaceous monocot. Potrykus I. *et al.* (1985) *Mol. Gen. Genet.* **199**; 183–188.

664. Stable transformation of soybean (*Glycine max*) by particle acceleration. McCabe D. E. *et al.* (1988) *BioTechnology* **6**; 923–926.

665. Transgenic plants of turfgrass (*Agrostis palustris* Huds.) from microprojectile bombardment of embryonic callus. Zhong H. *et al.* (1993) *Plant Cell Rep.* **13**; 1–6.

666. Herbicide resistant turfgrass (*Agrostis palustris* Huds.) by bolistic transformation. Hartman C. L. *et al.* (1994) *BioTechnology* **12**; 919–923.

667. Gene transfer into mouse lymphoma cells by electroporation in high electric fields. Neumann E. *et al.* (1982) *EMBO J.* **1**; 841–845.

668. Electric field induced cell to cell fusion. Zimmerman U. & Vienken J. (1983) *J. Membr. Biol.* **67**; 165–182.

669. Electric field-mediated DNA transfer: transient and stable gene expression in human and mouse lymphoid cells. Toneguzzo F. *et al.* (1986) *Mol. Cell. Biol.* **6**; 703–706.

670. Electroporation for the efficient transfection of mammalian cells with DNA. Chu G. *et al.* (1987) *Nucleic Acids Res.* **15**; 1311–1326.

671. High-efficiency transformation of mammalian cells by plasmid DNA. Chen C. & Okayama H. (1987) *Mol. Cell. Biol.* **7**; 2745–2752.

672. Lipofection: a highly efficient, lipid-mediated DNA-transfection procedure. Felgner P. L. *et al.* (1987) *Proc. Natl. Acad. Sci. USA* **84**; 7413–7417.

673. High efficiency transformation by direct microinjection of DNA into cultured mammalian cells. Capecchi M. R. (1980) *Cell* **22**; 479–488.

674. Integration and stable germline transmission of genes injected into mouse pronuclei. Gordon J. W. & Ruddle F. H. (1981) *Science* **214**; 1244–1246.

675. *In vivo* and *in vitro* gene transfer to mammalian somatic cells by particle bombardment. Yang N. S. *et al.* (1990) *Proc. Natl. Acad. Sci. USA* **87**; 9568–9572.

676. Introduction of foreign genes into tissues of living mice by DNA-coated microprojectiles. Williams R. S. *et al.* (1991) *Proc. Natl. Acad. Sci. USA* **88**; 2726–2730.

677. Chloroplast transformation in *Chlamydomonas* with high velocity microprojectiles. Boynton J. E. *et al.* (1988) *Science* **240**; 1534–1538.

678. Mitochondrial transformation in yeast by bombardment with microprojectiles. Johnston S. A. *et al.* (1988) *Science* **240**; 1538–1541.

679. High velocity microprojectiles for delivering nucleic acids into living cells. Klein T. M. *et al.* (1987) *Nature* **327**; 70–73.

680. Gene transfer to tumour-infiltrating lymphocytes and other mammalian somatic cells by microprojectile bombardment. Fitzpatrick-McElligott S. (1992) *BioTechnology* **10**; 1036–1040.

681. A general method for cloning eukaryotic structural gene sequences. Higuchi R. *et al.* (1976) *Proc. Natl. Acad. Sci. USA* **73**; 3146–3150.

682. Chromosome walking and jumping to isolate DNA from the *Ace* and *rosy* loci and *bithorax* complex in *Drosophila melanogaster*. Bender W. *et al.* (1983) *J. Mol. Biol.* **168**; 17–33.

683. Construction of a general human chromosome jumping library, with application to cystic fibrosis. Collins F. S. *et al.* (1987) *Science* **235**; 1040–1046.

684. Construction and use of human chromosome jumping libraries from *Not* I-digested DNA. Poustka A. *et al.* (1987) *Nature* **325**; 353–355.

685. Genome walking by single specific primer polymerase chain reaction – SSP-PCR. Shyamala V. & Ames G. F.-L. (1989) *Gene* **84**; 1–8.

686. A polymerase chain reaction (PCR) approach for constructing jumping and linking libraries. Kandpal R. P. *et al.* (1990) *Nucleic Acids Res.* **18**; 3081.

687. Directional cloning of DNA fragments at a large distance from an initial probe: a circularisation method. Collins F. S. & Weissman S. M. (1984) *Proc. Natl. Acad. Sci. USA* **81**; 6812–6816.

688. Use of yeast artificial chromosome clones for mapping and walking within human chromosome segment 18q21.3. Silverman G.A. *et al.* (1989) *Proc. Natl. Acad. Sci. USA* **86**; 7485–7489.

689. Yeast artificial chromosome libraries containing large inserts from mouse and human DNA. Larin Z. *et al.* (1991) *Proc. Natl. Acad. Sci. USA* **88**; 4123–4127.

690. Amplification of genomic sequences flanking transposable elements in host and heterologous plants. A tool for transposon tagging and genome characterisation. Earp D. J. *et al.* (1990) *Nucleic Acids Res.* **18**; 3271–3279.

691. Differential display of eukaryotic mRNAs by PCR. Liang P. & Pardee A. (1992) *Science* **257**; 967–971.

692. Colony hybridisation: a method for the isolation of cloned cDNAs that contain a specific gene. Grunstein M. & Hoggness D. S. (1975) *Proc. Natl. Acad. Sci. USA* **72**; 3961–3965.

693. Screening λgt recombinant clones by hybridisation to single plaques *in situ*. Benton W. D. & Davies R. W. (1977) *Science* **196**; 180–182.

694. The isolation of structural genes from libraries of eukaryotic DNA. Maniatis T. *et al.* (1978) *Cell* **15**; 687–701.

695. Plasmid screening at high colony density. Hanahan D. & Meselson M. (1980) *Gene* **10**; 63–67.

696. Use of synthetic oligonucleotides as hybridisation probes: isolation of cloned cDNA sequences for human β2-microglobulin. Suggs S. V. *et al.* (1981) *Proc. Natl. Acad. Sci. USA* **78**; 6613–6617.

697. Chemical synthesis and cloning of a poly(arginine)-coding gene fragment designed to aid polypeptide purification. Smith J. C. *et al.* (1984) *Gene* **32**; 321–327.

698. Base composition-independent hybridisation in tetramethyl-ammonium chloride: a method for oligonucleotide screening of highly complex gene libraries. Wood W. I. *et al.* (1985) *Proc. Natl. Acad. Sci. USA* **82**; 1585–1588.

699. *In situ* immunoassays for gene translation products in phage plaques and bacterial colonies. Skalka A. & Shapio L. (1976) *Gene* **1**; 65–79.

700. Immunological screening method to detect specific translation products. Broome S. & Gilbert W. (1978) *Proc. Natl. Acad. Sci. USA* **75**; 2746–2749.

701. Efficient isolation of genes by using antibody probes. Young R. A. & Davis R. W. (1983) *Proc. Natl. Acad. Sci. USA* **80**; 1194–1198.

702. Screening an expression library with a ligand probe: isolation and sequence of a cDNA corresponding to a brain calmodulin-binding protein. Sikela J. M. & Hahn W. (1987) *Proc. Natl. Acad. Sci. USA* **84**; 3038–3042.

703. Molecular cloning of the CD2 antigen, the T-cell erythrocyte receptor, by a rapid immunoselection procedure. Seed B. & Aruffo A. (1987) *Proc. Natl. Acad. Sci. USA* **84**; 3365–3369.

704. Expression cloning of the murine erythropoietin receptor. D'Andrea A. D. *et al.* (1989) *Cell* **57**; 277–285.

705. Direct clone characterisation from plaques and colonies by the polymerase chain reaction. Gussow D. & Clackson T. (1989) *Nucleic Acids Res.* **17**; 4000.

706. Enzymatic amplification of specific cDNA inserts from λgt11 libraries. Friedman K. D. *et al.* (1988) *Nucleic Acids Res.* **16**; 8718.

707. Cloning the differences between two complex genomes. Lisitsyn N. *et al.* (1993) *Science* **259**; 946.

708. Genomic mismatch scanning: a new approach to genetic linkage mapping. Nelson S. F. *et al.* (1993) *Nature Genet.* **4**; 11–17.

709. Finding similarities and differences among genomes. Lander E. S. (1993) *Nature Genet.* **4**, 5–10.

710. Direct isolation of polymorphic markers linked to a trait by genetically directed representational difference analysis. Lisitsyn N. *et al.* (1994) *Nature Genet.* **6**; 57–65

711. A genome wide search for human type 1 diabetes susceptibility genes. Davies J. L. *et al.* (1994) *Nature* **371**; 130–134

712. A procedure for *in vitro* amplification of DNA segments that lie outside the boundaries of known sequences. Triglia T. *et al.* (1988) *Nucleic Acids Res.* **16**; 8186.

713. Genetic applications of an inverse polymerase chain reaction. Ochman H. *et al.* (1988) *Genetics* **120**; 621–623.

714. DNA polymorphisms amplified by arbitrary primers are useful as genetic markers. Williams J. G. K. *et al.* (1990) *Nucleic Acids Res.* **18**; 6531–6535.

715. Fingerprinting genomes using PCR with arbitrary primers. Welsh J. & McClelland M. (1990) *Nucleic Acids Res.* **18**; 7213–7218.

716. Polymorphisms generated by arbitrarily primed PCR in the mouse: application to strain identification and genetic mapping. Welsh J. *et al.* (1990) *Nucleic Acids Res.* **19**; 303–306.

717. DNA amplification fingerprinting using arbitrary mini-hairpin oligonucleotide primers. Caetano-Anolles G. & Gresshoff P. M. (1994) *BioTechnology* **12**; 619–623.

718. Rapid identification of genetic variation and pathotype of *Leptosphaeria maculans* by Random Amplified Polymorphic DNA assay. Goodwin P. H. & Annis S. L. (1991) *Appl. Environ. Microbiol.* **57**; 2482–2486.

719. Estimating genomic distance from DNA sequence – location in cell nuclei by a random walk model. van den Engh *et al.* (1992) *Science* **257**; 1410.

720. A gel electrophoretic method for quantifying the binding of proteins to specific DNA regions. Garnier M. M. & Revzin A. (1981) *Nucleic Acids Res.* **9**; 3047–3060.

721. Hydroxyl radical footprinting: a high-resolution method for mapping protein–DNA contacts. Tullius T. D. *et al.* (1987) *Methods Enzymol.* **155**; 537–558.

722. Real time monitoring of DNA manipulation using biosensor technology. Nilsson P. *et al.* (1995) *Anal. Biochem.* **224**; 400–408.

723. Identification of regulatory elements of cloned genes with functional assays. Rosenthal N. (1987) *Methods Enzymol.* **152**; 704–720.

724. Recombinant genomes which express chloramphenicol acetyltransferase in mammalian cells. Gorman C. M. *et al.* (1982) *Mol. Cell. Biol.* **2**; 1044–1051.

725. Transient and stable expression of the firefly luciferase gene in plant cells and transgenic plants. Ow D. *et al.* (1986) *Science* **234**; 856–859.

726. Luciferase reporter gene vectors for the analysis of promoters and enhancers. Nordeen S. K. (1988) *BioTechniques* **6**; 454–458.

727. The GUS reporter gene system. Jefferson R. A. (1989) *Nature* **342**; 837–838.

728. Characterisation of enhancer elements in the long terminal repeat of moloney murine sarcoma virus. Laimins L. A. *et al.* (1984) *J. Virol.* **49**; 183–189.

729. Cell-specific expression of the rat insulin gene: evidence for the role of two distinct 5′ flanking regions. Edlund T. *et al.* (1985) *Science* **230**; 912–916.

730. The smallest known gene. Gonzalez-Pastor J. E. *et al.* (1994) *Nature* **369**; 281.

731. Genomic sequence databases. Waterman M. S. (1990) *Genomics* **6**; 700–701.

732. Molecular sequence accuracy and the analysis of protein coding regions. States D. J. & Botstein D. (1991) *Proc. Natl. Acad. Sci. USA* **88**; 5518–5522.

733. Isolation of sequences that span the Fragile-X and identification of a Fragile-X related CpG island. Heitz D. *et al.* (1991) *Science* **251**; 1236–1239.

734. R loop mapping of the 18S and 28S sequences in the long and short repeating units of *Drosophila melanogaster* rDNA. White R. L. & Hogness D. S. (1977) *Cell* **10**; 177–192.

735. Structural organisation of a 17 kb segment of the α2 collagen gene: evaluation by R-loop mapping. Schafer P. *et al.* (1980) *Nucleic Acids Res.* **8**; 2241–2249.

736. Translation maps of polyoma virus-specific RNA: analysis by two-dimensional nuclease S1 gel mapping. Favoloro J. *et al.* (1980) *Methods Enzymol.* **65**; 718–749.

737. Exon trapping. A genetic screen to identify candidate transcribed sequences in cloned mammalian genomic DNA. Duyk G. M. *et al.* (1990) *Proc. Natl. Acad. Sci. USA* **87**; 8995–8999.

738. Exon amplification. A strategy to isolate mammalian genes based on RNA splicing. Buckler A. J. *et al.* (1991) *Proc. Natl. Acad. Sci. USA* **88**; 1005–1009.

739. Individual-specific 'fingerprints' of human DNA. Jeffreys A. J. *et al.* (1985) *Nature* **316**; 76–79.

740. DNA 'fingerprints' and segregation analysis of multiple markers in human pedigrees. Jeffreys J. A. *et al.* (1986) *Am. J. Hum. Genet.* **39**; 11–24.

741. Characterisation of a panel of highly variable minisatellites cloned from human DNA. Wong Z. *et al.* (1987) *Ann. Hum. Genet.* **51**; 269–288.

742. The development of methods for the analysis of DNA extracted from forensic samples. Hopkins B. *et al.* (1989) *Technique* **1**; 96–102.

743. Forensic applications of DNA 'fingerprints'. Gill P. *et al.* (1985) *Nature* **318**; 577–579.

744. The analysis of hypervariable DNA profiles: problems associated with the objective determination of a probability of a match. Gill P. *et al.* (1990) *Hum. Genet.* **85**; 79–89.

745. The efficiency of multilocus DNA probes for individualisation and establishment of family relationships, determined from extensive casework. Jeffreys A. J. *et al.* (1991) *Am. J. Hum. Genet.* **48**; 824–840.

746. Identification of the skeletal remains of a murder victim by DNA analysis. Hagleberg E. *et al.* (1991) *Nature* **352**; 427–429.

747. Positive identification of an immigration test-case using human DNA fingerprints. Jeffreys A. J. *et al.* (1985) *Nature* **317**; 818–819.

748. Cross-contamination of human esophageal squamous carcinoma cell lines detected by DNA fingerprint analysis. Van Helden P. D. *et al.* (1988) *Cancer Res.* **48**; 5660–5662.

749. Use of hypervariable minisatellite DNA probes (33.15) for evaluating engraftment two or more years after bone marrow transplantation for aplastic anaemia. Weitzer J. N. *et al.* (1988) *Br. J. Haematol.* **70**; 91–99.

750. The use of locus-specific minisatellite probes to check engraftment following allogenic bone marrow transplantation for severe combined immunodeficiency disease. Katz F. E. *et al.* (1990) *Bone Marrow Transplant.* **5**; 199–204.

751. Assessment of clonality in gastrointestinal cancer by DNA fingerprinting. Fey M. F. *et al.* (1988) *J. Clin. Invest.* **82**; 1532–1537.

752. Differences in DNA fingerprints between remission and relapse in childhood acute lymphoblastic leukemia. Pakkala S. *et al.* (1988) *Leuk. Res.* **9**; 757–762.

753. Fingerprinting cell lines: use of human hypervariable DNA probes to characterise mammalian cell cultures. Thacker J. *et al.* (1988) *Som. Cell Mol. Genet.* **14**; 519–525.

754. Demographic study of a wild house sparrow population by DNA fingerprinting. Wetton J. H. *et al.* (1987) *Nature* **327**; 147–149.

755. DNA fingerprints applied to paternity analysis in apples. Nybom H. (1990) *Theoret. Appl. Genet.* **79**; 763–768.

756. Ancient DNA: extraction, characterisation, molecular cloning, and enzymatic amplification. Paabo S. (1989) *Proc. Natl. Acad. Sci. USA* **86**; 1939–1943.

757. Molecular cloning of Ancient Egyptian mummy DNA. Paabo S. (1985) *Nature* **314**; 644–645.

758. DNA sequences from the quagga, and extinct member of the horse family. Higuchi R. *et al.* (1984) *Nature* **312**; 282–284.

759. Mitochondrial DNA sequences from a 7000 year old brain. Paabo S. *et al.* (1988) *Nucleic Acids Res.* **16**; 9775–9787.

760. Chloroplast DNA sequence from a Miocene Magnolia species. Golenberg E. M. *et al.* (1990) *Nature* **344**; 656–658.

761. DNA typing from single hairs. Higuchi R. *et al.* (1988) *Nature* **332**; 543–546.

762. Mitochondrial DNA sequences in single hairs from a South African population. Vigilant L. *et al.* (1989) *Proc. Natl. Acad. Sci. USA* **86**; 9350–9354.

763. Mitochondrial DNA and human evolution. Cann R. L. *et al.* (1987) *Nature* **325**; 31–36.

764. The problem of our common mitochondrial mother. Kruger J. & Vogel F. (1991) *Hum. Genet.* **82**; 308–312.

765. Branching pattern in the evolutionary tree for human mitochondrial DNA. Rienzo A. D. & Wilson A. C. (1991) *Proc. Natl. Acad. Sci. USA* **88**; 1597–1601.

766. Zones of sharp genetic change in Europe are also linguistic boundaries. Barbujani G. & Sokal R. R. (1990) *Proc. Natl. Acad. Sci. USA* **87**; 1816–1819.

767. The phylogeny of hominoid primates, as indicated by DNA–DNA hybridisation. Sibley C. G. & Ahlquist J. E. (1984) *J. Mol. Evol.* **20**; 2–15.

768. Primate evolution at the DNA level and a clasification of hominoids. Goodman M. *et al.* (1990) *J. Mol. Evol.* **30**; 260–266.

769. Resolution of the African hominoid trichotomy by use of a mitochondrial gene sequence. Ruvolo M. *et al.* (1991) *Proc. Natl. Acad. Sci. USA* **88**; 1570–1574.

770. DNA phylogeny of the extinct marsupial wolf. Thomas R. H. *et al.* (1989) *Nature* **340**; 465–467.

771. A method for saturation mutagenesis of cloned DNA fragments. Myers R. M. *et al.* (1985) *Science* **229**; 242–247.

772. Regulatory mutants of simian virus 40: constructed mutants with base substitutions at the origin of viral replication. Shortle D. & Nathans D. (1979) *J. Mol. Biol.* **131**; 801–817.

773. *In vitro* mutagenesis of a circular DNA molecule using synthetic restriction sites. Heffron F. *et al.* (1978) *Proc. Natl. Acad. Sci. USA* **75**; 6012–6016.

774. Gap misrepair mutagenesis: efficient site-directed introduction of transition, transversion and frameshift mutations *in vitro*. Shortle D. *et al.* (1982) *Proc. Natl. Acad. Sci. USA* **79**; 1588–1592.

775. Site-specific mutagenesis by error-directed DNA synthesis. Zakour R. A. & Loeb L. A. (1982) *Nature* **295**; 708–710.

776. Enzymatic techniques for the isolation of random single-base substitutions *in vitro* at high frequency. Abarzua P. & Marians K. J. (1984) *Proc. Natl. Acad. Sci. USA* **81**; 2030–2034.

777. Receptor and antibody epitopes in human growth hormone identified by homolog-scanning mutagenesis. Cunningham B. C. *et al.* (1989) *Science* **243**; 1330–1336.

778. *Saccharomyces cerevisiae* CYC1 mRNA 5′ end positioning: analysis by *in vitro* mutagenesis using synthetic duplexes with random mismatch base repair. McNeil J. B. & Smith M. (1985) *Mol. Cell. Biol.* **5**; 3545–3551.

779. A method for unidirectional deletion mutagenesis with application to nucleotide sequencing and preparation of gene fusions. Barrack G. J. & Wolf R. E. (1986) *Gene* **49**; 119–128.

780. A general method of *in vitro* preparation and specific mutagenesis of DNA fragments: study of protein and DNA interactions. Higuchi R. *et al.* (1988) *Nucleic Acids Res.* **16**; 7351–7367.

781. Oligonucleotide directed mutagenesis of the human β-globin gene: a general method for producing specific point mutations in cloned DNA. Wallace R. B. *et al.* (1981) *Nucleic Acids Res.* **9**; 3647–3656.

782. Oligonucleotide-directed mutagenesis using M13-derived vectors: an efficient and general procedure for production of point mutations in any fragment of DNA. Zoller M. J. & Smith M. (1982) *Nucleic Acids Res.* **10**; 6487–6500.

783. Oligonucleotide-directed mutagenesis as a general and powerful method for studies of protein functions. Dalbadie-McFarland G. *et al.* (1982) *Proc. Natl. Acad. Sci. USA* **79**; 6409–6413.

784. Improved oligonucleotide site-directed mutagenesis using M13 vectors. Carter P. *et al.* (1985) *Nucleic Acids Res.* **13**; 4431–4443.

785. The gapped duplex DNA approach to oligonucleotide-directed mutation construction. Kramer W. *et al.* (1984) *Nucleic Acids Res.* **12**; 9441–9456.

786. The rapid generation of oligonucleotide-directed mutations at high frequency using phosphothioate-modified DNA. Taylor J. W. *et al.* (1985) *Nucleic Acids Res.* **13**; 8765–8785.

787. Cassette mutagenesis: an efficient method for generation of multiple mutations at defined sites. Wells J. A. *et al.* (1985) *Gene* **34**; 315–323.

788. Combinatorial cassette mutagenesis as a probe of the informational content of protein sequences. Reidhaar-Olsen J. F. & Sauer R. T. (1988) *Science* **241**; 53–57.

789. 'Sticky-feet' mutagenesis and its application to swapping antibody domains. Clackson T. & Winter G. (1989) *Nucleic Acids Res.* **17**; 10163–10170.

790. Oligonucleotide-directed misincorporation mutagenesis on single-stranded DNA templates. Singh M. *et al.* (1986) *Prot. Eng.* **1**; 75–76.

791. Site-specific mutagenesis using asymmetric polymerase chain reaction and single mutant primer. Perrin S. & Gilliland G. (1990) *Nucleic Acids Res.* **18**; 7433–7438.

792. PCR-based site-directed mutagenesis using primers with mismatched 3′ ends. Nassal M. & Rieger A. (1990) *Nucleic Acids Res.* **18**; 3077–3078.

793. A method for random mutagenesis of a defined DNA segment using a modified polymerase chain reaction. Leung D. W. *et al.* (1989) *Bio Technique* **1**; 11–15.

794. A dwarf mutant of *Arabidopsis* generated by T-DNA insertion mutagenesis. Feldmann K. A. *et al.* (1989) *Science* **243**; 1351–1354.

795. Floral homeotic mutations produced by transposon-mutagenesis in *Antirrhinum majus*. Carpenter R. & Coen E. S. (1990) *Genes Devel.* **4**; 1483–1493.

796. Direct expression in *E. coli* of a DNA sequence coding for human growth hormone. Goeddel D. V. *et al.* (1979) *Nature* **281**; 544–548.

797. Expression of chemically synthesized genes for human insulin. Goeddel D. V. *et al.* (1979) *Proc. Natl. Acad. Sci. USA* **76**; 106–110.

798. Secretion of human interferons by yeast. Hitzeman R. A. *et al.* (1983) *Science* **219**; 620–625.

799. Secretion cloning vectors in *Escherichia coli*. Ghrayeb J. *et al.* (1984) *EMBO J.* **3**; 2437–2442.

800. High level expression and *in vivo* processing of chimeric ubiquitin fusion proteins in *Saccharomyces cerevisae*. Sabin E. A. *et al.* (1989) *BioTechnology* **7**; 705–709.

801. Efficient, low-cost protein factories: expression of human adenosine deaminase in baculovirus infected insect larvae. Medin J. A. *et al.* (1990) *Proc. Natl. Acad. Sci. USA* **87**; 2760–2764.

802. An erythromycin derivative produced by targeted gene disruption in *Saccharopolyspora erythraea*. Weber J. M. *et al.* (1991) *Science* **252**; 114–117.

803. High frequency T-DNA-mediated gene tagging in plants. Koncz C. *et al.* (1989) *Proc. Natl. Acad. Sci. USA* **86**; 8467–8471.

804. Anti-sense RNA inhibition of polygalacturonase gene expression in transgenic tomatoes. Smith C. J. S. *et al.* (1988) *Nature* **334**; 724–726.

805. Expression of a truncated tomato polygalacturonase gene inhibits expression of the endogenous gene in transgenic plants. Smith C. J. S. *et al.* (1990) *Mol. Gen. Genet.* **224**; 477–481.

806. Reversible inhibition of tomato fruit senescence by antisense RNA. Oeller P. W. *et al.* (1991) *Science* **254**; 437–439.

807. Down-regulation of two non-homologous endogenous tomato genes with a single chimaeric sense gene construct. Seymour G. B. *et al.* (1993) *Plant Mol. Biol.* **23**; 1–9.

808. Transgenic tomato plants expressing the tomato yellow leaf curl virus capsid protein are resistant to virus. Kunik T. *et al.* (1994) *BioTechnology* **12**; 500–504.

809. Delay of disease development in transgenic plants that express the tobacco mosaic virus coat protein gene. Powell Abel P. *et al.* (1986) *Science* **232**; 738–743.

810. Plants transformed with tobacco mosaic virus non-structural gene sequence are resistant to virus. Golemboski D. B. *et al.* (1990) *Proc. Natl. Acad. Sci. USA* **87**; 6311–6315.

811. Induction of a highly specific anti-viral state in transgenic plants: implications for regulation of gene expression and virus resistance. Lindbo J. A. *et al.* (1993) *Plant Cell* **5**; 1749–1769.

812. Transgenic plants protected from insect attack. Vaeck M. *et al.* (1987) *Nature* **328**; 33–37.

813. Insect resistant cotton plants. Perlak F. J. *et al.* (1990) *BioTechnology* **8**; 939–943.

814. Engineering herbicide resistance in plants by expression of a detoxifying enzyme. De Block M. *et al.* (1987) *EMBO J.* **6**; 2513–2518.

815. Relocating a gene for herbicide tolerance: a chloroplast gene is converted into a nuclear gene. Cheung A. Y. *et al.* (1988) *Proc. Natl. Acad. Sci. USA* **85**; 391–395.

816. Herbicide resistance in transgenic plants expressing a bacterial detoxification gene. Stalker D. M. *et al.* (1988) *Science* **242**; 419–423.

817. Elevated levels of superoxide dismutase protect transgenic plants against ozone damage. Van Camp W. *et al.* (1994) *BioTechnology* **12**; 165–168.

818. Genetically engineered alteration in the chilling sensitivity of plants. Murata N. *et al.* (1992) *Nature* **356**; 710–713.

819. Transgenic pea seeds expressing the alpha-amylase inhibitor of the common bean are resistant to bruchid beetles. Shade R. E. *et al.* (1994) *BioTechnology* **12**; 793–799.

820. Enhanced protection against fungal attack by constitutive co-expression of chitinase and glucanase genes in transgenic tobacco. Zhu Q. *et al.* (1994) *BioTechnology* **12**; 807–812.

821. Production of antibodies in transgenic plants. Hiatt A. *et al.* (1989) *Nature* **342**; 76–78.

822. Production of correctly processed human serum albumin in transgenic plants. Sijmons P. C. *et al.* (1990) *BioTechnology* **8**; 217–221.

823. Genetic manipulation: guidelines issued. Norman C. (1976) *Nature* **262**; 2–4.

824. Establishment in culture of pluripotent cells from mouse embryos. Evans M. J. & Kaufman M. H. (1981) *Nature* **292**; 154–156.

825. Integration and stable germline transmission of genes injected into mouse pronuclei. Gordon J. W. & Ruddle F. H. (1981) *Science* **214**; 1244–1246.

826. Transgenesis by means of blastocyte-derived embryonic stem cell lines. Gossler A. *et al.* (1986) *Proc. Natl. Acad. Sci. USA* **83**; 9065–9069.

827. Germline transmission of genes introduced into cultured pluripotent cells by retroviral vector. Robertson E. *et al.* (1986) *Nature* **223**; 445–448.

828. Production of chimaeric mice containing embryonic stem (ES) cells carrying a homeobox *Hox 1.1* allele mutated by homologous recombination. Zimmer A. & Gruss P. (1989) *Nature* **338**; 150–153.

829. Variations of cervical vertebrae after expression of a *Hox 1.1* transgene in mice. Kessel M. *et al.* (1990) *Cell* **61**; 301–308.

830. Transgenic mice with inducible dwarfism. Borelli E. *et al.* (1989) *Nature* **339**; 538–541.

831. Cystic fibrosis in the mouse by targeted insertional mutagenesis. Dorin J. R. *et al.* (1992) *Nature* **359**; 211–215.

832. An animal model for cystic fibrosis made by gene targetting. Snouwaert J. N. *et al.* (1992) *Science* **257**; 1083–1088.

833. Defective epithelial chloride transport in a gene targetted mouse model of cystic fibrosis. Clarke L. L. *et al.* (1992) *Science* **257**; 1125–1128.

834. Cystic fibrosis mouse with intestinal obstruction. Colledge W. H. *et al.* (1992) *Lancet* **340**; 680.

835. The molecular basis of muscular dystrophy in the *mdx* mouse – a point mutation. Sicinski P. *et al.* (1989) *Science* **244**; 1578–1580.

836. New *mdx* mutation disrupts expression of muscle and non-muscle isoforms of dystrophin. Cox G. A. *et al.* (1993) *Nature Genet.* **4**; 87–93.

837. Development of venous occlusions in mice transgenic for the plasminogen activator inhibitor–1 gene. Erickson L. A. *et al.* (1990) *Nature* **346**; 74–76.

838. A potential animal model for Lesch–Nyhan syndrome through introduction of HPRT mutations in mice. Kuehn M. R. *et al.* (1987) *Nature* **326**; 295–298.

839. Retinoblastoma in transgenic mice. Windle J. J. *et al.* (1990) *Nature* **343**; 665–669.

840. Production of transgenic rabbits, sheep and pigs by microinjection. Hammer R. E. *et al.* (1985) *Nature* **315**; 680–683.

841. Generation of transgenic dairy cattle using *in vitro* embryo production. Krimpenfort P. *et al.* (1991) *BioTechnology* **9**; 844–847.

842. Generation of transgenic dairy cattle from transgene-analyzed and sexed embryos produced *in vitro*. Hyttinen J-H. *et al.* (1994) *BioTechnology* **12**; 606–609.

843. Sexing and detection of gene constructs in microinjected bovine blastocytes using the polymerase chain reaction. Horvat S. *et al.* (1993) *Transgenic Res.* **2**; 134–140.

844. A milk protein gene directs the expression of human tissue plasminogen activator cDNA to the mammary gland in transgenic mice. Pittius C. W. *et al.* (1988) *Proc. Natl. Acad. Sci. USA* **85**; 5874–5878.

845. Expression of human anti-haemophilic Factor IX in the milk of transgenic sheep. Clark A. J. *et al.* (1989) *BioTechnology* **7**; 487–492.

846. High level expression of active human alpha–1-antitrypsin in the milk of transgenic sheep. Wright G. *et al.* (1991) *BioTechnology* **9**; 830–834.

847. Transgenic production of a variant human tissue-type plasminogen activator in goat milk: generation of transgenic goats and analysis of expression. Ebert K. M. *et al.* (1991) *BioTechnology* **9**; 835–838.

848. Production of functional human haemoglobin in transgenic swine. Swanson M. E. *et al.* (1992) *BioTechnology* **10**; 557–559.

849. Direct gene analysis of chorionic villi: a possible technique for first trimester diagnosis of haemoglobinopathies. Williamson R. *et al.* (1981) *Lancet* **ii**; 1127.

850. Prenatal sex determination by DNA amplification from maternal peripheral blood. Lo Y.-M. *et al.* (1989) *Lancet* **ii**; 1363–1365.

851. Antenatal diagnosis of sickle-cell anaemia by DNA anlysis of amniotic fluid cells. Kan Y. W. & Dozy A. M. (1978) *Lancet* **ii**; 910–912.

852. Genetic analysis of DNA from single human oocytes: a model for preimplantation diagnosis of cystic fibrosis. Coutelle C. *et al.* (1989) *Br. Med. J.* **299**; 22–24.

853. Diagnosis of beta-thalassaemia by DNA amplification in single blastomeres from mouse preimplantation embryos. Holding C. & Monk M. (1989) *Lancet* **ii**; 532–535.

854. Pregnancies from biopsied human preimplantation embryos sexed by Y-specific DNA amplification. Handyside A. *et al.* (1990) *Nature* **344**; 768–770.

855. Birth of a normal girl after *in vitro* fertilisation and preimplantation diagnostic testing for cystic fibrosis. Handyside A. H. *et al.* (1992) *New Engl. J. Med.* **327**; 905–909.

856. DNA micro-extraction from dried blood spots on filter paper blots: potential applications to newborn screening. McCabe E. R. B. *et al.* (1987) *Hum. Genet.* **75**; 213–216.

857. Direct method for prenatal diagnosis and carrier detection in Duchenne/Becker muscular dystrophy using the entire dystrophin cDNA. Darras B. T. *et al.* (1988) *Am. J. Med. Genet.* **29**; 713–726.

858. Deletion screening of the Duchenne muscular dystrophy locus via multiple DNA amplification. Chamberlain J. S. *et al.* (1988) *Nucleic Acids Res.* **16**; 11141–11156.

859. Prenatal diagnosis of Duchenne muscular dystrophy: prospective linkage analysis and retrospective dystrophin cDNA analysis. Ward P. A. *et al.* (1989) *Am. J. Hum. Genet.* **44**; 270–281.

860. Determination of carrier status in Duchenne and Becker muscular dystrophies by quantitative PCR and allele specific oligonucleotides. Prior T. W. *et al.* (1990) *Clin. Chem.* **36**; 2113–2117.

861. Carrier detection and prenatal diagnosis in Duchenne and Becker muscular dystrophy families using dinucleotide repeats. Clemens P. R. *et al.* (1991) *Am. J. Hum. Genet.* **49**; 951–960.

862. Direct detection of dystrophin gene rearrangements by analysis of dystrophin mRNA in peripheral blood lymphocytes. Roberts R. G. *et al.* (1991) *Am. J. Hum. Genet.* **49**; 298–310.

863. Presymptomatic diagnosis of delayed-onset disease with linked DNA markers: the experience with Huntington's disease. Brandt J. *et al.* (1989) *J. Am. Med. Assoc.* **261**; 3108–3114.

864. Should we test children for 'adult' genetic diseases? Harper P. S. and Clarke A. (1990) *Lancet* **335**; 1206–1207.

865. The cystic fibrosis gene: medical and social implications for heterozygote detection. Wilfond B. S. & Fost N. (1990) *J. Am. Med. Assoc.* **263**; 2777–2783.

866. A polymorphic marker linked to cystic fibrosis is located to chromosome 7. Knowlton R. G. *et al.* (1985) *Nature* **318**; 380–382.

867. Localisation of cystic fibrosis locus to human chromosome 7. Wainwright B. J. *et al.* (1985) *Nature* **318**; 384–385.

868. A closely linked genetic marker for cystic fibrosis. White R. *et al.* (1985) *Nature* **318**; 382–384.

869. Identification of the cystic fibrosis gene: cloning and characterisation of complementary DNA. Riordan J. R. *et al.* (1989) *Science* **245**; 1066–1073.

870. Identification of the cystic fibrosis gene: chromosome walking and jumping. Rommens J. M. *et al.* (1989) *Science* **245**; 1059–1065.

871. Identification of the cystic fibrosis gene: genetic analysis. Kerem B. S. *et al.* (1989) *Science* **245**; 1073–1080.

872. Isolation of a new DNA marker in linkage disequilibrium with cystic fibrosis, situated between J3.11 (D7S8) and IRP. Estivill X. *et al.* (1989) *Am. J. Hum. Genet.* **44**; 704–710.

873. Expression and characterisation of the cystic fibrosis transmembrane conductance regulator. Gregory R. J. *et al.* (1990) *Nature* **347**; 382–386.

874. Chromosomal region of the cystic fibrosis gene in yeast artificial chromosomes: a model for human genome mapping. Green E. D. & Olson M. V. (1990) *Science* **250**; 94–98

875. Demonstration the CFTR is a chloride channel by alteration of its anion selectivity. Anderson M. P. *et al.* (1991) *Science* **253**; 202–205.

876. Purification and functional reconstitution of the cystic fibrosis transmembrane conductance regulator (CFTR). Bear C. E. *et al.* (1992) *Cell* **6** 809–811.

877. Expression of cystic fibrosis transmembrane conductance regulator corrects defective chloride channel regulatioin in cystic fibrosis airway epithelia cells. Rich D. P. *et al.* (1990) *Nature* **347**; 358–363.

878. Correction of the cystic fibrosis defect *in vitro* by retrovirus-mediated gene transfer. Drumm M. L. *et al.* (1990) *Cell* **62**; 1227–1233.

879. Expression of the cystic fibrosis gene in non-epithelial invertebrate cells produces a regulated anion conductance. Kartner N. *et al.* (1991) *Cell* **64**; 681–691.

880. Successful targetting of the mouse cystic fibrosis transmembrane conductance regulator gene in embryonal stem cells. Dorin J. R. *et al.* (1992) *Transgenic Res.* 1; 101–105.

881. Correction of the ion transport defect in cystic fibrosis transgenic mice b gene therapy. Hyde S. C. *et al.* (1993) *Nature* **362**; 250–255.

882. Cloning of a representative genomic library of the human X-chromosom after sorting by flow cytometry. Davies K. E. *et al.* (1981) *Nature* **293**; 374–37

883. Linkage relationship of a cloned DNA sequence on the short arm of the X-chromosome to Duchenne muscular dystrophy. Murray J. M. *et al.* (1982) *Nature* **300**; 69–71.

884. Specific cloning of DNA fragments absent from the DNA of a male patient with an X-chromosome deletion. Kunkel L. M. *et al.* (1985) *Proc. Natl. Acad. Sci. USA* **82**; 4778–4782.

885. Cloning of the breakpoint of an X:21 translocation associated with Duchenne muscular dystrophy. Ray P. N. *et al.* (1985) *Nature* **318**; 672–675.

886. Complete cloning of the Duchenne muscular dystrophy (DMD) cDNA an preliminary genomic organisation of the DMD gene in normal and affected individuals. Koenig M. *et al.* (1987) *Cell* **50**; 509–517.

887. Complete cloning of the Duchenne muscular dystrophy (DMD) cDNA an preliminary organisation of the DMD gene in normal and affected individuals. Koenig M. *et al.* (1987) *Cell* **50**; 509–517.

888. A cDNA clone from the Duchenne/Becker muscular dystrophy gene. Burghes A. H. M. *et al.* (1987) *Nature* **328**; 434–437.

889. The complete sequence of dystrophin predicts a rod-shaped cytoskeleta protein. Koenig M. *et al.* (1988) *Cell* **53**; 219–228.

890. Association of dystrophin-related protein with dystrophin-associated proteins in *mdx* mouse muscle. Matsumura K. *et al.* (1992) *Nature* **360**; 588–591.

891. Primary structure of dystrophin-related protein. Tinsley J. M. *et al.* (1992 *Nature* **360**; 591–593.

92. The structural and functional diversity of dystrophin. Ahn A. H. & Kunkel .M. (1993) *Nature Genet.* **3**; 283–291.

93. Association of dystrophin and an integral membrane glycoprotein. Campbell K. P. & Kahl S.D. (1989) *Nature* **338**; 259.

94. Analysis of deletions in DNA from patients with Becker and Duchenne muscular dystrophy. Kunkel L. M. *et al.* (1986) *Nature* **322**; 73–77.

95. Topography of the Duchenne muscular dystrophy (DMD) gene: FIGE and cDNA analysis of 194 cases reveals 115 deletions and 13 duplications. Den Dunnen J. T. *et al.* (1989) *Am. J. Hum. Genet.* **45**; 835–847.

96. Over-expression of dystrophin in transgenic *mdx* mice eliminates dystrophic symptoms without toxicity. Cox G. A. *et al.* (1993) *Nature* **364**; 25–728.

97. Completion of mouse embryogenesis requires both the maternal and paternal genomes. McGrath J. & Solter D. (1984) *Cell* **37**; 179–183.

98. Parental legacy determines methylation and expression of an autosomal transgene: a molecular mechanism for parental imprinting. Swain J. L. (1987) *Cell* **50**; 719–727.

99. A polymorphic DNA marker genetically linked to Huntington's disease. Gusella J. F. *et al.* (1983) *Nature* **306**; 234–238.

00. Chromosome jumping from D4S10 (G8) toward the Huntington's disease gene. Richards J. E. *et al.* (1988) *Proc. Natl. Acad. Sci. USA* **85**; 6437–6441.

01. Predictive testing for Huntington's disease with use of a linked DNA marker. Meissen G. J. (1988) *N. Engl. J. Med.* **318**; 535–542.

02. Identification of an *Alu* retrotransposition event in close proximity to a strong candidate gene for Huntington's disease. Goldberg Y. P. *et al.* (1993) *Nature* **362**; 370–373.

03. A novel gene containing a trinucleotide repeat that is expanded and unstable on Huntington's disease chromosomes. The Huntington's Disease Collaborative Research Group (1993) *Cell* **72**; 971–983.

04. The LDL receptor gene: a mosaic of exons shared with different proteins. Sudhof T. C. *et al.* (1985) *Science* **228**; 815–822.

05. Cassette of eight exons shared by genes for LDL receptor and EGF precursor. Sudhof T. C. *et al.* (1985) *Nature* **228**; 893–895.

06. A receptor mediated pathway for cholesterol homeostasis. Brown M. S. & Goldstein J. L. (1986) *Science* **232**; 34–47.

07. cDNA sequence of human apolipoprotein(a) is homologous to plasminogen. McLean J. W. *et al.* (1987) *Nature* **330**; 132–137.

08. Plasma Lp(a) concentration is inversely correlated with the ratio of Kringle IV/Kringle V encoding domains in the apo(a) gene. Gavish D. *et al.* (1989) *J. Clin. Invest.* **84**; 2021–2027.

09. Apolipoprotein(a) size heterogeneity is related to variable number of repeat sequences in its mRNA. Koschinsky M. *et al.* (1990) *Biochemistry* **29**; 640–644.

10. Variation in the size of human apolipoprotein(a) is due to a hypervariable region in the gene. Lindahl G. *et al.* (1990) *Hum. Genet.* **84**; 563–567.

11. Lipoprotein(a) is an independent risk factor for myocardial infarction at a young age. Sandkamp M. *et al.* (1990) *Clin. Chem.* **36**; 20–23.

912. Molecular basis of apolipoprotein(a) isoform size heterogeneity as revealed by pulse-field gel electrophoresis. Lackner C. *et al.* (1991) *J. Clin. Invest.* **87**; 2077–2086.

913. Sequence polymorphisms in the apolipoprotein (a) gene: evidence for dissociation between apolipoprotein(a) size and plasma Lipoprotein(a) levels. Cohen J. C. *et al.* (1993) *J. Clin. Invest.* **91**; 1630–1636.

914. Atherogenesis in transgenic mice expressing human apolipoprotein(a). Lawn R. M. *et al.* (1992) *Nature* **360**; 670–672.

915. Efficient expression of retroviral vector-transduced human low density lipoprotein (LDL) receptor in LDL receptor-deficient rabbit fibrobalsts *in vitro*. Miyanohara A. *et al.* (1988) *Proc. Natl. Acad. Sci. USA* **85**; 6538–6542.

916. Correction of the genetic defect in hepatocytes from the Watanabe heritable hyperlipidemic rabbit. Wilson J. M. *et al.* (1988) *Proc. Natl. Acad. Sci. USA* **85**; 4421–4425.

917. Nucleic acid hybridisation in viral diagnosis. Landry M. L. (1990) *Clin. Biochem.* **23**; 267–277.

918. Detection of human cytomegalovirus in clinical specimens by DNA–DNA hybridisation. Spector S. A. *et al.* (1984) *J. Infect. Dis.* **150**; 121–126.

919. Use of molecular probes to detect human cytomegalovirus and human immunodeficiency virus. Spector S. A. *et al.* (1989) *Clin. Chem.* **35**; 1581–1587.

920. Comparison of dot-blot DNA hybridisation and immediate early nuclear antigen production in cell culture for the rapid detection of human cytomegalovirus in urine. Morris D. J. *et al.* (1987) *J. Virol. Methods* **18**; 47–55.

921. A rapid and quantitative solution hybridisation method for HBV DNA in serum. Jalava T. *et al.* (1992) *J. Virol. Methods* **36**; 171–180.

922. Detection of hepatitis B virus DNA in serum with nucleic acid probes labelled with ^{32}P, biotin, alkaline phosphatase or sulphone. Valentine-Thon E. *et al.* (1991) *Mol. Cell. Probes* **5**; 299–305.

923. Typing *Chlamydia trachomatis* by detection of restriction fragment length polymorphism in the gene encoding the major outer membrane protein. Frost E. H. *et al.* (1991) *J. Infect. Dis.* **163**; 1103–1107.

924. Rapid genotyping of the *Chlamydia trachomatis* major outer membrane protein by the polymerase chain reaction. Sayada C. *et al.* (1991) *FEMS Microbiol. Lett.* **83**; 73–78.

925. Polymerase chain reaction amplification and *in situ* hybridisation for the detection of human β-lymphotropic virus. Buchbinder A. *et al.* (1988) *J. Virol. Methods* **21**; 191–197.

926. Competitive polymerase chain reaction assay for quantitation of HIV–1 DNA and RNA. Stieger M. *et al.* (1991) *J. Virol. Methods* **34**; 149–160.

927. Absolute quantitation of viremia in human immunodeficiency virus infection by competitive reverse transcription and polymerase chain reaction. Menzo S. *et al.* (1992) *J. Clin. Microbiol.* **30**; 1752–1757.

928. Quantitation of plasma human immunodeficiency virus type 1 RNA by competitive polymerase chain reaction. Scadden D. T. *et al.* (1992) *J. Infect. Dis.* **165**; 1119–1123.

929. High levels of HIV–1 in plasma during all stages of infection determined by competitive PCR. Piatak M. *et al.* (1993) *Science* **259**; 1749–1754.

930. Quantification of hepatitis B virus DNA by competitive amplification and hybridisation on microtitre plates. Jalava T. *et al.* (1993) *BioTechniques* **15**; 1–5.

931. Detection of hepatitis B virus DNA in human serum samples: use of digoxigenin-labelled oligonucleotides as modified primers for the polymerase chain reaction. Escarceller M. *et al.* (1992) *Anal. Biochem.* **206**; 36–42.

932. Direct and sensitive detection of a pathogenic protozoan, *Taxoplasma gondii*, by polymerase chain reaction. Burg J. L. *et al.* (1989) *J. Clin. Microbiol.* **27**; 1787–1792.

933. Use of polymerase chain reaction and rapid infectivity testing to detect *Treponema pallidum* in amniotic fluid, fetal and neonatal sera, and cerebrospinal fluid. Grimpel E. *et al.* (1991) *J. Clin. Microbiol.* **29**; 1711–1718.

934. Detection of cytomegalovirus infected cells by flow cytometry and fluorescence in suspension hybridisation (FLASH) using DNA probes labelled with biotin by the polymerase chain reaction. Link H. *et al.* (1992) *J. Virol. Methods* **37**; 143–148.

935. Detection and identification of multiple mycobacterial pathogens by DNA amplification in a single tube. Wilton S. & Cousins D. (1992) *PCR Methods Applic.* **1**; 269–273.

936. DNA amplification on induced sputum samples for diagnosis of *Pneumocystis carinii* pneumonia. Wakefield A. E. *et al.* (1991) *Lancet* **337**; 1378–1379.

937. Species-specific detection of *Legionella pneumophila* in water by DNA amplification and hybridisation. Starnbach M. N. *et al.* (1989) *J. Clin. Microbiol.* **27**; 1257–1261.

938. Detection of *Aspergillus fumigatus* by polymerase chain reaction. Spreadbury C. *et al.* (1993) *J. Clin. Microbiol.* **31**; 615–621.

939. Detection of *Trypanosoma cruzi* by DNA amplification using the polymerase chain reaction. Moser D. R. *et al.* (1989) *J. Clin. Microbiol.* **27**; 1477–1482.

940. Detection of mycobacterial DNA in sarcoidosis and tuberculosis with polymerase chain reaction. Saboor S. A. *et al.* (1992) *Lancet* **339**; 1012–1015.

941. Detection of *Borrelia burgdorferi* infection in *Ixodes dammini* ticks with the polymerase chain reaction. Presing D. H. *et al.* (1990) *J. Clin. Microbiol.* **28**; 566–572.

942. Detection of *Borrelia burgdorferi* using the polymerase chain reaction. Malloy D. C. *et al.* (1990) *J. Clin. Microbiol.* **28**; 1089–1093.

943. Isolation of a T-lymphotropic retrovirus from a patient at risk for Acquired Immune Deficiency Syndrome (AIDS). Barre-Sinoussi F. *et al.* (1983) *Science* **220**; 868–871.

944. Evolutionary origin of human and simian immunodeficiency viruses. Gojobori T. *et al.* (1990) *Proc. Natl. Acad. Sci. USA* **87**; 4108–4111.

945. Genetic organisation of a chimpanzee lentivirus related to HIV–1. Huet *et al.* (1990) *Nature* **345**; 356–359.

946. Detection, isolation, and continuous production of cytopathic retrovirus (HTLV-III) from patients with AIDS and pre-AIDS. Popovic M. *et al.* (1984) *Science* **224**; 497–500.

947. Immunodeficiency virus *rev*-trans-activator modulates the expression of the viral regulatory genes. Malim M. H. *et al.* (1988) *Nature* **335**; 181–183.

948. Nef protein of HIV–1 is a transcriptional repressor of HIV-LTR. Ahmad & Venkatesan S. (1988) *Science* **241**; 1481–1485.

949. HIV–1 Tat protein increases transcriptional initiation and stabilises elongation. Laspia M. F. *et al.* (1989) *Cell* **59**; 283–292.

950. Regulation by HIV *Rev* depends upon recognition of splice sites. Chang D. D. & Sharp P. A. (1989) *Cell* **59**; 789–795.

951. The HIV–1 *rev*-trans-activator acts through a structured target sequence to activate nuclear export of unspliced viral mRNA. Malim M. H. *et al.* (1989) *Nature* **338**; 254–257.

952. Importance of the *nef* gene for maintenance of high virus loads and development of AIDS. Kestler H. W. *et al.* (1991) *Cell* **65**; 651–662.

953. Phosphorylation of 3′-azido–3′-deoxythymidine and selective interaction 5′-triphosphate with human immunodeficiency virus reverse transcriptase. Furman P. A. *et al.* (1985) *Proc. Natl. Acad. Sci. USA* **83**; 8333–8337.

954. The efficacy of azidothymidine (AZT) in the treatment of patients with AIDS and AIDS-related complex: a double-blind, placebo-controlled trial. Fisch M. A. *et al.* (1987) *N. Engl. J. Med.* **317**; 185–191.

955. Multiple mutations in HIV–1 reverse transcriptase confer high-level resistance to zidovudine (AZT). Larder B. A. & Kemp S. D. (1989) *Science* **24**; 1155–1158.

956. Resistance to ddI and sensitivity to AZT induced by a mutation in HIV–reverse transcriptase. St Clais M. H. *et al.* (1991) *Science* **253**; 1557–1559.

957. Inhibition of HIV–1 protease in infected T-lymphocytes by synthetic peptide analogues. Meek T. D. *et al.* (1990) *Nature* **343**; 90–91.

958. Rational design of peptide-based HIV proteinase inhibitors. Roberts A. *et al.* (1990) *Science* **248**; 358–361.

959. Blocking of HIV–1 infectivity by a soluble, secreted form of the CD4 antigen. Smith D. H. *et al.* (1987) *Science* **238**; 1704–1707.

960. Soluble CD4 molecules neutralize human immunodeficiency virus type Traunecker A. *et al.* (1988) *Nature* **331**; 84–86.

961. Prevention of HIV–1 IIB infection in chimpanzees by CD4 immunoadhesin. Ward R. H. R. *et al.* (1991) *Nature* **352**; 434–436.

962. Elimination of infectious HIV from human T-cell cultures by synergistic action of CD4-*Psuedomonas* exotoxin and reverse transcriptase inhibitors. Ashorn P. *et al.* (1990) *Proc. Natl. Acad. Sci. USA* **87**; 8889–8893.

963. Immunoconjugates containing ricin A chain and either human anti-gp41 CD4 kill H9 cells infected with different isolates of HIV, but do not inhibit norm T or B cell function. Tioll M. A. *et al.* (1990) *AIDS* **3**; 609–614.

964. Protection of chimpanzees from infection by HIV–1 after vaccination wi recombinant glycoprotein gp120 but not gp160. Berman P. W. *et al.* (1990) *Nature* **345**; 622–625.

65. Conserved sequence and structural elements on the HIV–1 principal neutralizing determinant. LaRosa G. J. *et al.* (1990) *Science* **249**; 932–935.

66. A general role for the *lux* auto-inducer in bacterial cell signalling. Bainton N. J. *et al.* (1992) *Gene* **116**; 87–91.

67. Nucelotide sequence of the *lux*B gene of *Vibrio harveyi* and the complete amino acid sequence of the beta subunit of bacterial luciferase. Johnston T. C. *et al.* (1986) *J. Biol. Chem.* **261**; 4805–4811.

68. Nucleotide sequence of the *lux*A and *lux*B genes of the bioluminescent marine bacterium *Vibrio fischeri*. Foran D. R. & Brown W. M. (1988) *Nucleic Acids Res.* **16**; 777.

69. The complete nucleotide sequence of the *lux* regulon of *Vibrio fisheri* and the *lux*ABN region of the photobacterium *Leiognathi* and the mechanism of control of bacterial bioluminescence. Baldwin T. O. *et al.* (1989) *J. Biolum. Chemilumin.* **4**; 326–341

70. PCR-based genetic engineering of the *Vibrio harveyi lux* operon and the *Escherichia coli trp* operon provides for biochemically functional native and fused gene products. Hill P. J. *et al.* (1991) *Mol. Gen. Genet.* **226**; 41–48.

71. Highly bioluminescent *Bacillus subtilis* obtained through high level expression of a *lux*AB fusion gene. Jacobs M. *et al.* (1991) *Mol. Gen. Genet.* **30**; 251–256.

72. Increased neurovirulence associated with a single nucleotide change in a non-coding region of the Sabian type–3 poliovirus genome. Evans D. M. A. *et al.* (1984) *Nature* **314**; 548–550.

73. A recombinant live oral cholera vaccine. Kaper J. B. *et al.* (1984) *BioTechnology* **1**; 345–349.

74. Correlation between amount of virus with altered nucleotide sequence and the monkey test for acceptability of oral poliovirus vaccine. Chumakov K. M. *et al.* (1991) *Proc. Natl. Acad. Sci. USA* **88**; 199–203.

75. RNA sequence variants in live poliovirus vaccine and their relationship to neurovirulence. Chumakov K. M. *et al.* (1992) *J. Virol.* **66**; 966–970.

76. New antiviral strategy using capsid–nuclease fusion proteins. Natsoulis G. & Boeke J. D. (1991) *Nature* **352**; 632–635.

77. Safety and immunological response to a recombinant vaccinia virus vaccine expressing HIV envelope glycoprotein. Cooney E. L. *et al.* (1991) *Lancet* **337**; 567–572.

78. The safety and immunogenicity of a human immunodeficiency virus type 1 (HIV–1) recombinant gp160 candidate vaccine in humans. Dolin R. *et al.* (1991) *Ann. Intern. Med.* **114**; 119–127.

79. Induction of *Plasmodium falciparum* transmission-blocking antibodies by recombinant vaccinia virus. Kaslow D. C. *et al.* (1991) *Science* **252**; 130–1313.

80. Recombinant Pfs25 protein of *Plasmodium falciparum* elicits malaria transmisson-blocking immunity in experimental animals. Barr P. J. *et al.* (1991) *J. Exp. Med.* **174**; 1203–1208.

81. Production, purification and immunogenicity of a malaria transmission-blocking vaccine candidate: TBV25H expressed in yeast and purified using Nickel-NTA agarose. Kaslow D. C. & Shiloach J. (1994) *BioTechnology* **12**; 494–499.

982. A prototype recombinant vaccine against respiratory syncytial virus and parainfluenza virus type 3. Du R-P. *et al.* (1994) *BioTechnology* **12**; 813–818.

983. Idiotype/granulocyte-macrophage colony-stimulating factor fusion protein as a vaccine for B-cell lymphoma. Tao M-H. & Levy R. (1993) *Nature* **362**; 755–758.

984. Primary structure of α-subunit precursor of *Torpedo californica* acetylcholine receptor deduced from cDNA sequence. Noda M. *et al.* (1982) *Nature* **299**; 793–797.

985. Cloning and sequence analysis of calf cDNA and human genomic DNA encoding the α-subunit precursor of muscle acetylcholine receptor. Noda M. *et al.* (1983) *Nature* **305**; 818–823.

986. Primary structure of *Electrophus electricus* sodium channel deduced from cDNA sequence. Noda M. *et al.* (1984) *Nature* **312**; 121–127.

987. Hydrophobic organisation of membrane proteins. Rees D. C. *et al.* (1989) *Science* **245**; 510–513.

988. A simple method for displaying the hydropathic character of a protein. Kyte J. & Doolittle R. F. (1982) *J. Mol. Biol.* **157**; 105–132.

989. The hydrophobic moment detects periodicity in protein hydrophobicity. Eisenberg D. *et al.* (1984) *Proc. Natl. Acad. Sci. USA* **82**; 140–144.

990. Model for the structure of bacteriorhodopsin based upon high-resolution electron cryo-microscopy. Henderson R. *et al.* (1990) *J. Mol. Biol.* **213**; 899–929.

991. β2-adrenergic receptors: delineation of domains involved in effector coupling and ligand binding specificity. Kobilka B. K. *et al.* (1988) *Science* **240**; 1310–1316.

992. Predicted secondary structure for the *src* homology–3 domain. Benner S. A. *et al.* (1993) *J. Mol. Biol.* **229**; 295–305.

993. Conservation analysis and structure prediction of the SH2 family of phosphotyrosine binding domains. Russell R. B. *et al.* (1992) *FEBS Lett.* **304**; 15–20.

994. Crystal structure of the catalytic subunit of cyclic adenosine monophosphate dependent protein kinase. Knighton D. R. *et al.* (1991) *Science* **253**; 407–414.

995. Functionally distinct G proteins selectively couple different receptors to PI hydrolysis in the same cell. Ashkenazi A. *et al.* (1989) *Cell* **56**; 487–493.

996. G protein diversity: distinct class of alpha subunits is present in vertebrates and invertebrates. Strathman M. & Simon M. I. (1990) *Proc. Natl. Acad. Sci. USA* **87**; 9133–9117.

997. Membrane-associated GTPases in bacteria. March P. E. (1992) *Mol. Microbiol.* **6**; 1253–1257.

998. Evidence for epidermal growth factor (EGF)-induced intermolecular autophosphorylation of the EGF receptors in living cells. Honegger A. M. *et al.* (1990) *Mol. Cell. Biol.* **10**; 4035–4044.

999. Autophosphorylation of the PDGF receptor in the kinase insert region regulates interactions with cell proteins. Kazlauskas A. & Cooper J. A. (1989) *Cell* **58**; 1121–1133.

1000. Requirement for integration of signals from two distinct phosphorylation pathways for activation of MAP kinase. Anderson N. G. et al. (1990) Nature 334; 651–653.

1001. Signal transduction and transcriptional regulation by glucocorticoid receptor–lexA fusion proteins. Godowski P. J. et al. (1988) Science 241; 812–816.

1002. Differences between the ribonucleic acids of transforming and non-transforming avian tumour viruses. Duesberg P. H. & Vogt P. K. (1970) Proc. Natl. Acad. Sci. USA 67; 1673–1680.

1003. Transformation of rat embryo fibroblasts by cloned polyoma virus DNA fragments containing only part of the early region. Hassell J. A. et al. (1980) Proc. Natl. Acad. Sci. USA 77; 3978–3982.

1004. Nucleotide sequences related to the transforming gene of avian sarcoma virus are present in the DNA of uninfected vertebrates. Spector D. H. et al. (1978) Proc. Natl. Acad. Sci. USA 75; 4102–4106.

1005. DNA sequences homologous to vertebrate oncogenes are conserved in Drosophila melanogaster. Shilo B.-Z. & Weinberg R. A. (1981) Proc. Natl. Acad. Sci. USA 78; 6789–6792.

1006. Many tumours induced by the mouse mammary tumour virus contain a provirus integrated in the same region of the host genome. Nusse R. & Varmus H. E. (1982) Cell 31; 99–109.

1007. Three different human tumour cell lines contain different oncogenes. Murray M. J. et al. (1981) Cell 25; 355–361.

1008. Mechanism of activation of a human oncogene. Tabin C. J. et al. (1982) Nature 300; 143–149.

1009. Human c-myc oncogene is located on the region of chromosome 8 that is translocated in Burkitt lymphoma cells. Dalla-Favera R. et al. (1982) Proc. Natl. Acad. Sci. USA 79; 7824–7827.

1010. Translocation of the c-myc gene into the immunoglobulin heavy chain locus in human Burkitt lymphoma and murine plasmacytoma cells. Taub R. et al. (1982) Proc. Natl. Acad. Sci. USA 79; 7838–7841.

1011. Amplification of endogenous c-myc related sequences in a human myeloid leukemia cell line. Collins S. & Groudine M. (1982) Nature 299; 679–681.

1012. Altered transcription of the c-abl oncogene in K–562 and other chronic myelogenous leukaemia cells. Collins S. J. et al. (1984) Science 225; 72–74.

1013. Philadelphia chromosome breakpoints are clustered within a limited region, bcr, on chromosome 2. Groffen J. et al. (1984) Cell 36; 93–99.

1014. The neu oncogene: an erb-B related gene encoding a 185000Mr tumour antigen. Schechter A. L. et al. (1984) Nature 312; 513–516.

1015. Close similarity of epidermal growth factor receptor and v-erbB oncogene protein sequences. Downward J. et al. (1984) Nature 307; 521–527.

1016. Human proto-oncogene c-jun encodes a DNA-binding protein with structural and functional properties of transcription factor AP-1. Bohmann D. et al. (1987) Science 238; 1386–1392.

1017. Gene amplification on chromosome band 11q13 and oestrogen receptor status in breast cancer. Frantl V. et al. (1990) Eur. J. Cancer 26; 423–429.

1018. Deletions of a DNA sequence in retinoblastoma and mesenchymal tumours: organisation of the sequence and its encoded protein. Friend S. H. al. (1986) Proc. Natl. Acad. Sci. USA 84; 9059–9063.

1019. Cell specific inhibitory and stimulatory effects of Fos and Jun on transcriptional activation by nuclear receptors. Shemshedini L. et al. (1991) EMBO J. 10; 3839–3849.

1020. Evidence for the recessive nature of cellular immortality. Pereira-Smi O. M. & Smith J. R. (1983) Science 221; 964–966.

1021. The p53 proto-oncogene can act as a supressor of transformation. Finlay C. A. et al. (1989) Cell 57; 1083–1093.

1022. Mutations in the p53 gene occur in diverse human tumour types. Nig J. M. et al. (1989) Nature 342; 705–708.

1023. Germline p53 mutations in a familial syndrome of breast cancer, sarcomas, and other neoplasms. Malkin D. et al. (1990) Science 250; 1233–1238.

1024. Association of multiple copies of the N-Myc oncogene with rapid progression of neuroblastoma. Seeger R. C. et al. (1985) N. Engl. J. Med. 31 1111–1116.

1025. Human breast cancer: correlation of relapse and survival with amplification of the HER2/Neu oncogene. Slamon D. J. et al. (1987) Science 235; 177–182.

1026. Proto-oncogene abnormalities in human breast cancer: correlation wi anatomical features and clinical course of disease. Cline M. J. et al. (1990) J. Clin. Oncol. 5; 999–1006.

1027. Detection of two alternative bcr/abl mRNA junctions and minimal residual disease in Philadelphia chromosome positive chronic myelogenous leukaemia by polymerase chain recation. Lee M.-S. et al. (1989) Blood 73; 2165–2170.

1028. Detection of Philadelphia chromosome-positive cells by polymerase chain reaction following bone marrow transplant for chronic myelogenous leukaemia. Roth M. S. et al. (1989) Blood 74; 882–885.

1029. Non-isotopic SSCP and competitive PCR for DNA quantification: p53 breast cancer cells. Yap E. P. H. & McGee J. O. (1992) Nucleic Acids Res. 2 145.

1030. Highly specific in vivo tumour targetting by monovalent and divalent forms of 741F8 sFV, an anti-c-erbB–2 single chain Fv. Adams G. P. et al. (1993) Cancer Res. 53; 4026–4034.

1031. In vivo tumour targetting of a recombinant single chain antigen-bindin protein. Colcher D. et al. (1990) J. Natl. Cancer Inst. 82; 1191–1197.

1032. Rapid tumour penetration of a single chain Fv and comparison with other immunoglobulin forms. Yokota T. et al. (1992) Cancer Res. 52; 3402–3408.

1033. Use of tumour infiltrating lymphocytes and interleukin–2 in the immunotherapy of patients with metastatic melanoma. A preliminary report. Rosenberg S. A. et al. (1988) N. Engl. J. Med. 319; 1676–1680.

1034. Synthesis of acylhydrazido-substituted cephems – design of *Cephalosporin vinca* alkaloid prodrugs as substrates for an antibody-targetted enzyme. Jungheim L. N. *et al.* (1992) *J. Org. Chem.* **57**; 2334–2340.

1035. Genetic construction, expression and characterisation of a single-chain anti-carcinoma antibody fused to beta-lactamase. Goshorn S. C. *et al.* (1993) *Cancer Res.* **53**; 2123–2127.

1036. Suppression of tumourigenicity in human colon carcinoma cells by introduction of normal chromosomes 5 or 18. Tanaka K. *et al.* (1991) *Nature* **349**; 340–342.

1037. Rational design of receptor-specific variants of human growth hormone. Cunningham B. C. & Wells J. A. (1991) *Proc. Natl. Acad. Sci. USA* **88**; 3407–3411.

1038. Systematic evolution of ligands by exponential enrichment: RNA ligands to bacteriophage T4 DNA polymerase. Tuerk C. & Gold L. (1990) *Science* **249**; 505–510.

1039. Structure-based design of nonpeptide inhibitors specific for the human immunodeficiency virus 1 protease. DesJarlais R. L. *et al.* (1990) *Proc. Natl. Acad. Sci. USA* **87**; 6644–6648.

1040. Expression of human β1- and β2-adrenergic receptors in *E. coli* as a new tool for ligand screening. Marullo S. *et al.* (1989) *BioTechnology* **7**; 923–927.

1041. *In vitro* selection of RNA molecules that bind specific ligands. Ellington A. D.& Szostak J. W. (1990) *Nature* **346**; 818–822.

1042. Selection *in vitro* of single-stranded DNA moleclues that fold into specific ligand binding structures. Ellington A. D. & Szostak J. W. (1992) *Nature* **355**; 850–852.

1043. Screening for receptor ligands using large libraries of peptide linked to the C-terminus of the *lac* repressor. Cull M. G. *et al.* (1992) *Proc. Natl. Acad. Sci. USA* **89**; 1865–1869.

1044. Systematic evolution of ligands by exponential enrichment. RNA ligands to bacteriophage T4 DNA polymerase. Tuerk C. & Gold L. (1990) *Science* **249**; 386–390.

1045. Antibody-selectable filamentous fd phage vectors – affinity purification of target genes. Parmley S. F. & Smith G. P. (1988) *Gene* **73**; 305–318.

1046. Searching for peptide ligands with an epitope library. Scott J. K. & Smith G. P. (1990) *Science* **249**; 386–390.

1047. Random peptide libraries. A source of specific protein binding molecules. Devlin J. J. *et al.* (1990) *Science* **249**; 404–406.

1048. Peptides on phage: a vast library of peptides for identifying ligands. Cwirla S. E. *et al.* (1990) *Proc. Natl. Acad. Sci. USA* **87**; 6378–6382.

1049. Encoded combinatorial chemistry. Brenner S. & Lerner R. A. (1992) *Proc. Natl. Acad. Sci. USA* **89**; 5381–5383.

1050. Generation and screening of an oligonucleotide-encoded synthetic peptide library. Needels M. C. *et al.* (1993) *Proc. Natl. Acad. Sci. USA* **90**; 10700–10704.

1051. β-turns in proteins. Chou P. Y. & Fasman G. D. (1977) *J. Mol. Biol.* **115**; 135–175.

1052. Structure of proteins: packing of α-helices and pleated sheets. Chothia C. *et al.* (1977) *Proc. Natl. Acad. Sci. USA* **74**; 4130–4134.

1053. Prediction of chain turns in globular proteins on a hydrophobic basis. Rose G. D. (1987) *Nature* **272**; 586–590.

1054. A systematic approach to the comparison of protein structures. Remmington S. J. & Matthews B. W. (1980) *J. Mol. Biol.* **140**; 77–99.

1055. Rapid and sensitive protein similarity searches. Lippman D. J. & Pearson W. R. (1985) *Science* **227**; 1435–1441.

1056. Designing substrate specificity by protein engineering of electrostatic interactions. Wells J. A. *et al.* (1987) *Proc. Natl. Acad. Sci. USA* **84**; 1219–1223.

1057. DIROM: an experimental design interactive system for directed mutagenesis and nucleic acid engineering. Makarova K. S. *et al.* (1992) *CABIOS* **8**; 425–431.

1058. Design, synthesis and characterisation of a 34-residue polypeptide that interacts with nucleic acids. Gutte B. *et al.* (1979) *Nature* **281**; 650–655.

1059. Redesigning enzyme structure by site-directed mutagenesis. Winter G. *et al.* (1982) *Nature* **299**; 756–758.

1060. Engineering enzyme specificity by 'substrate-assisted catalysis'. Carter P. & Wells J. A. (1987) *Science* **237**; 394–399.

1061. Recruitment of substrate-specificity properties from one enzyme into a related one by protein engineering. Wells J. A. *et al.* (1987) *Proc. Natl. Acad. Sci. USA* **84**; 5167–5171.

1062. Rational scanning mutagenesis of a protein kinase identifies functional regions involved in catalytic and substrate interactions. Gibbs C. S. & Zoller M. J. (1991) *J. Biol. Chem.* **266**; 8923–8931.

1063. Disulphide bond engineered into T4 lysozyme: stabilisation of the protein toward thermal inactivation. Perry L. J. & Wetzel R. (1984) *Science* **226**; 555–557.

1064. Disulphide bonds and thermal stability in T4 lysozyme. Wetzel R. *et al.* (1988) *Proc. Natl. Acad. Sci. USA* **85**; 401–405.

1065. A large increase in enzyme–substrate affinity by protein engineering. Wilkinson A. J. *et al.* (1984) *Nature* **307**; 187–188.

1066. Engineering an enzyme by site-directed mutagenesis to be resistant to chemical oxidation. Estell D. A. *et al.* (1985) *J. Biol. Chem.* **260**; 6518–6521.

1067. Engineering subtilisin and its substrates for efficient ligation of peptide bonds in aqueous solution. Abrahamsen L. *et al.* (1991) *Biochemistry* **30**; 4151–4159.

1068. The *Tetrahymena* ribozyme acts like an RNA restriction endonuclease. Zaug A. J. *et al.* (1986) *Nature* **324**; 429–433.

1069. External guide sequences for an RNA enzyme. Forster A. C. & Altman S. (1990) *Science* **249**; 783–786.

1070. Binding and cleavage of nucleic acids by the hairpin ribozyme. Chowrira B. M. & Burke J. M. (1991) *Biochemistry* **30**; 8518–8522.

1071. Intermolecular exon ligation of the ribosomal RNA precursor of *Tetrahymena* – oligonucleotides can function as 5′ exons. Inoue T. *et al.* (1985) *Cell* **43**; 431–437.

1072. One binding site determines sequence specificity of *Tetrahymena* pre-rRNA self-splicing, trans-splicing and RNA enzyme activity. Been M. D. & Cech T. R. (1986) *Cell* **47**; 2017–216.

1073. Simple RNA enzymes with new and highly specific endoribonuclease activity. Haselhoff J. & Gerlach W. L. (1988) *Nature* **334**; 585–591.

1074. The molecular basis of antibody formation: a paradox. Dreyer W. J. & Bennet J. D. (1965) *Proc. Natl. Acad. Sci. USA* **54**; 864–869.

1075. A complete immunoglobulin gene is created by somatic recombination. Brack C. *et al.* (1978) *Cell* **15**; 1–14.

1076. An immunoglobulin heavy-chain variable region gene is generated from three segments of DNA; VH, D and JH. Early P. *et al.* (1980) *Cell* **19**; 981–992.

1077. An immunoglobulin heavy-chain gene is formed by at least two recombinational events. Davis M. M. *et al.* (1980) *Nature* **283**; 733–739.

1078. The repertoire of human germline VH sequences reveals about 50 groups of VH segments with different hypervariable loops. Tomlinson I. M. *et al.* (1992) *J. Mol. Biol.* **227**; 776–798.

1079. A directory of human germ-line Vκ segments reveals a strong bias in their usage. Cox J. P. *et al.* (1994) *Eur. J. Immunol.* **24**; 827–836.

1080. Rearrangement of immunoglobulin γ1-chain gene and mechanisms for class switching. Kataoka T. *et al.* (1980) *Proc. Natl. Acad. Sci. USA* **77**; 919–923.

1081. Introduced T cell receptor variable region gene segments recombine in pre-B cells: evidence that B and T cells use a common recombinase. Yancopoulos G. D. *et al.* (1986) *Cell* **44**; 251–259.

1082. The V(D)J recombination activating gene, RAG–1. Schatz D. G. *et al.* (1989) *Cell* **59**; 1035–1048.

1083. RAG–1 and RAG–2, adjacent genes that synergistically activate V(D)J recombination. Oettinger M. A. *et al.* (1990) *Science* **248**; 1517–1523.

1084. Structure of the human histocompatibility antigen, HLA-A2. Bjorkman P. J. *et al.* (1987) *Nature* **329**; 506–512.

1085. The SCID defect affects the final step of the immunoglobulin VDJ recombinase mechanism. Malynn B. A. *et al.* (1983) *Cell* **54**; 453–460.

1086. Rearrangement of antigen receptor genes is defective in mice with severe combined immunodeficiency. Schuler W. *et al.* (1986) *Cell* **46**; 963–972.

1087. Continuous culture of fused cells secreting antibody of predetermined specificity. Kohler G. & Milstein C. (1975) *Nature* **256**; 495–498.

1088. The functional expression of antibody Fv fragments in *E. coli*: improved vectors and a generally applicable purification technique. Skerra A. *et al.* (1991) *BioTechnology* **9**; 273–278.

1089. Targetting recombinant antibodies to the surface of *E. coli*: fusion to a peptidoglycan associated lipoprotein. Fuchs P. *et al.* (1991) *BioTechnology* **9**; 1369–1372.

1090. Phage antibodies: filamentous phage displaying antibody variable regions. McCafferty J. *et al.* (1990) *Nature* **348**; 552–554.

1091. By-passing immunization: human antibodies from synthetic repertoires of germline VH-gene segments re-arranged *in vitro*. Hoogenboom H. R. & Winter G. (1992) *J. Mol. Biol.* **227**; 381–388.

1092. *In vitro* assembly of repertoires of antibody chains by renaturation on the surface of phage. Figini M. *et al.* (1994) *J. Mol. Biol.* **239**; 68–78.

1093. Multi-subunit proteins on the surface of filamentous phage: methodologies for displaying antibody (Fab) heavy and light chains. Hoogenboom H. R. *et al.* (1991) *Nucleic Acids Res.* **19**; 4133–4137.

1094. Guiding the selection of human antibodies from phage display repertoires to a single epitope of an antigen. Jespers L. S. *et al.* (1994) *BioTechnology* **12**; 899–903.

1095. Human antibody fragments specific for human blood group antigens from phage display library. Marks J. D. *et al.* (1993) *BioTechnology* **11**; 1145–1149.

1096. Antibody fragments from a 'single pot' phage display library as immunochemical reagents. Nissim *et al.* (1994) *EMBO J.* **13**; 692–698.

1097. Cloning immunoglobulin variable domains for expression by the polymerase chain reaction. Orlandi R. *et al.* (1989) *Proc. Natl. Acad. Sci. USA* **86**; 3833–3837.

1098. Replacing the complementarity-determining regions in a human antibody with those from a mouse. Jones P. T. *et al.* (1986) *Nature* **321**; 522–525.

1099. Antibody framework residues affecting the conformation of the hypervariable loops. Foote J. & Winter G. (1992) *J. Mol. Biol.* **224**; 487–499.

1100. A genetically engineered murine/human chimeric antibody retains specificity for human tumour-associated antigen. Sahagan B. G. *et al.* (1986) *J. Immunol.* **137**; 1066–1074.

1101. Protein engineering of antibody binding sites: recovery of specific activity in an anti-digoxin single-chain Fv analogue produced in *E. coli*. Huston J. S. *et al.* (1988) *Proc. Natl. Acad. Sci. USA* **85**; 5879–5883.

1102. Single chain antigen-binding proteins. Bird R. E. *et al.* (1988) *Science* **242**; 423–426.

1103. A comparison of strategies to stabilize immunoglobulin Fv-fragments. Glockshuber R. *et al.* (1990) *Biochemistry* **29**; 1362–1367.

1104. A rapid method of cloning functional variable region antibody genes in *E. coli* as single chain immunotoxins. Chaudhary V. K. *et al.* (1990) *Proc. Natl. Acad. Sci. USA* **87**; 1066–1070.

1105. Mammalian cell expression of single-chain Fv (sFV) antibody proteins and their C-terminal fusions with interleukin-2 and other effector domains. Dora H. *et al.* (1994) *BioTechnology* **12**; 890–897.

1106. A bifunctional fusion protein containing Fc-binding fragment of staphylococcal protein A amino terminal to anti-digoxin single-chain Fv. Tai M-S. *et al.* (1990) *Biochemistry* **29**; 8024–8030.

1107. A recombinant single chain antibody interleukin-2 fusion protein. Savage P. *et al.* (1993) *Br. J. Cancer* **67**; 304–310.

1108. Regionally restricted development defects resulting from targetted disruption of the mouse homeobox gene *hox–1.5*. Chisaka O. & Capecchi M. R (1991) *Nature* **350**; 473–479.

1109. Site-directed mutagenesis by gene targetting in mouse embryo-derived stem cells. Thomas K. R. & Capecchi M. R. (1987) *Cell* **51**; 503–512.

1110. Targetted mutation of the *hprt* gene in mouse embryonic stem cells. Doetschman T. *et al.* (1988) *Proc. Natl. Acad. Sci. USA* **85**; 8583–8587.

1111. Disruption of the proto-oncogene *int–2* in mouse embryonic stem cells: a general strategy for targeting mutations to non-selectable genes. Mansour S. L. *et al.* (1988) *Nature* **336**; 348–352.

1112. Germline transmission of a disrupted β2-microglobulin gene produced by homologous recombination in embryonic stem cells. Zijlstra M. *et al.* (1989) *Nature* **342**; 435–438.

1113. Partial correction of murine hereditary growth disorder by germline incorporation of a new gene. Hammer R. E. *et al.* (1984) *Nature* **311**; 65–67.

1114. Human gene transfer: characterisation of human tumour-infiltrating lymphocytes as vehicles for retroviral-mediated gene transfer in man. Kasid A. *et al.* (1990) *Proc. Natl. Acad. Sci. USA* **87**; 473–477.

1115. Long-term correction of Parkinson's disease by gene therapy. Jiao S. *et al.* (1993) *Nature* **362**; 450–453.

1116. Grafting fibroblasts genetically modified to produce L-dopa in a rat model of Parkinson disease. Wolff J. A. *et al.* (1989) *Proc. Natl. Acad. Sci. USA* **86**; 9011–9014.

1117. An alternative approach to somatic gene therapy. St Louis D. & Verma I. M. (1988) *Proc. Natl. Acad. Sci. USA* **85**; 3150–3154.

1118. Retroviral gene transfer into primary hepatocytes: implications for genetic therapy of liver-specific functions. Ledley F. D. *et al.* (1987) *Proc. Natl. Acad. Sci. USA* **84**; 5335–5339.

1119. Expression of human Factor IX in rabbit hepatocytes by retrovirus-mediated gene transfer: potential for gene therapy of haemophilia B. Armentano D. *et al.* (1990) *Proc. Natl. Acad. Sci. USA* **87**; 6141–6145.

1120. Normal myoblast injections provide genetic treatment for murine dystrophy. Law P. K. *et al.* (1988) *Muscle & Nerve* **11**; 525–533.

1121. Conversion of *mdx* myofibres from dystrophin-negative to -positive by injection of normal myoblasts. Partridge T. A. *et al.* (1989) *Nature* **337**; 176–179.

1122. Dystrophin production induced by myoblast transfer therapy in Duchenne muscular dystrophy. Law P. K. *et al.* (1990) *Lancet* **336**; 114–115.

1123. Site-specific expression *in vivo* by direct gene transfer into the arterial wall. Nabel E. G. *et al.* (1990) *Science* **249**; 1285–1288.

1124. Introduction of a normal human chromosome 11 into a Wilms' tumour cell line controls its tumourigenic expression. Weissman B. E. *et al.* (1987) *Science* **236**; 175–180.

1125. The role and fate of DNA ends for homologous recombination in embryonic stem cells. Hasty P. *et al.* (1992) *Mol. Cell. Biol.* **12**; 1464–1474.

1126. Re-examination of gene targetting frequency as a function of the extent of homology between the targetting vector and the target locus. Deng C. & Capecchi M. R. (1992) *Mol. Cell Biol.* **12**; 3365–3371.

1127. Germline transmission and expression of a human-derived yeast artificial chromosome. Jakobovits A. *et al.* (1993) *Nature* **362**; 255–258.

1128. A yeast artificial chromosome covering the tyrosinase gene confers copy number-dependent expression in transgenic mice. Schedl A. *et al.* (1993) *Nature* **362**; 258–261.

1129. Short modified anti-sense oligonucleotides directed against the *Ha-ras* point mutation induce selective cleavage of the messenger RNA and inhibit T24 cell proliferation. Saison-Behmoaras T. *et al.* (1991) *EMBO J.* **10**; 1111–1118.

1130. Inhibition of Rous sarcoma viral DNA translation by a specific oligodeoxyribonucleotide. Stephenson M. L. & Zamecnik P. C. (1978) *Proc. Natl. Acad. Sci. USA* **75**; 285–288.

1131. A *c-myc* anti-sense oligodeoxynucleotide inhibits entry into S phase but does not progress from G0 to G1. Heikkila R. *et al.* (1987) *Nature* **328**; 445–449.

1132. Oligodeoxynucleoside phosphoramidates and phosphorothioates as inhibitors of human immunodeficiency virus. Agarwal S. J. *et al.* (1988) *Proc. Natl. Acad. Sci. USA* **85**; 7079–7083.

1133. Anti-sense RNA inhibits expression of membrane skeleton protein 4.1 during embryonic development of *Xenopus*. Giebelhaus D. H. *et al.* (1988) *Cell* **53**; 601–615.

1134. Anti-sense-mediated inhibition of *BCL2* proto-oncogene expression and leukemic cell growth: comparisons of phosphodiester and phosphorothioate oligodeoxynucleotides. Reed J. C. *et al.* (1990) *Cancer Res.* **50**; 6565–6570.

1135. Construction and analysis of monomobile DNA junctions. Chen J. H. *et al.* (1988) *Biochemistry* **27**; 6032–6038.

1136. Oligonucleoside methylphosphonates as anti-sense reagents. Miller P. S. (1991) *BioTechnology* **9**; 358–362.

1137. Oligonucleotide clamps arrest DNA synthesis on a single-stranded DNA target. Giovannangeli C. *et al.* (1993) *Proc. Natl. Acad. Sci. USA* **90**; 10013–10017.

1138. Centre d'Etude du Polymorphisme Humain (CEPH): collaborative mapping of the human genome. Dausset J. *et al.* (1990) *Genomics* **6**; 575–577.

1139. Toward a physical map of the genome of the nematode *Caenorhabditis elegans*. Coulson A. *et al.* (1986) *Proc. Natl. Acad. Sci. USA* **83**; 7821–7825.

1140. The physical map of the whole *E. coli* chromosome: application of a new strategy for rapid analysis and sorting of a large genomic library. Kohara Y. *et al.* (1987) *Cell* **50**; 495–508.

1141. A map of the distal region of the long arm of human chromosome 21 constructed by radiation hybrid mapping and pulsed-field gel electrophoresis. Burmeister M. *et al.* (1991) *Genomics* **9**; 19–30.

1142. The complete DNA sequence of yeast chromosome III. Oliver S. *et al.* (1991) *Nature* **357**; 38–46.

1143. The European project for sequencing the yeast genome. Goffeau A. & Vassarotti A. (1991) *Res. Microbiol.*

1144. A genetic map of the mouse with 4006 simple sequence length polymorphisms. Dietrich W. F. *et al.* (1994) *Nature Genet.* **7**; 220–230.

1145. The CEPH consortium linkage map of human chromosome 1. Dracopoli N. C. *et al.* (1991) *Genomics* **9**; 686–700.

1146. A 1st generation physical map of the human genome. Cohen D. *et al.* (1993) *Nature* **366**; 698–701.

1147. The 1993-94 Genethon human genetic linkage map. Gyapay G. *et al.* (1994) *Nature Genet.* **7**; 246–267.

1148. Microdissection of and microcloning from the short arm of human chromosome 2. Bates G. P. *et al.* (1986) *Mol. Cell. Biol.* **6**; 3826–3830.

1149. Cloning defined regions of the human genome by microdissection of banded chromosomes and enzymatic amplification. Ludecke H.-J. *et al.* (1989) *Nature* **338**; 348–350.

1150. Physical mapping of complex genomes by cosmid multiplex analysis. Evans G. A. & Lewis K. A. (1989) *Proc. Natl. Acad. Sci. USA* **86**; 5030–5034.

1151. High-resolution mapping of human chromosome 11 by *in situ* hybridisation with cosmid clones. Lichter P. *et al.* (1990) *Science* **247**; 64–69.

1152. Mapping of human chromosome Xq28 by two-colour fluorescence *in situ* hybridisation of DNA sequences to interphase nuclei. Trask B. J. *et al.* (1991) *Am. J. Hum. Genet.* **48**; 1–15.

1153. Construction of linkage maps with DNA markers for human chromosomes. White R. *et al.* (1985) *Nature* **313**; 101–105.

1154. Physical mapping of human chromosomes by repetitive sequence fingerprinting. Stallings R. L. *et al.* (1990) *Proc. Natl. Acad. Sci. USA* **87**; 6218–6222.

1155. Variable number of tandem repeat (VNTR) markers for human gene mapping. Nakamura Y. *et al.* (1987) *Science* **235**; 1616–1622.

1156. Systematic screening of yeast artificial-chromosome libraries by use of the polymerase chain reaction. Green E. D. & Olson M. V. (1990) *Proc. Natl. Acad. Sci. USA* **87**; 1213–1217.

1157. Radiation hybrid mapping: a somatic cell genetics method for constructing high-resolution maps of mammalian chromosomes. Cox D. R. *et al.* (1990) *Science* **250**; 245–250.

1158. A common language for physical mapping of the human genome. Olson M. V. *et al.* (1989) *Science* **245**; 1434–1435.

1159. Genomic fingerprints produced by PCR with consensus transfer-RNA gene primers. Welsh J. & McClelland M. (1991) *Nucleic Acids Res.* **19**; 861–866.

1160. Optimising restriction fragment fingerprinting methods for ordering large genomic libraries. Branscomb E. *et al.* (1990) *Genomics* **8**; 315–320.

1161. Ordering of cosmid clones covering the Herpes simplex virus type 1 (HSV–1) genome: a test case for fingerprinting by hybridisation. Graig A. G. *et al.* (1990) *Nucleic Acids Res.* **18**; 2653–2660.

1162. *Alu* polymerase chain reaction: a method for rapid isolation of human-specific sequences from complex DNA sources. Nelson D. L. *et al.* (1989) *Proc. Natl. Acad. Sci. USA* **86**; 6686–6690.

1163. *Drosophila* genome project: one-hit coverage in yeast artificial chromosomes. Ajioka J. W. *et al.* (1991) *Chromosoma* **100**; 495–509.

1164. Multiplex DNA sequencing. Church G. M. & Kieffer-Higgins S. (1988) *Science* **240**; 185–188.

1165. Genomic sequencing. Church G. M. & Gilbert W. (1984) *Proc. Natl. Acad. Sci. USA* **81**; 1991–1995.

1166. A novel method for nucleic acid sequence determination. Bains W. & Smith G. C. (1988) *J. Theoret. Biol.* **135**; 303–307.

1167. Complementary DNA sequencing: expressed sequence tags and the human genome project. Adams M. D. *et al.* (1991) *Science* **252**; 1651–1656.

1168. Sequencing of megabase-plus DNA by hybridisation: theory of the method. Drmanac R. *et al.* (1989) *Genomics* **4**; 114–128.

1169. An oligonucleotide hybridisation approach to DNA sequencing. Khrapko K. R. *et al.* (1989) *FEBS Lett.* **256**; 118–122.

1170. Analysing and comparing nucleic acid sequences by hybridisation to arrays of oligonucleotides – evaluation using experimental models. Southern E. M. *et al.* (1992) *Genomics* **13**; 1008–1017.

1171. Finding errors in DNA sequences. Posfai J. & Roberts R. J. (1992) *Proc. Natl. Acad. Sci. USA* **89**; 4698–4702.

1172. Characterisation of a 1.0 Mb YAC contig spanning two chromosome breakpoints related to Menkes disease. Tumer Z. *et al.* (1992) *Hum. Mol. Genet.* **1**; 483–489.

BIBLIOGRAPHY 2:
REVIEW ARTICLES

1. Mendel – now down to the molecular level. Fincham J. R. S. (1990) *Nature* **343**; 208–209.

2. Structure and function of nucleic acids as cell constituents. Chargaff E. (1951) *Fed. Proc.* **10**; 654–659.

3. Supercoiled DNA. Bauer W. R. *et al.* (1980) *Scientific American* **243**; 118–133.

4. DNA topoisomerases. Wang J. C. (1982) *Scientific American* **247**; 84–95.

5. A glimpse at chromosomal order. Gasser S. M. & Laemmli U. K. (1987) *Trends Genet.* **3**; 16–22.

6. Chromatin assembly *in vitro* and *in vivo*. Dilworth S. M. & Dingwall C. (1988) *BioEssays* **9**; 44–49.

7. The nucleosome. Kornberg R. D. & Klug A. (1981) *Scientific American* **244**; 48–60.

8. Nucleosome structure. McGhee J.D. & Falsenfeld G. (1980) *Annu. Rev. Biochem.* **49**; 1115–1156.

9. The chromatin domain as a unit of gene regulation. Goldman M. A. (1988) *BioEssays* **9**; 50–53.

10. Structural changes in nucleosomes during transcription: strip, split or flip? Thoma F. (1991) *Trends Genet.* **7**; 175–177.

11. Scaffold associated regions: cis-acting determinants of chromatin loops and functional domains. Laemmli U. K. *et al.* (1992) *Curr. Opin. Gen. Devel.* **2**; 275–285.

12. Structural basis of DNA–protein recognition. Brennan R. G. & Matthews B. W. (1989) *Trends Biochem. Sci.* **14**; 286–290.

13. DNA recognition by proteins with the helix-turn-helix motif. Harrison S. C. & Aggarwal A. K. (1990) *Annu. Rev. Biochem.* **59**; 933–969.

14. Structural studies of protein–nucleic acid interaction: the sources of sequence-specific binding. Steitz T. A. (1990) *Q. Rev. Biophys.* **23**; 205–280.

15. Protein–nucleic acid interactions in transcription: a molecular analysis. Von Hippel P. *et al.* (1984) *Annu. Rev. Biochem.* **53**; 389–446.

16. Inside a living cell. Goodsell D. S. (1991) *Trends Biochem. Sci.* **16**; 203–206.

17. Structure and function of the bacterial chromosome. Schmid M. B. *Trends Biochem. Sci.* **13**; 131–135.

18. Structure and function of ribosomal RNA. Brimacombe R. & Stiege W. (1985) *Biochem. J.* **229**; 1–17.

19. Molecular dissection of a transfer RNA and the basis for its identity. Hou Y.-M. *et al.* (1989) *Trends Biochem. Sci.* **14**; 233–237.

20. Centromeres of budding and fission yeasts. Clarke L. (1990) *Trends Genet.* **6**; 151–154.

21. Centromeres of mammalian chromosomes. Willard H. F. (1990) *Trends Genet.* **6**; 410–415.

22. Telomeres. Blackburn E. M. (1991) *Trends Biochem. Sci.* **16**; 378–381.

23. The structure and function of telomeres. Blackburn E. H. (1991) *Nature* **350**; 569–573.

24. The human telomere. Moyzis R. K. (1991) *Scientific American* **265**; 34–41.

25. The major histocompatability complex and human evolution. Klein J. *et al.* (1990) *Trends Genet.* **6**; 7–11.

26. Globin pseudogenes. Little P. F. R. (1982) *Cell* **28**; 683–684.

27. Mitochondria spring surprises. Hall B. D. (1979) *Nature* **282**; 129–130.

28. Chloroplast gene organisation. Umesono K. & Ozeki H. (1987) *Trends in Genetics* **3**; 281–287.

29. Split genes. Chambon P. (1981) *Scientific American* **244**; 48–59.

30. The discovery of split genes: a scientific revolution. Witkowski J. A. (1988) *Trends Biochem. Sci.* **13**; 110–113.

31. Intron phylogeny: a new hypothesis. Cavalier-Smith T. (1991) *Trends Genet.* **7**; 145–148.

32. Exons – original building blocks of proteins. Pathy L. (1991) *BioEssays* **13**; 187–192.

33. Overlapping genes. Normark S. *et al.* (1983) *Annu. Rev. Genet.* **17**; 499–525.

34. Trisomy in man. Hassold T. J. & Jacobs P. A. (1984) *Annu. Rev. Genet.* **18**; 69–98.

35. Chromosome translocations and human cancer. Croce C. M. & Klein G. (1985) *Scientific American* **252**; 54–60.

36. A chromosomal basis of lymphoid malignancy in man. Boehm T. & Rabbitts T. H. (1989) *Eur. J. Biochem.* **185**; 1–17.

37. Updating the theory of mutation. Drake J. W. *et al.* (1983) *American Scientist* **71**; 621–630.

38. Spontaneous mutation. Drake J. W. (1991) *Annu. Rev. Genet.* **25**; 125–146.

39. Inducible repair of DNA. Howard-Flanders P. (1981) *Scientific American* **245**; 56–64.

40. DNA mismatch correction. Modrich P. (1987) *Annu. Rev. Biochem.* **56**; 435–466.

41. Mechanisms and biological effects of mismatch repair. Modrich P. (1991) *Annu. Rev. Genet.* **25**; 229–253.

42. DNA repair at the level of the gene. Bohr V. A. & Wassermann K. (1988) *Trends Biochem. Sci.* **13**; 429–433.

43. DNA repair enzymes. Sancar A. & Sancar G. B. (1988) *Annu. Rev. Biochem.* **57**; 29–67.

44. Mutagenesis and inducible responses to deoxyribonucleic acid damage in *E. coli*. Walker G. C. (1984) *Microbiol. Rev.* **48**; 60–93.

45. Genetics of DNA repair in bacteria. Ossanna N. *et al.* (1986) *Trends Genet.* **2**; 55–58.

46. Classes of aminoacyl-tRNA synthetases and the establishment of the genetic code. Schimmel P. (1991) *Trends Biochem. Sci.* **16**; 1–3.

47. The genetic code. Crick F. F. C. (1962) *Scientific American* **207**; 66–74.

48. The genetic code II. Nirenberg M. W. (1963) *Scientific American* **208**; 80–94.

49. The genetic code III. Crick F. H. C. (1966) *Scientific American* **215**; 55–62.

50. The transfer RNA identity problem: a search for rules. Saks M. E. *et al.* (1994) *Science* **236**; 191–197.

51. UGA: a split personality in the genetic code. Hatfield D. & Diamond A. (1993) *Trends Genet.* **9**; 69–70.

52. Recent evidence for evolution of the genetic code. Osawa S. *et al.* (1992) *Microbiol. Rev.* **56**; 229–264.

53. The interaction of *E. coli* RNA polymerase with promoters. Bujord H. (1980) *Trends Biochem. Sci.* **5**; 274–278.

54. The *E. coli* Rho protein. Bear D. G. & Peabody D. S. (1988) *Trends Biochem. Sci.* **13**; 343–347.

55. How RNA polymerase II terminates transcription in higher eukaryotes. Proudfoot N. J. (1989) *Trends Biochem. Sci.* **14**; 105–110.

56. RNA polymerase II: sub-unit structure and function. Woychik N. J. & Young R. A. (1990) *Trends Biochem. Sci.* **15**; 347–351.

57. Advances in RNA polymerase II transcription. Zawel L. & Reinberg D. (1992) *Curr. Opin. Cell Biol.* **4**; 488–495.

58. Initiation of transcription by RNA polymerase II; a multistep process. Zawel L. & Reinberg D. (1993) *Prog. Nucleic Acid Res. Mol. Biol.* **44**; 67–108.

59. Transcriptional activation; a complex puzzle with few easy pieces. Tjian R. & Maniatis T. (1994) *Cell* **77**; 5–8.

60. Capping of eukaryotic mRNAs. Shatkin A. J. (1976) *Cell* **9**; 645–653.

61. How the messenger got its tail. Wickens M. (1990) *Trends Biochem. Sci.* **15**; 277–281.

62. Transcription termination and 3′ processing; the end is in site. Birnstiel M. L. *et al.* (1985) *Cell* **41**; 349–359.

63. Autoregulated instability of tubulin mRNAs: a novel eukaryotic regulatory mechanism. Cleveland D. W. (1988) *Trends Biochem. Sci.* **13**; 339–343.

64. Splicing messenger RNA precursors: branch sites and lariat RNAs. Padgett R. A. *et al.* (1985) *Trends Biochem. Sci.* **10**; 154–157.

65. The role of the small nuclear ribonucleoprotein particles in pre-mRNA splicing. Maniatis T. & Reed R. (1987) *Nature* **325**; 673–678.

66. Spliceosomes and snurposomes. Gall J. (1991) *Science* **252**; 1499–1500.

67. RNA editing – a novel genetic phenomenon. Simpson L. (1990) *Science* **250**; 512–513.

68. RNA editing in trypanosomes: is there a message? Benne R. (1990) *Trends Genet.* **6**; 177–181.

69. RNA editing: what's in a mechanism? Hoffman M. (1991) *Science* **147**; 1462–1465.

70. Apolipoprotein B mRNA editing: a new tier for the control of gene expression. Hodges P. & Scott J. (1992) *Trends Biochem. Sci.* **17**; 77–81.

71. RNA as an enzyme. Cech T. R. (1986) *Scientific American* **255**; 64–75.

72. Catalytic RNA and the origin of genetic systems. Lamond A. I. & Gibson T. J. (1990) *Trends Genet.* **6**; 145–149.

73. Self-splicing of group I introns. Cech T. R. (1990) *Annu. Rev. Biochem.* **59**; 543–568.

74. Small catalytic RNAs. Symons R. H. (1992) *Annu. Rev. Biochem.* **61**; 641–671.

75. Processing of prokaryotic ribonucleic acid. Gegenheimer P. & Apirion D. (1981) *Microbiol. Rev.* **45**; 502–541.

76. Transfer RNA modification. Bjork G. R. *et al.* (1987) *Annu. Rev. Biochem.* **56**; 263–287.

77. DNA methylation and gene activity. Ceder H. (1988) *Cell* **53**; 3–4.

78. Eukaryotic DNA methylation and gene expression. Weissbach A. *et al.* (1989) *Curr. Topics Cell. Regul.* **30**; 1–21.

79. Alternative splicing in the control of gene expression. Smith C. W. J. *et al.* (1989) *Annu. Rev. Genet.* **23**; 527–577.

80. Regulation of inducible and tissue-specific gene expression. Maniatis T. *et al.* (1987) *Science* **236**; 1237–1245.

81. Gene regulation by steroid hormones. Beato M. (1989) *Cell* **56**; 335–344.

82. Promoter specificity and modulation of RNA polymerase II transcription. Saltzman A. G. & Weinmann R. (1989) *FASEB J.* **3**; 1723–1733.

83. Modularity in promoters and enhancers. Dynan W. S. (1989) *Cell* **58**; 1–4.

84. Transcription enhancer sequences: a novel regulatory element. Dynan W. & Tjian R. (1982) *Trends Biochem. Sci.* **7**; 124–127.

85. Enhancers and eukaryotic gene transcription. Serfling E. *et al.* (1985) *Trends Genet.* **1**; 224–230.

86. Enhancer function in viral and cellular gene regulation. Marriott S. J. & Brady J. N. (1989) *Biochim. Biophys. Acta* **989**; 97–100.

87. UARs and enhancers: common mechanism of transcriptional activation in yeast and mammals Guarente L. (1988) *Cell* **52**; 303–305.

88. Control of gene activity in higher eukaryotic cells by prokaryotic regulatory elements. Gossen M. *et al.* (1994) *Trends Biotech.* **12**; 58–62.

89. Genetic repressors. Ptashne M. & Gilbert W. (1970) *Scientific American* **222**; 36–44.

90. Attenuation in the control of expression of bacterial operons. Yanofsky C. (1981) *Nature* **289**; 751–758.

91. Attenuation in amino acid biosynthetic operons. Yanofsky C. & Kolter R. (1982) *Annu. Rev. Genet.* **16**; 113–134.

92. The lactose-operon controlling elements. A complex paradigm. Reznikoff W. S. (1992) *Mol. Microbiol.* **6**; 2419–2422.

93. Archaeal rRNA operons. Garrett R. A. *et al.* (1991) *Trends Biochem. Sci.* **16**; 22–26.

94. How eukaryotic transcriptional activators work. Ptashne M. (1988) *Nature* **335**; 683–689.

95. Eukaryotic transcriptional regulatory proteins. Johnson P. F. & McKnight S. L. (1989) *Annu. Rev. Biochem.* **58**; 797–839.

96. 'Zinc fingers': a novel protein motif for nucleic acid recognition. Klug A. & Rhodes D. (1987) *Trends Biochem. Sci.* **12**; 464–469.

97. Helix-turn-helix, zinc finger, and leucine zipper motifs for eukaryotic transcriptional regulatory proteins. Struhl K. (1989) *Trends Biochem. Sci.* **14**; 137–140.

98. Molecular zippers in gene regulation. McKnight S. L. (1991) *Scientific American* **264**; 54–64.

99. Steroid hormone receptors: biochemistry, genetics and molecular biology. Gehring U. (1987) *Trends Biochem. Sci.* **12**; 399–402.

100. The steroid and hormone receptor superfamily. Evans R. M. (1988) *Science* **240**; 889–895.

101. Involvement of RNA in the synthesis of proteins. Watson J. D. (1963) *Science* **140**; 17–26.

102. The ribosome. Lake J. A. (1981) *Scientific American* **245**; 56–69.

103. Three dimensional crystals of ribosomal particles. Yonath A. (1984) *Trends Biochem. Sci.* **9**; 227–230.

104. The initiation of protein synthesis. Hunt T. (1980) *Trends Biochem. Sci.* **5**; 178–181.

105. Cap recognition and the entry of mRNA into the protein synthesis initiation cycle. Rhoads R. E. (1988) *Trends Biochem. Sci.* **13**; 52–56.

106. The elongation step of protein biosynthesis. Clark B. (1980) *Trends Biochem. Sci.* **5**; 207–210.

107. Peptide chain termination. Caskey C. T. (1980) *Trends Biochem. Sci.* **5**; 234–237.

108. Identifying the peptidyl transferase centre. Garrett R. A. & Wooley P. (1982) *Trends Biochem. Sci.* **7**; 385–386.

109. Life at the end of the ribosome tunnel. Eisenberg H. (1987) *Trends Biochem. Sci.* **12**; 207–208.

110. Translational control in mammalian cells. Hershey J. W. B. (1991) *Annu. Rev. Biochem.* **60**; 717–755.

111. The proteins. Doolittle R. F. (1985) *Scientific American* **216**; 80–94

112. The protein folding problem. Richards F. M. (1991) *Scientific American* **264**; 34–41.

113. Recognition of nascent polypeptides for targetting and folding. Landry S. J. & Gierasch L. M. (1991) *Trends Biochem. Sci.* **16**; 159–163.

114. Multiple mechanisms of protein insertion into and across membranes. Wickner W. T. & Lodish H. F. (1985) *Science* **230**; 400–407.

115. How proteins enter the nucleus. Silver P. (1991) *Cell* **64**; 489–497.

116. 'Zip codes' direct intracellular protein tyrosine phosphatases to the correct cellular 'address'. Mauro L. J. & Dixon J. E. (1994) *Trends Biochem. Sci.* **19**; 151–155.

117. The cell cycle. Mazia D. (1974) *Scientific American* **230**; 54–64.

118. The cell cycle: spinning further afield. Marx J. (1991) *Science* **252**; 1490–1492.

119. Cell cycle advances. Strange C. (1992) *BioScience* **42**; 252–256.

120. The synaptonemal complex in genetic segregation. von Wettstein D. et al. (1984) *Annu. Rev. Genet.* **18**; 331–414.

121. Motor proteins of cytoplasmic microtubules. Vallee R. B. & Shpetner H. S. (1990) *Annu. Rev. Biochem.* **59**; 909–932.

122. The mitotic spindle. McIntosh J. R. & McDonald K. L. (1989) *Scientific American* **261**; 48–56.

123. Universal control mechanism regulating onset of M-phase. Nurse P. (1990) *Nature* **344**; 503–508.

124. Animal cell cycles and their control. Norbury C. & Nurse P. (1992) *Annu. Rev. Biochem.* **61**; 441–470.

125. Arresting developments in cell-cycle control. Pines J. (1994) *Trends Biochem. Sci.* **19**; 143–146.

126. What controls the cell cycle. Murray A. W. & Kirschner M. W. (1991) *Scientific American* **264**; 56–63.

127. Apoptosis and its role in human disease. Barr P. J. & Tomei L. D. (1994) *BioTechnology* **12**; 487–493.

128. Apoptosis and disease. Carson D. A. & Ribeiro J. M. (1993) *Lancet* **341**; 1251–1254.

129. Genetic recombination. Stahl F. (1987) *Scientific American* **256**; 53–63.

130. Genetic recombination in bacteria: a discovery account. Lederberg J. (1987) *Annu. Rev. Genet.* **21**; 23–46.

131. Homologous recombination in mammalian cells. Bollag R. J. et al. (1989) *Annu. Rev. Genet.* **23**; 199–225.

132. The conjugation system of F-like plasmids. Willets N. & Skurray R. (1980) *Annu. Rev. Genet.* **14**; 47–76.

133. Transduction in bacteria. Zinder N. D. (1958) *Scientific American* **199**; 38–43.

134. Plasmids that mobilise bacterial chromosomes. Holloway B. W. (1979) *Plasmid* **2**; 1–19.

135. Models of DNA transposition. Bukhari A. I. (1981) *Trends Biochem. Sci.* **6**; 56–60.

136. IS elements and transposons. Starlinger P. (1980) *Plasmid* **3**; 241–259.

137. Transposable genetic elements. Cohen S. N. & Shapiro J. A. (1980) *Scientific American* **242**; 36–45.

138. Transpositional recombination in prokaryotes. Grindley N. D. F. & Reed R. R. (1985) *Annu. Rev. Biochem.* **54**; 863–896.

139. Molecular mechanisms regulating *Drosophila* P element transposition. Rio D. C. (1990) *Annu. Rev. Genet.* **24**; 543–578.

140. Ty: a retroelement moving forward. Kingman A. J. & Kingman S. M. (1988) *Cell* **53**; 333–335.

140. Non-viral retrosposons: genes, pseudogenes, and transposable elements generated by the reverse flow of genetic information. Weiner A. M. et al. (1986) *Annu. Rev. Biochem.* **55**; 631–661.

141. Control of antigen gene expression in African trypanosomes. Pays E. & Steinert M. (1988) *Annu. Rev. Genet.* **22**; 107–126.

142. Eukaryotic transposable elements and genome evolution. Finnegan D. J. (1989) *Trends Genet* **5**; 103–107.

143. The synthesis of DNA. Kornberg A. (1968) *Scientific American* **219**; 64–78.

144. RNA-directed DNA synthesis. Temin H. (1972) *Scientific American* **226**; 24–33.

145. Discontinuous DNA replication. Ogawa T. & Okazaki T. (1980) *Annu. Rev. Biochem.* **49**; 421–457.

146. DNA replication. Kornberg A. (1984) *Trends Biochem. Sci.* **9**; 122–124.

147. Prokaryotic DNA replication systems. Nossal N.G. (1983) *Annu. Rev. Biochem.* **53**; 581–615.

148. Single-stranded DNA binding proteins required for DNA replication. Chase J. W. & Williams K. R. (1986) *Annu. Rev. Biochem.* **55**; 103–136.

149. The high fidelity of DNA duplication. Radman M. & Wagner R. (1988) *Scientific American* **259**; 24–30.

150. Eukaryotic chromosome replication requires both α and δ polymerases. Blow J. (1989) *Trends Genet.* **5**; 134–136.

151. Embryology gets down to the molecular level. Cherfas J. (1990) *Science* **250**; 33–35.

152. The bithorax complex: control of segmental identity. Peifer M. *et al.* (1987) *Genes Devel.* **1**; 891–898.

153. The egg came first, of course. Manseau L. J. & Schupbach T. (1989) *Trends Genet.* **5**; 400–405.

154. The molecular genetics of embryonic pattern formation in *Drosophila*. Ingham P. W. (1988) *Nature* **335**; 25–34.

155. Molecular and genetic organisation of the *Antennapedia* gene complex of *Drosophila melanogaster*. Kaufman T. C. *et al.* (1990) *Adv. Genet.* **27**; 309–362.

156. Determination of anteroposterior polarity in *Drosophila*. Nusslein-Volhard C. *et al.* (1987) *Science* **238**; 1675–1681.

157. The segmentation and homeotic gene network in early *Drosophila* development. Scott M. P. & Carroll S. B. (1987) *Cell* **51**; 689–698.

158. Role of gap genes in early *Drosophila* development. Gaul U. & Jackle J. (1990) *Adv. Genet.* **27**; 239–275.

159. Gap genes and gradients – the logic behind the gaps. Hulskamp M. & Tautz D. (1991) *BioEssays* **13**; 261–268.

160. What determines the specificity of action of *Drosophila* homeodomain proteins? Hayashi S. & Scott M. P. (1990) *Cell* **63**; 883–894.

161. Receptor tyrosine kinases: genetic evidence for their role in *Drosophila* and mouse development. Pawson T. & Bernstein A. (1990) *Trends Genet.* **6**; 350–356.

162. Smart genes. Beardsley T. (1991) *Scientific American* **265**; 72–81.

163. Pattern formation during animal development. Melton D. A. (1991) *Science* **252**; 234–241.

164. *HOX* and *HOM*: homologous gene clusters in insects and vertebrates. Akam M. (1989) *Cell* **57**; 347–349.

165. Murine developmental control genes. Kessel M. & Gruss P. (1990) *Science* **249**; 374–379.

166. Genetic control of cell lineage during nematode development. Sternberg P. W. & Horvitz H. R. (1984) *Annu. Rev. Genet.* **18**; 489–524.

167. The structure of the homeodomain and its functional implications. Gehring W. J. *et al.* (1990) *Trends Genet.* **6**; 323–329.

168. Deciphering the Hox code: clues to patterning brancial regions of the head. Hunt P. & Krumlauf R. (1991) *Cell* **66**; 1075–1078.

169. Imprinting a determined state into the chromatin of *Drosophila*. Paro R. (1990) *Trends Genet.* **6**; 416–421.

170. Mechanisms of heritable gene repression during development of *Drosophila*. Paro R. (1993) *Curr. Opin. Cell Biol.* **5**; 999–1005.

171. Silencers, silencing and heritable transcriptional states. Laurenson P. & Rine J. (1992) *Microbiol. Rev.* **56**; 543–560.

172. Building a bacterial virus. Wood W. B. & Edgar R. S. (1967) *Scientific American* **217**; 60–74.

173. The assembly of a virus. Butler P. J. G. & Klug A. (1978) *Scientific American* **239**; 52–59.

174. Lysogeny. Lwoff A. (1953) *Bacteriol. Rev.* **17**; 269–337.

175. How viruses insert their DNA into the DNA of the host cell. Campbell A. M. (1976) *Scientific American* **235**; 102–113.

176. The prions progress. Weissmann C. (1991) *Nature* **349**; 569–571.

177. A 'unified theory' of prion propagation. Weissmann C. (1991) *Nature* **352**; 679–683.

178. Restriction and modification enzymes and their recognition sequences. Roberts R. J. (1984) *Nucleic Acids Res.* **12**; 167–191.

179. Recognition sequences of restriction endonucleases and methylases – a review. Kessler C. *et al.* (1985) *Gene* **33**; 1–102.

180. Restriction and modification systems. Wilson G. G. & Murray N. E. (1991) *Annu. Rev. Genet.* **25**; 585–627.

181. Terminal deoxynucleotidyl transferase. Bollum F. J. (1974) In: *The Enzymes*, Vol 10. P. D. Boyer (ed).

182. T4 RNA ligase. Uhlenbeck O. C. & Gumport R. I. (1982) In: *The Enzymes*, Vol. 15. P. D. Boyer (ed).

183. Reverse transcriptase. Verma I. M. (1981) In: *The Enzymes*, Vol. 14. P. D. Boyer (ed).

184. Eukaryotic DNA polymerases. Wang T. S.-F. (1991) *Annu. Rev. Biochem.* **60**; 513–552.

185. RNA polymerase II. Yound R. A. (1991) *Annu. Rev. Biochem.* **60**; 689–715.

186. The isolation of messenger RNA. Taylor J. M. (1979) *Annu. Rev. Biochem.* **48**; 681.

187. Viewing molecules with scanning tunelling microscopy and atomic force microscopy. Edstrom R. D. *et al.* (1990) *FASEB J.* **4**; 3144–3151.

188. Pulsed field gel electrophoresis. A techniques for fractionating large DNA molecules. Anand R. (1986) *Trends Genet.* **2**; 278–283.

189. Pulsed field gel electrophoresis. Lai E. *et al.* (1989) *BioTechniques* **7**; 34–42.

190. Biotech's new nanotools. Edington S. M. (1994) *BioTechnology* **12**; 468–471.

191. Analytical strategies for the use of DNA probes. Matthews J. A. & Kricka L. J. (1988) *Anal. Biochem.* **169**; 1–25.

192. Nucleic acid hybridisation in plant virus diagnosis and characterisation. Hull R. & Al-Hakim A. (1988) *Trends Biotech.* **6**; 213–218.

193. Microbial diagnosis by nucleic acid sandwich hybridisation. Palva M. & Rankin M. (1985) *Clin. Lab. Med.* **5**; 475–490.

194. Detection of nucleic acid hybridisation using surface plasmon resonance. Schwarz T. *et al.* (1991) *Trends Biotech.* **9**; 339–340.

195. The reverse transcriptase. Verma I. M. (1977) *Biochem. Biophys. Acta* **473**; 1–38.

196. Reverse transcription. Varmus H. E. (1987) *Scientific American* **257**; 48–54.

197. Sequencing end-labelled DNA with base-specific cleavage. Maxam A. M & Gilbert W. (1980) *Meth. Enzymol.* **65**; 499–560.

198. DNA sequencing by hybridisation – a megasequencing method and a diagnostic tool? Mirzabekov A. D. (1994) *Trends Biotech.* **12**; 27–32.

199. The polymerase chain reaction. White T. J. *et al.* (1989) *Trends Genet.* **5**; 185–189.

200. Recent advances in the polymerase chain reaction. Erlich H. A. *et al.* (1991) *Science* **252**; 1643–1651.

201. Quantitative or semi-quantitative PCR: reality versus myth. Ferre F. (1992) *PCR Methods Applic.* **2**; 1–9.

202. Message amplification phenotyping (MAPPing) – principles, practice and potential. Larrick J. W. (1992) *Trends Biotech.* **10**; 146–152.

203. PCR analysis of DNA sequences in single cells: single sperm gene mapping and genetic disease diagnosis. Arnheim N. *et al.* (1990) *Genomics* **8**; 415–419.

204. The polymerase chain reaction: amplifying our options. Peter J. B. (1991) *Rev. Infect. Dis.* **13**; 166–171.

205. Replicatable RNA reporters. Kramer F. R. & Lizardi P. M. (1989) *Nature* **339**; 401–402.

206. Branched DNA signal amplification. Does bDNA represent post-PCR amplification technology? Urdea M. S. (1994) *BioTechnology* **12**; 926–928.

207. Use of T7 RNA polymerase to direct the expression of cloned genes. Studier F. & Moffat B. (1990) *Methods Enzymol.* **185**; 10–89.

208. Synthetic oligonucleotide probes deduced from amino acid sequence data. Theoretical and practical considerations. Lathe R. (1985) *J. Mol. Biol.* **183**; 1–12.

209. Use of nucleic acid probes in genetic tests. Dawson D. B. (1990) *Clin. Biochem.* **23**; 279–285.

210. Chemical DNA synthesis and recombinant DNA studies. Itakura K. & Riggs A. D. (1980) *Science* **209**; 1401–1405.

211. Gene synthesis machines. DNA chemistry and its uses. Caruthers M. H. (1985) *Science* **230**; 281–285.

212. Can DNA mimics improve on the real thing? Flam F. (1993) *Science* **262**; 1647–1649.

213. New photolabelling and crosslinking methods. Brunner J. (1993) *Annu. Rev. Biochem.* **62**; 483–514.

214. The filamentous phage (Ff) as vectors for recombinant DNA – a review. Zinder N. D. & Boeke J. D. (1982) *Gene* **19**; 1–10.

215. Cloning high molecular weight DNA fragments by the bacteriophage P1 system. Sternberg N. L. (1992) *Trends Genet.* **8**; 11–16.

216. Plasmids. Novick R. P. (1980) *Scientific American* **243**; 76–90.

217. The molecule of infectious drug resistance. Clowes R. C. (1973) *Scientific American* **228**; 18–27.

218. Plasmid vector pBR322 and its special-purpose derivatives. Balbas P. *et al.* (1986) *Gene* **50**; 3–40.

219. Cloning in yeast: an appropriate scale for mammalian genomes. Cooke H. (1987) *Trends Genet* **3**; 173–174.

220. Yeast artificial chromosomes – tools for mapping and analysis of complex genomes. Schlessinger D. (1990) *Trends Genet.* **6**; 248–258.

221. Yeast artificial chromosomes (YACs) and the analysis of complex genomes. Anand R. (1992) *Trends Biotech.* **10**; 35–40.

222. Isolating vector-insert junctions from yeast artificial chromosomes. Silverman G. A. (1993) *PCR Methods Applic.* **3**; 141–150.

223. Ordering up big MACs. Huxley C. *et al.* (1994) *BioTechnology* **12**; 586–590.

224. YACs, BACs, PACs, and MACs: artificial chromosomes as research tools. Monaco A. P. & Larin Z. (1994) *Trends Biotech.* **12**; 280–286.

225. Transfer and function of T-DNA genes from *Agrobacterium* Ti and Ri plasmids in plants. Zambryski P. *et al.* (1989) *Cell* **56**; 193–201.

226. T-DNA of the *Agrobacterium* Ti and Ri plasmids. Bevan M. W. & Chilton M. D. (1983) *Annu. Rev. Genet.* **16**; 375–384.

227. Form and function of retroviral vectors. Varmus H. E. (1982) *Science* **216**; 812–820.

228. Retroviral vectors for introduction of genes into mammalian cells. Eglitis M. A. & Anderson W. F. (1988) *BioTechniques* **6**; 608–614.

229. Gene expression using retroviral vectors. Tolstoshev P. & Anderson W. F. (1990) *Curr. Opin. Biotech.* **1**; 55–61.

230. Geminiviruses, genes and vectors. Davies J. W. & Stanley J. (1989) *Trends Genet.* **5**; 77–81.

231. Development of adenovirus vectors for the expression of heterologous genes. Berkner K. L. (1988) *BioTechniques* **6**; 619–629.

232. Mammalian cell expression. Gorman C. M. (1990) *Curr. Opin. Biotech.* **1**; 36–43.

233. Insect baculoviruses: powerful gene expression vectors. Miller L. K. (1989) *BioEssays* **11**; 91–95.

234. Baculovirus expression of alkaline phosphatase as a reporter gene for evaluation of production, glycosylation and secretion. Davis T. R. *et al.* (1992) *BioTechnology* **10**; 1148–1150.

235. Altering genotype and phenotype by DNA-mediated gene transfer. Pellicer A. *et al.* (1980) *Science* **209**; 414–1422.

236. Electroporation of eukaryotes and prokaryotes: a general approach to the introduction of macromolecules into cells. Shigekawa K. & Dower W. J. *BioTechniques* **6**; 742–751.

237. Bolistic transformation: microbes to mice. Johnston S. A. (1990) *Nature* **346**; 776–777.

238. Transformation of microbes, plants and animals by particle bombardment. Klein T. M. *et al.* (1992) *BioTechnology* **10**; 286–291.

239. Liposome mediated gene transfer. Mannino R. J. & Gould-Fogerite S. (1988) *BioTechniques* **6**; 682–690.

240. Retroviral-mediated gene transfer. McLachlin J. R. *et al.* (1990) *Prog. Nucleic Acids Res. Mol. Biol.* **38**; 91–135.

241. Jumping libraries and linking libraries: the next generation of molecular tools in mammalian genetics. Poustka A. & Lehrach H. (1986) *Trends Genet.* **2**; 174–179.

242. Construction and characterisation of band-specific DNA libraries. Ludecke H.-J. *et al.* (1990) *Hum. Genet.* **84**; 512–516.

243. PCR amplification techniques for chromosome walking. Rosenthal A. (1992) *Trends Biotech.* **10**; 44–48.

244. Chromosome microtechnology: microdissection and microcloning. Gruelich K. O. (1992) *Trends Biotech.* **10**; 48–53.

245. Positional cloning. Lets not call it reverse anymore. Collins F. S. (1992) *Nature Genet.* **1**; 3–6.

246. Fishing for complements. Finding genes by direct selection. Lovett M. (1994) *Trends Genet.* **10**; 352–357.

247. Patenting life. Barton J. H. (1991) *Scientific American* **264**; 40–46.

248. Chromosome mapping with DNA markers. White R. & Lolouel J.-M. (1988) *Scientific American* **258**; 40–48.

249. From linked marker to gene. Wicking C. & Williamson B. (1991) *Trends Genet.* **7**; 288–293.

250. The pluses of subtraction. Myers R. M. (1993) *Science* **259**; 942–946.

251. Fast tracks to disease genes. Aldhous P. (1994) *Science* **265**; 2008–2010.

252. Genome scanning methods. Brown P. O. (1994) *Curr. Opin. Genet. Devel.* **4**; 366–371.

253. Transposons as tools for the isolation of plant genes. Balcells L. *et al.* (1991) *Trends Biotech.* **9**; 31–36.

254. Using RAPD markers for crop improvement. Waugh R. & Powell W. (1992) *Trends Biotech.* **10**; 186–191.

255. Visualising gene expression with luciferase fusions. Schauer A. T. (1988) *Trends Biotech.* **6**; 23–27.

256. The history of genetic sequence databases. Smith T. F. (1990) *Genomics* **6**; 701–707.

257. Searching through sequence databases. Doolittle R. F. (1990) *Methods Enzymol.* **183**; 99–110.

258. CpG-rich islands and the function of DNA methylation. Bird A. P. (1986) *Nature* **321**; 209–213.

259. Splice junctions, branch point sites, and exons: sequence statistics, identification and application to genome projects. Senapathy P. *et al.* (1990) *Methods Enzymol.* **183**; 252–278.

260. DNA fingerprinting takes the witness stand. Marx J. L. (1988) *Science* **240**; 1616–1618.

261. DNA fingerprinting. Debenham P. G. (1991) *J. Pathol.* **164**; 101–106.

262. Minisatellite repeat coding as a digital approach to DNA typing. Jeffereys A. J. *et al.* (1991) *Nature* **354**; 204–209.

263. Probing identity: the changing face of DNA fingerprinting. Debenham P. G. (1992) *Trends Biotech.* **10**; 96–102.

264. Ancient DNA: still busy after death. Cherfas J. (1991) *Science* **253**; 1354–1356.

265. Ancient DNA and the polymerase chain reaction. The emerging field of molecular archeology. Paabo S. *et al.* (1989) *J. Biol. Chem.* **264**; 9709–9712.

266. Mitochondrial DNA. Grivell L. A. (1983) *Scientific American* **248**; 60–73.

267. On the molecular evolutionary clock. Zuckerland E. (1987) *J. Mol. Evolution* **26**; 34–46.

268. The future of DNA–DNA hybridisation studies. Diamond J. M. (1990) *J. Mol. Evolution* **30**; 196–201.

269. The geneology of some recently evolved vertebrate proteins. Doolittle R. F. (1985) *Trends Biochem. Sci.* **10**; 233–237.

270. In the beginning. Horgan J. (1991) *Scientific American* **264**; 116–125.

271. Origin of life – facing up to the physical setting. Pace N. R. (1991) *Cell* **65**; 531–533.

272. Strategies and applications of *in vitro* mutagenesis. Botstein D. & Shortle D. (1985) *Science* **229**; 1193–1201.

273. *In vitro* mutagenesis. Smith M. (1985) *Annu. Rev. Genet.* **19**; 423–462.

274. Towards engineering proteins by site-directed incorporation *in vivo* of non-natural amino acids. Ibba M. & Hennecke H. (1994) *BioTechnology* **12**; 678–682.

275. Expanding the genetic lexicon: incorporating non-standard amino acids into proteins by ribosome-based synthesis. Benner S. A. (1994) *Trends Biotech.* **12**; 158–163.

276. Genetic engineering of bacteria from managed and natural habitats. Lindow S. E. *et al.* (1989) *Science* **244**; 1300–1307.

277. Heterologous gene expression in filamentous fungi. Saunders G. *et al.* (1989) *Trends Biotech.* **7**; 283–287.

278. Designing microbial systems for gene expression in the field. de Lorenzo V. (1994) *Trends Biotech.* **12**; 365–371.

279. The cellulosome – a treasure trove for biotechnology. Bayer E. A. *et al.* (1994) *Trends Biotech.* **12**; 379–386.

280. Drug synthesis by genetically engineered microorganisms. Hutchinson C. H. (1994) *BioTechnology* **12**; 375–380.

281. Techniques in plant molecular biology. Walden R. & Schell J. (1990) *Eur. J. Biochem.* **192**; 563–576.

282. Plant biotechnology. Fraley R. & Schell J. (1991) *Curr. Opin. Biotech.* **2**; 145–210.

283. Saying it with genes: molecular flower breeding. Lindow S. E. *et al.* (1989) *Trends Biotech.* **7**; 148–153.

284. Insect control with genetically engineered crops. Brunke K. J. & Meeusen R. L. (1991) *Trends Biotech.* **9**; 197–200.

285. Engineering herbicide tolerance into plants. Oxtoby E. & Hughes M. A. (1990) *Trends Biotech.* **8**; 61–65.

286. Approaches to nonconventional control of plant virus diseases. Hull R. & Davies J. W. (1992) *Crit. Rev. Plant Sci.* **11**; 17–33.

287. Genetically engineering plants for crop improvement. Gasser C. S. & Fraley R. T. (1989) *Science* **244**; 1293–1299.

288. Foreign gene expression in transgenic cereals. McElroy D. & Brettell R. I. S. (1994) *Trends Biotech.* **12**; 62–68.

289. Genetic engineering for plant oils: potential and limitations. Battey J. F. *et al.* (1989) *Trends Biotech.* **7**; 122–125.

290. Modifying oilseed crops for non-edible products. Murphy D. J. (1992) *Trends Biotech.* **10**; 84–87.

291. Is pursuing improved canola an unctuous aim? Kidd G. (1993) *BioTechnology* **11**; 448–449.

292. Ethically sensitive genes and the consumer. Aldridge S. (1994) *Trends Biotech.* **12**; 71–72.

293. Do transgenic crops pose ecological risks? Fox J. L. (1994) *BioTechnology* **12**; 127–128.

294. Transgenic animals. Jaenisch R. (1989) *Science* **240**; 1468–1474.

295. Genetic engineering of livestock. Pursel V. G. et al. (1989) *Science* **244**; 1281–1288.

296. Improvement of wool production through genetic engineering. Rodgers G. E. (1990) *Trends Biotech.* **8**; 6–11.

297. Transgenic fish. Chen T. T. & Powers D. A. (1990) *Trends Biotech.* **8**; 209–215.

298. Transgenic fish: safe to eat? Berkowitz D. B. & Kryspin-Sorensen I. (1994) *BioTechnology* **12**; 247–252.

299. Genetic ablation in transgenic mice. Berstein A. & Breitman M. (1989) *Mol. Biol. Med.* **6**; 523–530.

300. Dissecting multistep tumourigenesis in transgenic mice. Hanahan D. (1988) *Annu. Rev. Genet.* **22**; 479–521.

301. Cystic fibrosis: a welcome animal model. Collins F. S. & Wilson J. M. (1992) *Nature* **358**; 708–709.

302. Transgenic animals – production of foreign proteins in milk. Henninghausen L. *et al.* (1990) *Curr. Opin. Biotech.* **1**; 74–78.

303. Transgenic animals as bioproducers of therapeutic proteins. Janne J. *et al.* (1992) *Ann. Med.* **24**; 273–280.

304. Disease diagnosis by recombinant DNA methods. Caskey C. T. (1987) *Science* **236**; 1223–1229.

305. Chorionic villus sampling. Goldberg J.D. & Golbus M. S. (1988) *Adv. Hum. Genet.* **17**; 1–25.

306. Diagnosis of genetic disorders at the DNA level. Antonarakis S. E. (1989) *N. Engl. J. Med.* **320**; 153–163.

307. Diagnosis by gene amplification. Kazazian H. H. (1989) *J. Lab. Clin. Med.* **114**; 95–96.

308. DNA diagnosis: molecular techniques and automation. Landegren U. *et al.* (1988) *Science* **242**; 229–237.

309. Cystic fibrosis. Molecular biology and therapeutic implications. Collins F. S. (1992) *Science* **256**; 774–779.

310. Dystrophin-related muscular dystrophies. Witkowski J. A. (1989) *J. Child Neurol.* **4**; 251–271.

311. Genomic imprinting and genetic disorders in man. Reik W. (1989) *Trends Genet.* **5**; 331–336.

312. The end is in sight for Huntington disease? Pritchard C. *et al.* (1991) *Am. J. Hum Genet.* **49**; 1–6.

313. Huntington's disease. The end of the beginning. Little P. (1993) *Nature* **362**; 408–409.

314. The pathogenesis of atherosclerosis. Ross R. (1986) *N. Engl. J. Med.* **314**; 488–494.

315. Apolipoproteins and lipoproteins of human plasma: significance in health and disease. Kostner G. M. (1983) *Adv. Lipid Res.* **20**; 1–43.

316. How LDL receptors influence cholesterol and atherosclerosis. Brown M. S. & Goldstein J. L. (1984) *Scientific American* **251**; 58–66.

317. Receptor-mediated endocytosis: concepts emerging from the LDL receptor system. Goldstein J. L. *et al.* (1985) *Annu. Rev. Cell Biol.* **1**; 1–39.

318. High density lipoprotein: the clinical implications of recent studies. Gordon D. J. & Rifkind B. M. (1989) *N. Engl. J. Med.* **321**; 1311–1316.

319. Genetics of plasma lipoprotein (a) concentrations. Boerwinkle E. (1992) *Curr. Opin. Lipidol.* **3**; 128–136.

320. Lipoproteins and heart disease. Breslow J. (1994) *BioTechnology* **12**; 365–370.

321. DNA probes for the diagnosis of intestinal infection. Char S. & Farthing M. J. G. (1991) *Gut* **32**; 1–3.

322. AIDS pathogenesis. Montagnier L. (1991) *The Biochemist* **13**; 3–7.

323. Karposi's sarcoma and the acquired immunodeficiency syndrome. Volberding P. A. (1989) *Med. Clin. N. Am.* **70**; 665–675.

324. Animal models for AIDS. Letvin N. L. (1990) *Immunol. Today* **11**; 322–326.

325. The molecular biology of human immunodeficiency virus type 1 infections. Greene W. C. (1991) *N. Engl. J. Med.* **324**; 308–317.

326. Functions of the auxillary gene products of the human immunodeficioency virus type 1. Cullen B. R. & Greene W. C. (1990) *Virology* **178**; 1–5.

327. Regulation of HIV gene expression by RNA–protein interaction. Rosen C. A. (1991) *Trends Genet.* **7**; 9–14.

328. The role of CD4 in normal immunity and HIV infection. Lifson J. D. & Engleman E. G. (1989) *Immunol. Rev.* **109**; 93–117.

329. Apoptosis in AIDS. Gougeon M-L. & Montagnier L. (1994) *Science* **260**; 1269–1270.

330. Molecular targets for AIDS therapy. Mitsuya H. *et al.* (1990) *Science* **249**; 1533–1544.

331. The 3-D structure of HIV-1 proteinase and the design of antiviral agents for the treatment of AIDS. Blundell T. L. *et al.* (1990) *Trends Genet.* **15**; 425–430.

332. Molecular biology of bacterial bioluminsecence Meighen E. A. (1991) *Microbiol. Rev.* **55**; 123–142.

333. *In vivo* bioluminescence: new potentials for microbiology. Stewart G. S. A. B. (1990) *Lett. Appl. Microbiol.* **10**; 1–8.

334. *In vivo* bioluminescence – a cellular reporter for research and industry. Jassim S. A. A. *et al.* (1990) *J. Biolumin. Chemilumin.* **5**; 115–122.

335. *lux* genes and the applications of bacterial luminescence. Stewart G. S. A. B. & Williams P. (1992) *J. Gen. Microbiol.* **138**; 1289–1300.

336. From Jenner to genes – the new vaccines. Brown F. (1990) *Lancet* **335**; 587–590.

337. Approaches to HIV vaccine design. Bolognesi D. P. (1990) *Trends Biotech.* **8**; 40–45.

338. Design and trials of AIDS vaccines. Sonigo P. *et al.* (1990) *Immunol. Today* **11**; 465–471.

339. Acetylcholine receptor structure, function and evolution. Stroud R. M. & Finer-Moore J. (1985) *Annu. Rev. Cell Biol.* **1**; 317–351.

340. The bacterial photosynthetic reaction centre as a model for membrane proteins. Rees D. C. *et al.* (1989) *Annu. Rev. Biochem.* **58**; 607–633.

341. The structure of proteins in biological membranes. Unwin N. & Henderson R. (1984) *Scientific American* **250**; 78–95.

342. Identifying nonpolar transbilayer helices in amino acid sequences of membrane proteins. Engelman D. M. *et al.* (1986) *Annu. Rev. Biophys. Chem.* **15**; 321–353.

343. The prediction of transmembrane protein sequences and their confromations: an evaluation. Fasman G. D. & Gilbert W. A. (1990) *Trends Biochem. Sci.* **15**; 89–95.

344. G proteins: transducers of receptor-generated signals. Gilman A. G. (1987) *Annu. Rev. Biochem.* **56**; 615–649.

345. The GTPase superfamily: a molecular switch for diverse cell functions. Bourne H. R. *et al.* (1990) *Nature* **348**; 125–132.

346. The GTPase superfamily: conserved structure and molecular mechanism. Bourne H. R. *et al.* (1991) *Nature* **349**; 117–127.

347. Signal transduction by the platelet-derived growth factor receptor. Williams L. T. (1989) *Science* **243**; 1564–1570.

348. Signal transduction by receptors with tyrosine kinase activity. Ullrich A. & Sclessinger J. (1990) *Cell* **61**; 203–212.

349. cAMP-dependent protein kinase: framework for a diverse family of regulatory enzymes. Taylor S. S. *et al.* (1990) *Annu. Rev. Biochem.* **59**; 971–1005.

350. Protein-tyrosine phosphatases. Walton K. M. & Dixon J. E. (1993) *Annu. Rev. Biochem.* **62**; 101–120.

351. From epinephrine to cyclic AMP. Levitzki A. (1988) *Science* **241**; 800–806.

352. The cycling of calcium as an intracellular messenger. Rasmussen H. (1989) *Scientific American* **261**; 66–73.

353. Inositol phosphates and cell signalling. Berridge M. J. & Irvine R. F. (1990) *Nature* **341**; 197–205.

354. Pseudosubstrates turn off protein kinases. Hardie G. (1988) *Nature* **335**; 592–593.

355. The molecular genetics of cancer. Bishop J. M. (1987) *Science* **235**; 305–311.

356. Oncogenes. Bishop J. M. (1982) *Scientific American* **246**; 80–92.

357. Viral oncogenes. Bishop J. M. (1985) *Cell* **42**; 23–38.

358. Oncogene amplification in neoplastic development and progression of human cancers. Schwab M. (1990) *Crit. Rev. Oncogen.* **2**; 35–51.

359. Oncogenes and signal transduction. Cantley L. C. *et al.* (1991) *Cell* **64**; 281–302.

360. Growth factors in development, transformation and tumourigenesis. Cross M. & Dexter T. M. (1991) *Cell* **64**; 271–280.

361. Oncogenic conversion by regulatory changes in transcription factors. Lewin B. (1991) *Cell* **64**; 303–312.

362. Cooperation among oncogenes. Hunter T. (1991) *Cell* **64**; 249–270.

363. Oncogenes, anti-oncogenes and the molecular basis of multistep carcinogenesis. Weinberg R. A. (1989) *Cancer Res.* **49**; 3713–3721.

364. Molecular themes in oncogenesis. Bishop J. M. (1991) *Cell* **64**; 235–248.

365. Human tumour supressor genes. Stanbridge E. J. (1990) *Annu. Rev. Genet.* **24**; 615–657.

366. The p53 tumour supressor gene. Levine A. J. *et al.* (1991) *Nature* **351**; 453–456.

367. *Ras* oncogenes in human breast cancer: a review. Boss J. L. (1989) *Cancer Res.* **49**; 4862–4869.

368. Exploiting altered glycosylation patterns in cancer: progress and challenges in diagnosis and therapy. Taylor-Papadimitriou J. & Epenetos A. (1994) *Trends Biotech.* **12**; 227–236.

369. Activation of prodrugs by antibody–enzyme conjugates – a new approach to cancer therapy. Senter P. D. (1990) *FASEB J.* **4**; 188–193.

370. Construction and chemotherapeutic potential of carboxypeptidase-A monoclonal antibody conjugates. Esswein A. *et al.* (1991) *Adv. Enzyme Reg* **31**; 3–12.

371. Taxol: the chemistry and structure–activity relationships of a novel anticancer agent. Kingston D. G. I. (1994) *Trends Biotech.* **12**; 222–227.

372. Tumour targeting: activation of prodrugs by enzyme–monoclonal antibody conjugates. Huennekens F. M. (1994) *Trends Biotech.* **12**; 234–239

373. Antibody structure-based design of pharmacological agents. Dougall W. C. *et al.* (1994) *Trends Biotech.* **12**; 372–379.

374. Receptor screening and the search for new pharmaceuticals. Hodgsor J. (1992) *BioTechnology* **10**; 973–980.

375. Combinatorial chemistry – applications of light-directed chemical synthesis. Jacobs J. W. & Fodor S. P. A. (1994) *Trends Biotech.* **12**; 19–26.

376. *In vitro* selection from protein and peptide libraries. Clackson T. & Wel J. A. (1994) *Trends Biotech.* **12**; 173–184.

377. Epitope discovery using peptide libraries displayed on phage. Cortese R. *et al.* (1994) *Trends Biotech.* **12**; 262–267.

378. Anti-sense oligonucleotides as anti-viral agents. Agrawal S. (1992) *Trends Biotech.* **10**; 152–158.

379. Secretion of peptides from *E. coli*: a production system for the pharmaceutical industry. Josephson S. & Bishop R. (1988) *Trends Biotech.* **6** 218–224.

380. Principles that determine the structure of proteins. Chothia C. (1984) *Annu. Rev. Biochem.* **53**; 537–572.

381. New molecular biology methods for protein engineering. Zoller M. J. (1991) *Curr. Opin. Struct. Biol.* **1**; 605–610.

382. Genetic approaches to protein structure and function: point mutations as modifiers of protein function. Medynski D. (1992) *BioTechnology* **10**; 1002–1006.

383. Structural and genetic analysis of protein stability. Matthews B. W. (1993) *Annu. Rev. Biochem.* **62**; 139–160.

384. Engineering surface loops of proteins – a preferred strategy for obtaining new enzyme function. El Hawrani A. S. *et al.* (1994) *Trends Biotech.* **12**; 207–214.

385. Tinkering with enzymes. What are we learning? Knowles J. R. (1987) *Science* **236**; 1252–1258.

386. The regulation of variable region gene assembly. Yancopoulos G. & Alt F. (1986) *Annu. Rev. Immunol.* **4**; 339–368.

387. Regulation of immunoglobulin gene rearrangement and expression. Tausig M. J. *et al.* (1989) *Immunol. Today* **10**; 143–146.

388. Immunoglobulin class switching: molecular and cellular analysis. Esser C. & Radbruch A. (1990) *Annu. Rev. Immunol.* **8**; 717–735.

389. The structure of CD4 and CD8 genes. Littman D. R. (1987) *Annu. Rev. Immunol.* **5**; 561–584.

390. The immunoglobulin superfamily – domains for cell surface recognition. Williams A. F. & Barclay A. N. (1988) *Annu. Rev. Immunol.* **6**; 381–406.

391. Monoclonal antibodies. Milstein C. (1980) *Scientific American* **243**; 66–74.

392. Antibodies from *Escherichia coli*. Pluckthun A. (1990) *Nature* **347**; 497–498.

393. Making antibodies by phage display technology. Winter G. *et al.* (1994) *Annu. Rev. Immunol.* **12**; 433–455.

394. Man-made antibodies. Winter G. & Milstein C. (1991) *Nature* **349**; 293–299.

395. Humanized antibodies. Winter G. & Harris W. J. (1993) *Trends Pharmacol. Sci.* **14**; 139–143.

396. The genetic engineering of monoclonal antibodies. Owens R. J. & Young R. J. (1994) *J. Immunol. Methods* **168**; 149–165.

397. Protein engineering of single-chain Fv analogues and fusion proteins. Huston J. S. *et al.* (1991) *Methods Enzymol.* **203**; 46–88.

398. Medical application of single chain antibodies. Huston J. S. *et al.* (1993) *Int. Rev. Immunol.* **10**; 195–217.

399. Monoclonal and engineered antibodies for human parenteral clinical use: regulatory considerations. Manohar V. & Hoffman T. (1992) *Trends Biotech.* **10**; 305–309.

400. Extending the chemistry of enzymes and abzymes. Hilvert D. (1991) *Trends Biotech.* **9**; 11–15.

401. Antibody design: beyond the natural limits. Ress A. R. *et al.* (1994) *Trends Biotech.* **12;** 199–206.

402. Catalytic antibodies: perusing combinatorial libraries. Posner B. *et al.* (1994) *Trends Biochem. Sci.* **19**; 145–150.

403. Catalytic antibodies: a critical assessment. Tawfik D. S. *et al.* (1994) *Mol. Biotechnol.* **1**; 87–103.

404. Gene transfer into primates and prospects for gene therapy in humans. Cornetta K. *et al.* (1988) *Prog. Nucleic Acids Res.* **36**; 311–322.

405. Altering the genome by homologous recombination. Capecchi M. R. (1989) *Science* **244**; 1288–1292.

406. Specific regulation of gene expression by anti-sense, sense and anti-gene nucleic acids. Helene C. & Toulme J.-J. (1990) *Biochim. Biophys. Acta* **1049**; 99–125.

407. Oligonucleotide therapeutics. Cohen J. S. (1992) *Trends Biotech.* **10**; 87–91.

408. Ribozymes and their medical implications. Cech T. R. (1988) *J. Am. Med. Assoc.* **260**; 3030–3034.

409. Mammalian artificial chromosomes. Brown W. R. A. (1992) *Curr. Opin. Gen. Devel.* **2**; 479–486.

410. Gene therapy for metabolic disorders. Kay M. A. & Eoo S. L. C. (1994) *Trends Genet.* **10**; 253–257.

411. New directions in molecular medicine. Karp J. E. & Broder S. (1994) *Cancer Res.* **54**; 653–665.

412. Therapeutic approaches to Alzheimer's disease. Schehr R. S. (1994) *BioTechnology* **12**; 140–144.

413. Progress towards gene therapy for HIV infection. Yu M. *et al.* (1994) *Gene Ther.* **1**; 13–26.

414. Gene therapy for cancer. Cilver K. W. & Blaese R. M. (1994) *Trends Genet.* **10**; 174–178.

415. Gene therapy for neurological disorders. Freidmann T. (1994) *Trends Genet.* **10**; 21–214.

416. Gene therapy for infectious disease: the AIDS model. Gilboa E. & Smith C. (1994) *Trends Genet.* **10**; 139–144.

417. The ethics of gene therapy. Walters L. (1986) *Nature* **320**; 225–227.

418. *Arabidopsis*, a useful weed. Myerowitz E. M. (1989) *Cell* **56**; 263–9.

419. YACs and the *C. elegans* genome. Coulson A. *et al.* (1991) *BioEssays* **13**; 413–417.

420. Toward cloning and mapping the genome of *Drosophila*. Merriam J. *et al.* (1991) *Science* **254**; 221–225.

421. The Human Genome Organisation: history, purposes and membership. McKusick V. A. (1989) *Genomics* **5**; 385–387.

422. Genome sequence analysis: scientific objectives and practical strategies. Venter J. C. *et al.* (1992) *Trends Biotech.* **10**; 8–11.

423. The human genome project: under an international ethical microscope. Knoppers B. M. & Chadwick R. (1994) *Science* **265**; 2035–2036.

424. Genome mapping the 'easy' way. Hodgson J. (1994) *BioTechnology* **12**; 581–584.

425. Analyzing and sorting human chromosomes. Green D. K. (1990) *J. Microsc.* **159**; 237–244.

426. DNA sequencing, automation and the human genome. Trainor G. L. (1990) *Anal. Chem.* **62**; 418–426.

427. The human genome project: prospects and implications for clinical medicine. Green E. D. & Waterson R. H. (1991) *J. Am. Med. Assoc.* **266**; 1966–1975.

428. Assessing mapping progress in the human genome project. Cox D. R. *et al.* (1994) *Science* **265**; 2031–2032.

BIBLIOGRAPHY 3:
SUGGESTED FURTHER READING

1. *The Origins of Genetics: A Mendel Source Book.* Stern C. & Sherwood E. R. (eds). Freeman, San Francisco (1966).

2. *Heritage from Mendel.* R. E. Brink (ed.). University of Wisconsin Press, Madison (1967).

3. *Origins of Mendelism.* R. Olby. University of Chicago Press, London (1985).

4. *History of Genetics.* H. Stubbe. M.I.T. Press, Cambridge, MA (1972).

5. *Looking at Chromosomes.* J. McLeish & B Snoad. Macmillan, New York (1958).

6. *Thomas Hunt Morgan; The Man and His Science.* Allen G. E. Princeton University Press, New Jersey (1978).

7. *Thomas Hunt Morgan: Pioneer of Genetics.* I. Shine & S. Wrobel. University Press of Kentucky, Lexington (1976).

8. *The Transforming Principle: Discovering that Genes are made of DNA.* M. McCarty. Norton, London (1985).

9. *Dealing With Genes: The Language of Heredity.* P. Berg & M. Singer. Blackwell Scientific Publishers, Oxford (1992).

10. *The Pyrimidines.* D. J. Brown. Interscience Press, New York (1962).

11. *The Chemistry of Nucleosides and Nucleotides.* A. M. Michelson. Academic Press, New York (1963).

12. *Handbook of Biochemistry and Molecular Biology*, 3rd Edition. *Nucleic Acids*, Vol. I (1975), and Vol. II (1976). G. D. Fasman (ed.). CRC Press.

13. *The Biochemistry of the Nucleic Acids*, 11th Edition. R. L. P. Adams *et al.* Chapman and Hall, London (1992).

14. *The Double Helix.* J. D. Watson. Atheneum, New York (1968).

15. *The Double Helix: A New Critical Edition.* G. S. Stent (ed.). Weidenfield & Nicolson, London (1981).

16. *Rosalind Franklin and DNA.* A. Sayre. W. W. Norton, New York (1975).

17. *The Path to the Double Helix.* R. Olby. Macmillan, London (1974).

18. *A Century of DNA: A History of the Discovery of the Structure and Function of the Genetic Substance.* F. H. Portugal & S. Cohen. M.I.T. Press, Cambridge, MA (1977).

19. *The DNA Story.* J. D. Watson & J. Tooze. W. H. Freeman & Co., San Francisco (1981).

20. *The Eighth Day of Creation.* H. F. Judson. Jonathan Cape, London (1979).

21. *What Mad Pursuit: A Personal View of Scientific Discovery.* F. C. Crick. Basic Books, New York (1988).

22. *Chromosomes: Eukaryotic, Prokaryotic and Viral.* K. W. Adolph. CRC Press (1989).

23. *Cold Spring Harbor Symp. Quant. Biol.*, Vol. 42: *Chromatin.* Cold Spring Harbor Laboratory Press, New York (1978).

24. *Cold Spring Harbor Symp. Quant. Biol.*, Vol. 47: *Structures of DNA.* Cold Spring Harbor Laboratory Press, New York (1983).

25. *Heterochromatin: Molecular and Structural Aspects.* R. Verma (ed.). Cambridge University Press, Cambridge (1988).

26. *Nucleosomes.* P. M. Wassarman & R. D. Kornberg (eds). *Methods in Enzymology*, Vol. 170 (1989).

27. *DNA–Protein interactions.* A. A. Travers. Chapman and Hall, London (1990).

28. *Protein–Nucleic Acid Interaction. Topics in Molecular and Structural Biology*, Vol 10. W. Saenger & U. Heinemann (eds). Macmillan Press, London (1989).

29. *Cytogenetics.* C.P. Swanson, T. Mertz & W. J. Young. Prentice-Hall, New Jersey (1967).

30. *DNA and Chromosomes.* E. J. Dupraw. Holt, Rinehart & Winston, New York (1970).

31. *The Bacterial Chromosome.* K. Drlica & M. Riley (eds). American Society for Microbiology, Washington DC (1990).

32. *Structure and Function of Plant Genomes.* O. Ciferri & L. Dure (eds). Plenum Press, New York (1983).

33. *The Genome of* Drosophila melanogaster. D. L. Lindsley & G. G. Zimm. Academic Press, San Diego (1992).

34. *Organelle Heredity.* N. W. Gillham. Raven Press, New York (1978).

35. *Cytoplasmic Genes and Organelles.* R. Sager. Academic Press, New York (1972).

36. *Genetics: A Molecular Approach*, 2nd Edition. T. A. Brown. Chapman and Hall, London (1992).

37. *Molecular Biology of the Cell*, 3rd Edition. B. Alberts, D. Bray, J. Lewis, M. Raff, K. Roberts & J. D. Watson. Garland Press, New York (1995).

38. *The Molecular Biology of the Gene*, 4th Edition. J. D. Watson *et al.* Benjamin/Cummings (1987).

39. *Molecular Cell Biology*, 2nd Edition. J. Darnell, H. Lodish and D. Baltimore. Scientific American Books, New York (1990).

40. *Chromosome Structure and Function.* J. P. Gustafson & R. Appels (eds). Plenum Press, New York (1988).

41. *Lampbrush Chromosomes.* Vol. 36, *Molecular Biology, Biochemistry, and Biophysics.* H. G. Callan. Springer-Verlag, New York (1986).

42. *Chromosomes. A Synthesis.* R. Wagner, M. P. Maguire & R. L. Stalling. J. Wiley & Sons, Chichester (1993).

43. *Human Chromosomes; Manual of Basic Techniques.* R. S. Varma & A. Babu. Pergamon Press, New York (1989).

44. *Flow Cytometry and Sorting*, 2nd Edition. M. R. Melamed, T. Lindmo & M. L. Mendelsohn (eds). Wiley-Liss, New York (1990).

45. *Chromosome Banding.* A. T. Sumner. Unwin Hyman, Cambridge, MA (1990).

46. *Genetics, Evolution and Man.* W. F. Bodmer & L.L. Cavalli-Sforza. W.H. Freeman & Co., New York (1976).

47. *Fundamentals of Molecular Evolution.* W. H. Li & D. Gaur. Sinauer, Sunderland, MA (1991).

48. *The Search for Eve.* M. H. Brown. Harper & Row, New York (1990).

49. *Gene Rearrangements.* Frontiers in Molecular Biology Series. B. D. Hames & D. M. Glover (eds). IRL Press, Oxford (1989).

50. *Gene Structure and Transcription.* T. Beebee & J. Burke. IRCL Press, Oxford (1988).

51. *Clinical Atlas of Human Chromosomes.* J. deGrouchy & C. Turleau. J. Wiley & Sons, New York (1984).

52. *Aneuploidy: Etiology and Mechanisms.* V. L. Dellarco, P. E. Voytek & A. Hollaender. Plenum Press, New York (1985).

53. *The Cytogenetics of Mammalian Autosomal Rearrangements.* A. Daniel (ed.). Alan R. Liss, Inc., New York (1988).

54. *Chromosomal Variation in Man: A Catalogue of Chromosomal Variants and Anomalies*, 5th Edition. D. S. Borgaonkar. Alan R. Liss, Inc., New York (1989).

55. *Chromosomal Anomalies and Prenatal Development: An Atlas.* D. Warburton, J. Byrne & N. Canki. Oxford University Press, Oxford (1990).

56. *Chromosomes and Cancer.* J. German (ed.). J. Wiley & Sons, New York (1974).

57. *Catalog of Chromosome Aberrations in Cancer*, 4th Edition. F. Mitelman. Wiley-Liss, New York (1991).

58. *The Molecular Basis of Mutation.* J. W. Drake. Holden-Day, San Francisco (1970).

59. *Mutation Research.* C. Auerbach. Chapman & Hall, London (1976).

60. *Chemical Mutagens: Principles and Methods for Their Detection*, Vol. 7. F. J. deSerres & A. Hollaender (eds). Plenum Press, New York (1982).

61. *Mechanisms of Cellular Transformation by Carcinogenic Agents.* D. Grunberger & S. P. Goff (eds). Pergamon Press, New York (1987).

62. *Analysis of Human Genetic Linkage.* J. Ott. Johns Hopkins University Press, Baltimore, MD (1985).

63. *General and Quantitative Genetics.* A. B. Chapman (ed.). Elsevier Science Publishers, Amsterdam (1985).

64. *Basic Concepts in Population, Quantitative and Evolutionary Genetics.* J. F. Crow. W. H. Freeman & Co., New York (1986).

65. *Introduction to Quantitative Genetics*, 3rd Edition. D. S. Falconer. Longman, London (1989).

66. *Human Genetics: Problems and Approaches*, 2nd Edition. F. Vogel & A. G. Motulsky. Springer-Verlag, Heidelberg (1986).

67. *An Introduction to Genetic Analysis*, 5th Edition. A. J. F. Griffiths, J. H. Miller, D. T. Suzuki, R. C. Lewontin & W. M. Gelbart. W. H. Freeman & Co., New York (1993).

68. *Mendelian Inheritance in Man: Catalogs of Autosomal Dominant, Autosomal Recessive, and X-linked Phenotypes*, 9th Edition. V. A. McKusick. Johns Hopkins University Press, Baltimore, MD (1990).

69. *Inborn Errors of Metabolism.* A. E. Garrod. Hodder & Stoughton, London (1909).

70. *Cold Spring Harbor Symp. Quant. Biol.*, Vol. 31: *The Genetic Code.* Cold Spring Harbor Laboratory Press, New York (1967).

71. *Signs of Life: The Language and Meanings of DNA.* R. Pollack. Viking, London (1994).

72. *Gene Transcription: A Practical Approach.* B. D. Hames & S. J. Higgins (eds). IRL Press, Oxford (1993).

73. *RNA Polymerase.* R. Losick & M. Chamberlain (eds). Cold Spring Harbor Laboratory Press, New York (1976).

74. *Transcription and Splicing.* Frontiers in Molecular Biology Series. B. D. Hames & D. M. Glover (eds). IRL Press, Oxford (1988).

75. *Gene Regulation: A Eukaryotic Perspective.* D. Latchman. Unwin Hyman, London (1990).

76. *Gene Function in Prokaryotes.* J. Beckwith, J. Davies & J. A. Gallant. Cold Spring Harbor Laboratory Press, New York (1983).

77. *Gene Expression*, Vol. 1: *Bacterial Genomes.* B. Lewin. J. Wiley & Sons, New York (1977).

78. *Gene Expression*, Vol. 3: *Plasmids and Phages.* B. Lewin. J. Wiley & Sons, New York (1977).

79. *The Molecular Biology of the Yeast* Saccharomyces: *Metabolism and Gene Expression.* J. N. Strathern, E. W. Jones & J. R. Broach (eds). Cold Spring Harbor Laboratory Press, New York (1982).

80. *The Control of Human Retrovirus Gene Expression.* B. R. Franza Jr., B. R. Cullen & F. Wong-Staal (eds). Cold Spring Harbor Laboratory Press, New York (1988).

81. *The Operon*, 2nd Edition. J. H. Miller & W. S. Reznikoff (eds). Cold Spring Harbor Laboratory Press, New York (1980).

82. *The Lactose Operon.* J. Beckwith & D. Zipser. Cold Spring Harbor Laboratory Press, New York (1970).

83. *The Statute Within. An Autobiography.* F. Jacob. Basic Books, New York (1988).

84. *A Genetic Switch: Gene Control and Phage Lambda*, 2nd Edition. M. Ptashne. Cell Press & Blackwell Scientific Publications, Cambridge, MA (1992).

85. *Transcription Factors: A Practical Approach.* D. S. Latchman (ed.). IRL Press, Oxford (1993).

86. *Ribosomes.* M. Nomura, A. Tissieres & P. Lengyel (eds). Cold Spring Harbor Laboratory Press, New York (1974).

87. *The Ribosome: Structure, Function and Evolution.* W. E. Hill *et al.* (eds) American Society for Microbiology, Washington DC (1990).

88. *The Mechanics of Inheritance*, 2nd Edition. F. W. Stahl. Prentice-Hall, New Jersey (1969).

89. *Cell Division and Heredity.* R. Kemp. Edward Arnold, London (1970).

90. *Mitosis: Molecules and Mechanisms.* J. S. Hyams & B. R. Brinkley (eds). Academic Press, San Diego (1989).

91. *Apoptosis: The Molecular Basis of Cell Death.* L. D. Tomei & F. O. Cope (eds). Cold Spring Harbor Laboratory Press, New York (1991).

92. *Apoptosis II: The Molecular Basis of Apoptosis in Disease.* L. D. Tomei & F. O. Cope (eds). Cold Spring Harbor Laboratory Press, New York (1994).

93. *Protein Phosphorylation: A Practical Approach.* D. G. Hardie (ed.). IRL Press, Oxford (1993).

94. *Genetic Recombination.* R. Kaucherlapti & G. R. Smith (eds). American Society of Microbiology, Washington DC (1988).

95. *Genetic Recombination: Thinking About It in Phage and Fungi.* F. W. Stahl. W.H. Freeman & Co., New York (1979).

96. *RNA Tumour Viruses*. R. Weiss, N. Teich, H. Varmus & J. Coffin (eds). Cold Spring Harbor Laboratory Press, New York (1985).

97. *Sexuality and the Genetics of Bacteria*. F. Jacob & E. Wollman. Academic Press, New York (1961).

98. *Escherichia coli and Salmonella typhimurium. Cellular and Molecular Biology*. F. C. Neidhardt (ed.). American Society for Microbiology, Washington DC (1987).

99. *Genetics of Bacterial Diversity*. D. A. Hopwood & K. F. Chater (eds). Academic Press, London (1989).

100. *The Emergence of Bacterial Genetics*. T. D. Brock. Cold Spring Harbor Laboratory Press, New York (1990).

101. *The Genetics of Bacteria and Their Viruses*, 2nd Edition. W. Hayes. J. Wiley & Sons, New York (1968).

102. *Bacterial and Bacteriophage Genetics*. E. A. Birge. Springer-Verlag, New York (1988).

103. *DNA Replication*, 2nd Edition. A. Kornberg & T. Baker. W.H. Freeman & Co., New York (1992).

104. *DNA Replication*. R. L. P. Adams. IRL Press, Oxford (1991).

105. *The Selfish Gene*. R. Dawkins. Oxford University Press (1989).

106. *DNA Insertion Elements, Plasmids and Episomes*. A. I. Bukhari, J. A. Shapiro & S. L. Adhya (eds). Cold Spring Harbor Laboratory Press, New York (1977).

107. *Mobile Genetic Elements*. J. A. Shapiro (ed.). Academic Press, New York (1983).

108. *The Discovery and Characterisation of Transposable Elements: The Collected Papers of Barbara McClintock*. B. McClintock. Garland Publishing, New York (1987).

109. *Mobile DNA*. D. E. Berg & M. M. Howe (eds). American Society for Microbiology, Washington DC (1989).

110. *Molecular Mechanisms in Cellular Growth and Differentiation*. A. R. Bellve & H. J. Vogel (eds). Academic Press, San Diego (1991).

111. *Genetic Variations of Drosophila melanogaster*. D. L. Lindsley & E. H. Grell. Carnegie Institute of Washington, Washington DC (1972).

112. *The Genetics and Biology of Drosophila*. M. Ashburner & E. Novitski (eds). Academic Press, New York (1976).

113. *The Making of a Fly*. P. A. Lawrence. Blackwell Scientific Publications, Oxford (1992).

114. *Developmental Genetics of Higher Organisms*. G. M. Malacinski (ed.). Macmillan, New York (1988).

115. *Genetic Analysis of Animal Development*, 2nd Edition. A. S. Wilkins. J. Wiley & Sons, New York. (1992).

116. *The Bacteriophages*, Vol 2. R. Calender (ed.). Plenum Press, New York (1988).

117. *Phage and the Origins of Molecular Biology*. J. Cairns, G. S. Stent & J. D. Watson (eds). Cold Spring Harbor Laboratory Press, New York (1966).

118. *Lambda II*. R. W. Hendrix, J. W. Roberts, F. W. Stahl & R. A. Weisberg. Cold Spring Harbor Laboratory Press, New York (1983).

119. *Recombinant DNA Technology and Applications*. C. Ho, A. Prokop & R. Baipai (eds). McGraw-Hill, New York (1990).

120. *Recombinant DNA,* 2nd Edition. J. D. Watson, M. Gilman, J. Witkowski & M. Zoller. W.H. Freeman & Co., New York (1992).

121. *Gene Cloning: An Introduction*, 3rd Edition. T. A. Brown. Chapman & Hall, London (1995).

122. *Principles of Gene Manipulation*, 5th Edition. R. W. Old & S. B. Primrose. Blackwell Scientific Publications, Oxford (1994).

123. *Genetic Engineering*. S. M. Kingsman & A. J. Kingsman. Blackwell Scientific Publications, Oxford. (1988).

124. *The Encyclopedia of Molecular Biology*. J. Kendrew (ed.). Blackwell Scientific Publications, Oxford. (1994).

125. *Enzymes of Molecular Biology*. M. M. Burrell (ed.). Humana Press Inc., New York (1993).

126. *Current Protocols in Molecular Biology*. F. A. Ausubel et al. (eds). Greene Publishing/Wiley Interscience, New York (1989).

127. *Molecular Cloning: A Laboratory Manual*, 2nd Edition. J. Sambrook, E. Fritsch & T. Maniatis. Cold Spring Harbor Laboratory Press, New York (1989).

128. *DNA Cloning: A Practical Approach*. D. M. Glover (ed.). IRL Press, Oxford (1985).

129. *Electrophoresis of Large DNA Molecules: Theory and Applications*. E. Lai & B. W. Birren (eds). Cold Spring Harbor Laboratory Press, New York (1990).

130. *Pulsed-Field Gel Electrophoresis. Protocols, Methods and Theories*. M. Burmeister & L. Ulanovsky (eds). Humana Press Inc., New York (1992).

131. *Nucleic Acid Hybridisation. A Practical Approach*. B. D. Hames & S. J. Higgins (eds). IRL Press, Oxford (1985).

132. *Immunocytochemistry – Modern Methods and Applications*. J. M. Polak & S. Van Noorden. Wright, Bristol (1986).

133. In situ *Hybridisation – Principles and Practice*. J. M. Polak & J. O. D. McGee (eds). Oxford Univeristy Press, Oxford (1990).

134. In situ *Hybridisation. Application to Neurobiology*. K. L. Valentino, J. H. Eberwine & J. D. Brachas (eds). Oxford University Press, Oxford (1987).

135. *Techniques of Autoradiography*. A. W. Rogers. Elsevier Science Publishers, Amsterdam (1979).

136. *PCR Technology – Principles and Applications for DNA Amplification*. H. A. Erlich (ed.). Stockton Press, New York (1989).

137. *PCR Protocols – A Guide to Methods and Applications*. M. A. Innis, D. H. Gelfand, J. J. Sninsky & T. J. White (eds). Academic Press, London (1990).

138. *PCR: A Practical Approach*. M. J. McPherson, P. Quirke & G. R. Taylor (eds). Oxford Univeristy Press, Oxford (1992).

139. *PCR Protocols: Current Methods and Applications*. B. A. White (ed.). Humana Press Inc., New York (1993).

140. *DNA Probes*, 2nd Edition. G. H. Keller & M. M. Manak (eds). Stockton Press, New York (1993).

141. *Nucleic Acid and Monoclonal Antibody Probes*. B. Swaminathan & G. Prakesh (eds). Dekkert, New York (1989).

142. *Genetic Alchemy*. S. Krimsky. MIT Press, Cambridge, MA (1983).

143. *Oligonucleotide Synthesis: A Practical Approach*. M. Gait (ed.). IRL Press, Oxford (1984).

144. *Invisible Frontiers: The Race to Synthesise a Human Gene*. S. S. Hall. Tempus, Washington DC (1989).

145. *Oligonucleotides and Analogues: A Practical Approach*. F. Eckstein (ed.). IRL Press, Oxford (1991).

146. *Bioluminescence and Chemiluminescence: Current Status*. P. E. Stanley & L. J. Kricka (eds). J. Wiley & Sons, Chichester (1991).

147. *Chemiluminescence Principles and Applications in Biology and Medicine*. A. K. Campbell. Ellis Horwood, Chichester (1988).

148. *Plasmids*. P. Broda. W.H. Freeman & Co., New York (1979).

149. *Bacterial Plasmids*. K. Hardy (ed.). American Society for Microbiology, Washington, DC (1986).

150. *The Molecular Biology of the Yeast* Saccharomyces: *Life Cycle and Inheritance*. J. N. Strathern, E. W. Jones & J. R. Broach (eds). Cold Spring Harbor Laboratory Press, New York (1981).

151. *Molecular Biology of Fission Yeast*. A. Nasim, P. Young & B. F. Johnson (eds). Acadmeic Press, New York (1989).

152. *Guide to Yeast Genetics and Molecular Biology*. C. Guthrie & G. R. Finks (eds). Academic Press, New York (1991).

153. *The Molecular Biology of Tumour Viruses*. J. Tooze (ed.). Cold Spring Harbor Laboratory Press, New York (1973).

154. *Eukaryotic Viral Vectors*. Y. Gluzman. Cold Spring Harbor Laboratory Press, New York (1982).

155. *The Baculovirus Expression System*. L. A. King & R. D. Possee (eds). Chapman and Hall, New York.

156. *Plant Cell and Tissue Culture*. A. Stafford & G. Warren (eds). Open University Press, Oxford (1991).

157. *Cell Fusion*. A. E. Sowers (ed.). Plenum Press, New York (1987).

158. *Transgenesis. Applications of Gene Transfer*. J. A. H. Murray (ed.). J. Wiley & Sons, Chichester (1992).

159. *Transgenesis Techniques. Principles and Protocols*. D. Murphy & D. A. Carter (eds). Humana Press, New York (1993).

160. *Genetic Maps*. S. J. Olsen (ed.). Cold Spring Harbor Press, New York (1984).

161. *Genome Analysis. A Practical Approach*. K.E. Davies (ed.). IRL Press, Oxford (1988).

162. *Genetic and Physical Mapping I: Genome Analysis*. K. E. Davies & S. M. Tilghman (eds). Cold Spring Harbor Laboratory Press, New York (1990).

163. *Genome*. J. E. Bishop & M. Waldholz. Simon & Schuster, New York (1990).

164. *Of URFs and ORFs; A Primer on How to Analyse Derived Amino Acid Sequences*. R. F. Doolittle. University Science Books, Mill Valley, CA (1987).

165. *Molecular Evolution: Computer Analysis of Protein and Nucleic Acid Sequences*. R. F. Doolittle (ed.). *Methods in Enzymology*, Vol. 183 (1990).

166. *DNA Technology and Forensic Science*. J. Ballantyne *et al.* (eds). Cold Spring Harbor Laboratory Press, New York (1989).

167. *DNA Fingerprinting: An Introduction*. L. T. Kirby. Stockton Press, New York (1990).

168. *DNA Fingerprinting: Approaches and Applications*. T. Burke, G. Dolf, A. J. Jeffreys & R. Wolff (eds). Birkhauser Verlag AG, Berlin (1991).

169. *Molecular Biology and Biotechnology*, 3rd Edition. J. M. Walker & E. B. Gingold (eds). Royal Society of Chemistry, London (1993).

170. *Purification and Analysis of Recombinant Proteins*. R. Seetharam & S. K. Sahrma (eds). Marcel Dekker, New York (1991).

171. *Monitoring Genetically Manipulated Microorganisms in the Environment*. C. Edwards (ed.). J. Wiley & Sons, New York (1993).

172. *Assessing Ecological Risks in Biotechnology*. L. R. Ginzburg (ed.). Butterworth-Heinemann, London (1991).

173. *Preconception and Preimplantation Diagnosis of Human Genetic Disease*. R. G. Edwards (ed.). Cambridge University Press, New York (1993).

174. *Essential Medical Genetics*, 3rd Edition. J. M. Connor & M. A. Ferguson-Smith. Blackwell Scientific Publications, Oxford (1991).

175. *Techniques in Diagnostic Human Biochemical Genetics: A Laboratory Manual*. F. A. Hommes (ed.). Wiley-Liss Inc., New York (1991).

176. *Elements of Medical Genetics*. A. E. H. Emery. Churchill Livingstone, Edinburgh (1983).

177. *Human Genetic Disease. A Practical Approach*. K. E. Davies (ed.). IRL Press, Oxford (1986).

178. *Principles of Medical Genetics*. T. D. Gelehrter & F. S. Collins. Williams & Wilkins, Baltimore, Maryland (1990).

179. *Molecular and Antibody Probes in Diagnosis*. M. R. Walker & R. Rapley (eds). J. Wiley & Sons, Chichester (1993).

180. *The New Genetics and Clinical Practice*, 3rd Edition. D. J. Weatherall. Oxford University Press, Oxford (1990).

181. *Molecular Diagnostics*. R. Rapley & M. R. Walker (eds). Blackwell Scientific Publications, Oxford (1993).

182. *The Metabolic Basis of Inherited Disease*, 6th Edition. C. R. Scriver, A. L. Beaudet, W. S. Sly & D. Valle (eds). McGraw-Hill Inc., New York (1989).

183. *Texbook of Biochemistry With Clinical Correlations*, 3rd Edition. T. M. Devlin (ed.). Wiley-Liss Inc., New York (1992).

184. *Cystic Fibrosis*. P. Goodfellow (ed.). Oxford University Press, Oxford (1989).

185. *Duchenne Muscular Dystrophy*. A. E. Emery. Oxford University Press, Oxford (1988).

186. *Cholesterol Metabolism, LDL and the LDL Receptor*. N. B. Myant. Acadmeic Press, San Diego (1990).

187. *Plasma Lipoproteins: New Comprehensive Biochemistry*. A. M. Gotto Jr. (ed.). Elsevier Science Publishers, Amsterdam (1987).

188. *Microbial Genetics*. D. Freifelder. Jones & Bartlett, Boston, MA (1987).

189. *Fungal Genetics*, 4th Edition. J. R. S. Finchman, P. R. Day & A. Radford. Blackwell Scientific Publications, Oxford (1979).

190. *Molecular Biology of Mycobacteria*. J. J. McFadden (ed.). Academic Press, London (1990).

191. *Laboratory Diagnosis of Infectious Diseases. Principles and Practices.* E. H. Lennette, P. Halonen & F. A. Murphy (eds). Springer-Verlag, Berlin (1988).

192. *Gene Probes for Bacteria.* A. J. L. Macario & E. C. de Marcario (eds). Academic Press, London (1990).

193. *Molecular Methods for Microbial Identification and Typing.* K. J. Towner & A. Cockayne. Chapman & Hall, London (1993).

194. *Vaccine Design.* F. Brown *et al.* J. Wiley & Sons, Chichester (1993).

195. *Cold Spring Harbor Symp. Quant. Biol.*, Vol. 53: *Signal Transduction.* Cold Spring Harbor Laboratory Press, New York (1988).

196. *Signal Transduction: A Practical Approach.* G. M. Milligan (ed.). IRL Press, Oxford (1992).

197. *The Molecular Basis of Cancer.* P. B. Farmer & J. M. Walker (eds). Croom Helm, Beckenham, Kent (1985).

198. *Cancer Cells: DNA Tumour Viruses: Control of Gene Expression and Regulation.* M. Botchan, T. Grodzicker & P. A. Sharp (eds). Cold Spring Harbor Laboratory Press, New York (1986).

199. *Oncogenes.* G. Cooper. Jones & Bartlett, Boston, MA (1990).

200. *Oncogenes, Genes, and Growth Factors.* G. Gurroff (ed.). Wiley Interscience, New York (1987).

201. *Genes and Cancer.* Carney D. & Sikora K. (eds). J. Wiley & Sons, Chichester (1990).

202. *Cellular and Molecular Biology of Mammary Cancer.* D. Medina, W. Kidwell, G. Heppner & E. Anderson (eds). Plenum Press, New York (1988).

203. *Molecular Diagnostics of Human Cancer.* M. Furth & M. Greaves (eds). Cold Spring Harbor Laboratory Press, New York (1989).

204. *Cancer: Science and Society.* J. Cairns. W.H. Freeman & Co., New York (1978).

205. *Drug Carrier Systems.* F. H. Roerdinck & A. M. Kroons (eds). J. Wiley & Sons, Chichester (1989).

206. *Development of Target-Oriented Anticancer Drugs.* Y-C. Cheng (ed.). Raven Press, New York (1983).

207. *Prospect for Antisense Nucleic Acid Therapy of Cancer and AIDS.* E. Wickstrom (ed.). Wiley-Liss Inc., New York (1991).

208. *Oligodeoxynucleotides – Antisense Inhibitors of Gene Expression.* J. S. Cohen (ed.). Macmillan Press, London (1989).

209. *Gene Regulation: Biology of Antisense RNA and DNA.* R. P. Erickson & J. G. Izant. Raven Press, New York (1992).

210. *Antisense RNA and DNA.* Murray J. (ed.). J. Wiley & Sons, Chichester (1992).

211. *Introduction to Protein Structure.* C. Branden & J. Tooze. Garland Publishing, New York (1991).

212. *Proteins: Structures and Molecular Properties*, 2nd Edition. T. E. Creighton. W. H. Freeman & Co., New York (1993).

213. *Protein Engineering: Approaches to the Manipulation of Protein Folding.* S. A. Narang (ed.). Butterworth Press, Boston, MA (1990).

214. *Molecular Immunology. Frontiers in Molecular Biology Series.* B. D. Hames & D. M. Glover (eds). IRL Press, Oxford (1988).

215. *Cold Spring Harbor Symp. Quant. Biol.*, Vol. 54: *Immunological Recognition.* Cold Spring Harbor Laboratory Press, New York (1989).

216. *Molecular Genetics of Immunoglobulins.* F. Calabi & M. S. Neuberger (eds). Elsevier Science Publishers, Amsterdam (1987).

217. *Immunoglobulin Genes.* T. Honjo, F. W. Alt & T. H. Rabbitts (eds). Academic Press, London (1989).

218. *Monoclonal Antibodies.* K. Sikora & H. M. Smedley. Blackwell Scientific Publications, Oxford (1984).

219. *Therapeutic Monoclonal Antibodies.* C. A. K. Borrebaeck & J. W. Larrie (eds). Stockton Press, New York (1990).

220. *Human Gene Therapy.* E. K. Nichols. Harvard University Press, Cambridge, MA (1988).

221. *Myoblast Transfer Therapy.* R. C. Griggs & G. Karparti (eds). Plenum Press, New York (1990).

222. *Evolution at the Molecular Level.* R. K. Selander, A. G. Clark & T. S. Wittam (eds). Sinauer Associates, Sunderland, MA (1991).

223. *The Human Genome.* T. Strachan. BIOS Scientific Publishers, Oxford (1992).

224. *Travelling Around the Human Genome: An* in situ *Investigation.* B. Jordan. John Libbey Eurotext (1993).

225. *The Human Genome Project: Deciphering the Blueprint of Heredity.* N. G. Cooper (ed.). University Science Books, Mill Valley, CA (1994).

226. *The Gene-Splicing Wars: Reflections on the Recombinant DNA Controversy.* R. A. Zilinskas & B. K. Zimmerman. Macmillan, New York (1986).

227. *Genetics, Law and Social Policy.* P. Reilly. Harvard University Press, Cambridge, MA (1977).

228. *In the Name of Eugenics.* D. Kelves. Alfred A. Knopf, New York (1985).

229. *Proceed with Caution: Predicting Genetic Risks in the Recombinant DNA Era.* N. A. Holtzman. Johns Hopkins University Press, Baltimore, MD (1989).

INDEX

325